Generalized Riccati Theory and Robust Control: A Popov Function Approach

Generalized Riccati Theory and Robust Control

A Popov Function Approach

Vlad Ionescu
Faculty of Automatic Control And Computers
Polytechnic University of Bucharest
Bucharest, Romania

Cristian Oară
Faculty of Automatic Control And Computers
Polytechnic University of Bucharest
Bucharest, Romania

Martin Weiss
Faculty of Applied Physics
Eindhoven University of Technology
Eindhoven, The Netherlands

JOHN WILEY & SONS
Chichester · New York · Weinheim · Brisbane · Singapore · Toronto

National 01243 779777
International (+44) 1243 779777

e-mail (for orders and customer service enquiries): cs-books@wiley.co.uk

Visit our Home Page on http://www.wiley.co.uk
or
http://www.wiley.com

Other Wiley Editorial Offices

John Wiley & Sons, Inc., 605 Third Avenue,
New York, NY 10158-0012, USA

WILEY-VCH GmbH, Pappelallee 3,
D-69469 Weinheim, Germany

Jacaranda Wiley Ltd, 33 Park Road, Milton,
Queensland 4064, Australia

John Wiley & Sons (Asia) Pte Ltd, 2 Clementi Loop #02-01,
Jin Xing Distripark, Singapore 129809

John Wiley & Sons (Canada) Ltd, 22 Worcester Road,
Rexdale, Ontario M9W 1L1, Canada

British Library Cataloguing in Publication Data

A catalogue record for this book is available from the British Library.

ISBN 0 471 97147 2

Produced from camera-ready-copy supplied by the authors.
Printed and bound in Great Britain by Antony Rowe Ltd, Chippenham, Wilts.
This book is printed on acid-free paper responsibly manufactured from sustainable forestry, in which at least two trees are planted for each one used in paper production.

To our families, in love and gratitude

Contents

Preface

This book is a comprehensive presentation of an original approach – which we call the Popov function approach – to the theory of algebraic Riccati equation and its applications to the robust control of linear dynamical systems. This is the first monograph to give a systematic account on the algebraic Riccati equation with an indefinite sign quadratic term, and which achieves the same level of generality and in-depth details for both the continuous–time and the discrete–time settings. Although the Popov function approach has been developed during the last few years in a series of contributions by the authors, it has notwithstanding been strongly influenced and stimulated by older and modern developments in control theory.

Ever since it emerged in the solution to the quadratic optimal control and filtering problems during the 1960s in the works of Kalman [Kal60, Kal64], the algebraic Riccati equation has played a crucial role and has been a central topic in control theory. The fact that Riccati equations provide a nice theoretical framework, together with sound and stable numerical algorithms, is probably the best explanation for the current popularity of Riccati techniques in control theory. The influential paper of Willems [Wil71] in early 1970s, and later the introduction of the two–Riccati approach for the H^∞–control problem in the paper by Doyle, Glover, Khargonekar, and Francis [DGKF89], increased dramatically the range of control applications of the algebraic Riccati theory. These two papers have also extended the class of equations interesting from a practical point of view, with equations featuring indefinite sign free and quadratic terms. Consideration of singular LQ–problems as in Silverman [Sil76], or singular spectral factorizations as in Chen and Francis [CF89], or the singular H^∞–control problem as in Stoorvogel [Sto92], led to a further enlargement of the class of "interesting" Riccati equations. In parallel with the theoretical achievements, the development of stable, efficient, and reliable algorithms for computing solutions to Riccati equations, which began with the seminal papers of Laub [Lau79] and Van Dooren [Van81], made of Riccati theory an effective and computer implementable design tool for engineers. All these successive advances resulted in an explosive number of contributions, revealing various aspects of the Riccati equation and steadily increasing our knowledge of these equations. Many of the general results on Riccati equations have been gathered in article collections such as Bittanti, Laub, and Willems [BLW91], or in monographs such as Lancaster and Rodman [LR95], while applications of these results to robust and more general modern control systems have been covered by Anderson and Moore [AM71], Kwakernaak and Sivan [KS72], Kailath [Kai80], Francis [Fra87], Green and Limebeer [GL95], Zhou, Doyle and Glover [ZDG95], or Kimura [Kim96]. Extensions of these topics to the class of infinite dimensional systems have been recently accomplished by Curtain and Zwart [CZ95], and Feintuch [Fei98].

The literature gives the impression of great variety, in which the underlying connections between various results are often difficult to retrieve. In general, the theoretical advancements in Riccati theory can be classified as broad–based mathematical results that give existence conditions, classifications, computable formulas, and numerical algorithms for various solutions to the Riccati equation, and results that relate the Riccati equation to the control theoretical problems. Of course, the importance of the Riccati equation relies on both aspects of the theory. Nevertheless, the gap between the general mathematical results and the way in which they may be applied in control is sometimes important, and not infrequently developments in control systems have been the driving force in advancing the theory.

The Popov function approach that we present in this book tries to fill this gap and has the feature that it enables a unifying treatment of many of the results in both divisions of the Riccati equation theory. In this way hidden connections between theoretical system properties, computational algorithms, and rather abstract mathematical results become transparent.

As we have already pointed out, a particular feature of this book is that it presents for the first time a comprehensive theory of Riccati equations with an indefinite sign quadratic term. This newly born theory proves not to be a trivial extension of the definite sign case, as it requires many new technical tools and rich extensions of results, previously known only in the context of the positivity theory originally developed by Popov [Pop66]. Another trait of the monograph that singles it out from other references, in both the fields of algebraic Riccati equation and robust control systems, is that every topic is treated in parallel – first in the continuous–time and then in the discrete–time setting and, mostly important, at the same broad level of generality. A manifest aim of the book is that every reader should come to the conclusion that there is nothing more hairy, tricky, or bizarre, in the behavior of discrete–time systems than in the behavior of continuous–time systems, provided one defines and handles the relevant counterpart notions and results.

There have been numerous contributions to Riccati equation theory over the years covering rather distinctive directions and having particular motivations, and it is not possible to give a comprehensive exposé in the quite limited space of a book. Therefore, some important aspects of the theory of algebraic Riccati equations – for example the geometric description of the solutions, comparison theorems, existence and characterization of all Hermitian and definite solutions, structure–preserving algorithms, and many others – have been intentionally left out. The interested reader is referred to the recent comprehensive monograph by Lancaster and Rodman [LR95] for theoretical aspects, and to Mehrmann [Meh91] for numerical algorithms. Primarily, we have focused on those constituents of the theory that point to applications in systems theory. These parts of the theory are mainly related to certain particular solutions (positive, stabilizing, antistabilizing, semistabilizing, strong stabilizing) rather than to the class of all solutions to the Riccati equation. Yet, as already mentioned, we have paid special attention to the case of Riccati equation with an indefinite sign quadratic term, and we have enriched the class of well–known Riccati equations with new Riccati–like equations (as Riccati systems, or Riccati equations in reverse–time) so as to cover in their full generality the scope of applications that we present. All these features of the book have reduced to a minimum the possible overlapping with the single other monograph dedicated to the algebraic Riccati equation by Lancaster and Rodman [LR95], which concentrates more on global properties of the class of all solutions for the case of the definite sign quadratic term.

The prime reason that drove us towards writing this book was to extend to the indefinite sign and singular cases the Positivity Theory developed by Popov in the early 1960s [Pop66], so as to become applicable to the new problems that have emerged in the last decade from the theory of control systems. In this respect, it will be most rewarding for us if the reader considers this book as a continuation of Popov's work.

The background prerequisites assumed of any reader are linear algebra and advanced matrix theory, linear differential and difference equations and operator theory, both at a modest level, and linear systems theory and optimal and robust control, both at graduate level. The intended readership consists of graduate and postgraduate students and researchers in electrical engineering and applied mathematics who work in linear systems theory, with emphasis on optimal and robust control.

Structure of the book and main contributions

The book encompasses in a unified framework many and various results in control systems theory that are related to the Riccati equations, and also facts related to the theory and computation of the solutions of these equations. The book is built up of three main parts. Part I introduces and outlines in two chapters some background as well as more advanced results in linear algebra, with emphasis on general matrix theory, rational matrix functions and the theory of realizations, and linear dynamical systems. The purpose of this first part is to ensure a fair degree of self–containedness and to establish the notation and notions for the rest of the exposition. The treatment is by no means exhaustive and, as many of the results are classical, we omit most of the proofs and provide suitable references instead. Proofs are given only for those results which are either new or for which we were not able to find an appropriate source.

Chapter 1 gives, in increasing complexity, a short outline of canonical forms for matrices, regular and general matrix pencils, linear matrix equations that play a central role in system theory: Lyapunov, Sylvester, and Stein equations, and some relevant aspects of rational matrix functions theory. A particular focus is on invariant subspaces of matrices, and their extensions to the case of regular and general pencils, as they provide both theoretical tools for investigating and numerically sound algorithms for computing solutions to Riccati equations. For rational matrices we review briefly the Smith–McMillan forms, structural elements including structure at infinity, and realization theory in both the proper and improper cases.

Chapter 2 contains a short review of several basic facts in linear systems theory that will be extensively used in the sequel. Again, the emphasis is on results that are not usually encountered in the literature, such as dichotomy, L^2 input–output maps, associated Toeplitz and Hankel operators, continuity of the H^∞–norm. As elsewhere in the book, we not only treat in the same depth and generality the continuous and discrete–time cases, but also give attention to the distinctive features of the discrete–time case that are usually ruled out in the literature by restrictive assumptions.

Part II, entitled Generalized Riccati Theory, is the heart of the monograph. It embodies a comprehensive presentation of general results on the Riccati equation under assumptions of the matrix coefficients relaxed as far as possible, allowing for the inclusion of indefinite sign and singular cases. Not infrequently, we consider extensions to settings so general that even the Riccati equation is no longer defined as such, and has to be replaced by an appropriate substitute. The approach to the Riccati equation is made from several standpoints: time domain, frequency domain, and state–space. This leads to characterizations in terms of input–output operator, transfer function matrix, and transmission pencil, respectively, all associated with the underlying Hamiltonian system. These quite different perspectives, that point to areas of mathematics as diverse as operator theory, functional analysis, and linear algebra, are strongly connected and closely interwoven in the applications to systems and robust control given in Part III.

Part II consists of four chapters as follows. In Chapter 3 we introduce the basic setting for our approach to the Riccati equation theory. The central notion here is the Popov triplet, for which we define a special equivalence transformation and with which we associate a number of mathematical objects that will play various parts later in the book. In continuous–time, several of these objects are: the Hamiltonian system, the Popov function, the extended Hamiltonian pencil, the Riccati equation, and the Kalman–Popov–Yakubovich system. The main results in this chapter are giving the effect of the equivalence transformation for Popov triplets on the associated objects as well as the relations between these objects. In particular, the equivalence relation for Popov triplets enables an elegant exposition of the general Riccati equation theory. Particular attention is paid to finding the appropriate discrete–time counterparts of the continuous–time notions, as these will be used throughout the book.

Chapters 4, 5, and 6 represent the core of our approach as they contain the main existence results for the stabilizing solution and other related solutions to the Riccati equation as well as explicit computational formulas. Chapter 4 gives the system theoretic connections of the solution to the Riccati equation and the approach is successively in time–domain and frequency domain. To briefly explain the major results of this chapter, we confine ourselves for the moment to the continuous–time case only. The main result in time–domain says that the algebraic Riccati equation has a stabilizing solution if and only if the Toeplitz of the input–output operator of the underlying Hamiltonian system is invertible. The same result, translated to the frequency domain, reads as the algebraic Riccati equation has a stabilizing solution if and only if the associated Popov function is antianalytic factorizable or, under mild conditions, if and only if a certain signature condition on the Popov function holds. Actually, we show that the signature condition on the Popov function is equivalent to the existence of a stabilizing solution to the Kalman–Popov–Yakubovich system in J–form, which is a more handy and appropriate way of writing Riccati equations as it inherits from the Popov function useful information about the signature. This feature is not present in Riccati equations and becomes even more striking in the discrete–time setting. In particular, it emerges that the solutions to most problems in Part III can be written in compact form in terms of Kalman–Popov–Yakubovich systems.

Chapter 5 is primarily concerned with the computational aspects of the theory. In continuous–time, the main result shows that the Riccati equation has a solution if and only if the associated extended Hamiltonian pencil has a stable deflating subspace satisfying an invertibility condition called disconjugacy and, if a solution exists, we give computable formulas for the solution expressed in terms of a basis matrix for the deflating subspace. Incidentally, it turned out that this result provides a convenient tool for deriving, in a systematic and simple way, useful relations between solutions to various Riccati equations involved in certain problems in Part III.

Chapter 6 presents an extension of the theory in Chapter 5 to so-called Riccati systems. These are a generalization of the Riccati equation to cope with the case in which the underlying Popov function is not invertible, and are related to the solution of singular optimal control problems and various factorizations. Numerical algorithms are discussed in detail, and appropriate hints for the computation of various solutions to Riccati equations and systems are provided.

The generality and the unifying aspects of the Riccati theory that we present become clear in these last three aforementioned chapters. The wide applicability of the book is enabled by the very general assumptions under which we work (practically, no assumptions on the matrix coefficients of the Riccati equation). The unifying character is rendered by the natural duality Popov function–extended Hamiltonian pencil which are nothing but two equivalent representations of the same Hamiltonian system. On the basis of the above, it should be clear at this point why the Popov function – which is just the transfer function of the Hamiltonian system – describes in a natural manner the input–output behavior in time and frequency domain, whereas the Hamiltonian pencil – which is but the transmission pencil of the same Hamiltonian system – is more suited for computational purposes or for describing the structural properties. The book attempts to reaffirm the primordial role of the Popov function that was introduced in connection with the positivity theory back in the 1960s in Popov's celebrated monograph [Pop66]. Most of the material in Part II is the original contribution of the authors and appears here for the first time.

Part III includes five chapters which comprise mathematical problems from the realm of systems theory and robust control. The solution to all these problems is obtained by applying the tools developed in Part II.

In Chapter 7 we obtain new straightforward proofs for or extensions of several classical results in systems theory: the bounded real lemma, convergence of a family of Riccati equations, normalized coprime factorizations, the Small Gain Theorem, quadratic index with constrained dynamics, and spectral and inner–outer factorizations.

Chapter 8 contains the Nehari four block problem for which we provide the class of all solutions in the suboptimal case. Our method of proof, and the construction of the solutions, is novel, and has the advantage of simplicity and versatility over the classical all–pass embedding techniques. The versatility pertains to the potential of our method to be applied to the more general Adamjan–Arov–Krein problem, or to obtain the class of optimal (as opposed to suboptimal) solutions to the four block Nehari problem. Due to space limitations, we have not pursued these advances here.

Chapter 9 deals with the optimal H^2–control problem, for which we construct an optimal solution by employing a new method of reducing successively the original problem to simpler ones.

Chapter 10 is devoted to the H^∞–control problem. The two–Riccati criterion for the existence of a suboptimal H^∞–controller and the parameterization of the solutions are derived by using a new method of proof emanated by the results in Chapter 4. In contrast to most of the other approaches presented in the literature, we do not make the usual simplifying assumptions about the problem. Nonetheless, the derivation is reasonably short, and the resulting state–space formulas for the class of solutions, in both the continuous and discrete–time setting, are the most general available so far.

Finally, we solve in Chapter 11 the robust stabilization problem in the cases in which the uncertainties act either on normalized coprime factors or have multiplicative representation. For the problem of robust stabilization with normalized coprime factor uncertainties, we obtain an optimal (as opposed to suboptimal) solution which is amenable to numerical algorithms. The numerical features include the reduced order of the solution and the removal of the large pole–zeros that usually lead to an ill–conditioned computation of the solution near the optimum. The method pursued here is to reduce the robust stabilization problem to a suboptimal H^∞–control problem and to make a singular perturbation analysis. Concerning the theoretical importance, it emerged that the singular perturbation technique conveniently supplements the Popov function approach which can handle, generally speaking, the suboptimal cases. We also present a solution to the robust stabilization problem with multiplicative uncertainties, which turns out to be a singular H^∞–control problem. We use again limiting arguments to reduce the problem at hand to a regular H^∞–control problem.

Each chapter ends with notes and references pointing to the sources that were influential for the authors. At the same time, we place the topics in a short historical perspective, and provide glimpses into closely related results. In view of our limited knowledge and space, when facing the towering number of publications in the field, our comments on and references to works that advanced the field will not be exhaustive. Here is the right place to apologize to all those whose names or works were unintentionally omitted.

Acknowledgments

The main research results contained in this book have been obtained jointly by all three authors, partially while the last two were PhD students under the supervision of the first. The task of writing this book has been mainly accomplished by the two first–named authors, based on a draft version completed by them in 1996.

The second author gratefully acknowledges the support provided by the Alexander von Humboldt Foundation, Bonn, during the writing of this book. He is also indebted to Professor Paul Van Dooren and Dr. Andras Varga for many interesting and stimulating discussions, and for inviting him to spend extensive research periods at Université Catholique de Louvain and Deutsches Zentrum für Luft– und Raumfahrt, DLR–Oberpfaffenhofen.

Our deepest appreciation goes to Professor Aristide Halanay who for several decades enthusiastically headed the Seminar of Differential Equations and Dynamical Systems at the University of Bucharest, and who recently passed away. This seminar was the place where the authors presented and checked many of the results of their research. The authors are especially grateful for his penetrating criticism and generous advice on many aspects of the theory presented in the book.

Above all, we are grateful to our families for their love, patience and understanding.

Vlad Ionescu
Cristian Oară
Martin Weiss

Acronyms, Notation, and Symbols

Acronyms: Riccati Theory

Σ	Popov triplet
F_{Ricc}	Riccati feedback
$D_{\Sigma}(X)$	dissipation matrix associated with the Popov triplet Σ
$\mathbf{\Pi}_{\Sigma}$	Popov function associated with the Popov triplet Σ
EHP(Σ)	extended Hamiltonian pencil associated with the Popov triplet Σ
ESP(Σ)	extended symplectic pencil associated with the Popov triplet Σ
CTARE(Σ)	continuous–time algebraic Riccati equation associated with the Popov triplet Σ
DTARE(Σ)	discrete–time algebraic Riccati equation associated with the Popov triplet Σ
CTARS(Σ)	continuous–time algebraic Riccati system associated with the Popov triplet Σ
DTARS(Σ)	discrete–time algebraic Riccati system associated with the Popov triplet Σ
GCTARS(Σ)	generalized continuous–time algebraic Riccati system associated with the Popov triplet Σ
GDTARS(Σ)	generalized discrete–time algebraic Riccati system associated with the Popov triplet Σ
KPYS(Σ, J)	Kalman–Popov–Yakubovich system associated with the Popov triplet Σ and the sign matrix J
KSPYS(Σ, J)	Kalman–Szego–Popov–Yakubovich system associated with the Popov triplet Σ and the sign matrix J
CTHS(Σ)	continuous–time Hamiltonian system associated with the Popov triplet Σ
DTHS(Σ)	discrete–time Hamiltonian system associated with the Popov triplet Σ
ITHS(Σ)	inverse–time (discrete) Hamiltonian system associated with the Popov triplet Σ

Acronyms: Systems Theory and Robust Control

TFM	transfer function matrix
LQP	linear quadratic problem
4–NP	four block Nehari problem
IDEP	impulse disturbance estimation problem
IDFP	impulse disturbance feedforward problem
IOEP	impulse output estimation problem
DAP	disturbance attenuation problem
DE	disturbance estimation
DF	disturbance feedforward
OE	output estimation

Miscellaneous

$:=$	the left–hand side is defined by the expression in the right–hand side
$=:$	the right–hand side is defined by the expression in the left–hand side
	end of proof
$B \leftarrow A$	store the value A in B (seen as a computer variable)
$u \equiv y$	u is identical to y

Numbers and Sets

$\epsilon \downarrow a$	ϵ tends to a with values greater than a
$\mathbb{R}$	field of real numbers
$\mathbb{C}$	field of complex numbers or complex plane
$\mathbb{F}$	field, either $\mathbb{R}$ or $\mathbb{C}$
$\mathbb{N}$	set of natural numbers
$\mathbb{Z}$	set of integers
$\mathbb{C}_-$	open left–half plane
$\mathbb{C}_+$	open right–half plane
$\mathbb{C}_0$	imaginary axis
$\mathbb{D}$	unit disk
$\partial\mathbb{D}$	unit circle
$\mathbb{M}$	closure of the set $\mathbb{M}$

Operators

$\mathcal{R}^{-1}$	bounded inverse of $\mathcal{R}$ (for boundedly invertible $\mathcal{R}$)
$\mathcal{R}^*$	adjoint of $\mathcal{R}$
$\Vert\mathcal{R}\Vert$	operator norm of $\mathcal{R}$
$\Vert\mathcal{R}\Vert_H$	Hankel norm of $\mathcal{R}$
$\mathcal{R} = \mathcal{R}^* >> 0$	$\mathcal{R}$ is coercive: $\exists\delta > 0$, such that $\langle \mathcal{R}u, u\rangle_X \geq \delta \Vert u\Vert_X^2$, $\forall u \in X$
$\mathbb{H}_\mathcal{G}$ ($\widehat{\mathbb{H}}_\mathcal{G}$)	causal (anticausal) Hankel operator of $\mathcal{G}$
$\mathbb{T}_\mathcal{G}$ ($\widehat{\mathbb{T}}_\mathcal{G}$)	causal (anticausal) Toeplitz operator of $\mathcal{G}$
d	differential operator
σ	shift operator: $(\sigma x)_k := x_{k+1}$
Ω	time–reversal operator: $(\Omega x)_k := x_{-k}$
$\dot{x}(t)$	derivative of $x(t)$
$\delta(t)$	unit (or Dirac) impulse

Rational Matrices and Systems

$\mathbb{F}(\lambda)$	set of rational matrices with coefficients in $\mathbb{F}$
$\mathbb{F}^{p\times m}(\lambda)$	set of $p \times m$ rational matrices with coefficients in $\mathbb{F}$
$\mathbb{F}^{m}(\lambda)$	vector space of columns with m rational elements with coefficients in $\mathbb{F}$
$\left[\begin{array}{cc} A & B \\ C & D \end{array}\right](\lambda)$	$:= C(\lambda I - A)^{-1}B + D$
$\left[\begin{array}{c\|c} A & B \\ C & D \end{array}\right]^{a}(\lambda)$	$:= C(\frac{1}{\lambda}I - A)^{-1}B + D$
$\mathcal{P}(\mathbf{G})$	union of poles of the rational matrix $\mathbf{G}$
$\mathcal{Z}(\mathbf{G})$	union of zeros of the rational matrix $\mathbf{G}$
$\mathbf{G}^{*}(\lambda)$	adjoint or conjugate: $:= \mathbf{G}^{T}(-\lambda)$ (continuous–time) $:= \mathbf{G}^{T}(\frac{1}{\lambda})$ (discrete–time)
RL^{∞} $(RL^{\infty}_{p\times m})$	set of $(p \times m)$ rational matrices with elements analytic on either $\mathbb{C}_0$ or $\partial\mathbb{D}$
RH^{∞}_{+} $(RH^{\infty}_{+,p\times m})$	set of $(p \times m)$ rational matrices with elements analytic on either $\mathbb{C}_{+} \cup \overline{\mathbb{C}}_0$ or $\overline{\mathbb{C}} \setminus \mathbb{D}$
RH^{∞}_{-} $(RH^{\infty}_{-,p\times m})$	set of $(p \times m)$ rational matrices with elements analytic on either $\mathbb{C}_{-} \cup \overline{\mathbb{C}}_0$ or $\overline{\mathbb{D}}$
LLFT$(\mathbf{T},\mathbf{K})$	lower linear fractional transformation of $\mathbf{T}$ with $\mathbf{K}$
ULFT$(\mathbf{T},\mathbf{K})$	upper linear fractional transformation of $\mathbf{T}$ with $\mathbf{K}$
LFT$(\mathbf{T},\mathbf{K})$	shorthand for LLFT$(\mathbf{T},\mathbf{K})$
$\mathbf{T} \otimes \mathbf{K}$	Redheffer product of $\mathbf{T}$ with $\mathbf{K}$

Spaces of Functions

$L^{2}(\mathbb{M})$	Hilbert space of $\mathbb{F}$–valued matrix (vector or scalar) functions $f(t)$ for which $\int_{\mathbb{M}} \text{Trace}\,[f^{*}(t)f(t)]\mathrm{d}t < \infty$; $\mathbb{M}$ is either $\mathbb{Z}$, $(-\infty,\infty)$, $\mathbb{C}_0$, or $\partial\mathbb{D}$
$L^{\infty}(\mathbb{M})$	Banach space of $\mathbb{F}$–valued matrix functions $F(t)$ essentially bounded on $\mathbb{M}$: $\sup_{\mathbb{M}} \sigma(F(t)) < \infty$; $\mathbb{M}$ is either $\mathbb{C}_0$ or $\partial\mathbb{D}$
$\|\|\cdot\|\|_2$	norm in space $L^{2}(\mathbb{M})$
$\|\|\cdot\|\|_\infty$	norm in space $L^{\infty}(\mathbb{M})$
$L^{2,m}$	shorthand for $L^{2,m}(-\infty,\infty)$: Hilbert space of norm–square Lebesgue integrable on $(-\infty,\infty)$, $\mathbb{F}^{m}$–valued functions
$L^{2,m}_{+}$	closed subspace of $L^{2,m}$ of functions with support in $[0,\infty)$
$L^{2,m}_{-}$	closed subspace of $L^{2,m}$ of functions with support in $(-\infty,0]$
$\ell^{2,m}$	shorthand for $L^{2,m}(\mathbb{Z})$: Hilbert space of doubly infinite norm–square summable, $\mathbb{F}^{m}$–valued sequences
$\ell^{2,m}_{+}$	closed subspace of $\ell^{2,m}$ of sequences with positive support $k \geq 0$
$\ell^{2,m}_{-}$	closed subspace of $\ell^{2,m}$ of sequences with negative support $k \leq -1$
P^{r}_{+}	orthogonal projection of $L^{2,r}$ (or $\ell^{2,r}$) onto $L^{2,r}_{+}$ (or $\ell^{2,r}_{+}$)
P^{r}_{-}	orthogonal projection of $L^{2,r}$ (or $\ell^{2,r}$) onto $L^{2,r}_{-}$ (or $\ell^{2,r}_{-}$)
$\langle\cdot,\cdot\rangle$	inner product (the Hilbert space is implicit)
$\langle\cdot,\cdot\rangle_X$	inner product in Hilbert space X

Vector Spaces and Matrices

$\mathbb{F}^m$	vector space of columns with m elements in $\mathbb{F}$
$\langle x, y\rangle$	standard scalar product of $x, y \in \mathbb{F}^m$
$\|\|u\|\|$	Euclidean norm of $u \in \mathbb{F}^m$
$\dim \mathcal{V}$	dimension of the space $\mathcal{V}$
$\mathcal{V}+\mathcal{W}$, $(\mathcal{V}\oplus\mathcal{W})$	(orthogonal) direct sum of subspaces $\mathcal{V}$ and $\mathcal{W}$
$\mathbb{F}^{m\times n}$	set of $m \times n$ matrices with elements in $\mathbb{F}$
I_n, I	$n \times n$ identity matrix
0, O	zero matrix (not necessarily square)
$\star$	irrelevant (block) element of a matrix
$\Lambda(A)$	union of the eigenvalues of A
$\sigma(A)$	largest singular value of A
$\mathcal{X}_-(A)$	stable invariant subspace of A
$\mathcal{X}_+(A)$	antistable invariant subspace of A
$\Pi_-(A)$	orthogonal projection onto the stable subspace of A
$\Pi_+(A)$	orthogonal projection onto the antistable subspace of A
$A\|_{\mathcal{V}}$	restriction of A to its invariant subspace $\mathcal{V}$
$\Lambda(A)\|_{\mathcal{V}}$	restriction of the spectrum of A to its invariant subspace $\mathcal{V}$
$\nu_c(A)$, $\pi_c(A)$, $\delta_c(A)$	number of eigenvalues of A with strictly negative, strictly positive, and null real parts
$\nu_d(A)$, $\pi_d(A)$, $\delta_d(A)$	number of eigenvalues of A with modulus less than, greater than, and equal to 1
A^T	transpose of A
A^{-1}	inverse of A (for invertible A)
$A^{\dagger}$	pseudoinverse of A
A^*	complex conjugate transpose of A
A^{-T}	shorthand for $(A^{-1})^T$
A^{-*}	shorthand for $(A^{-1})^*$
A^P	pertranspose of A
$A \geq B$	$A - B$ is positive semidefinite (for hermitian A and B)
$A > B$	$A - B$ is positive definite (for hermitian A and B)
$A^{\frac{1}{2}}$	positive semidefinite hermitian square–root of A (for hermitian positive semidefinite A)
$\det(A)$	determinant of A
$\mathrm{Trace}\,(A)$	trace of A
$\mathrm{Ker}\,A$	kernel (or null space) of A
$\mathrm{Im}\,A$	image (or range) of A
$\mathrm{rank}\,A$	rank of A
$\rho(A)$	spectral radius of A
$\mathrm{sgn}\,(A)$	sign matrix of A (for hermitian A)
$\|\|A\|\|$	operator–induced norm of A
$\mathrm{diag}\,(A_1,\ldots,A_k)$	diagonal matrix with diagonal elements $A_1,\ldots,A_k$
$\lambda M - N$	matrix pencil
$\Lambda(M,N)$	union of the generalized eigenvalues of $\lambda M - N$
$(\lambda M - N)\|_{\mathcal{V}}$	restriction of $\lambda M - N$ to its deflating subspace $\mathcal{V}$
$\Lambda(M,N)\|_{\mathcal{V}}$	restriction of $\Lambda(M,N)$ to the deflating subspace $\mathcal{V}$

Part I

General Matrix Theory and Linear Dynamical Systems

1

Matrices, Matrix Pencils, and Rational Matrix Functions

We start with an outline of the chapter. In Sections 1.1–1.3 we review the eigenvalue problem, the Jordan canonical form, and the invariant subspace of a constant square matrix to show then, in Sections 1.6–1.8, how these concepts extend to the generalized eigenvalue problem, the Weierstrass canonical form, and the deflating subspace, all introduced for a matrix pencil that satisfies a regularity assumption. We take a further step in Sections 1.9–1.11 and discuss for a general pencil (possibly singular) the generalized eigenstructure problem, the Kronecker canonical form, and introduce the notion of proper deflating subspace as an extension to the general pencil case of the concept of deflating subspace of a regular pencil. In Section 1.4 we study several properties of pairs of matrices, like controllability and observability, which play a central role in Section 1.5 where we investigate properties of linear matrix equations: Sylvester, Stein, and their symmetric versions, the Lyapunov and the symmetric Stein equations. Further we introduce several important results from the theory of rational matrices, for example the structural elements, realization theory, and their mutual connections. In Section 1.12 we define the poles, zeros, and the left and right minimal indices of an arbitrary rational matrix. We recall that any rational matrix has a canonical form called the Smith–McMillan form in terms of which we define the orders of poles and zeros. We recall also the local Smith–McMillan form which provides complete information about poles and zeros at any single prescribed point in the closed complex plane. A very useful way to represent a rational matrix is in terms of a realization. We review in some detail the realization theory for rational matrices which take finite values at infinity in Section 1.13, and discuss several aspects of realization theory in the case of general rational matrices as needed in the sequel. One of the most important features of a realization is the possibility to read off the structural elements of a rational matrix from the eigenstructure of two matrix pencils associated with a particular realization. We make precise these ideas in Section 1.14.

1.1 The eigenvalue problem

We start with the eigenvalue problem for a square matrix, and show further its connection with the Jordan canonical form and with the notion of invariant subspace.

By $\mathbb{F}$ we denote either $\mathbb{C}$ or $\mathbb{R}$, $\mathbb{F}^n$ stands for the vector space of columns with n elements in $\mathbb{F}$, and $\mathbb{F}^{p\times m}$ is the set of $p \times m$ matrices with elements in $\mathbb{F}$.

Let $N \in \mathbb{F}^{n\times n}$. In a natural way N represents a linear transformation from $\mathbb{F}^n$ into $\mathbb{F}^n$. The problem of solving for $\lambda \in \mathbb{C}$ the degree n polynomial equation

$$\chi(\lambda) := \det(\lambda I_n - N) = 0 \tag{1.1}$$

is called the *eigenvalue problem*, and $\chi(\lambda)$ is called the *characteristic polynomial* of the matrix N. Due to the fundamental theorem of the algebra there always exist n complex roots (counting multiplicities) of the characteristic polynomial, called the *eigenvalues* of N. The union of the n eigenvalues is called the *spectrum* of N and is denoted by $\Lambda(N)$. For an eigenvalue λ_0 of N, there always exists a nonzero vector $x \in \mathbb{C}^n$ – called an *eigenvector* of N – such that $Nx = \lambda_0 x$.

If $\lambda_0 = 0$ is an eigenvalue of N then N is called *singular.* A nonsingular matrix has always an inverse, and it is also called *invertible.* Precisely, if $N \in \mathbb{F}^{n\times n}$ is nonsingular, there exists a unique matrix in $\mathbb{F}^{n\times n}$, called the inverse of N and denoted N^{-1}, such that $N^{-1}N = NN^{-1} = I_n$ (the identity matrix of order n).

For square matrices we introduce the following relation of equivalence called *similarity.* Two matrices $N, \widetilde{N} \in \mathbb{F}^{n\times n}$ are called *similar* if there exists an invertible matrix $T \in \mathbb{F}^{n\times n}$ such that

$$TNT^{-1} = \widetilde{N}. \tag{1.2}$$

A similarity transformation can be associated with a change of the basis in $\mathbb{F}^n$. Thus any $x \in \mathbb{F}^n$ changes to $\widetilde{x} = Tx$, and N and $\widetilde{N}$ represent the same linear transformation from $\mathbb{F}^n$ into $\mathbb{F}^n$ with respect to the two different choices of a basis.

1.2 The Jordan canonical form of a square matrix

For matrices in $\mathbb{F}^{n\times n}$ we discuss the Jordan canonical form induced by similarity.

The complex Jordan canonical form

For $\mathbb{F} = \mathbb{C}$ the similarity equivalence (1.2) induces a canonical form on the set of $n \times n$ matrices with elements in $\mathbb{C}$ called the *complex Jordan canonical form* as follows. Let $N \in \mathbb{C}^{n\times n}$ and let $\{\lambda_1, \lambda_2, \ldots, \lambda_k\}$ be the set of k distinct eigenvalues of N. Then there always exists an invertible $T \in \mathbb{C}^{n\times n}$ such that

$$TNT^{-1} = N_J \tag{1.3}$$

where

$$N_J := \operatorname{diag}(N_{11}(\lambda_1), N_{22}(\lambda_2), \ldots, N_{kk}(\lambda_k)), \tag{1.4}$$

$$N_{ii}(\lambda_i) := \operatorname{diag}(J_{s_1^{(i)}}(\lambda_i), J_{s_2^{(i)}}(\lambda_i), \ldots, J_{s_{h_i}^{(i)}}(\lambda_i)), \tag{1.5}$$

and $J_s(\lambda_i)$ is an $s \times s$ elementary matrix of the form

$$J_s(\lambda_i) := \begin{bmatrix} \lambda_i & 1 & & \\ & \lambda_i & \ddots & \\ & & \ddots & 1 \\ & & & \lambda_i \end{bmatrix}, \tag{1.6}$$

called an *elementary Jordan block.*

The set of distinct eigenvalues $\{\lambda_1, \lambda_2, \ldots, \lambda_k\}$ together with the set of block sizes $\{s_j^{(i)},\ i = 1, \ldots, k,\ j = 1, \ldots, h_i\}$ completely determine the complex Jordan canonical form (up to a permutation of the blocks on the main diagonal). For a given eigenvalue λ_i, the sizes $s_j^{(i)}$, $(j = 1, \ldots, h_i)$ are called *elementary Jordan indices* (or partial multiplicities) of the eigenvalue λ_i. The positive integer h_i is called the *geometric multiplicity* of λ_i. The sum $n_i = s_1^{(i)} + \cdots + s_{h_i}^{(i)}$ is called the *algebraic multiplicity* of λ_i and it is the multiplicity of λ_i as a root of the characteristic polynomial.

The real Jordan canonical form

Let now N be a matrix with real elements. A matrix $N \in \mathbb{R}^{n \times n}$ does not necessarily have n real eigenvalues, yet it has n complex eigenvalues. The non–real eigenvalues always occur in complex conjugated pairs. This makes the treatment of canonical forms under real similarity transformations somehow different. Let $\{\lambda_1, \lambda_2, \ldots, \lambda_r, \lambda_{r+1}, \ldots, \lambda_k, \overline{\lambda}_{r+1}, \ldots, \overline{\lambda}_k\}$ be the spectrum of N, where $\lambda_1, \ldots, \lambda_r$, are the real eigenvalues and $\lambda_{r+1}, \ldots, \lambda_k, \overline{\lambda}_{r+1}, \ldots, \overline{\lambda}_k$, are the non–real eigenvalues that occur in complex conjugated pairs. Thus if we restrict the similarity transformation T in (1.3) to have real elements as well, we obtain the *real Jordan canonical form* N_J of N, where N_J is as in (1.4), $N_{ii}(\lambda_i)$ has the form (1.5)–(1.6) for real eigenvalues λ_i $(i = 1, \ldots, r)$ while for nonreal eigenvalues $\lambda_i = \alpha_i + j\beta_i$ $(\alpha_i, \beta_i \in \mathbb{R}, \beta_i \neq 0,$ $i = r+1, \ldots, k)$, we have

$$N_{ii}(\lambda_i) := \operatorname{diag}\,(J_{s_1^{(i)}}(\alpha_i, \beta_i), J_{s_2^{(i)}}(\alpha_i, \beta_i), \ldots, J_{s_{h_i}^{(i)}}(\alpha_i, \beta_i)), \tag{1.7}$$

and $J_s(\alpha_i, \beta_i)$ is a $2s \times 2s$ elementary matrix of the form

$$J_s(\alpha_i, \beta_i) := \begin{bmatrix} K(\alpha_i,\beta_i) & I_2 & & & \\ & K(\alpha_i,\beta_i) & I_2 & & \\ & & K(\alpha_i,\beta_i) & \ddots & \\ & & & \ddots & I_2 \\ & & & & K(\alpha_i,\beta_i) \end{bmatrix} \tag{1.8}$$

called an *elementary real Jordan block*, where

$$K(\alpha_i, \beta_i) := \begin{bmatrix} \alpha_i & \beta_i \\ -\beta_i & \alpha_i \end{bmatrix}.$$

It is clear from the above discussion that two similar matrices always have the same Jordan canonical form.

1.3 Invariant subspace of a square matrix

We introduce now the concept of invariant subspace of a matrix and discuss further its different characterizations.

Definition 1.3.1 *Let $A \in \mathbb{F}^{n \times n}$. The linear space $\mathcal{V} \subset \mathbb{F}^n$ is called an* invariant subspace *of A if*

$$A\mathcal{V} \subset \mathcal{V}. \tag{1.9}$$

The simplest examples of an invariant subspace are the eigenvectors. It is easy to see that if $\mathcal{V}$ is an invariant subspace of dimension ℓ, then there exists a similarity matrix U, partitioned as

$$U = [\underbrace{U_1}_{\ell} \quad \underbrace{U_2}_{n-\ell}], \tag{1.10}$$

with $\mathcal{V} = \operatorname{Im} U_1$, such that

$$U^{-1}AU = \begin{bmatrix} A_{11} & A_{12} \\ O & A_{22} \end{bmatrix} \begin{matrix} \}\ell \\ \}n-\ell \end{matrix} \tag{1.11}$$

with column blocks of widths ℓ and $n-\ell$,

and $\Lambda(A) = \Lambda(A_{11}) \cup \Lambda(A_{22})$. In this new coordinate system $\mathcal{V}$ is represented by

$$\mathcal{V} = \operatorname{Im} \begin{bmatrix} I_\ell \\ O \end{bmatrix}. \tag{1.12}$$

The matrix A and its spectrum *restricted to the space* $\mathcal{V}$ are defined by

$$A|_\mathcal{V} := A_{11},$$
$$\Lambda(A)|_\mathcal{V} := \Lambda(A_{11}),$$

respectively. Conversely, if (1.11) holds for a nonsingular U as in (1.10), then $\mathcal{V} = \operatorname{Im} U_1$ is an invariant subspace of A and $\Lambda(A)|_\mathcal{V} = A_{11}$.

Consider a partition of the complex plane in two disjoint sets

$$\mathbb{C} = \mathbb{C}_g \cup \mathbb{C}_b$$

(the good and the bad one). An invariant subspace $\mathcal{V}$ will be called $\mathbb{C}_g$*–invariant* if the spectrum of $\Lambda(A)|_\mathcal{V}$ lies in $\mathbb{C}_g$.

If $\Lambda_1 \cup \Lambda_2 = \Lambda(A)$ is a partition of the spectrum of $A \in \mathbb{F}^{n\times n}$ then there exists an invariant subspace $\mathcal{V}$ of A such that $\Lambda(A)|_\mathcal{V} = \Lambda_1$. This follows immediately from the Jordan canonical form and the considerations above. If A has real elements, and Λ_1 is a symmetric set (with respect to the imaginary axis), then we can choose $\mathcal{V} \subset \mathbb{R}^n$. Furthermore, if $\Lambda_1 \cap \Lambda_2 = \emptyset$ then $\mathcal{V}$ is unique.

The following result gives a useful characterization of the invariant subspace in terms of an associated basis matrix.

Theorem 1.3.2 *Let* $A \in \mathbb{F}^{n\times n}$.

1. If $\mathcal{V} = \operatorname{Im} V \subset \mathbb{F}^n$ *is an* ℓ*–dimensional invariant subspace of* A*, where* $V \in \mathbb{F}^{n\times\ell}$ *is a basis matrix for* $\mathcal{V}$*, then there exists a matrix* $S \in \mathbb{F}^{\ell\times\ell}$ *such that*

$$AV = VS. \tag{1.13}$$

Moreover, $S = A|_\mathcal{V}$.

2. Conversely, if $AV = VS$ *holds for a certain basis matrix* $V \in \mathbb{F}^{n\times\ell}$ *and a matrix* $S \in \mathbb{F}^{\ell\times\ell}$*, then* $\mathcal{V} = \operatorname{Im} V$ *is an invariant subspace of* A *and* $S = A|_\mathcal{V}$.

Proof. It follows easily from the definition of the invariant subspace of a matrix.

1.4 Controllability and observability

Controllability and observability are important structural properties in systems theory. In the case of linear systems they admit purely algebraic characterizations in terms of the system matrices. In this section we take a purely algebraic point of view and we introduce controllability and observability as properties of pairs of matrices.

The characterizations of controllability and observability given in the following definition are usually known as the Popov–Belevitch–Hautus (PBH) tests.

Definition 1.4.1 *Let* $\mathbb{M}$ *be a set in* $\mathbb{C}$.
Controllability *The ordered pair of matrices* (A, B), *with* $A \in \mathbb{F}^{n\times n}$, $B \in \mathbb{F}^{n\times m}$, *is called* $\mathbb{M}$–controllable *if*

$$\operatorname{rank} \begin{bmatrix} \lambda I_n - A & -B \end{bmatrix} = n, \quad \forall \lambda \in \mathbb{M},$$

and is called simply controllable *if* $\mathbb{M} = \mathbb{C}$.
Observability *The ordered pair of matrices* (C, A), *with* $A \in \mathbb{F}^{n\times n}$, $C \in \mathbb{F}^{p\times n}$, *is called* $\mathbb{M}$–observable *if*

$$\operatorname{rank} \begin{bmatrix} \lambda I_n - A \\ C \end{bmatrix} = n, \quad \forall \lambda \in \mathbb{M},$$

and is called simply observable *if* $\mathbb{M} = \mathbb{C}$.

We also say that *a set* $\mathbb{M}$ *is controllable (observable)* with respect to a certain matrix pair if the respective pair is $\mathbb{M}$–controllable (observable). In the particular cases in which $\mathbb{M} = \mathbb{C}_+ \cup \mathbb{C}_0$ (or $\mathbb{M} = \mathbb{C} \setminus \mathbb{D}$) the pair (A, B) is called *c–controllable* (or *d–controllable*). Accordingly, we shall use also the terms c–observable and d–observable. The c– and d–controllability are related to the stabilizability of continuous and discrete–time dynamical systems, respectively. Similarly, c– and d–observability are related to the detectability of continuous and discrete–time systems, respectively. These connections will be investigated further in Chapter 2.

Controllability and observability are dual properties in the following sense. The pair (A, B) is controllable if and only if the pair (B^*, A^*) is observable, and the pair (C, A) is observable if and only if the pair (A^*, C^*) is controllable. The following theorem gives equivalent characterizations of controllability.

Theorem 1.4.2 *Let* (A, B) *be a pair of matrices, with* $A \in \mathbb{F}^{n\times n}$, $B \in \mathbb{F}^{n\times m}$.

1. *The following assertions are equivalent:*
   (a) *The pair* (A, B) *is controllable.*
   (b) $\operatorname{rank} \begin{bmatrix} B & AB & \dots & A^{n-1}B \end{bmatrix} = n$.
   (c) *For any nonempty set* $\mathbb{M} \subset \mathbb{C}$ *there exists a matrix* $F \in \mathbb{C}^{m\times n}$ *such that* $\Lambda(A+BF) \subset \mathbb{M}$. *Moreover, if* $\mathbb{F} = \mathbb{R}$ *and the set* $\mathbb{M}$ *is symmetric (with respect to the imaginary axis) then the matrix* F *can be chosen with real elements.*
2. *Let* $\mathbb{M}$ *be a given set in* $\mathbb{C}$ *such that* $\mathbb{M}_c := \mathbb{C} \setminus \mathbb{M}$ *is nonempty. The following assertions are equivalent:*
   (a) *The pair* (A, B) *is* $\mathbb{M}$*–controllable.*

(b) *There exists a matrix* $F \in \mathbb{C}^{m\times n}$ *such that* $\Lambda(A+BF) \subset \mathbb{M}_c$. *Moreover, if* $\mathbb{F} = \mathbb{R}$ *and* $\mathbb{M}_c$ *contains a symmetric set then* F *can be chosen with elements in* $\mathbb{R}$.

By duality we can formulate a similar result for observability.

Theorem 1.4.3 *Let* (C, A) *be a pair of matrices, with* $A \in \mathbb{F}^{n\times n}$, $C \in \mathbb{F}^{p\times n}$.

1. *The following assertions are equivalent:*
   (a) *The pair* (C, A) *is observable.*
   (b)
$$\operatorname{rank}\begin{bmatrix} C \\ CA \\ \vdots \\ CA^{n-1} \end{bmatrix} = n.$$
   (c) *For any nonempty set* $\mathbb{M} \subset \mathbb{C}$ *there exists a matrix* $K \in \mathbb{C}^{n\times p}$ *such that* $\Lambda(A+KC) \subset \mathbb{M}$. *Moreover, if* $\mathbb{F} = \mathbb{R}$ *and the set* $\mathbb{M}$ *is symmetric (with respect to the imaginary axis) then the matrix* K *can be chosen with real elements.*
2. *Let* $\mathbb{M}$ *be a given set in* $\mathbb{C}$ *such that* $\mathbb{M}_c := \mathbb{C}\setminus\mathbb{M}$ *is nonempty. The following assertions are equivalent:*
   (a) *The pair* (C, A) *is* $\mathbb{M}$*–observable.*
   (b) *There exists a matrix* $K \in \mathbb{R}^{n\times p}$ *such that* $\Lambda(A+KC) \subset \mathbb{M}_c$. *Moreover, if* $\mathbb{F} = \mathbb{R}$ *and* $\mathbb{M}_c$ *contains a symmetric set then* K *can be chosen with elements in* $\mathbb{R}$.

The next results shows that any matrix pair (A, B) contains, in a certain sense, a sub–pair that is controllable.

Theorem 1.4.4 (Kalman's controllable form) *Let* (A, B) *be a pair of matrices, with* $A \in \mathbb{F}^{n\times n}$, $B \in \mathbb{F}^{n\times m}$. *Then there exists an invertible* $T \in \mathbb{F}^{n\times n}$ *such that*

$$\widetilde{A} := TAT^{-1} = \begin{bmatrix} A_1 & A_{12} \\ O & A_2 \end{bmatrix}, \quad \widetilde{B} := TB = \begin{bmatrix} B_1 \\ O \end{bmatrix},$$

where the pair (A_1, B_1) *is controllable, with* $A_1 \in \mathbb{F}^{n_1\times n_1}$ *and* $B_1 \in \mathbb{F}^{n_1\times m}$, *and the space*

$$\mu_c(A, B) := \operatorname{Im} T^{-1}\begin{bmatrix} I_{n_1} \\ O \end{bmatrix}$$

is the controllable space *of* (A, B) *defined as the smallest invariant subspace of* A *which contains* $\operatorname{Im} B$.

The previous theorem suggests the following definition: the space

$$\mu_{uc}(A, B) := \operatorname{Im} T^{-1}\begin{bmatrix} O \\ I_{n_2} \end{bmatrix}$$

where $n_2 := n - n_1$, is called the *uncontrollable space* of the pair (A, B).

A dual result holds for observability as shown in the following theorem.

Theorem 1.4.5 (Kalman's observable form) *Let (C, A) be a pair of matrices, with $A \in \mathbb{F}^{n\times n}$, $C \in \mathbb{F}^{p\times n}$. Then there exists an invertible $T \in \mathbb{F}^{n\times n}$ such that*

$$\widetilde{A} := TAT^{-1} = \begin{bmatrix} A_1 & A_{12} \\ O & A_2 \end{bmatrix}, \quad \widetilde{C} := CT^{-1} = \begin{bmatrix} O & C_2 \end{bmatrix},$$

where the pair (C_2, A_2) is observable, with $A_2 \in \mathbb{F}^{n_2\times n_2}$ and $C_2 \in \mathbb{F}^{p\times n_2}$, and the space

$$\nu_{uo}(C, A) := \operatorname{Im} T^{-1} \begin{bmatrix} I_{n_1} \\ O \end{bmatrix}$$

is the unobservable space *of (C, A) defined as the largest invariant subspace of A which is included in* $\operatorname{Ker} C$.

The previous theorem suggests the following definition: the space

$$\nu_o(C, A) := \operatorname{Im} T^{-1} \begin{bmatrix} O \\ I_{n_2} \end{bmatrix}$$

where $n_2 := n - n_1$, is called the *observable space* of the pair (C, A).

Other important equivalent characterizations of controllability and observability are connected with certain properties of linear dynamical systems, either in continuous or discrete–time, and will be revealed in Chapter 2.

1.5 Linear matrix equations

Several linear matrix equations, such as the Sylvester and the Stein equations, and their symmetric versions called the Lyapunov and the symmetric Stein equations, play a central role in the algebraic Riccati equation theory and, in general, in linear systems theory. We present in this section solvability conditions for, and some remarkable properties of these equations.

Recall first that a matrix A is called *symmetric* if $A = A^T$ and *hermitian* if $A = A^*$. If A is hermitian then all its eigenvalues are real and we introduce the *sign matrix of* A as

$$\operatorname{sgn}(A) := \operatorname{diag}(I_{n_-}, I_{n_+}, O_{n_0}) \tag{1.14}$$

where n_-, n_+, and n_0 are the number of negative, positive and zero eigenvalues of A, respectively, and O_{n_0} stands for an $n_0 \times n_0$ null matrix. A matrix having the form in the right–hand side of (1.14) is called a *sign matrix.* Notice that the sign matrix is uniquely defined for a given hermitian matrix.

Sylvester and Stein equations

Theorem 1.5.1 (Sylvester and Stein equations) *Let $A \in \mathbb{F}^{n\times n}$, $B \in \mathbb{F}^{m\times m}$, and $C \in \mathbb{F}^{n\times m}$.*

1. The Sylvester equation

$$AX + XB + C = 0 \tag{1.15}$$

has a unique solution $X \in \mathbb{F}^{n\times m}$ if and only if

$$\lambda + \mu \neq 0, \quad \text{for any } \lambda \in \Lambda(A),\ \mu \in \Lambda(B). \tag{1.16}$$

2. The Stein equation

$$AXB - X + C = 0 \tag{1.17}$$

has a unique solution $X \in \mathbb{F}^{n\times m}$ *if and only if*

$$\lambda\mu \neq 1, \quad \text{for any } \lambda \in \Lambda(A),\ \mu \in \Lambda(B). \tag{1.18}$$

If the condition (1.16) (or (1.18)) does not hold then the Sylvester equation (respectively the Stein equation) may have no solution or an infinity of solutions, depending on the free term C.

Lyapunov and symmetric Stein equations

If $B = A^*$ and $Q := C = C^*$ then the Sylvester equation is known as the *Lyapunov equation* and takes the form

$$AX + XA^* + Q = 0, \tag{1.19}$$

whereas the Stein equation is called the *symmetric Stein equation* (or the *discrete–time Lyapunov equation*) and takes the form

$$AXA^* - X + Q = 0. \tag{1.20}$$

The following result is a direct consequence of Theorem 1.5.1.

Theorem 1.5.2 (Lyapunov and symmetric Stein equations) *Let* $A \in \mathbb{F}^{n\times n}$ *and* $Q = Q^* \in \mathbb{F}^{n\times n}$.

1. The Lyapunov equation

$$AX + XA^* + Q = 0 \tag{1.21}$$

has a unique solution $X \in \mathbb{F}^{n\times n}$ *if and only if*

$$\lambda + \mu \neq 0, \quad \text{for any } \lambda, \mu \in \Lambda(A). \tag{1.22}$$

Moreover, if X *is unique, then it is hermitian.*

2. The symmetric Stein equation

$$AXA^* - X + Q = 0 \tag{1.23}$$

has a unique solution $X \in \mathbb{F}^{n\times n}$ *if and only if*

$$\lambda\mu \neq 1, \quad \text{for any } \lambda, \mu \in \Lambda(A). \tag{1.24}$$

Moreover, if X *is unique, then it is hermitian.*

We give now a general result relating the inertia of the solution X to the inertia of the matrix A from which many important, well–known theorems on the Lyapunov and symmetric Stein equations follow as simple corollaries.

Given a square matrix $A \in \mathbb{F}^{n\times n}$, we denote by $\nu_c(A)$, $\pi_c(A)$, and $\delta_c(A)$, the number of eigenvalues of A with strictly negative, strictly positive, and null real parts, respectively, and by $\nu_d(A)$, $\pi_d(A)$, and $\delta_d(A)$, the number of eigenvalues of A with modulus less than, greater than, and equal to 1, respectively. The triplet of integers $(\nu_c(A),\pi_c(A),\delta_c(A))$ is called the *c–inertia* and the triplet of integers $(\nu_d(A),\pi_d(A),\delta_d(A))$ is called the *d–inertia* of the matrix A. Clearly,

$$\nu_c(A) + \pi_c(A) + \delta_c(A) = \nu_d(A) + \pi_d(A) + \delta_d(A) = n.$$

Theorem 1.5.3 *Let $A, Q, X \in \mathbb{F}^{n\times n}$, with $X = X^*$ and $Q = Q^* \geq 0$.*

1. Assume A, Q, X satisfy the Lyapunov equation

$$AX + XA^* + Q = 0.$$

Then we have:

(a) If $\delta_c(A) = 0$ then $\nu_c(A) \geq \pi_c(X)$ and $\pi_c(A) \geq \nu_c(X)$.
(b) If $\delta_c(X) = 0$ then $\nu_c(X) \geq \pi_c(A)$ and $\pi_c(X) \geq \nu_c(A)$.

2. Assume A, Q, X satisfy the symmetric Stein equation

$$AXA^* - X + Q = 0.$$

Then we have:

(a) If $\delta_d(A) = 0$ then $\nu_d(A) \geq \pi_c(X)$ and $\pi_d(A) \geq \nu_c(X)$.
(b) If $\delta_c(X) = 0$ then $\nu_c(X) \geq \pi_d(A)$ and $\pi_c(X) \geq \nu_d(A)$.

Assume that $Q = Q^* \geq 0$ is given in the form $Q = BB^*$. If the pair (A, B) is controllable, the inequalities in Theorem 1.5.3 can be strengthened to equalities.

Theorem 1.5.4 *Let $A, X \in \mathbb{F}^{n\times n}$, $B \in \mathbb{F}^{n\times m}$, $X = X^*$, and assume the pair (A, B) is controllable.*

1. If A, X, B satisfy the Lyapunov equation

$$AX + XA^* + BB^* = 0$$

then we have $\delta_c(A) = 0$, $\delta_c(X) = 0$, $\nu_c(A) = \pi_c(X)$, $\pi_c(A) = \nu_c(X)$.

2. If A, X, B satisfy the symmetric Stein equation

$$AXA^* - X + BB^* = 0$$

then we have $\delta_d(A) = 0$, $\delta_c(X) = 0$, $\nu_d(A) = \pi_c(X)$, $\pi_d(A) = \nu_c(X)$.

If we relax the controllability assumption for the pair (A, B) we obtain the following result that will be used extensively in the sequel.

Theorem 1.5.5 *Let $A, X \in \mathbb{F}^{n\times n}$, $B \in \mathbb{F}^{n\times m}$, and $X = X^*$.*

1. If the pair (A, B) is c-controllable, and A, X, B satisfy the Lyapunov equation

$$AX + XA^* + BB^* = 0$$

then we have $\delta_c(A) = 0, \nu_c(A) = \pi_c(X) + \delta_c(X), \pi_c(A) = \nu_c(X)$.

2. If the pair (A, B) is d-controllable and A, X, B satisfy the symmetric Stein equation

$$AXA^* - X + BB^* = 0$$

then we have $\delta_d(A) = 0, \nu_d(A) = \pi_c(X) + \delta_c(X), \pi_d(A) = \nu_c(X)$.

If in Theorem 1.5.4 and Theorem 1.5.5 we replace A with A^* and B with C^* we obtain corresponding dual versions. Under certain circumstances, the kernel of X satisfying the Lyapunov (or symmetric Stein) equation is related in a precise way to the controllable space of the pair (A, B). However, we shall only need the following dual result.

Theorem 1.5.6 *Let $A, X \in \mathbb{F}^{n\times n}$, $C \in \mathbb{F}^{p\times n}$, $X = X^*$ and let $\nu_{uo}(C, A)$ be the unobservable space of the pair (C, A).*

1. *Assume A, X, C satisfy the Lyapunov equation*

$$A^*X + XA + C^*C = 0. \tag{1.25}$$

   Then we have:

   (a) $\operatorname{Ker} X \subset \nu_{uo}(C, A)$.
   (b) *If the Lyapunov equation (1.25) has a unique solution X then we also have $\nu_{uo}(C, A) \subset \operatorname{Ker} X$.*

2. *Assume A, X, C satisfy the symmetric Stein equation*

$$A^*XA - X + C^*C = 0. \tag{1.26}$$

   Then we have:

   (a) $\operatorname{Ker} X \subset \nu_{uo}(C, A)$.
   (b) *If the symmetric Stein equation (1.26) has a unique solution X then we have $\nu_{uo}(C, A) \subset \operatorname{Ker} X$.*

1.6 The generalized eigenvalue problem: the regular case

We turn now our attention to matrix pencils and discuss first the generalized eigenvalue problem, the Weierstrass canonical form, and the notion of deflating subspace of a matrix pencil that satisfies a regularity assumption. Further, we shall generalize all these concepts to the case of a general pencil.

Let M and N be $m \times n$ matrices with elements in $\mathbb{F}$. The first–order matrix polynomial $\lambda M - N$ in the indeterminate λ is called a *matrix pencil* or, briefly, *pencil.*

We study first matrix pencils $\lambda M - N$ that are *regular*, i.e., which are square $(n \times n)$ and have a non–vanishing determinant $\det(\lambda M - N) \not\equiv 0$. A pencil which is not regular is called *singular.* In particular, if M or N are invertible then the pencil $\lambda M - N$ is regular. However, we are interested here in more general regular matrix pencils for which M and N are both singular.

Given a regular matrix pencil $\lambda M - N$, with $M, N \in \mathbb{F}^{n\times n}$, the problem of solving for $\lambda \in \mathbb{C}$ the polynomial equation

$$\chi(\lambda) := \det(\lambda M - N) = 0 \tag{1.27}$$

is called the *generalized eigenvalue problem.* Since M may be singular, the *characteristic polynomial* $\chi(\lambda)$ has degree $n_f \leq n$. The n_f roots of $\chi(\lambda)$ are called the *finite generalized eigenvalues* of $\lambda M - N$. We say that $\lambda = \infty$ is a generalized eigenvalue of $\lambda M - N$ if $\lambda = 0$ is a generalized eigenvalue of the *reciprocal pencil* $\lambda N - M$. The multiplicity n_∞ of the infinite generalized eigenvalue is by definition the multiplicity of $\lambda = 0$ as a generalized eigenvalue of $\lambda N - M$. It follows easily that $n_\infty = n - n_f$. Therefore, an $n \times n$ regular pencil $\lambda M - N$ always has n generalized eigenvalues (finite and infinite) which form the *spectrum* denoted $\Lambda(M, N)$. For a generalized eigenvalue λ_0 there always exists a nonzero vector $x \in \mathbb{C}^n$ – called a *generalized eigenvector* – such that

$$\begin{cases} Nx = \lambda_0 Mx & \text{if } \lambda_0 \text{ is finite,} \\ Mx = 0 & \text{if } \lambda_0 \text{ is infinite.} \end{cases}$$

1.7 The Weierstrass canonical form of a regular matrix pencil

We introduce first an equivalence relation for general matrix pencils, and we study in this section its properties in the particular case of regular pencils.

Two matrix pencils $\lambda M - N$ and $\lambda \widetilde{M} - \widetilde{N}$, with $M, N, \widetilde{M}, \widetilde{N} \in \mathbb{F}^{m\times n}$ are called *(strictly) equivalent* if there exist two constant invertible matrices $Q \in \mathbb{F}^{m\times m}$, $Z \in \mathbb{F}^{n\times n}$, such that

$$Q(\lambda M - N)Z = \lambda \widetilde{M} - \widetilde{N}. \tag{1.28}$$

For $\mathbb{F} = \mathbb{C}$ the strict equivalence relation (1.28) induces a canonical form on the set of regular $n \times n$ pencils called the *complex Weierstrass canonical form.* Thus, for any regular matrix pencil $\lambda M - N$, with $M, N \in \mathbb{C}^{n\times n}$, there always exist two invertible matrices $Q, Z \in \mathbb{C}^{n\times n}$ such that

$$Q(\lambda M - N)Z = \lambda M_W - N_W$$

where

$$\lambda M_W - N_W := \operatorname{diag}(\lambda M_\infty - I_{n_\infty}, \lambda I_{n_f} - N_f), \tag{1.29}$$

N_f is in the complex Jordan canonical form (1.4), M_∞ is nilpotent and it is in the Jordan canonical form

$$M_\infty := \operatorname{diag}(J_{s_1^\infty}(0), J_{s_2^\infty}(0), \ldots, J_{s_{h_\infty}^\infty}(0)) \tag{1.30}$$

and $J_s(0)$ is an $s \times s$ elementary (nilpotent) Jordan block (1.6) with eigenvalue $\lambda = 0$. Note that the finite generalized eigenvalues of the pencil $\lambda M - N$ are the eigenvalues of the matrix N_f. Therefore, we use N_f to define the concepts of partial, algebraic and geometric multiplicities of a finite generalized eigenvalue of $\lambda M - N$. Next, we define the partial, algebraic and geometrical multiplicities of the infinite eigenvalue of $\lambda M - N$ as the corresponding multiplicities of the zero eigenvalue of M_∞. Thus, for $\lambda = \infty$ the partial multiplicities are s_i^∞ $(i = 1, \ldots, h_\infty)$, the algebraic multiplicity is n_∞ and satisfies $n_\infty = \sum_{i=1}^{h_\infty} s_i^\infty$, and the geometric multiplicity is h_∞. The set of generalized eigenvalues together with their partial multiplicities completely determines the Weierstrass canonical form of a regular pencil (up to a permutation of the diagonal blocks on the main diagonal).

If M is invertible then there are no infinite generalized eigenvalues and the Weierstrass canonical form reduces to the Jordan canonical form of the matrix $M^{-1}N$.

Consider now M and N having elements in $\mathbb{R}$. If we restrict the class of transformation matrices Q and Z to have real elements as well we obtain the *real Weierstrass canonical form* (1.29) where now N_f is in the real Jordan canonical form.

1.8 Deflating subspace of a regular matrix pencil

We discuss now the generalization to the regular pencil case of the notion of invariant subspace of a matrix.

Definition 1.8.1 *Let $\lambda M - N$ be a regular pencil, with $M, N \in \mathbb{F}^{n\times n}$. The linear space $\mathcal{V} \subset \mathbb{F}^n$ is called a* deflating subspace *of $\lambda M - N$ if*

$$\dim(M\mathcal{V} + N\mathcal{V}) = \dim\mathcal{V}. \tag{1.31}$$

The simplest examples of a deflating subspace are the generalized eigenvectors. For the case $M = I_n$, the definition of a deflating subspace reduces to that of an invariant subspace of N since $\dim(\mathcal{V} + N\mathcal{V}) = \dim\mathcal{V}$ is equivalent to $N\mathcal{V} \subset \mathcal{V}$. Now let $\mathcal{V} \subset \mathbb{F}^n$ denote an arbitrary space (not necessarily deflating) of dimension ℓ and define

$$\mathcal{W} := M\mathcal{V} + N\mathcal{V}. \tag{1.32}$$

Then, since $\lambda M - N$ is regular, one can easily prove that we have in general

$$k := \dim\mathcal{W} \geq \dim\mathcal{V} = \ell, \tag{1.33}$$

with equality taking place only if $\mathcal{V}$ is a deflating subspace. Indeed, for any basis matrices Z_1 and Q_1 of $\mathcal{V}$ and $\mathcal{W}$, respectively, we can write

$$\operatorname{rank} Q_1 = \operatorname{rank} \begin{bmatrix} MZ_1 & NZ_1 \end{bmatrix} = \operatorname{rank} \begin{bmatrix} MZ_1 & (-\lambda_0 M + N)Z_1 \end{bmatrix} \geq \operatorname{rank} Z_1$$

for any λ_0 for which $\det(\lambda_0 M - N) \neq 0$. Hence (1.33) follows. Construct the invertible matrices Q and Z, partitioned as

$$Z = [\; \underbrace{Z_1}_{\ell} \quad \underbrace{Z_2}_{n-\ell} \;], \qquad Q = [\; \underbrace{Q_1}_{k} \quad \underbrace{Q_2}_{n-k} \;] \tag{1.34}$$

such that

$$\mathcal{V} = \operatorname{Im} Z_1, \qquad \mathcal{W} = \operatorname{Im} Q_1. \tag{1.35}$$

Then it follows from (1.32) that $M\mathcal{V} \subset \mathcal{W}$ and $N\mathcal{V} \subset \mathcal{W}$, and thus

$$Q^{-1}(\lambda M - N)Z = \begin{bmatrix} \lambda M_{11} - N_{11} & \lambda M_{12} - N_{12} \\ O & \lambda M_{22} - N_{22} \end{bmatrix} \begin{matrix} \}\, k \\ \}\, n-k, \end{matrix} \tag{1.36}$$

with the first block column of width ℓ and the second of width $n - \ell$, where $k \geq \ell$. In this new coordinate system $\mathcal{V}$ and $\mathcal{W}$ are represented by

$$\mathcal{V} = \operatorname{Im} \begin{bmatrix} I_\ell \\ O \end{bmatrix}, \qquad \mathcal{W} = \operatorname{Im} \begin{bmatrix} I_k \\ O \end{bmatrix}. \tag{1.37}$$

If $\mathcal{V}$ is now a deflating subspace, then $\dim\mathcal{W} = k = \ell = \dim\mathcal{V}$, M_{ii} and N_{ii} are square for $i = 1, 2$, and we define the *map and the spectrum of the pencil restricted to the space $\mathcal{V}$* as

$$(\lambda M - N)|_{\mathcal{V}} := \lambda M_{11} - N_{11}, \tag{1.38}$$

$$\Lambda(M,N)|_{\mathcal{V}} := \Lambda(M_{11}, N_{11}). \tag{1.39}$$

Consider a partition of the complex plane in two symmetric disjoint sets

$$\mathbb{C} = \mathbb{C}_g \cup \mathbb{C}_b \tag{1.40}$$

(the good and the bad one). A deflating subspace $\mathcal{V}$ will be $\mathbb{C}_g$–deflating if $\Lambda(M,N)|_{\mathcal{V}}$ lies in $\mathbb{C}_g$. We have the following result.

Proposition 1.8.2 *Let $\lambda M - N$ be a regular pencil, with $M, N \in \mathbb{F}^{n\times n}$.*

1. If $\mathcal{V} \subset \mathbb{F}^n$ is an ℓ–dimensional deflating subspace of $\lambda M - N$ and $\mathcal{W} := M\mathcal{V} + N\mathcal{V}$ then there exist invertible matrices

$$Z = [\ \underbrace{Z_1}_{\ell}\quad \underbrace{Z_2}_{n-\ell}\ ], \qquad Q = [\ \underbrace{Q_1}_{\ell}\quad \underbrace{Q_2}_{n-\ell}\ ] \tag{1.41}$$

with $\mathcal{V} = \operatorname{Im} Z_1$ and $\mathcal{W} = \operatorname{Im} Q_1$, and $Q, Z \in \mathbb{F}^{n\times n}$, such that

$$Q^{-1}(\lambda M - N)Z \;=\; \left[\begin{array}{cc} \lambda M_{11} - N_{11} & \lambda M_{12} - N_{12} \\ O & \lambda M_{22} - N_{22} \end{array}\right] \begin{array}{l} \}\,\ell \\ \}\,n-\ell, \end{array} \tag{1.42}$$

with the first block column of width ℓ and the second of width $n-\ell$,

where the diagonal pencils $\lambda M_{ii} - N_{ii}$ $(i = 1,2)$ are regular with $\Lambda(M_{11}, N_{11}) \cup \Lambda(M_{22}, N_{22}) = \Lambda(M,N)$ and $\Lambda(M,N)|_{\mathcal{V}} = \Lambda(M_{11}, N_{11})$.

2. Conversely, if (1.42) holds for the invertible matrices Q and Z as in (1.41), then $\mathcal{V} = \operatorname{Im} Z_1$ is a deflating subspace of $\lambda M - N$, $\mathcal{W} = \operatorname{Im} Q_1$ where $\mathcal{W} := M\mathcal{V} + N\mathcal{V}$, and $\Lambda(M,N)|_{\mathcal{V}} = \Lambda(M_{11}, N_{11})$.

Proof. It follows immediately from (1.34)–(1.36), written for $k = \ell$. ■

Corollary 1.8.3 *Let $\lambda M - N$ be a regular pencil with coefficients in $\mathbb{F}$ and let $\Lambda_1 \cup \Lambda_2 = \Lambda(M,N)$ be a partition of its spectrum. Then there exists a deflating subspace $\mathcal{V} \subset \mathbb{C}^n$ such that*

$$\Lambda(M,N)|_{\mathcal{V}} = \Lambda_1.$$

If M and N have real elements, and Λ_1 is a symmetric set, then we can choose $\mathcal{V} \subset \mathbb{R}^n$. If $\Lambda_1 \cap \Lambda_2 = \emptyset$, then $\mathcal{V}$ is unique.

Proof. The first part is a simple consequence of the Weierstrass canonical form and of Proposition 1.8.2. For the uniqueness we refer to Stewart [Ste72]. ■

We give now a useful characterization of a deflating subspace in terms of an associated basis matrix. We deal only with deflating subspaces with a finite spectrum, i.e., deflating subspaces $\mathcal{V}$ for which $\Lambda(M,N)|_{\mathcal{V}}$ has only finite elements, because this is the only case that will be used in the sequel. The general case is more technical and may be found in Oară and Van Dooren [OV97].

Theorem 1.8.4 *Let $\lambda M - N$ be a regular pencil, with $M, N \in \mathbb{F}^{n\times n}$.*

1. If $V \in \mathbb{F}^{n\times \ell}$ is a basis matrix for an ℓ-dimensional deflating subspace $\mathcal{V}$ with finite spectrum, then there exists a matrix $S \in \mathbb{F}^{\ell\times\ell}$ such that

$$MVS = NV, \tag{1.43}$$

and $\lambda I - S$ is equivalent to $(\lambda M - N)|_{\mathcal{V}}$.

2. Conversely, if $MVS = NV$ holds for a basis matrix $V \in \mathbb{F}^{n\times\ell}$ and a matrix $S \in \mathbb{F}^{\ell\times\ell}$ then $\mathcal{V} = \operatorname{Im} V$ is a deflating subspace of $\lambda M - N$ and $\lambda I - S$ is equivalent to $(\lambda M - N)|_{\mathcal{V}}$.

Proof. 1. Let $\mathcal{W} := M\mathcal{V} + N\mathcal{V}$ and let W be a basis matrix for it. Since $\mathcal{V}$ is a deflating subspace there exist invertible matrices $Z = \begin{bmatrix} V & \overline{V} \end{bmatrix}$ and $Q = \begin{bmatrix} W & \overline{W} \end{bmatrix}$ such that

$$Q^{-1}(\lambda M - N)Z = \begin{bmatrix} \lambda M_{11} - N_{11} & \lambda M_{12} - N_{12} \\ O & \lambda M_{22} - N_{22} \end{bmatrix} \tag{1.44}$$

and $\Lambda(M,N)|_{\mathcal{V}} = \Lambda(M_{11}, N_{11})$ contains only finite elements. Thus M_{11} is invertible and the conclusion follows by taking $S := M_{11}^{-1}N_{11}$ in (1.43).

2. We rewrite (1.43) as

$$\begin{bmatrix} MV & NV \end{bmatrix} \begin{bmatrix} S \\ -I \end{bmatrix} = 0. \tag{1.45}$$

This shows that the $n \times 2\ell$ matrix $\begin{bmatrix} MV & NV \end{bmatrix}$ is of rank less or equal to ℓ which implies that $\dim \mathcal{W} \leq \dim \mathcal{V} = \ell$, where $\mathcal{W} := M\mathcal{V} + N\mathcal{V}$. Combining this with (1.33) shows that $\dim \mathcal{W} = \dim \mathcal{V}$ and therefore $\mathcal{V}$ is a deflating subspace.

Finally, we show that $\lambda I - S$ is equivalent to $(\lambda M - N)|_{\mathcal{V}}$. Since $\mathcal{V}$ is a deflating subspace, there exist invertible matrices $Z = \begin{bmatrix} V & \overline{V} \end{bmatrix}$ and $Q = \begin{bmatrix} W & \overline{W} \end{bmatrix}$ such that (1.44) holds, where $\mathcal{V} = \operatorname{Im} V$, $\mathcal{W} = \operatorname{Im} W$, and $\Lambda(M,N)|_{\mathcal{V}} = \Lambda(M_{11}, N_{11})$. In this new coordinate system (1.43) becomes

$$\begin{bmatrix} M_{11} & M_{12} \\ O & M_{22} \end{bmatrix} \begin{bmatrix} I \\ O \end{bmatrix} S = \begin{bmatrix} N_{11} & N_{12} \\ O & N_{22} \end{bmatrix} \begin{bmatrix} I \\ O \end{bmatrix}$$

or, equivalently,

$$M_{11}S = N_{11}. \tag{1.46}$$

Since $\lambda M_{11} - N_{11}$ is regular it follows from (1.46) that M_{11} is invertible and therefore $S = M_{11}^{-1}N_{11}$. The conclusion follows from $\lambda M_{11} - N_{11} = (\lambda M - N)|_{\mathcal{V}}$. ■

Relation (1.43) generalizes the characterization (1.13) of an invariant subspace of a matrix to that of a deflating subspace of a regular pencil. Indeed, if $M = I$ we get $VS = NV$ which shows that $\mathcal{V} = \operatorname{Im} V$ is an invariant subspace of N.

1.9 The generalized eigenvalue problem: the general case

We extend now all the results stated so far for regular pencils to general pencils.

We start by describing the structural elements for an arbitrary $m \times n$ pencil $\lambda M - N$ with elements in $\mathbb{F}$. Recall that a singular pencil is a pencil that is not regular, i.e., which is either non–square ($m \neq n$), or is square and has a vanishing determinant $\det(\lambda M - N) \equiv 0$.

It is easy to realize that $\lambda M - N$ has constant rank for all but a finite number of values $\lambda \in \mathbb{C}$ for which it has a smaller rank. The *normal rank* r of a pencil $\lambda M - N$ is defined as the rank of $\lambda M - N$ for almost all $\lambda \in \mathbb{C}$. For a regular pencil we have $m = n = r$. If $\nu_r := n - r > 0$ then the pencil $\lambda M - N$ has a *right singular structure.* If $\nu_\ell := m - r > 0$ then the pencil has a *left singular structure.* A regular pencil has neither a left nor a right singular structure.

We define the *characteristic polynomial* $\chi(\lambda)$ of a pencil $\lambda M - N$, with $M, N \in \mathbb{F}^{m\times n}$, as the greatest common divisor of the polynomials $R_i(\lambda)$, where $\{R_i(\lambda)\}$ is the set of all minors of order r of $\lambda M - N$ (seen as a polynomial matrix). The problem of solving for $\lambda \in \mathbb{C}$ the polynomial equation $\chi(\lambda) := \chi_0 + \chi_1\lambda + \ldots + \chi_{n_f}\lambda^{n_f} = 0$, with $\chi_{n_f} \neq 0$, is called the *generalized eigenvalue problem* for the pencil $\lambda M - N$. The n_f roots of $\chi(\lambda)$ are called the *finite generalized eigenvalues* of $\lambda M - N$. We say that $\lambda = \infty$ is a generalized eigenvalue of $\lambda M - N$ if $\lambda = 0$ is a generalized eigenvalue of the reciprocal pencil $\lambda N - M$ or, equivalently, if $\operatorname{rank} M < r$. The multiplicity n_∞ of the infinite generalized eigenvalue $\lambda = \infty$ of $\lambda M - N$ is by definition the multiplicity of $\lambda = 0$ as a generalized eigenvalue of $\lambda N - M$. The union of $n_f + n_\infty$ generalized eigenvalues of $\lambda M - N$ (finite and infinite) forms the spectrum of $\lambda M - N$, which is denoted by $\Lambda(M, N)$.

1.10 The Kronecker canonical form of a general matrix pencil

Recall that two pencils $\lambda M - N$ and $\lambda\widetilde{M} - \widetilde{N}$, with $M, N, \widetilde{M}, \widetilde{N} \in \mathbb{F}^{m\times n}$, are called *strictly equivalent* if there exist two invertible matrices $Q \in \mathbb{F}^{m\times m}$ and $Z \in \mathbb{F}^{n\times n}$ such that

$$Q(\lambda M - N)Z = \lambda\widetilde{M} - \widetilde{N}. \tag{1.47}$$

For $\mathbb{F} = \mathbb{C}$ the strict equivalence relation (1.47) induces a canonical form on the set of $m\times n$ pencils called the *complex Kronecker canonical form.* Thus, for any matrix pencil $\lambda M - N$, with $M, N \in \mathbb{C}^{m\times n}$, there always exist two invertible matrices $Q \in \mathbb{C}^{m\times m}$ and $Z \in \mathbb{C}^{n\times n}$ such that

$$Q(\lambda M - N)Z = \lambda M_{KR} - N_{KR},$$

where

$$\lambda M_{KR} - N_{KR} := \begin{bmatrix} L_{\epsilon_1} & & & & & & & \\ & \ddots & & & & & & \\ & & L_{\epsilon_{\nu_r}} & & & & & \\ & & & \lambda M_\infty - I_{n_\infty} & & & & \\ & & & & \lambda I_{n_f} - N_f & & & \\ & & & & & L^T_{\eta_1} & & \\ & & & & & & \ddots & \\ & & & & & & & L^T_{\eta_{\nu_\ell}} \end{bmatrix}, \tag{1.48}$$

N_f is a matrix in the Jordan canonical form (1.4), M_∞ is a nilpotent matrix in Jordan canonical form, and L_k ($k \geq 0$) denotes the bidiagonal $k \times (k+1)$ pencil

$$L_k := \begin{bmatrix} \lambda & -1 & & \\ & \ddots & \ddots & \\ & & \lambda & -1 \end{bmatrix}.$$

Note that k may be zero corresponding to either a column or a row of zeros in the pencil (1.48). The Kronecker structure is completely determined by the regular part and the singular part as follows.

The generalized eigenvalues (finite and infinite) of the pencil $\lambda M - N$ are the generalized eigenvalues of the regular pencil

$$\lambda \begin{bmatrix} M_\infty & O \\ O & I_{n_f} \end{bmatrix} - \begin{bmatrix} I_{n_\infty} & O \\ O & N_f \end{bmatrix}$$

(also called the *regular part* of $\lambda M - N$). Therefore, we use this regular pencil to define the concepts of partial, algebraic and geometric multiplicities of the generalized eigenvalues of $\lambda M - N$.

The *singular part* of the pencil is given by the right and left singular Kronecker structure as follows. The $\epsilon_i \times (\epsilon_i + 1)$ blocks L_{ϵ_i}, $(i = 1, \ldots, \nu_r)$, are the right elementary Kronecker blocks, and $\epsilon_i \geq 0$ are called the *right Kronecker indices.* The $(\eta_j + 1) \times \eta_j$ blocks ${L_{\eta_j}}^T$, $(j = 1, ..., \nu_\ell)$, are the left elementary Kronecker blocks, and $\eta_j \geq 0$ are called the *left Kronecker indices.* Notice that ϵ_i and η_j can be zero.

From the Kronecker canonical form, the normal rank of $\lambda M - N$ can be expressed as

$$r = n_r + n_\infty + n_f + n_\ell, \tag{1.49}$$

where $n_r := \sum_{i=1}^{\nu_r} \epsilon_i$ and $n_\ell := \sum_{j=1}^{\nu_\ell} \eta_j$. If $\lambda M - N$ is regular, there are no Kronecker indices (i.e., the pencils $L_{\epsilon_i}, L_{\eta_j}^T$ are void) and the Kronecker canonical form reduces to the Weierstrass canonical form.

Consider now M and N having elements in $\mathbb{R}$. If we restrict the class of transformation matrices Q and Z to have real elements as well, we obtain the *real Kronecker canonical form* (1.48) where now N_f is in the real Jordan canonical form (see Section 1.2).

1.11 Proper deflating subspace of a general matrix pencil

We begin with the definition.

Definition 1.11.1 *Let $\lambda M - N$ be a matrix pencil, with $M, N \in \mathbb{F}^{m \times n}$. The linear space $\mathcal{V} \subset \mathbb{F}^n$ of dimension ρ is called* deflating (to the right) *for this matrix pencil if for a basis matrix $V \in \mathbb{F}^{n \times \rho}$ of $\mathcal{V}$, there exists a matrix $S \in \mathbb{F}^{\rho \times \rho}$ such that*

$$MVS = NV. \tag{1.50}$$

The deflating subspace $\mathcal{V}$ is additionally called proper *if MV has full column rank.*

Notice that the definition of a deflating subspace coincides formally with the characterization (1.43) of a deflating subspace (with finite spectrum) of a regular pencil. Furthermore, if the pencil $\lambda M - N$ is regular and (1.50) holds, then it is easy to see that $\mathcal{V}$ is automatically proper. However, the singularity of the pencil makes the properties and the computation of proper deflating subspaces far more intricate.

We define now deflating subspaces to the left.

Definition 1.11.2 *Let $\lambda M - N$ be a matrix pencil, with $M, N \in \mathbb{F}^{m\times n}$. The linear space $\mathcal{W} \subset \mathbb{F}^m$ of dimension ρ is called* deflating to the left *if $\mathcal{W}$ is a deflating subspace to the right of $\lambda M^T - N^T$. Equivalently, $\mathcal{W}$ is a deflating subspace to the left if for a row basis matrix $W \in \mathbb{F}^{\rho\times m}$ of $\mathcal{W}$ there exists a matrix $T \in \mathbb{F}^{\rho\times\rho}$ such that*

$$TWM = WN. \tag{1.51}$$

The deflating subspace to the left $\mathcal{W}$ is additionally called proper *if WM has full row rank.*

We give now two general results about proper deflating subspaces. We shall make intensive use of the notation introduced in the previous section in connection with the Kronecker canonical form.

Consider a partition of the complex plane in two disjoint sets

$$\mathbb{C} = \mathbb{C}_g \cup \mathbb{C}_b \tag{1.52}$$

(the good and the bad one). A deflating subspace $\mathcal{V}$ will be called $\mathbb{C}_g$*-deflating* if the spectrum of S in (1.50) lies in $\mathbb{C}_g$.

Let n_g and n_b be the number of finite generalized eigenvalues (multiplicity counted) in $\mathbb{C}_g$ and $\mathbb{C}_b$, respectively. Clearly, we have $n_f = n_g + n_b$.

By using a mere permutation of rows and columns in (1.48) it is not difficult to realize that there exist two invertible matrices Q and Z such that

$$Q(\lambda M - N)Z = \begin{bmatrix} \lambda I_{n_r} - A_r & -B & & & & \\ & & \lambda I_{n_g} - N_g & & & \\ & & & \lambda I_{n_b} - N_b & & \\ & & & & \lambda M_\infty - I_{n_\infty} & \\ & & & & & \lambda I_{n_\ell} - A_\ell \\ & & & & & -C \end{bmatrix}. \tag{1.53}$$

Here (A_r, B) is a controllable pair, B is an $n_r \times \nu_r$ matrix, N_g contains the finite generalized eigenvalues in $\mathbb{C}_g$, N_b contains the finite generalized eigenvalues in $\mathbb{C}_b$, M_∞ is a nilpotent matrix which determines the infinite generalized eigenvalues, (C, A_ℓ) is an observable pair, and C is an $\nu_\ell \times n_\ell$ matrix. If M, N, have real elements and the set $\mathbb{C}_g$ is symmetric, then all the matrices in (1.53) can be chosen with real elements as well.

Theorem 1.11.3 *Let $\lambda M - N$ be an arbitrary matrix pencil, with $M, N \in \mathbb{F}^{m\times n}$, and let (1.52) be a disjoint partition of the complex plane. Then $\lambda M - N$ has a proper $\mathbb{C}_g$-deflating subspace to the right (left) if and only if $n_r + n_g > 0$ ($n_\ell + n_g > 0$). In this* *case, there exists a proper $\mathbb{C}_g$–deflating subspace $\mathcal{V}$ of dimension n_r+n_g $(n_\ell+n_g)$ and any such subspace cannot have a larger dimension. If M and N have real elements, and $\mathbb{C}_g$ is a symmetric set, then we can choose $\mathcal{V} \subset \mathbb{R}^n$.*

Proof. It is enough to prove the result for right deflating subspaces.

"Only if". Let $\mathcal{V}$ be a proper $\mathbb{C}_g$–deflating subspace to the right and let V be a basis matrix for $\mathcal{V}$. Then MV has full column rank and there exists a square matrix S with the spectrum in $\mathbb{C}_g$ such that $MVS = NV$. Let Q and Z be two invertible matrices such that $Q(\lambda M - N)Z$ is in the form (1.53). We get

$$QMZ\,Z^{-1}V\,S = QNZ\,Z^{-1}V. \tag{1.54}$$

Now write $Z^{-1}V$ partitioned conformably to (1.53), i.e.,

$$Z^{-1}V = \begin{bmatrix} V_{r1} \\ V_{r2} \\ V_g \\ V_b \\ V_\infty \\ V_\ell \end{bmatrix} \begin{matrix} \}n_r \\ \}\nu_r \\ \}n_g \\ \}n_b \\ \}n_\infty \\ \}n_\ell \end{matrix} .$$

Writing (1.54) explicitly, we obtain

$$N_b V_b = V_b S, \quad V_\infty = M_\infty V_\infty S, \quad A_\ell V_\ell = V_\ell S, \quad C V_\ell = 0.$$

From these relations we shall prove successively that $V_\ell = 0$, $V_b = 0$, and $V_\infty = 0$. Since (C, A_ℓ) is an observable pair, there exists K such that $\Lambda(A_\ell + KC) \subset \mathbb{C}_b$. Hence we have $(A_\ell + KC)V_\ell = V_\ell S$. Since $\Lambda(S) \cap \Lambda(A_\ell + KC) = \Lambda(S) \cap \Lambda(N_b) = \emptyset$, we deduce that $V_\ell = 0$ and $V_b = 0$. We shall deduce now that $V_\infty = 0$. For this purpose, remember that M_∞ is nilpotent and let $k \in \mathbb{N}$ such that $M_\infty^k = 0$. Multiplying successively to the left $V_\infty = M_\infty V_\infty S$ with decreasing powers $k-1, \ldots, 1$ of M_∞ gives $M_\infty^{k-1} V_\infty = \cdots = V_\infty = 0$. Since $\mathcal{V}$ is proper, MV has full column rank and also

$$QMV = \begin{bmatrix} V_{r1} \\ V_g \\ O \end{bmatrix} \begin{matrix} \}n_r \\ \}n_g \\ \}s \end{matrix} , \tag{1.55}$$

where $s := n_b + n_\infty + n_\ell + \nu_\ell$. Consequently $n_r + n_g > 0$ and $\dim \mathcal{V} \leq n_r + n_g$. This justifies also the last assertion in the statement.

"If". Let Q and Z be two invertible matrices such that (1.53) holds. Assume now $n_r + n_g > 0$. We shall construct a proper $\mathbb{C}_g$–deflating subspace $\mathcal{V}$ of dimension $n_r + n_g$.

If $n_r > 0$ choose F such that $\Lambda(A_r + BF) \subset \mathbb{C}_g$. This is possible since (A_r, B) is a controllable pair. Let

$$\overline{V} = \begin{bmatrix} I_{n_r} & O \\ F & O \\ O & I_{n_g} \\ O & O \end{bmatrix} \begin{matrix} \\ \\ \\ \}t \end{matrix} , \tag{1.56}$$

where $t := n_b + n_\infty + n_\ell$. The matrix $\overline{V}$ has full column rank and, therefore, $\overline{\mathcal{V}} = \operatorname{Im} \overline{V}$ has dimension $n_r + n_g$. Then (1.50) is satisfied if we take

$$S = \begin{bmatrix} A_r + BF & O \\ O & N_g \end{bmatrix},$$

and $V = Z\overline{V}$. Notice that $\Lambda(S) \subset \mathbb{C}_g$ and

$$QMV = \begin{bmatrix} I_{n_r} & O \\ O & I_{n_g} \\ O & O \end{bmatrix} \}v$$

has full column rank which shows that MV has full column rank too. Here we have denoted $v := t + \nu_r$.

If $n_r = 0$, then (1.50) holds for

$$V := Z \begin{bmatrix} I_{n_g} \\ O \end{bmatrix}$$

and $S := N_g$. In this case,

$$QMV = \begin{bmatrix} I_{n_g} \\ O \end{bmatrix}$$

has again full column rank and therefore MV has full column rank.

Hence $\mathcal{V}$ =ImV is a proper $\mathbb{C}_g$–deflating subspace of dimension $n_r + n_g$.

Finally, if M and N have real elements and $\mathbb{C}_g$ is a symmetric set, then in the above construction of $\mathcal{V}$ we start from (1.53), which holds now with real matrices, and we choose the matrix F to have real elements as well, which is always possible due to Theorem 1.4.2. Further, the entire construction can be repeated such that all the intervening matrices have real elements, and the resulting $\mathcal{V}$ will be in $\mathbb{R}^n$. ■

Following Theorem 1.11.3 we call a proper $\mathbb{C}_g$–deflating subspace of dimension $n_r + n_g$, *maximal.* As we shall see in Chapter 6, certain properties of maximal proper deflating subspaces are connected with existence conditions and computational procedures for the solutions to the Riccati systems. However, from the proof of Theorem 1.11.3, maximal proper deflating subspaces with specified spectrum are not unique, in general. Fortunately, this will not affect the possibility of finding the solution to the Riccati systems. In this respect, we shall show that finding *one* maximal proper deflating subspace is enough to decide whether a solution exists or not and, when it exists, to compute it. The following result will be instrumental in establishing this fact.

Theorem 1.11.4 *Let $\lambda M - N$ be an $m \times n$ pencil with $n_r + n_g > 0$. Let $\mathcal{V}$ be an arbitrary maximal proper $\mathbb{C}_g$–deflating subspace and V be a basis matrix of $\mathcal{V}$. Then* Im MV *is independent of $\mathcal{V}$.*

Proof. Let Q and Z be the matrices that bring the pencil to the form of (1.53). From (1.55) it follows that the matrix

$$\begin{bmatrix} V_{r1} \\ V_g \end{bmatrix} \tag{1.57}$$

has full column rank. Since the dimension of $\mathcal{V}$ equals $n_r + n_g$, it follows that (1.57) is square, and therefore, invertible. Now

$$\operatorname{Im} MV = \operatorname{Im} Q^{-1} \begin{bmatrix} V_{r1} \\ V_g \\ O \end{bmatrix} = \operatorname{Im} Q^{-1} \begin{bmatrix} I_{n_r+n_g} \\ O \end{bmatrix}$$

which is clearly independent of the particular choice of $\mathcal{V}$. ■

1.12 Structural elements of a rational matrix

Throughout the book we view a $p \times m$ rational matrix function $\mathbf{G}(\lambda)$ as a matrix whose elements are rational functions with coefficients in $\mathbb{F}$, and of the complex variable λ. More precisely,

$$\mathbf{G}(\lambda) = [g_{ij}(\lambda)]_{i,j}^{p,m} = \left[\frac{a_{ij}(\lambda)}{b_{ij}(\lambda)}\right]_{i,j=1}^{p,m} \tag{1.58}$$

where $a_{ij}(\lambda)$ and $b_{ij}(\lambda)$ are scalar polynomials with coefficients in $\mathbb{F}$, with neither of $b_{ij}(\lambda)$ identical zero. By $\mathbb{F}(\lambda)$ we denote the set of all rational matrices with coefficients in $\mathbb{F}$, and by $\mathbb{F}^{p\times m}(\lambda)$ the subset of $\mathbb{F}(\lambda)$ of $p \times m$ matrices. Alternatively, we shall interpret $\mathbb{F}^{m}(\lambda)$ as the space of columns with m rational elements. Occasionally, as we shall see in the next chapter, we assimilate the rational matrix $\mathbf{G}(\lambda)$ with the transfer–function matrix of a linear time–invariant continuous–time or discrete–time system, and λ is either s or z, the complex variables appearing in the Laplace or Z–transform, respectively, according to the type of the system.

It is easy to show that $\mathbf{G}(\lambda)$ has constant rank for all but a finite number of points $\lambda \in \mathbb{C}$ for which it has smaller rank. The *normal rank* of $\mathbf{G}$ is by definition the rank of $\mathbf{G}(\lambda)$ for almost all $\lambda \in \mathbb{C}$, and is denoted $\text{rank}_{\mathbb{F}(\lambda)}\mathbf{G}$. A rational matrix which is square (say $p \times p$) and has full normal rank p is called nonsingular (or invertible), otherwise it is called singular. Clearly, a nonsingular rational matrix has an inverse which is a rational matrix as well.

For an arbitrary rational matrix we define the following structural elements that play an important role in systems theory: poles and zeros together with their multiplicities, minimal bases for the left and right null spaces together with the associated minimal indices. We start with poles and zeros.

Poles and zeros, and the Smith–McMillan forms

The following theorem is central in the theory of rational matrices.

Theorem 1.12.1 (Smith–McMillan form) *Let $\mathbf{G}(\lambda)$ be a $p \times m$ rational matrix, with coefficients in $\mathbb{F}$, and having normal rank r. Then there exists a $p \times p$ polynomial matrix $\mathbf{U}(\lambda)$ and an $m \times m$ polynomial matrix $\mathbf{V}(\lambda)$, both with coefficients in $\mathbb{F}$ and constant nonzero determinants (independent of λ), that bring $\mathbf{G}(\lambda)$ to the* Smith–McMillan form

$$\mathbf{S}(\lambda) = \mathbf{U}(\lambda)\mathbf{G}(\lambda)\mathbf{V}(\lambda), \tag{1.59}$$

where

$$\mathbf{S}(\lambda) := \text{diag}\left\{\frac{\epsilon_1(\lambda)}{\eta_1(\lambda)}, \frac{\epsilon_2(\lambda)}{\eta_2(\lambda)}, \ldots, \frac{\epsilon_r(\lambda)}{\eta_r(\lambda)}, 0, \ldots, 0\right\}, \tag{1.60}$$

the polynomials $\epsilon_i(\lambda)$, $\eta_i(\lambda)$ are monic (have the highest degree coefficient equal to 1), are pairwise coprime for $i = 1, \ldots, r$, and satisfy the divisibility properties

$$\begin{array}{rcl} \epsilon_i(\lambda) & | & \epsilon_{i+1}(\lambda), \\ \eta_{i+1}(\lambda) & | & \eta_i(\lambda), \end{array} \quad i = 1, \ldots, r-1. \tag{1.61}$$

Moreover, the polynomials $\epsilon_i(\lambda)$ and $\eta_i(\lambda)$ are uniquely defined by $\mathbf{G}(\lambda)$, and are called elementary divisors *of $\mathbf{G}$.*

Define

$$\begin{aligned} z(\lambda) &:= \epsilon_1(\lambda)\cdots\epsilon_r(\lambda), \\ p(\lambda) &:= \eta_1(\lambda)\cdots\eta_r(\lambda). \end{aligned} \tag{1.62}$$

Then the n_z (n_p) roots of $z(\lambda)$ ($p(\lambda)$) are by definition *the finite zeros (poles)* of $\mathbf{G}$. The order of the zero (pole) λ_0 of $\mathbf{G}(\lambda)$ is by definition its *multiplicity* as a root of $z(\lambda)$ ($p(\lambda)$). The *partial orders* of the zero (pole) λ_0 of $\mathbf{G}(\lambda)$ are by definition the nonzero multiplicities of λ_0 as a root of $\epsilon_i(\lambda)$ ($\eta_i(\lambda)$), for $i = 1, \ldots, r$. We say that ∞ is a zero (pole) of $\mathbf{G}$ if 0 is a zero (pole) of the rational matrix $\widehat{\mathbf{G}}(\lambda) = \mathbf{G}(\frac{1}{\lambda})$, and we define the orders of the (zero) pole $\lambda = \infty$ of $\mathbf{G}(\lambda)$ as the orders of the pole $\lambda = 0$ of $\widehat{\mathbf{G}}(\lambda)$. The union of all zeros and the union of all poles of $\mathbf{G}$ are denoted by $\mathcal{Z}(\mathbf{G})$ and $\mathcal{P}(\mathbf{G})$, respectively, (including infinity, and repeating elements according to their orders). By definition, the *McMillan degree* of $\mathbf{G}$ is the sum of the orders of all its poles (finite and infinite). Thus the McMillan degree of a rational matrix equals its number of poles (counting orders and including infinity).

When only one point λ_0 is of interest, a customary alternative to the Smith–McMillan form is the local Smith–McMillan form that displays only the orders of λ_0 as a zero and as a pole.

Theorem 1.12.2 (Local Smith–McMillan form) *Let* $\mathbf{G}$ *be a* $p \times m$ *rational matrix with coefficients in* $\mathbb{F}$*, having normal rank* r*, and let* $\lambda_0 \in \mathbb{C}$*. Then there exists a* $p \times p$ *rational matrix* $\mathbf{U}(\lambda)$ *and an* $m \times m$ *rational matrix* $\mathbf{V}(\lambda)$*, both nonsingular and without poles and zeros at* λ_0*, that bring* $\mathbf{G}(\lambda)$ *to the* local Smith–McMillan form around λ_0,

$$\mathbf{S}_\ell(\lambda) = \mathbf{U}(\lambda)\mathbf{G}(\lambda)\mathbf{V}(\lambda), \tag{1.63}$$

where

$$\mathbf{S}_\ell(\lambda) := \operatorname{diag}\left\{(\lambda-\lambda_0)^{k_1}, (\lambda-\lambda_0)^{k_2}, \ldots, (\lambda-\lambda_0)^{k_r}\right\}, \tag{1.64}$$

and

$$k_1 \le k_2 \le \ldots \le k_r$$

are integers. Moreover, the strictly positive (or strictly negative) integers k_i *are the partial orders of* λ_0 *as a zero (or pole) of* $\mathbf{G}$.

The local Smith–McMillan form around $\lambda_0 = \infty$ is defined as the local Smith–McMillan form of $\mathbf{G}(\frac{1}{\lambda})$ around $\lambda_0 = 0$.

Minimal bases for the left and right null spaces

For a rational matrix $\mathbf{G} \in \mathbb{F}^{p\times m}(\lambda)$ of normal rank r the right null space over $\mathbb{F}^m(\lambda)$ has dimension $m - r$ and the corresponding left null space over $\mathbb{F}^p(\lambda)$ has dimension $p - r$. We introduce now the notion of left and right minimal indices of a rational matrix. To this end we briefly recall several key ideas from the theory of minimal bases.

Given any subspace of $\mathbb{F}^m(\lambda)$, we can always construct a polynomial basis for it. The *degree of a polynomial vector* is defined to be the highest degree among its entries. The columns of a polynomial matrix $\mathbf{P}(\lambda)$ are said to form a *minimal basis* for the space spanned by them if:

1. $\mathbf{P}(\lambda)$ has full column rank for all finite λ, and
2. $\mathbf{P}(\lambda)$ is *column reduced* (or column proper), i.e., denoting the degree of the ith column of $\mathbf{P}(\lambda)$ by n_i, and constructing the constant matrix P_n whose ith column is the vector coefficient of λ^{n_i} in the ith column of $\mathbf{P}(\lambda)$, we have P_n to be of full column rank.

The matrix P_n is termed the *highest-order column coefficient* matrix. Equivalently, a polynomial basis is minimal if the sum of the degrees of the polynomial vectors that form it is as small as possible. The set of degrees of the vectors in any minimal basis for a given space is invariant with respect to unimodular column operations, and these degrees are termed the *minimal indices* (or the *Kronecker indices*) of the space. By definition, the sets of *left and right minimal indices of a rational matrix* are the sets of minimal indices of the left and right null space, respectively.

For an arbitrary rational matrix there is an interesting relation between its structural elements that will be insightful for the problems treated in the sequel, as shown below. Let n_r and n_ℓ be the sum of its right and left minimal indices, respectively, and let n_z and n_p be its number of zeros and poles (counting orders and including infinity), respectively. Then we have:

$$n_p = n_z + n_r + n_\ell.$$

1.13 Realization theory for rational matrices

Let $\mathbf{G} \in \mathbb{F}^{p\times m}(\lambda)$ be given as in (1.58). The rational matrix $\mathbf{G}$ is called *proper* if $\mathbf{G}(\lambda)$ is finite at infinity or, equivalently, the degrees of the polynomials a_{ij} are pairwise less than or equal to the degrees of the polynomials b_{ij}, otherwise it is called *improper.* If $\mathbf{G}(\lambda)$ takes the value 0 at infinity it is called *strictly proper*, and if $\mathbf{G}(\lambda)$ is proper and has a proper inverse then it is called *biproper.* Notice that a proper rational matrix has no poles at infinity and thus its McMillan degree is given simply by the degree of the polynomial $p(\lambda)$ in (1.62). We shall present first the case in which $\mathbf{G}$ is proper for which realization theory is a well understood subject and which will be used most frequently in this book. Further, we discuss very briefly realization theory for general (possibly improper) rational matrices which is a more recent achievement and is considerably more intricate.

1.13.1 The proper case

A realization of the proper rational matrix $\mathbf{G} \in \mathbb{F}^{p\times m}(\lambda)$ is a representation of the form

$$\mathbf{G}(\lambda) = D + C(\lambda I - A)^{-1}B, \tag{1.65}$$

where $A \in \mathbb{F}^{n\times n}$, $B \in \mathbb{F}^{n\times m}$, $C \in \mathbb{F}^{p\times n}$, and $D \in \mathbb{F}^{p\times m}$, and which holds for every $\lambda \notin \Lambda(A)$. We call the positive integer n the *order* (or the *dimension*) of the realization (1.65). Notice that $D = \lim_{\lambda\to\infty} \mathbf{G}(\lambda)$. Alternatively, we shall use for rational matrices $\mathbf{G}$ the following quite popular notation, that emphasizes also one of its realizations,

$$\mathbf{G} = \left[\begin{array}{c|c} A & B \\ \hline C & D \end{array}\right]. \tag{1.66}$$

Note that the right–hand side of (1.66) is a rational matrix, not to be confused with the block matrix

$$\begin{bmatrix} A & B \\ C & D \end{bmatrix}.$$

The next theorem is crucial in systems theory.

Theorem 1.13.1 (Realization of rational matrices) *Any proper rational matrix $\mathbf{G} \in \mathbb{F}^{p\times m}(\lambda)$ has a realization (1.65) of a certain order n, with the matrices A, B, C, D having elements in $\mathbb{F}$.*

A realization of a rational matrix is far from being unique, and it is easy to see, starting from a given realization, that there exist realizations of arbitrary larger dimensions. Therefore, it is natural to determine realizations which have an order as small as possible – called *minimal realizations* – and to ask to what extent such minimal realizations are unique. The following key theorem provides an answer to both these questions, and gives a characterization of minimality in terms of the controllability and observability of the pairs of matrices (A, B) and (C, A), respectively.

Theorem 1.13.2 (Minimal realizations) *Let $\mathbf{G} \in \mathbb{F}^{p\times m}(\lambda)$ be a proper rational matrix of McMillan degree n. Then we have:*

1. Any realization (1.65) of $\mathbf{G}(\lambda)$ has an order greater or equal to n and there exists a realization of minimal order n, with coefficients in $\mathbb{F}$, called a minimal realization.

2. If

$$\mathbf{G} = \left[\begin{array}{c|c} A & B \\ \hline C & D \end{array}\right] = \left[\begin{array}{c|c} \widetilde{A} & \widetilde{B} \\ \hline \widetilde{C} & \widetilde{D} \end{array}\right]$$

are two minimal realizations with elements in $\mathbb{F}$, then

$$D = \widetilde{D}$$

and there exists a unique invertible matrix $T \in \mathbb{F}^{n\times n}$, called a similarity transformation, *such that*

$$\widetilde{A} = TAT^{-1}, \quad \widetilde{B} = TB, \quad \widetilde{C} = CT^{-1}. \tag{1.67}$$

3. If

$$\mathbf{G} = \left[\begin{array}{c|c} A & B \\ \hline C & D \end{array}\right]$$

is an arbitrary realization of order k, then there exists an invertible $k \times k$ matrix T such that

$$\widetilde{A} := TAT^{-1} = \begin{bmatrix} A_1 & A_{12} & A_{13} & A_{14} \\ O & A_2 & A_{23} & A_{24} \\ O & O & A_3 & A_{34} \\ O & O & O & A_4 \end{bmatrix}, \quad \widetilde{B} := TB = \begin{bmatrix} B_1 \\ B_2 \\ O \\ O \end{bmatrix},$$
$$\widetilde{C} := CT^{-1} = \begin{bmatrix} O & C_2 & O & C_4 \end{bmatrix}$$

where the pair (A_2, B_2) is controllable, the pair (C_2, A_2) is observable, and

$$\mathbf{G} = \left[\begin{array}{c|c} A_2 & B_2 \\ \hline C_2 & D \end{array}\right].$$

4. A realization

$$\mathbf{G} = \left[\begin{array}{c|c} A & B \\ \hline C & D \end{array}\right]$$

is minimal if and only if the pair (A, B) *is controllable and the pair* (C, A) *is observable.*

1.13.2 The general case

Let $\mathbf{G} \in \mathbb{F}^{p\times m}(\lambda)$ be a general rational matrix. It is clear that if $\mathbf{G}$ is not proper then $\mathbf{G}$ cannot be represented as in (1.65) since this would imply $\mathbf{G}(\infty) = D$ is finite. There are several alternatives to represent an arbitrary (possibly improper) rational matrix, each with specific advantages and drawbacks. The most common representation is

$$\mathbf{G}(\lambda) = D + C(\lambda E - A)^{-1}B, \tag{1.68}$$

where $A, E \in \mathbb{F}^{n\times n}$, $B \in \mathbb{F}^{n\times m}$, $C \in \mathbb{F}^{p\times n}$, $D \in \mathbb{F}^{p\times m}$, $\lambda E - A$ is a regular pencil, and which holds for any $\lambda \notin \Lambda(E, A)$. We call the positive integer n the *order* (or the *dimension*) of the realization (1.68). As in the proper case, to denote an arbitrary rational matrix simultaneously with one of its realizations, we shall use the notation

$$\mathbf{G} = \left[\begin{array}{c|c} \lambda E - A & -B \\ \hline C & D \end{array}\right]. \tag{1.69}$$

Notice the sign convention used in (1.69). Any rational matrix has a representation of the form (1.68). Indeed, let

$$\mathbf{G}(\lambda) = \widetilde{\mathbf{G}}(\lambda) + \widehat{\mathbf{G}}(\lambda),$$

where $\widetilde{\mathbf{G}}(\lambda)$ is the proper part and $\widehat{\mathbf{G}}(\lambda)$ is the polynomial part of $\mathbf{G}(\lambda)$. Then

$$\overline{\mathbf{G}}(\lambda) := \frac{1}{\lambda}\widehat{\mathbf{G}}(\frac{1}{\lambda})$$

is strictly proper and we may write realizations as in (1.65) for the proper rational matrices $\widetilde{\mathbf{G}}$ and $\overline{\mathbf{G}}$ as

$$\widetilde{\mathbf{G}} = \left[\begin{array}{c|c} \widetilde{A} & \widetilde{B} \\ \hline \widetilde{C} & \widetilde{D} \end{array}\right], \quad \overline{\mathbf{G}} = \left[\begin{array}{c|c} \overline{A} & \overline{B} \\ \hline \overline{C} & O \end{array}\right], \tag{1.70}$$

having orders $\widetilde{n}$ and $\overline{n}$, respectively. Then a realization of $\mathbf{G}$ in the form (1.68) is

$$\mathbf{G} = \left[\begin{array}{cc|c} \lambda I_{\widetilde{n}} - \widetilde{A} & O & -\widetilde{B} \\ O & I_{\overline{n}} - \lambda\overline{A} & -\overline{B} \\ \hline \widetilde{C} & \overline{C} & D \end{array}\right] \tag{1.71}$$

and has the order $\widetilde{n} + \overline{n}$. The principal drawback of the realizations (1.68) is that their minimal possible order is greater than the McMillan degree of $\mathbf{G}$, unless $\mathbf{G}$ is proper. A remedy to this is to use realizations in which the "B" or "C" matrices are replaced by matrix pencils as explained further.

Consider any additive decomposition of $\mathbf{G}$ as

$$\mathbf{G}(\lambda) = \mathbf{G}_p(\lambda) + \mathbf{G}_i(\lambda), \tag{1.72}$$

where $\mathbf{G}_p$ has no poles at infinity and $\mathbf{G}_i$ has no poles at zero. Then $\mathbf{G}_p$ is proper, $\mathbf{G}_i$ contains the polynomial (or improper part), and

$$\widetilde{\mathbf{G}}_i(\lambda) := \mathbf{G}_i(\frac{1}{\lambda})$$

is proper. Then we may write realizations as in (1.65) for the proper rational matrices $\mathbf{G}_p$ and $\widetilde{\mathbf{G}}_i$ as

$$\begin{array}{rcl} \mathbf{G}_p(\lambda) & = & D_p + C_p(\lambda I - A_p)^{-1}B_p, \\ \widetilde{\mathbf{G}}_i(\lambda) & = & D_i + C_i(\lambda I - A_i)^{-1}B_i, \end{array} \tag{1.73}$$

having orders n_p and n_i, respectively. Then we have

$$\mathbf{G}(\lambda) = D_p + D_i + C_p(\lambda I - A_p)^{-1}B_p + \lambda C_i(I - \lambda A_i)^{-1}B_i$$

which is a realization of the form

$$\mathbf{G}(\lambda) = D + (C_1 + \lambda C_2)(\lambda E - A)^{-1}B, \tag{1.74}$$

which we also denote as

$$\mathbf{G} = \left[\begin{array}{c|c} \lambda E - A & -B \\ \hline C_1 + \lambda C_2 & D \end{array}\right]. \tag{1.75}$$

In our case we obtain a realization (1.75) with the particular coefficients

$$\mathbf{G} = \left[\begin{array}{cc|c} \lambda I_{n_p} - A_p & O & -B_p \\ O & I_{n_i} - \lambda A_i & -B_i \\ \hline C_p & \lambda C_i & D_p + D_i \end{array}\right],$$

and of order $n := n_p + n_i$. Clearly, it is always possible to arrange that, in the decomposition (1.72), the McMillan degrees of $\mathbf{G}_p$ and $\mathbf{G}_i$ add up to the McMillan degree of $\mathbf{G}$. As a consequence, the realization (1.75) has the smallest possible order n among all realizations of this type, and n equals the McMillan degree of $\mathbf{G}$. We shall see in subsequent chapters that the representations (1.75) arise naturally in the study of discrete–time systems with mixed direct and inverse–time evolutions.

A realization of the form (1.69) or of the form (1.75) is called a *descriptor realization* to distinguish it from the *standard realization* (1.66). For descriptor realizations we can develop a similar theory as for standard realizations including the characterization of minimality in terms of suitable defined controllability and observability properties, and the relation between two minimal realizations. We do not pursue these developments here as they are not needed in the sequel. We only mention that the standard realization (1.66), and the descriptor realizations (1.69) and (1.75), are particular cases of the following general realization of a rational matrix due to Rosenbrock [Ros70]

$$\mathbf{G}(\lambda) = \mathbf{W}(\lambda) + \mathbf{V}(\lambda)\mathbf{T}^{-1}(\lambda)\mathbf{U}(\lambda) = \left[\begin{array}{c|c} \mathbf{T}(\lambda) & -\mathbf{U}(\lambda) \\ \hline \mathbf{V}(\lambda) & \mathbf{W}(\lambda) \end{array}\right] \tag{1.76}$$

where $\mathbf{T}(\lambda)$, $\mathbf{U}(\lambda)$, $\mathbf{V}(\lambda)$, and $\mathbf{W}(\lambda)$ are matrix polynomials, with $\mathbf{T}(\lambda)$ square and invertible.

A general realization (1.76) is called controllable (observable) if

$$\begin{bmatrix} \mathbf{T}(\lambda) & -\mathbf{U}(\lambda) \end{bmatrix}, \qquad (\begin{bmatrix} \mathbf{T}(\lambda) \\ \mathbf{V}(\lambda) \end{bmatrix}), \tag{1.77}$$

seen as a rational matrix has no zeros (finite or infinite). Similarly as for constant matrices we define, for a set $\mathbb{M} \subset \overline{\mathbb{C}}$, the notion of $\mathbb{M}$–controllability ($\mathbb{M}$–observability), with the significance that the respective pair (1.77) has no zeros in $\mathbb{M}$.

1.14 Structural elements in terms of realizations for a rational matrix

We have introduced so far several important structural elements of a rational matrix, and we have shown that with each rational matrix we can associate a realization in standard or in descriptor form. A natural question is whether or not and to what extent can we determine the structural elements starting from a particular realization. It turns out that if we construct a minimal realization (i.e., a realization of smallest possible dimension) of the form

$$\mathbf{G}(\lambda) = D + C(\lambda E - A)^{-1}B, \tag{1.78}$$

then all the structural elements of $\mathbf{G}$ (poles and zeros, their orders, and left and right minimal indices) can be read off and through from the eigenstructure of the *pole pencil*

$$\lambda E - A \tag{1.79}$$

and of the *transmission pencil*

$$\lambda E_T - A_T := \lambda \begin{bmatrix} E & O \\ O & O \end{bmatrix} - \begin{bmatrix} A & B \\ C & D \end{bmatrix} \tag{1.80}$$

associated with the realization (1.78). Similarly, we can develop a theory in terms of minimal realizations of the form (1.75) where now the pole pencil is defined as in (1.79) and the transmission pencil is defined as

$$\lambda E_T - A_T := \lambda \begin{bmatrix} E & O \\ -C_2 & O \end{bmatrix} - \begin{bmatrix} A & B \\ C_1 & D \end{bmatrix}. \tag{1.81}$$

The focus will be on proper rational matrices for which we can use standard realizations with $E = I$, while the improper case will only be briefly mentioned.

Theorem 1.14.1 *Let $\mathbf{G} \in \mathbb{F}^{p\times m}(\lambda)$ be a proper rational matrix of McMillan degree n and normal rank r. Let*

$$\mathbf{G} = \left[\begin{array}{c|c} A & B \\ \hline C & D \end{array}\right] \tag{1.82}$$

be a realization of order k with coefficients in $\mathbb{F}$, and let

$$\lambda I - A, \quad \lambda E_T - A_T := \begin{bmatrix} \lambda I - A & -B \\ -C & -D \end{bmatrix}$$

be the pole pencil and the transmission pencil, respectively, associated with this particular realization.

1. For any realization (1.82) we have:

(a) **Normal rank.** *The normal ranks of* $\mathbf{G}$ *and* $\lambda E_T - A_T$ *are related as*

$$\operatorname{rank}_{\mathbb{F}(\lambda)} \mathbf{G} = \operatorname{rank}_{\mathbb{F}(\lambda)} (\lambda E_T - A_T) - k.$$

(b) **Poles.** *The union of poles* $\mathcal{P}(\mathbf{G})$ *of* $\mathbf{G}$ *is included in the union of eigenvalues of* A,

$$\mathcal{P}(\mathbf{G}) \subset \Lambda(A).$$

(c) **Zeros.** *The union of zeros* $\mathcal{Z}(\mathbf{G})$ *(finite and infinite) of* $\mathbf{G}$ *is included in the union of generalized eigenvalues of* $\lambda E_T - A_T$,

$$\mathcal{Z}(\mathbf{G}) \subset \Lambda(E_T, A_T).$$

2. For a minimal realization (1.82) we have in addition:

(a) **Poles.** *The poles of* $\mathbf{G}$ *are the eigenvalues of* A. *The partial orders of the poles of* $\mathbf{G}$ *are pairwise equal to the partial multiplicities of the eigenvalues of* A.

(b) **Finite zeros.** *The finite zeros of* $\mathbf{G}$ *are the finite generalized eigenvalues of* $\lambda E_T - A_T$. *The partial orders of the finite zeros of* $\mathbf{G}$ *are pairwise equal to the partial multiplicities of the finite generalized eigenvalues of* $\lambda E_T - A_T$.

(c) **Infinite zeros.** *The partial orders of the zero at infinity of* $\mathbf{G}$ *are pairwise in excess with 1 to the partial multiplicities of the infinite generalized eigenvalue of* $\lambda E_T - A_T$.

(d) **Left minimal indices.** *The minimal indices to the left of* $\mathbf{G}$ *are pairwise equal to the left Kronecker indices of* $\lambda E_T - A_T$.

(e) **Right minimal indices.** *The minimal indices to the right of* $\mathbf{G}$ *are pairwise equal to the right Kronecker indices of* $\lambda E_T - A_T$.

For rational matrices represented by descriptor realizations of type (1.69) or (1.75) a similar result holds. However, we shall only need the following weaker result.

Theorem 1.14.2 *Let* $\mathbf{G} \in \mathbb{F}^{p\times m}(\lambda)$ *be a rational matrix with McMillan degree* n *and normal rank* r. *Associate with* $\mathbf{G}$ *a realization of the form (1.69) (or of the form (1.75)), of order* k, *and define the pole pencil as in (1.79) and the transmission pencil as in (1.80) (or as in (1.81)). Then we have:*

1. **Normal rank.** *The normal ranks of* $\mathbf{G}$ *and* $\lambda E_T - A_T$ *are related as*

$$\operatorname{rank}_{\mathbb{F}(\lambda)} \mathbf{G} = \operatorname{rank}_{\mathbb{F}(\lambda)} (\lambda E_T - A_T) - k.$$

2. **Poles.** *The union of poles* $\mathcal{P}(\mathbf{G})$ *(finite and infinite, counting multiplicities) of* $\mathbf{G}$ *is included in the union of generalized eigenvalues of* $\lambda E - A$,

$$\mathcal{P}(\mathbf{G}) \subset \Lambda(E, A).$$

3. **Zeros.** *The union of zeros* $\mathcal{Z}(\mathbf{G})$ *(finite and infinite, counting multiplicities) of* $\mathbf{G}$ *is included in the union of generalized eigenvalues of* $\lambda E_T - A_T$,

$$\mathcal{Z}(\mathbf{G}) \subset \Lambda(E_T, A_T).$$

Notes and references

The eigenvalue problem and the Jordan canonical form are the central issues of matrix algebra, and can be found in any standard text covering these topics. Excellent references are Gohberg, Lancaster and Rodman [GLR82, GLR83, GLR86], Horn and Johnson [HJ85, HJ91], and Golub and Van Loan [GV89] for computational aspects. A thorough treatment of the generalized eigenvalue problem, and of the Weierstrass and Kronecker canonical forms is given by Gantmacher [Gan59]. The notion of deflating subspace of a regular pencil is due to Stewart [Ste72]. Numerical procedures for computing deflating subspaces of regular pencils were developed by Van Dooren [Van81]. The extension of deflating subspaces to the case of general pencils (possibly singular) comprises two directions: the reducing subspaces which were introduced by Van Dooren [Van83] from a geometrical viewpoint, and the proper deflating subspaces which were introduced by Ionescu, Oară, and Weiss [IW93a, Oar93]. A detailed study of proper deflating subspaces can be found in Ionescu and Oară [Oar94, Oar95, IO96d]. Oară and Van Dooren [Oar95, OV97] have shown that reducing and proper deflating subspaces are different characterizations of the same concept, and have developed numerically sound algorithms for their computation. Controllability and observability, and their different characterizations can be found in any standard book on multivariable systems. Very good and comprehensive references are Kailath [Kai80], and Wonham [Won85]. Sylvester and Stein equations are fundamental for matrix algebra and appear in innumerable texts, for example in Gantmacher [Gan59], or in Horn and Johnson [HJ91]. Lyapunov and symmetric Stein equations are a central technical tool for many results in linear systems theory. The results given here are the most refined versions available and may be found in or easily adapted from Ostrowski and Schneider [OS62], and Bittanti and Colaneri [BC86] (see also Glover [Glo84]). The theory of rational matrices as reviewed in this chapter appears in many books on linear systems, for example in Belevitch [Bel68], Rosenbrock [Ros70], Kailath [Kai80] and Gohberg, Lancaster and Rodman [GLR86]. However, most of the texts consider in detail only finite poles and zeros, and leave out the important case of poles and zeros at infinity. In a very general setting, Rosenbrock [Ros70] shows that the orders of the finite poles and zeros of a rational matrix are isomorphic with the orders of the finite poles and zeros of any irreducible (sort of controllable plus observable) polynomial state–space realization of the rational matrix. The proof that this isomorphism extends to the pole and zero at infinity in the particular case of descriptor state–space representations of type (1.68) was given by Verghese, Van Dooren and Kailath [VVK79]. There appear firstly also the connections between the left and right minimal indices of a singular rational matrix and the left and right Kronecker indices of the transmission pencil associated with a particular realization. The realization theory for improper rational matrices together with a comprehensive theory of generalized state–space systems is developed in Verghese and Kailath [VLK81]. The minimal indices of rational vector spaces were investigated in full generality by Forney [For75].

2

Linear Dynamical Systems

This chapter is devoted to a brief review of several basic notions and facts in linear systems theory that will be extensively used in the sequel. As elsewhere in this book, we treat the continuous and discrete–time cases in parallel. Section 2.1 is devoted to the notion of the continuous–time linear system as used in this book, and to several related concepts like stability, dichotomy, input–output norms, continuity, controllability and observability Gramians, balancing transformations, and interconnections of systems. In Sections 2.2–2.4 we introduce successively several notions like L^2–evolutions, input–output, Hankel and Toeplitz operators, and study their features in various instances.

Discrete–time systems are introduced in Section 2.5, and the discrete–time counterpart of the results in Sections 2.1–2.4 are presented in Sections 2.5–2.8.

A. The continuous–time case

2.1 Linear systems

Throughout this monograph we make the leading assumptions that all systems are linear, finite–dimensional, time–invariant, either continuous or discrete–time. The most handy means of describing the dynamical behavior of a system satisfying all these assumptions is the *state–space representation.* A *continuous–time state–space system* is given by the equations

$$\begin{aligned} \dot{x}(t) &= Ax(t) + Bu(t), \\ y(t) &= Cx(t) + Du(t), \end{aligned} \tag{2.1}$$

where $t \in \mathbb{R}$, $x(t) \in \mathbb{F}^n$ is the *state*, $u(t) \in \mathbb{F}^m$ is the *input*, $y(t) \in \mathbb{F}^p$ is the *output* of the system, and A, B, C, D are constant matrices of appropriate dimensions with the entries in $\mathbb{F}$. The matrix A is usually called the *state matrix* of the system. Often, we shall refer to (2.1) by indicating the quadruple (A, B, C, D). Almost anywhere in this book we make the common assumption in systems theory that the matrices (A, B, C, D) have real elements, and the functions u, x, y of the real variable t take real values as well. Thus from now on, if not otherwise stated, we take $\mathbb{F} = \mathbb{R}$.

The input–output behavior of the system (2.1) is conveniently described by the *transfer function matrix* (TFM) which is the proper $p \times m$ rational matrix function

$$\mathbf{G}(s) := C(sI - A)^{-1}B + D = \left[\begin{array}{c|c} A & B \\ \hline C & D \end{array}\right](s). \tag{2.2}$$

Notice that $\mathbf{G}(s)$ is obtained by taking the Laplace transform in (2.1) and making explicit $y(s)$ as a function of $u(s)$ in the form

$$y(s) = \mathbf{G}(s)u(s) \tag{2.3}$$

(where now $y(s)$ and $u(s)$ are viewed as the Laplace transforms of $y(t)$ and $u(t)$, respectively). Starting with an *input–output representation* (2.3), where $\mathbf{G}$ is a rational matrix given entrywise as in (1.58), we can always arrive to the state–space representation (2.1) by writing down a standard realization (A, B, C, D) of $\mathbf{G}$ (see Theorem 1.13.1). Of course, we can always arrange for the realization to be minimal, as Theorem 1.13.2 shows. Notice that if $\mathbf{G}$ is improper, then $\mathbf{G}$ cannot be the TFM of a state–space system of type (2.1). Summarizing the above discussion, to any state–space representation (2.1) corresponds an input–output representation (2.3) where $\mathbf{G}$ is the proper rational matrix given in (2.2), and to any input–output representation (2.3), with $\mathbf{G}$ a proper rational matrix, corresponds a state–space representation (2.1). From now on, whenever we shall refer to a system we understand the state–space representation (2.1), or the input–output representation (2.3).

2.1.1 *Stable, antistable, and dichotomic systems*

The system (2.1) is called *stable* if all the eigenvalues of A are in $\mathbb{C}_-$. The system is called *antistable* if all the eigenvalues of A are in $\mathbb{C}_+$. The system is called *dichotomic* if A has no eigenvalues on the imaginary axis. By a slight but common terminology abuse, we shall say that A is stable, antistable, or dichotomic if it is the state matrix of a stable, antistable, or dichotomic system, respectively.

After a linear change of variables of the form

$$\widetilde{x} := Tx, \tag{2.4}$$

where T is an invertible matrix, the transformed state vector $\widetilde{x}$ satisfies

$$\begin{aligned} \dot{\widetilde{x}}(t) &= \widetilde{A}\widetilde{x}(t) + \widetilde{B}u(t), \\ y(t) &= \widetilde{C}\widetilde{x}(t) + \widetilde{D}u(t), \end{aligned} \tag{2.5}$$

where

$$\widetilde{A} := TAT^{-1}, \quad \widetilde{B} := TB, \quad \widetilde{C} := CT^{-1}, \quad \widetilde{D} := D. \tag{2.6}$$

Formulas (2.6) imply that a change of variables (2.4) (also called a *similarity transformation*, or an *equivalence transformation*) does affect neither the eigenvalues of A, nor the associated TFM. Therefore, stability, antistability, dichotomy, and the TFM, are invariant under coordinate changes. By a coordinate transformation, a dichotomic system can be decomposed into a stable and antistable part in the following sense. If the system is dichotomic, there exists a coordinate transformation T such that

$$\widetilde{A} = TAT^{-1} = \begin{bmatrix} A_- & O \\ O & A_+ \end{bmatrix} \begin{matrix} \}n_- \\ \}n_+ \end{matrix} \tag{2.7}$$

where A_- is square and stable and A_+ is square and antistable. With such a transformation it is convenient to write explicitly the other system matrices as

$$\widetilde{B} = TB = \begin{bmatrix} B_- \\ B_+ \end{bmatrix} \begin{matrix} \}n_- \\ \}n_+ \end{matrix}, \quad \widetilde{C} = CT^{-1} = \begin{bmatrix} \underbrace{C_-}_{n_-} & \underbrace{C_+}_{n_+} \end{bmatrix}. \tag{2.8}$$

Such a transformation is not unique. However, the subspaces

$$\mathcal{X}_-(A) := \operatorname{Im} T^{-1} \begin{bmatrix} I_{n_-} \\ O \end{bmatrix}, \quad \mathcal{X}_+(A) := \operatorname{Im} T^{-1} \begin{bmatrix} O \\ I_{n_+} \end{bmatrix} \tag{2.9}$$

are independent of T and are called *the stable* and *the antistable subspace* of A, respectively. It is easy to see that they are invariant subspaces of A that are complementary, i.e.,

$$\mathcal{X}_-(A) \oplus \mathcal{X}_+(A) = \mathbb{R}^n,$$

and that the projection on these subspaces are given by the formulas

$$\Pi_- = T^{-1} \begin{bmatrix} I_{n_-} & O \\ O & O \end{bmatrix} T, \quad \Pi_+ = T^{-1} \begin{bmatrix} O & O \\ O & I_{n_+} \end{bmatrix} T. \tag{2.10}$$

2.1.2 Controllability, stabilizability, and duals

If the pair of matrices (A, B) is controllable then we say that the system (2.1) is *controllable.* A system (2.1) is called *stabilizable* if the pair (A, B) is c–controllable. The system (2.1) is said *antistabilizable* if the pair $(-A, B)$ is c–controllable. Equivalently, as Theorem 1.4.2 shows, a system is stabilizable (or antistabilizable) if and only if there exists $F \in \mathbb{R}^{m \times n}$ such that $A + BF$ is stable (or antistable). In this case, the matrix F is called a *stabilizing (antistabilizing)* feedback. By a common abuse of terminology we shall also talk, with obvious meaning, about stabilizable and antistabilizable pairs of matrices (A, B). Clearly, a controllable pair is stabilizable, and also antistabilizable. Conversely, if a pair is simultaneously stabilizable and antistabilizable, then it is controllable.

We also have the following dual notions. If (C, A) is an observable pair, then we say that the system (2.1) is observable. The system (2.1) is said to be *detectable* (*antidetectable*) if the pair (C, A) is c–observable (the pair $(C, -A)$ is c–observable). Correspondingly, we shall talk about detectable and antidetectable pairs of matrices (C, A). Similarly, an observable pair is detectable, and also antidetectable, and if a pair is simultaneously detectable and antidetectable, then it is observable.

2.1.3 Spaces of rational matrices, norms, and continuity

Let $L^\infty(\mathbb{C}_0)$ denote the Banach space of matrix–valued complex functions $G(t)$ that are (essentially) bounded on $\mathbb{C}_0$, with the so–called $L^\infty(\mathbb{C}_0)$*–norm,*

$$||G||_\infty := \sup_{t \in \mathbb{C}_0} \overline{\sigma}(G(t)) \tag{2.11}$$

where $\overline{\sigma}$ denotes the maximum singular value. The rational subspace of $L^\infty(\mathbb{C}_0)$ denoted by $RL^\infty_{p\times m}$ consists of all $p\times m$ proper rational matrices with elements bounded on the imaginary axis (with no poles on the imaginary axis). Also $RH^\infty_{+,p\times m}$ and $RH^\infty_{-,p\times m}$ will denote the spaces of proper rational matrices with elements analytic in $\mathbb{C}_+ \cup \mathbb{C}_0$, respectively on $\mathbb{C}_- \cup \mathbb{C}_0$. Whenever the dimensions of the matrices follow from the context we shall simply speak about matrices in RL^∞, RH^∞_+, or RH^∞_-. If a rational matrix $\mathbf{G}$ belongs to one of these rational matrix sets and has an inverse that belongs to the same set we say that $\mathbf{G}$ is a *unit* in that set.

If A is dichotomic, then the TFM $\mathbf{G}$ given by (2.2) is in RL^∞. If A is stable, then $\mathbf{G}$ is in RH^∞_+, and if A is antistable, then $\mathbf{G}$ is in RH^∞_-.

A system which has a TFM in RH^∞_+ is called *input–output stable.* Thus, the class of input–output stable systems coincides with RH^∞_+, and a stable system (2.1) (having the matrix A stable) is always input–output stable. However, input–output stability does not imply in general stability of the system (2.1). To make the distinction between these two concepts more striking we sometimes term a stable system (2.1) as *internally stable.* Internal stability is equivalent to input–output stability under the assumption that the realization (2.2) is stabilizable and detectable. The main concern in systems theory is internal stability (as opposed to input–output stability), as it guarantees that to any input (signal) of bounded energy injected anywhere into the system, all the resulting signals will be of bounded energy as well.

The same discussion can be repeated with respect to dichotomy and antistability and the corresponding spaces RL^∞ and RH^∞_-. Thus a system (2.1) is dichotomic if and only if it is $\mathbb{C}_0$–controllable, $\mathbb{C}_0$–observable, and its TFM (2.2) is in RL^∞. Similarly, a system (2.1) is antistable if and only if it is antistabilizable, antidetectable, and its TFM (2.2) belongs to RH^∞_-.

Notice that (2.11) can be written equivalently for the rational matrix $G \equiv \mathbf{G}$ as

$$\|\mathbf{G}\|_\infty = \sup_{\omega\in\mathbb{R}} \overline{\sigma}(\mathbf{G}(j\omega)) = \sup_{\omega\in\mathbb{R}} \|\mathbf{G}(j\omega)\| \qquad (2.12)$$

where $\|\cdot\|$ stands for the operator–induced norm of the matrix. As we shall see in Theorem 2.3.2, $\|\mathbf{G}\|_\infty$ is the induced norm of the so–called input–output operator of the system (2.1). The H^∞*-norm* is defined for rational matrices $\mathbf{G} \in RH^\infty_+$ and has the same expression (2.12).

We need the following classical result due to Rellich (see for example Kato [Kat66]) that we mention here without proof.

Theorem 2.1.1 (Rellich's theorem) *Let the $n \times n$ matrix $M(x) = M^*(x)$ depending continuously on the real parameter x ranging in the interval I. Then there exist real continuous functions $\mu_1, \ldots, \mu_n$ on I, such that $\mu_n(x) \leq \ldots \leq \mu_1(x)$, for every x in I, and $\Lambda(M(x)) = \{\mu_i(x) : i = 1, \ldots, n\}$.*

We have the following simple consequence.

Corollary 2.1.2 *For $\mathbf{G} \in RL^\infty_{p\times m}$, $\overline{\sigma}(\mathbf{G}(j\omega))$ depends continuously on $\omega \in \mathbb{R}$, is bounded and attains its maximum for $\omega \in [-\infty, \infty]$, where by definition $\mathbf{G}(j\infty) = \mathbf{G}(-j\infty) = D$.*

Proof. The proof follows immediately from Rellich's Theorem and the fact that $\mathbf{G}$ is proper and has no poles on the imaginary axis. ∎

The following result relates convergence of state–space realizations to the convergence of the respective TFMs.

Theorem 2.1.3 *Assume that the sequence of linear systems*

$$\mathbf{G}_k = \left[\begin{array}{c|c} A_k & B_k \\ \hline C_k & D_k \end{array}\right], \quad k \in \mathbb{N}, \tag{2.13}$$

is such that

$$A_k \to A, B_k \to B, C_k \to C \text{ and } D_k \to D \text{ as } k \to \infty,$$

and A and A_k are all stable, and $\mathbf{G}(s) = C(sI - A)^{-1}B + D \equiv 0$. Then

$$\lim_{k\to\infty} ||\mathbf{G}_k||_\infty = 0.$$

This result can be straightforwardly deduced from its discrete–time counterpart, which in turn is easier to prove. Therefore we postpone the proof of this theorem until after the proof of Theorem 2.5.1.

The *adjoint (or conjugate)* of a TFM $\mathbf{G}$ of a continuous–time system is defined as

$$\mathbf{G}^*(s) := \mathbf{G}^T(-s),$$

such that, for each real ω,

$$\mathbf{G}^*(j\omega) = \mathbf{G}^T(-j\omega) = [\mathbf{G}(j\omega)]^*.$$

A realization of $\mathbf{G}^*$ is given by

$$\mathbf{G} = \left[\begin{array}{c|c} -A^T & -C^T \\ \hline B^T & D^T \end{array}\right]. \tag{2.14}$$

Let $L^2(\mathbb{M})$ denote the Hilbert space of matrix (or vector or scalar) complex–valued functions $G(t)$ for which

$$\int_{\mathbb{M}} \text{Trace}\,[G^*(t)G(t)]\text{d}t < \infty, \tag{2.15}$$

and where $\mathbb{M}$ stands for either $\mathbb{Z}$, $(-\infty, \infty)$, $\mathbb{C}_0$, or $\partial\mathbb{D}$. If the set $\mathbb{M}$ results from the context, we shall omit it and write simply L^2.

In particular, $L^2(\mathbb{C}_0)$ denotes the Hilbert space of matrix (or vector or scalar) complex–valued functions $G(t)$ for which the so–called $L^2(\mathbb{C}_0)$*–norm*

$$||G||_2 := \{\frac{1}{2\pi}\int_{-\infty}^{\infty} \text{Trace}\,[G^*(j\omega)G(j\omega)]\text{d}\omega\}^{\frac{1}{2}} \tag{2.16}$$

is finite. If $G \equiv \mathbf{G}$ is a rational matrix, then the $L^2(\mathbb{C}_0)$–norm is finite only if $\mathbf{G}$ is strictly proper. The H^2*–norm* is defined for rational matrices $\mathbf{G} \in RH_+^\infty$ and has the same expression (2.16).

2.1.4 Gramians and Hankel singular values

Assume that A is stable. Then we have from Theorem 1.5.2 that the following Lyapunov equations

$$AP + PA^T + BB^T = 0 \tag{2.17}$$

and

$$A^TQ + QA + C^TC = 0 \tag{2.18}$$

have unique positive semidefinite solutions P and Q in $\mathbb{R}^{n\times n}$. P is called the *causal controllability Gramian* associated with the pair of matrices (A, B) and Q is called the *causal observability Gramian* associated with the pair of matrices (C, A). This terminology is justified by the fact that $P > 0$ if and only if the system is controllable, and similarly $Q > 0$ if and only if the system is observable, as Theorems 1.5.6 and 1.5.4 show.

It is easy to check that they can also be expressed as

$$P = \int_0^\infty e^{At}BB^Te^{A^Tt}\mathrm{d}t \ , \qquad Q = \int_0^\infty e^{A^Tt}C^TCe^{At}\mathrm{d}t. \tag{2.19}$$

Let $\Lambda(PQ) = \Lambda(QP) = \{\sigma_1^2, \ldots, \sigma_n^2\}$. As both P and Q are positive semidefinite, we have that σ_i, $(i = 1, \ldots, n)$, are non–negative. Those values σ_i which are non–zero depend only on the TFM $\mathbf{G}(s)$ and are called the *Hankel singular values* of the system.

After a coordinate transformation $\widetilde{x} = Tx$ in the system (2.1), it is easy to see that the Gramians change according to the formulas

$$\widetilde{P} = TPT^T, \quad \widetilde{Q} = T^{-T}QT^{-1}. \tag{2.20}$$

According to the next result, a particular coordinate change can be performed to make the Gramians equal and diagonal,

Theorem 2.1.4 *Assume that the system (2.1) is stable. Then there exists a coordinate change $\widetilde{x} = Tx$ such that the Gramians become equal and diagonal,*

$$\widetilde{P} = \widetilde{Q} = \begin{bmatrix} \sigma_1 & & \\ & \ddots & \\ & & \sigma_n \end{bmatrix}, \qquad (\sigma_1 \geq \ldots \geq \sigma_n).$$

A coordinate transformation T as in the previous theorem is called a *balancing transformation* and the system (2.1) in the new coordinates is said to be in a *balanced form*. Let us denote by $(\widetilde{A}, \widetilde{B}, \widetilde{C}, D)$ the system matrices of (2.1) in balanced form. The following fact is of importance in model reduction.

Theorem 2.1.5 *Assume that the system (2.1) is stable and let T be a balancing transformation and $\widetilde{A} = TAT^{-1}$ be the state matrix of the system in balanced form. If there exists an integer i such that the Hankel singular values of the system (2.1) satisfy $\sigma_i > \sigma_{i+1}$, and A_1 and A_2 are the upper–left $i \times i$ corner, respectively the lower–right $(n-i) \times (n-i)$ corner of $\widetilde{A}$, then A_1 and A_2 are both stable.*

In the case that A is antistable, we can introduce the *anticausal controllability Gramian* $\widehat{P}$ associated with the pair of matrices (A, B) and the *anticausal observability Gramian* $\widehat{Q}$ associated with the pair of matrices (C, A) as the unique solutions (see Theorem 1.5.2) of the Lyapunov equations

$$A\widehat{P} + \widehat{P}A^T - BB^T = 0 \tag{2.21}$$

and

$$A^T\widehat{Q} + \widehat{Q}A - C^TC = 0. \tag{2.22}$$

They can also be expressed as

$$\widehat{P} = \int_{-\infty}^{0} e^{At}BB^Te^{A^Tt}\mathrm{d}t \ , \quad \widehat{Q} = \int_{-\infty}^{0} e^{A^Tt}C^TCe^{At}\mathrm{d}t, \tag{2.23}$$

and enjoy similar properties as those mentioned in the causal case. In fact, it is easy to see that $\widehat{P}$ is the causal controllability Gramian associated with the pair $(-A, B)$, whereas $\widehat{Q}$ is the observability Gramian associated with the pair $(C, -A)$.

Assume that $D = 0$. If A is stable, the $L^2(\mathbb{C}_0)$–norm of the TFM $\mathbf{G}$ can be computed using the formulas

$$||\mathbf{G}||_2 = [\text{Trace}\,(B^TQB)]^{\frac{1}{2}} = [\text{Trace}\,(CPC^T)]^{\frac{1}{2}}. \tag{2.24}$$

Similarly, if A is antistable,

$$||\mathbf{G}||_2 = [\text{Trace}\,(B^T\widehat{Q}B)]^{\frac{1}{2}} = [\text{Trace}\,(C\widehat{P}C^T)]^{\frac{1}{2}}. \tag{2.25}$$

2.1.5 Connections of systems

Several connections of systems will play a particular role in the sequel: parallel, series, closed–loop, linear fractional transformations, and Redheffer product. Assuming that the given systems are expressed in terms of realizations, it will be handy to have formulas for a realization of the resulting system. However, we shall see that these connections, or complex combinations of them, may lead to a resulting system with a TFM that is no longer proper. Thus, it may happen that the resulting system has no state–space representation of type (2.1) even if both systems to start with have one. This discussion leads to the following definition: a certain connection of state–space systems is called *well-posed* if the resulting system is a state–space system as well. Thus, well–posedness automatically implies that the TFM of the resulting system is proper.

As a simple example that clarifies this point, we take the inverse of a system. The *inverse* of a system (2.1) exists provided the matrix D is invertible, and in this case the inverse system is given by

$$\begin{array}{rcl} \dot{x}(t) & = & (A - BD^{-1}C)x(t) + BD^{-1}y(t), \\ u(t) & = & -D^{-1}Cx(t) + D^{-1}y(t), \end{array} \tag{2.26}$$

and has the TFM given by

$$\mathbf{G}^{-1} = \left[\begin{array}{c|c} A - BD^{-1}C & BD^{-1} \\ \hline -D^{-1}C & D^{-1} \end{array}\right]. \tag{2.27}$$

The invertibility of D is not necessary for the existence of the inverse of the TFM $\mathbf{G}(s)$ (seen as a rational matrix), and is made solely to ensure that $\mathbf{G}^{-1}$ is proper and therefore is the TFM of a state–space system of type (2.1).

Parallel connection

Consider two systems,

$$\begin{aligned} \dot{x}_1(t) &= A_1x_1(t) + B_1u_1(t), \\ y_1(t) &= C_1x_1(t) + D_1u_1(t), \end{aligned} \tag{2.28}$$

and

$$\begin{aligned} \dot{x}_2(t) &= A_2x_2(t) + B_2u_2(t), \\ y_2(t) &= C_2x_2(t) + D_2u_2(t), \end{aligned} \tag{2.29}$$

having the TFMs

$$\mathbf{G}_1 = \left[\begin{array}{c|c} A_1 & B_1 \\ \hline C_1 & D_1 \end{array}\right], \quad \mathbf{G}_2 = \left[\begin{array}{c|c} A_2 & B_2 \\ \hline C_2 & D_2 \end{array}\right], \tag{2.30}$$

respectively. If the two systems have the same number of inputs and outputs, respectively, then we define the *parallel connection* of $\mathbf{G}_1$ with $\mathbf{G}_2$ as the system with the input u and output $y := y_1 + y_2$ obtained by setting $u_1 \equiv u_2 = u$. The connection is always well–posed and the TFM of the resulting system is

$$\mathbf{G}(s) := \mathbf{G}_1(s) + \mathbf{G}_2(s), \tag{2.31}$$

which has a realization given by

$$\mathbf{G} = \left[\begin{array}{cc|c} A_1 & O & B_1 \\ O & A_2 & B_2 \\ \hline C_1 & C_2 & D_1 + D_2 \end{array}\right]. \tag{2.32}$$

Serial connection

Consider two systems (2.28) and (2.29), having the TFMs given in (2.30). If the number of outputs of $\mathbf{G}_2$ is equal to the number of inputs of $\mathbf{G}_1$ then we define the *serial* connection of $\mathbf{G}_1$ with $\mathbf{G}_2$ (in this order) as the system with input $u := u_2$ and output $y := y_1$, obtained by setting $u_1 \equiv y_2$. The connection is always well–posed and the TFM of the resulting system is

$$\mathbf{G}(s) := \mathbf{G}_1(s)\mathbf{G}_2(s), \tag{2.33}$$

which has a realization given by

$$\mathbf{G} = \left[\begin{array}{cc|c} A_1 & B_1C_2 & B_1D_2 \\ O & A_2 & B_2 \\ \hline C_1 & D_1C_2 & D_1D_2 \end{array}\right]. \tag{2.34}$$

Feedback connection

Consider two systems (2.28) and (2.29), having the TFMs given in (2.30). If the number of inputs and outputs of $\mathbf{G}_2$ is equal to the number of outputs and inputs of $\mathbf{G}_1$, respectively, then we define the *feedback connection* of $\mathbf{G}_1$ with $\mathbf{G}_2$ (in this order) as the system with input u and output $y := y_1$ obtained by setting $u_1 \equiv u + y_2$, $u_2 \equiv y_1$. The connection is well–posed if and only the matrix

$$\begin{bmatrix} I & D_1 \\ D_2 & I \end{bmatrix} \tag{2.35}$$

is invertible. In this case the TFM of the resulting *feedback system* is

$$\mathbf{G}(s) := \mathbf{G}_1(s)(I - \mathbf{G}_2(s)\mathbf{G}_1(s))^{-1}, \tag{2.36}$$

and has a realization given by

$$\mathbf{G} = \left[\begin{array}{cc|c} A_1 + B_1 D_2 \widehat{S}^{-1} C_1 & B_1 S^{-1} C_2 & B_1 S^{-1} \\ B_2 \widehat{S}^{-1} C_1 & A_2 + B_2 D_1 S^{-1} C_2 & B_2 D_1 S^{-1} \\ \hline \widehat{S}^{-1} C_1 & \widehat{S}^{-1} D_1 C_2 & D_1 S^{-1} \end{array}\right], \tag{2.37}$$

where we have denoted $\widehat{S} := I - D_1 D_2$ and $S := I - D_2 D_1$ which are both invertible due to the invertibility of the matrix in (2.35).

Remark 2.1.6 For parallel, series, and feedback connections, the two realizations (2.30) are not necessarily of the the same order. The resulting realizations (2.32), (2.34), and (2.37) are not necessarily minimal even the original realizations (2.30) are.

Linear fractional transformations

These connections can be viewed as an extension to systems (or, more exactly, to rational matrix functions) of the well–known *linear fractional transformation (LFT)* (or Moebius transformation),

$$f(\lambda) = \frac{a + b\lambda}{c + d\lambda},$$

where a, b, c, d are all scalars.

Consider two systems given by

$$\begin{aligned} \dot{x} &= Ax + B_1 u_1 + B_2 u_2, \\ y_1 &= C_1 x + D_{11} u_1 + D_{12} u_2, \\ y_2 &= C_2 x + D_{21} u_1 + D_{22} u_2, \end{aligned} \tag{2.38}$$

and

$$\begin{aligned} \dot{x}_K &= A_K x_K + B_K u_K, \\ y_K &= C_K x_K + D_K u_K, \end{aligned} \tag{2.39}$$

having the TFMs

$$\mathbf{T} = \begin{bmatrix} \mathbf{T}_{11} & \mathbf{T}_{12} \\ \mathbf{T}_{21} & \mathbf{T}_{22} \end{bmatrix} = \left[\begin{array}{c|cc} A & B_1 & B_2 \\ \hline C_1 & D_{11} & D_{12} \\ C_2 & D_{21} & D_{22} \end{array}\right] \tag{2.40}$$

and

$$\mathbf{K} = \left[\begin{array}{c|c} A_K & B_K \\ \hline C_K & D_K \end{array}\right], \tag{2.41}$$

respectively. A partitioned system of type (2.40) will be called as a *generalized system.* If $\mathbf{T}_{22}$ has the number of inputs and outputs equal to the numbers of outputs and inputs of $\mathbf{K}$, respectively, then we define the *lower linear fractional transformation (LLFT)* of $\mathbf{T}$ with $\mathbf{K}$ (in this order) as the system with input $u := u_1$ and output $y := y_1$ obtained by setting $u_2 \equiv y_K$ and $u_K \equiv y_2$. The connection is well–posed if and only if the matrix

$$\begin{bmatrix} I & D_{22} \\ D_K & I \end{bmatrix} \tag{2.42}$$

is invertible and in this case the TFM of the resulting system is defined as

$$\mathbf{T}_R = \mathrm{LLFT}(\mathbf{T}, \mathbf{K}) := \mathbf{T}_{11} + \mathbf{T}_{12}\mathbf{K}(I - \mathbf{T}_{22}\mathbf{K})^{-1}\mathbf{T}_{21}, \tag{2.43}$$

and has a realization given by

$$\mathbf{T}_R = \left[\begin{array}{cc|c} A + B_2\widetilde{S}^{-1}D_K C_2 & B_2\widetilde{S}^{-1}C_K & B_1 + B_2\widetilde{S}^{-1}D_K D_{21} \\ B_K S^{-1} C_2 & A_K + B_K S^{-1} D_{22} C_K & B_K S^{-1} D_{21} \\ \hline C_1 + D_{12} D_K S^{-1} C_2 & D_{12}\widetilde{S}^{-1}C_K & D_{11} + D_{12} D_K S^{-1} D_{21} \end{array}\right], \tag{2.44}$$

where we have denoted $S := I - D_{22}D_K$ and $\widetilde{S} := I - D_K D_{22}$ which are both invertible due to the invertibility of the matrix in (2.42). Notice that $D_K S^{-1} = \widetilde{S}^{-1} D_K$ and $S^{-1}D_{22} = D_{22}\widetilde{S}^{-1}$ as directly follows from the matrix identity $U(I - VU)^{-1} = (I - UV)^{-1}U$ which holds for any matrices U and V of compatible dimensions.

Analogously, consider two systems (2.38) and (2.39) having the TFMs (2.40) and (2.41), respectively. If $\mathbf{T}_{11}$ has the number of inputs and outputs equal to the number of outputs and inputs of $\mathbf{K}$, respectively, we define the *upper linear fractional transformation (ULFT)* of $\mathbf{T}$ with $\mathbf{K}$ (in this order) as the system with input $u := u_2$ and output $y := y_2$ obtained by setting $u_1 \equiv y_K$ and $u_K \equiv y_1$. The connection is well–posed if and only the matrix

$$\begin{bmatrix} I & D_{11} \\ D_K & I \end{bmatrix} \tag{2.45}$$

is invertible and in this case the TFM of the resulting system is defined as

$$\mathbf{T}_R = \mathrm{ULFT}(\mathbf{T}, \mathbf{K}) := \mathbf{T}_{22} + \mathbf{T}_{21}\mathbf{K}(I - \mathbf{T}_{11}\mathbf{K})^{-1}\mathbf{T}_{12}, \tag{2.46}$$

and has a realization given by

$$\mathbf{T}_R = \left[\begin{array}{cc|c} A + B_1\widetilde{T}^{-1}D_K C_1 & B_1\widetilde{T}^{-1}C_K & B_2 + B_1\widetilde{T}^{-1}D_K D_{12} \\ B_K T^{-1} C_1 & A_K + B_K T^{-1} D_{11} C_K & B_K T^{-1} D_{12} \\ \hline C_2 + D_{21} D_K T^{-1} C_1 & D_{21}\widetilde{T}^{-1}C_K & D_{22} + D_{21} D_K T^{-1} D_{12} \end{array}\right], \tag{2.47}$$

where we have denoted $T := I - D_{11}D_K$ and $\widetilde{T} := I - D_K D_{11}$ which are both invertible due to the invertibility of the matrix in (2.45). Notice that $D_K T^{-1} = \widetilde{T}^{-1} D_K$ and $T^{-1}D_{11} = D_{11}\widetilde{T}^{-1}$.

In most cases, and when there is no danger of confusion, we shall use the term LFT to denote the LLFT.

Redheffer product

This connection is an extension of the LFTs. Consider two generalized systems (2.38) and

$$\begin{aligned} \dot{x}_K &= A_K x_K + B_{K1} u_{K1} + B_{K2} u_{K2}, \\ y_{K1} &= C_{K1} x_K + D_{K11} u_{K1} + D_{K12} u_{K2}, \\ y_{K2} &= C_{K2} x_K + D_{K21} u_{K1} + D_{K22} u_{K2}, \end{aligned} \tag{2.48}$$

having the TFMs given by (2.40) and

$$\mathbf{K} = \begin{bmatrix} \mathbf{K}_{11} & \mathbf{K}_{12} \\ \mathbf{K}_{21} & \mathbf{K}_{22} \end{bmatrix} = \left[\begin{array}{c|cc} A_K & B_{K1} & B_{K2} \\ \hline C_{K1} & D_{K11} & D_{K12} \\ C_{K2} & D_{K21} & D_{K22} \end{array}\right], \tag{2.49}$$

respectively. If the number of inputs and outputs of $\mathbf{T}_{22}$ is equal to the number of outputs and inputs of $\mathbf{K}_{11}$, respectively, then we define the *Redheffer product* of $\mathbf{T}$ with $\mathbf{K}$ (in this order) as the system with input $u := \begin{bmatrix} u_1 \\ u_{K2} \end{bmatrix}$ and output $y := \begin{bmatrix} y_1 \\ y_{K2} \end{bmatrix}$ obtained by setting $u_2 \equiv y_{K1}$ and $u_{K1} \equiv y_2$. The connection is well–posed if and only if the matrix

$$\begin{bmatrix} I & D_{22} \\ D_{K11} & I \end{bmatrix} \tag{2.50}$$

is invertible and in this case the TFM of the resulting generalized system is defined as

$$\begin{aligned} \mathbf{T}_R &= \mathbf{T} \otimes \mathbf{K} \\ &:= \begin{bmatrix} \mathbf{T}_{11} + \mathbf{T}_{12}\mathbf{K}_{11}(I - \mathbf{T}_{22}\mathbf{K}_{11})^{-1}\mathbf{T}_{21} & \mathbf{T}_{12}(I - \mathbf{K}_{11}\mathbf{T}_{22})^{-1}\mathbf{K}_{12} \\ \mathbf{K}_{21}(I - \mathbf{T}_{22}\mathbf{K}_{11})^{-1}\mathbf{T}_{21} & \mathbf{K}_{22} + \mathbf{K}_{21}\mathbf{T}_{22}(I - \mathbf{K}_{11}\mathbf{T}_{22})^{-1}\mathbf{K}_{12} \end{bmatrix}, \end{aligned} \tag{2.51}$$

and has a realization given by

$$\mathbf{T}_R = \left[\begin{array}{cc|cc} A + B_2\widetilde{S}^{-1}D_{K11}C_2 & B_2\widetilde{S}^{-1}C_{K1} & B_1 + B_2\widetilde{S}^{-1}D_{K11}D_{21} & B_2\widetilde{S}^{-1}D_{K12} \\ B_{K1}S^{-1}C_2 & A_K + B_{K1}S^{-1}D_{22}C_{K1} & B_{K1}S^{-1}D_{21} & B_{K2} + B_{K1}S^{-1}D_{22}D_{K12} \\ \hline C_1 + D_{12}D_{K11}S^{-1}C_2 & D_{12}\widetilde{S}^{-1}C_{K1} & D_{11} + D_{12}D_{K11}S^{-1}D_{21} & D_{12}\widetilde{S}^{-1}D_{K12} \\ D_{K21}S^{-1}C_2 & C_{K2} + D_{K21}S^{-1}D_{22}C_{K1} & D_{K21}S^{-1}D_{21} & D_{K22} + D_{K21}S^{-1}D_{22}D_{K12} \end{array}\right] \tag{2.52}$$

where we have denoted $S := I - D_{22}D_{K11}$ and $\widetilde{S} := I - D_{K11}D_{22}$ which are both invertible due to the invertibility of the matrix in (2.50).

Remark 2.1.7 A remark similar to Remark 2.1.6 can be made for the LFTs and Redheffer product. Note that if in (2.40) we have $\mathbf{T}_{22}(\infty) = D_{22} = 0$ then the matrices (2.42) and (2.50) are automatically invertible, and the LFTs and Redheffer product of $\mathbf{T}$ with $\mathbf{K}$ are automatically well–posed.

2.2 State evolutions corresponding to L^2 inputs and associated operators

The purpose of this section is to introduce some important particular solutions of the linear differential system

$$\dot{x} = Ax + Bu, \tag{2.53}$$

where $A \in \mathbb{R}^{n\times n}, B \in \mathbb{R}^{n\times m}$, in the case in which u is an $L^{2,m}(-\infty,\infty)$ function. Here $L^{2,m}(-\infty,\infty)$ – or, briefly, $L^{2,m}$ – denotes the Hilbert space of norm–square Lebesgue integrable on $(-\infty,\ \infty)$, $\mathbb{C}^m$–valued functions. If $u \in L^{2,m}$ then

$$||u||_2 := (\int_{-\infty}^{\infty} ||u(t)||^2 \mathrm{d}t)^{\frac{1}{2}} < \infty$$

is the L^2*–norm* of u. Let $L_+^{2,m}$ and $L_-^{2,m}$ be the closed subspaces of $L^{2,m}$ consisting of those functions whose support is in $[0,\ \infty)$ and $(-\infty, 0]$, respectively. Clearly, $L^{2,m} = L_+^{2,m} \oplus L_-^{2,m}$. If the dimension is clear from the context we shall use also the notation L^2 for $L^{2,m}$. In what follows we shall refer usually to L^2 real–valued functions. The most common solution to (2.53) is the *initial–value solution.* For an arbitrary ξ in $\mathbb{R}^n$ and u in $L_+^{2,m}$, we denote by $x^{\xi,u}$ the solution to (2.53) that satisfies $x(0) = \xi$ and that is given by the variation of constants formula

$$x^{\xi,u}(t) = e^{At}\xi + \int_0^t e^{A(t-\tau)}Bu(\tau)\mathrm{d}\tau, \quad t \geq 0. \tag{2.54}$$

However, we shall study other solutions of (2.53) as well. We introduce them in this section and give their main properties. In particular, we give, whenever possible, an operator expression for these solutions.

We shall successively consider the cases in which A is stable, antistable, dichotomic, and arbitrary.

2.2.1 The case: A stable

If A is stable, the initial value solution $x^{\xi,u}$ is in $L_+^{2,n}$ for each $\xi \in \mathbb{R}^n$ and $u \in L_+^{2,m}$. We introduce the following linear operators:

$$\Phi : \mathbb{R}^n \to L_+^{2,n}\ , \quad (\Phi\xi)(t) := e^{At}\xi, \quad t \geq 0, \tag{2.55}$$

and

$$\mathcal{L} : L_+^{2,m} \to L_+^{2,n}\ , \quad (\mathcal{L}u)(t) := \int_0^t e^{A(t-\tau)}Bu(\tau)\mathrm{d}\tau, \quad t \geq 0. \tag{2.56}$$

It is easy to check that Φ and $\mathcal{L}$ are bounded operators. We can write

$$x^{\xi,u} = \Phi\xi + \mathcal{L}u. \tag{2.57}$$

Let us now consider the case of an input u in $L^{2,m}$ (defined on the entire real axis).

Theorem 2.2.1 *Let A be stable. Then, for each u in $L^{2,m}$, the equation (2.53) has a unique solution x_e^u in $L^{2,n}$, which is given by*

$$x_e^u(t) = \int_{-\infty}^t e^{A(t-\tau)}Bu(\tau)\mathrm{d}\tau\ , \quad t \in \mathbb{R}. \tag{2.58}$$

Moreover, this solution is absolutely continuous.

Proof. Since A is stable, we can find $\rho > 1$ and $\sigma > 0$ such that

$$||e^{At}|| \leq \rho e^{-\sigma t} \tag{2.59}$$

for each $t \geq 0$. Using (2.59), it is easy to show that the integral in (2.58) is convergent. Now we prove that the function $x_e^u : \mathbb{R} \to \mathbb{R}^n$ defined by (2.58) is in $L^{2,n}$ and it is absolutely continuous. We can estimate

$$\begin{aligned}
||x_e^u(t)||^2 &\leq \left[\int_{-\infty}^{t} \rho e^{-\sigma(t-\tau)}||B||||u(\tau)||\mathrm{d}\tau\right]^2 \\
&= \rho^2||B||^2\left[\int_{-\infty}^{t} e^{-\frac{\sigma}{2}(t-\tau)}e^{-\frac{\sigma}{2}(t-\tau)}||u(\tau)||\mathrm{d}\tau\right]^2
\end{aligned}$$

and using the Cauchy–Schwartz inequality

$$\begin{aligned}
&\leq \rho^2||B||^2\int_{-\infty}^{t} e^{-\sigma(t-\tau)}\mathrm{d}\tau\int_{-\infty}^{t} e^{-\sigma(t-\tau)}||u(\tau)||^2\mathrm{d}\tau \\
&= \rho^2||B||^2\int_{0}^{\infty} e^{-\sigma\theta}\mathrm{d}\theta\int_{-\infty}^{t} e^{-\sigma(t-\tau)}||u(\tau)||^2\mathrm{d}\tau \\
&= \frac{\rho^2||B||^2}{\sigma}\int_{-\infty}^{t} e^{-\sigma(t-\tau)}||u(\tau)||^2\mathrm{d}\tau.
\end{aligned}$$

Further, we have for every $T_1, T_2 \in \mathbb{R}$,

$$\begin{aligned}
\int_{T_1}^{T_2}||x_e^u(t)||^2\mathrm{d}t &\leq \frac{\rho^2||B||^2}{\sigma}\int_{T_1}^{T_2}\int_{-\infty}^{t} e^{-\sigma(t-\tau)}||u(\tau)||^2\mathrm{d}\tau\mathrm{d}t \\
&= \frac{\rho^2||B||^2}{\sigma}\int_{-\infty}^{T_2}\int_{\max(T_1,\tau)}^{T_2} e^{-\sigma(t-\tau)}||u(\tau)||^2\mathrm{d}\tau\mathrm{d}t \\
&= \frac{\rho^2||B||^2}{\sigma}\int_{-\infty}^{T_2}||u(\tau)||^2\left[\int_{\max(T_1,\tau)}^{T_2} e^{-\sigma(t-\tau)}\mathrm{d}t\right]\mathrm{d}\tau \\
&= \frac{\rho^2||B||^2}{\sigma}\int_{-\infty}^{T_2}||u(\tau)||^2\left[\int_{T_2-\tau}^{\max(T_1-\tau,0)} e^{-\sigma s}\mathrm{d}s\right]\mathrm{d}\tau \\
&\leq \frac{\rho^2||B||^2}{\sigma}\int_{-\infty}^{T_2}||u(\tau)||^2\left[\int_{0}^{\infty} e^{-\sigma s}\mathrm{d}s\right]\mathrm{d}\tau \\
&= \left(\frac{\rho||B||}{\sigma}||u||_2\right)^2.
\end{aligned}$$

Taking $T_1 \to -\infty$ and $T_2 \to \infty$, we obtain

$$||x_e^u||_2 \leq \frac{\rho||B||}{\sigma}||u||_2 \tag{2.60}$$

for each $u \in L^{2,m}$, and thus x_e^u is norm–square integrable. The fact that x_e^u is absolutely continuous follows directly from (2.58).

It is easy to check that (2.58) gives a solution of (2.53), by simply taking the derivative (almost everywhere) of both sides of the equality.

To prove uniqueness, it is enough to show that $x \equiv 0$ is the unique solution in $L^{2,n}$ of $\dot{x} = Ax$. Suppose this is not true. Then we can write the solution as $x(t) = e^{A(t-\tau)}x(\tau)$ for any τ such that $x(\tau) \neq 0$. Since A is stable, $\int_{-\infty}^{\infty} ||x(t)||^2 dt$ is unbounded, so x is not in L^2 as assumed. ■

From the proof of Theorem 2.2.1, in particular from (2.60), it follows that we can define the bounded linear operator $\mathcal{L}_e$ from $L^{2,m}$ to $L^{2,n}$, given by

$$(\mathcal{L}_e u)(t) := \int_{-\infty}^{t} e^{A(t-\tau)}Bu(\tau)d\tau , \quad t \in \mathbb{R}. \tag{2.61}$$

Accordingly, we have

$$x_e^u = \mathcal{L}_e u \tag{2.62}$$

and

$$||\mathcal{L}_e|| \leq \frac{\rho||B||}{\sigma}$$

as (2.60) shows. If we denote by $P_{\pm}^r$, the usual orthogonal projections of $L^{2,r}$ onto $L_{\pm}^{2,r}$, it immediately follows that

$$\mathcal{L} = P_+^n \mathcal{L}_e P_+^m = \mathcal{L}_e P_+^m, \tag{2.63}$$

where $\mathcal{L}$ and $\mathcal{L}_e$ are defined in (2.56) and (2.61), respectively.

2.2.2 The case: A antistable

The following result follows from Theorem 2.2.1 by a time–reversal transformation.

Theorem 2.2.2 *Let A be antistable. Then, for each u in $L^{2,m}$, the equation (2.53) has a unique solution in $L^{2,n}$, which is given by*

$$x_e^u(t) = -\int_t^{\infty} e^{A(t-\tau)}Bu(\tau)d\tau , \quad t \in \mathbb{R}. \tag{2.64}$$

Moreover, this solution is absolutely continuous.

Analogously, we can define for this case the bounded linear operator $\mathcal{L}_e$ from $L^{2,m}$ to $L^{2,n}$ given by

$$(\mathcal{L}_e u)(t) := -\int_t^{\infty} e^{A(t-\tau)}Bu(\tau)d\tau , \quad t \in \mathbb{R}, \tag{2.65}$$

such that (2.62) continues to be valid.

We shall also need the restriction of x_e^u to the positive half axis. Therefore, for $u \in L_+^{2,m}$ we introduce

$$x^u(t) = \begin{cases} -\int_t^{\infty} e^{A(t-\tau)}Bu(\tau)d\tau, & t \geq 0, \\ 0, & t < 0. \end{cases} \tag{2.66}$$

If we introduce the operator $\mathcal{L}$ in this case through the relation (2.66), we can write

$$x^u = \mathcal{L}u \tag{2.67}$$

where (2.63) changes into

$$\mathcal{L} = P_-^n \mathcal{L}_e P_-^m. \tag{2.68}$$

2.2.3 The case: A dichotomic

By combining Theorem 2.2.1 and Theorem 2.2.2, we obtain the following result.

Theorem 2.2.3 *Let A be dichotomic. Then, for each u in $L^{2,m}$, the equation (2.53) has a unique solution in $L^{2,n}$, which is given by*

$$x_e^u(t) = \int_{-\infty}^{t} e^{A(t-\tau)}\Pi_- Bu(\tau)\mathrm{d}\tau \; - \int_t^{\infty} e^{A(t-\tau)}\Pi_+ Bu(\tau)\mathrm{d}\tau \;, \quad t \in \mathbb{R}, \tag{2.69}$$

where Π_- and Π_+ are the projections (2.10) of $\mathbb{R}^n$ on the stable and antistable subspaces of A, respectively.

Proof. Consider the state transformation T that realizes the decomposition (2.7) and (2.8), i.e.,

$$\widetilde{A} = TAT^{-1} = \begin{bmatrix} A_- & O \\ O & A_+ \end{bmatrix}, \quad \widetilde{B} = TB = \begin{bmatrix} B_- \\ B_+ \end{bmatrix},$$

with $A_- \in \mathbb{R}^{n_- \times n_-}$ stable and $A_+ \in \mathbb{R}^{n_+ \times n_+}$ antistable. By denoting the new state vector by

$$\widetilde{x} = Tx = \begin{bmatrix} x_- \\ x_+ \end{bmatrix} \begin{matrix} \}n_- \\ \}n_+ \end{matrix}, \tag{2.70}$$

the system (2.53) is decomposed into the following subsystems

$$\dot{x}_- = A_- x_- + B_- u, \tag{2.71}$$

and

$$\dot{x}_+ = A_+ x_+ + B_+ u. \tag{2.72}$$

Applying Theorem 2.2.1 to (2.71) and Theorem 2.2.2 to (2.72), together with the formulas (2.10), the conclusion follows. ■

Precisely as in the previous two cases, we define the bounded linear operator

$$\mathcal{L}_e : L^{2,m} \to L^{2,n}, \quad x_e^u := \mathcal{L}_e u. \tag{2.73}$$

We also define

$$\mathcal{L} : L_+^{2,m} \to L_+^{2,n}, \quad \mathcal{L} := P_+^n \mathcal{L}_e P_+^m,$$

and

$$x^u := \mathcal{L}u.$$

2.2.4 The case: A arbitrary

As we have seen, in this case we can define the initial value solution $x^{\xi,u}$ through (2.54). In general, $x^{\xi,u}$ is not in $L_+^{2,n}$. It is interesting to consider the set of those input functions u that make $x^{\xi,u}$ norm–square integrable. These inputs can be regarded as stabilizing the system in open–loop. We introduce

$$\mathcal{U}_\xi := \{u : u \in L_+^{2,m} \text{ such that } x^{\xi,u} \in L_+^{2,n}\}. \tag{2.74}$$

If A is stable then $\mathcal{U}_\xi = L_+^{2,m}$. In general, the following result holds.

Lemma 2.2.4 *Assume that* (A, B) *is stabilizable and* F *is a stabilizing feedback. Let* $\xi \in \mathbb{R}^n$ *and* $u \in L_+^{2,m}$. *Then* $u \in \mathcal{U}_\xi$ *if and only if* $u = Fx_F^{\xi,v} + v$, *where* v *is arbitrary in* $L_+^{2,m}$ *and* $x_F^{\xi,v}$ *is the unique solution in* $L_+^{2,n}$ *of*

$$\dot{x}_F = (A + BF)x_F + Bv\ , \quad x_F(0) = \xi.$$

In the same vein, we may consider x^u given by (2.66) for those u for which the integral converges. Again, we may consider the set of inputs that make x^u a norm–square integrable function

$$\widehat{\mathcal{U}} := \{u : u \in L_+^{2,m} \text{ such that } x^u \in L_+^{2,n}\}. \tag{2.75}$$

This set is completely characterized by the following result.

Lemma 2.2.5 *Assume that* (A, B) *is antistabilizable and* F *is an antistabilizing feedback. Then* $u \in \widehat{\mathcal{U}}$ *if and only if* $u = Fx_F^v + v$, *where* v *is arbitrary in* $L_+^{2,m}$ *and* x_F^v *is the unique solution in* $L_+^{2,n}$ *of*

$$\dot{x}_F = (A + BF)x_F + Bv.$$

2.3 The L^2 input–output operator

Let us consider again the linear system (2.1),

$$\begin{array}{rcl} \dot{x}(t) & = & Ax(t) + Bu(t), \\ y(t) & = & Cx(t) + Du(t). \end{array} \tag{2.76}$$

We shall assume throughout this section that A is dichotomic and u is arbitrary in $L^{2,m}$. Then by (2.73), we have $x = x_e^u = \mathcal{L}_e u$. Accordingly, we have $y = (C\mathcal{L}_e + D)u$ and the output y is in $L^{2,p}$. Thus we deduce that the system (2.76) defines a linear bounded input–output operator from $L^{2,m}$ to $L^{2,p}$, whose expression follows easily from (2.69) as

$$(\mathcal{G}u)(t) = \int_{-\infty}^{t} Ce^{A(t-\tau)}\Pi_- Bu(\tau)\mathrm{d}\tau \ - \int_{t}^{\infty} Ce^{A(t-\tau)}\Pi_+ Bu(\tau)\mathrm{d}\tau \ + Du(t). \tag{2.77}$$

An alternative expression for $\mathcal{G}$ can be given by applying a state coordinate transformation T that realizes the decomposition (2.7), (2.8), into a stable and an antistable part. As a coordinate change does not affect the relationship between the input and the output of the system, we can write further

$$(\mathcal{G}u)(t) = \int_{-\infty}^{t} C_- e^{A_-(t-\tau)}B_- u(\tau)\mathrm{d}\tau \ - \int_{t}^{\infty} C_+ e^{A_+(t-\tau)}B_+ u(\tau)\mathrm{d}\tau \ + Du(t). \tag{2.78}$$

About the invertibility of the input–output operator we can prove the following result.

Theorem 2.3.1 *Assume that* D *is invertible and* $A - BD^{-1}C$ *is dichotomic. Then the input–output operator* $\mathcal{G}$ *is invertible and its inverse is the input–output operator of the system*

$$\begin{array}{rcrcl} \dot{x}(t) & = & (A - BD^{-1}C)x(t) & + & BD^{-1}y(t), \\ u(t) & = & -D^{-1}Cx(t) & + & D^{-1}y(t). \end{array} \tag{2.79}$$

Proof. It is a matter of simple substitutions to prove that the vector–valued functions u, x and y in L^2 satisfy (2.76) if and only if they satisfy (2.79). From the definition of the input–output operator, the assertion of the theorem follows immediately. ∎

Since A is dichotomic, the TFM $\mathbf{G}(s) = C(sI - A)^{-1}B + D$ of (2.76) belongs to $RL^{\infty}_{p\times m}$. If $y \in L^{2,p}$ and $u \in L^{2,m}$ are such that

$$y = \mathcal{G}u,$$

then their Fourier transforms $\widehat{y}$ and $\widehat{u}$ are elements in $L^2(\mathbb{C}_0)$ and

$$\widehat{y}(j\omega) = \mathbf{G}(j\omega)\widehat{u}(j\omega),$$

for almost every ω.

The following is a standard result.

Theorem 2.3.2 *If the system (2.76) is dichotomic, then the norm of its input–output operator equals the $L^{\infty}(\mathbb{C}_0)$–norm of its TFM, i.e.,*

$$||\mathcal{G}|| = ||\mathbf{G}||_{\infty}. \tag{2.80}$$

2.4 Hankel and Toeplitz operators

In the previous section we have associated with a dichotomic system a bounded linear operator $\mathcal{G}$ from $L^{2,m}$ to $L^{2,p}$. We introduce now other two important operators acting on half axis Lebesgue spaces L^2_- and L^2_+.

We denote by $P^r_{\pm}$, the usual orthogonal projections of $L^{2,r}$ onto $L^{2,r}_{\pm}$.

Definition 2.4.1 *Let $\mathcal{G}$ be a bounded linear operator from $L^{2,m}$ to $L^{2,p}$. We define:*

1. *The* causal Hankel operator *associated with $\mathcal{G}$,*

$$\mathbb{H}_{\mathcal{G}} := P^p_+\mathcal{G}|_{L^{2,m}_-}.$$

2. *The* anticausal Hankel operator *associated with $\mathcal{G}$,*

$$\widehat{\mathbb{H}}_{\mathcal{G}} := P^p_-\mathcal{G}|_{L^{2,m}_+}.$$

3. *The* causal Toeplitz operator *associated with $\mathcal{G}$,*

$$\mathbb{T}_{\mathcal{G}} := P^p_+\mathcal{G}|_{L^{2,m}_+}.$$

4. *The* anticausal Toeplitz operator *associated with $\mathcal{G}$,*

$$\widehat{\mathbb{T}}_{\mathcal{G}} := P^p_-\mathcal{G}|_{L^{2,m}_-}.$$

Remark 2.4.2 Since, generally, we are dealing with L^2–operators defined by linear systems, the restrictions to $L^2_{\pm}$ which appear in Definition 2.4.1 claim the specification of the appropriate initial value. For the sake of simplicity we omit the initial conditions and make further a slight abuse by using the following modified definitions:

$$\mathbb{H}_{\mathcal{G}} := P^p_+\mathcal{G}P^m_-, \quad \widehat{\mathbb{H}}_{\mathcal{G}} := P^p_-\mathcal{G}P^m_+, \quad \mathbb{T}_{\mathcal{G}} := P^p_+\mathcal{G}P^m_+, \quad \widehat{\mathbb{T}}_{\mathcal{G}} := P^p_-\mathcal{G}P^m_-.$$

Remark 2.4.3 Let $\mathcal{L}_e$ and $\mathcal{L}$ be the operators introduced in Section 2.2. Then (2.63) and (2.68) express the fact that $\mathcal{L}$ is either the causal Toeplitz operator or the anticausal Toeplitz operator associated with $\mathcal{L}_e$, i.e., $\mathcal{L} = \mathbb{T}_{\mathcal{L}_e}$, or $\mathcal{L} = \widehat{\mathbb{T}}_{\mathcal{L}_e}$.

For the rest of this book these operators will be associated with input–output operators $\mathcal{G}$ of linear systems

$$\begin{aligned} \dot{x}(t) &= Ax(t) + Bu(t), \\ y(t) &= Cx(t) + Du(t). \end{aligned} \tag{2.81}$$

In this case it is customary to call the corresponding TFM the *symbol* of the Toeplitz (Hankel) operator. We shall sometimes indicate a Toeplitz operator by its symbol, like $\mathbb{T}_{\mathbf{G}}$, instead of by the corresponding input–output operator, like $\mathbb{T}_{\mathcal{G}}$.

Proposition 2.4.4 *Let $\mathcal{G}$ be a bounded linear operator from $L^{2,m}$ to $L^{2,p}$. Then:*

1. $\mathbb{T}_{\mathcal{G}} = \mathcal{G}P_+^m$ *if and only if* $\widehat{\mathbb{H}}_{\mathcal{G}} = 0$.
2. $\widehat{\mathbb{T}}_{\mathcal{G}} = \mathcal{G}P_-^m$ *if and only if* $\mathbb{H}_{\mathcal{G}} = 0$.

Proof. We have

$$\mathcal{G}P_+^m = (P_+^p + P_-^p)\mathcal{G}P_+^m = P_+^p\mathcal{G}P_+^m + P_-^p\mathcal{G}P_+^m = \mathbb{T}_{\mathcal{G}} + \widehat{\mathbb{H}}_{\mathcal{G}}.$$

From this, point 1 follows immediately. Point 2 can be similarly proved. ■

The next result relates stability to the annulation of the anticausal Hankel operator.

Theorem 2.4.5 *Assume the system (2.81) is dichotomic and let $\mathcal{G}$ be its L^2 input–output operator. If A is stable, then $\widehat{\mathbb{H}}_{\mathcal{G}} = 0$.*

Conversely, if $\widehat{\mathbb{H}}_{\mathcal{G}} = 0$ and, additionally, (A, B) is controllable and (C, A) is observable, then A is stable.

Proof. Assume A is stable. Then the contribution of the antistable part of the system in (2.78) vanishes and the input–output operator becomes

$$(\mathcal{G}u)(t) = \int_{-\infty}^{t} Ce^{A(t-\tau)}Bu(\tau)d\tau + Du(t). \tag{2.82}$$

If we use in this expression $u(t) = 0$, for $t < 0$, it is clear that $(\mathcal{G}u)(t) = 0$, for $t < 0$. This means that $\widehat{\mathbb{H}}_{\mathcal{G}} := P_-^p\mathcal{G}P_+^m = 0$.

Conversely, assume now that $\widehat{\mathbb{H}}_{\mathcal{G}} = 0$ and that the system is controllable and observable. If the system were not stable, we could perform a change of coordinates that realizes the decomposition (2.7), (2.8). Then the realization of the system can be written as

$$\mathbf{G} = \left[\begin{array}{cc|c} A_- & O & B_- \\ O & A_+ & B_+ \\ \hline C_- & C_+ & D \end{array}\right]$$

with $A_- \in \mathbb{R}^{n_- \times n_-}$ stable and $A_+ \in \mathbb{R}^{n_+ \times n_+}$ antistable. Moreover, the pairs $(A_\pm, B_\pm)$ and $(C_\pm, A_\pm)$ are respectively, controllable and observable. If we apply (2.78) for the input $u_+ = P_+u$, then it is easy to see that

$$(\widehat{\mathbb{H}}_{\mathcal{G}}u)(t) = -\int_0^{\infty} C_+e^{A_+(t-\tau)}B_+u_+(\tau)d\tau, \; t \le 0,$$

or, equivalently,

$$(\widehat{\mathbb{H}}_{\mathcal{G}}u)(t) = C_+e^{A_+t}\xi_{u_+}, \tag{2.83}$$

where

$$\xi_{u_+} = -\int_0^\infty e^{-A_+\tau}B_+u_+(\tau)\mathrm{d}\tau. \tag{2.84}$$

If $(\widehat{\mathbb{H}}_{\mathcal{G}}u)(t) = 0$ for every t, it follows from the observability of (C_+, A_+) that $\xi_{u_+} = 0$ for every $u_+ \in L_+^{2,m}$. If we put in (2.84) $u_+(\tau) = -B_+^Te^{-A_+^T\tau}z$ for an arbitrary $z \in \mathbb{R}^{n_+}$, we get

$$0 = \int_0^\infty e^{-A_+\tau}B_+B_+^Te^{-A_+^T\tau}z\mathrm{d}\tau = \widehat{P}z$$

where $\widehat{P}$ is the anticausal controllability Gramian associated with the pair (A_+, B_+) and given in (2.23). However, since (A_+, B_+) is controllable, $\widehat{P}$ must be non–singular. This means that $n_+ = 0$, and thus the system is stable. ■

Remark 2.4.6 The dual of Theorem 2.4.5 can be obtained by replacing stability with antistability and the anticausal Hankel operator with the causal Hankel operator.

The following definition is suggested by Theorem 2.4.5.

Definition 2.4.7 *A bounded linear operator $\mathcal{G}$ from $L^{2,m}$ to $L^{2,p}$ is called* causal *if $\widehat{\mathbb{H}}_{\mathcal{G}} = 0$ and* anticausal *if $\mathbb{H}_{\mathcal{G}} = 0$.*

We shall say that the linear system (2.81) is (anti)causal if its input–output operator $\mathcal{G}$ is (anti)causal.

According to Proposition 2.4.4, the causality of $\mathcal{G}$ is equivalent to

$$\mathbb{T}_{\mathcal{G}} = \mathcal{G}P_+^m,$$

whereas the anticausality is equivalent to

$$\widehat{\mathbb{T}}_{\mathcal{G}} = \mathcal{G}P_-^m.$$

According to Theorem 2.4.5, a stable system is causal, whereas a causal system that is controllable and observable, is stable.

If the system (2.81) is causal and $u \in L_-^{2,m}$, then from (2.82) we deduce

$$(\mathbb{H}_{\mathcal{G}}u)(t) = (P_+\mathcal{G}u)(t) = Ce^{At}\int_{-\infty}^0 e^{-A\tau}Bu(\tau)\mathrm{d}\tau. \tag{2.85}$$

Let us define the *causal controllability operator* $\Psi : L_-^{2,m} \to \mathbb{R}^n$ as

$$\Psi u := \int_{-\infty}^0 e^{-A\tau}Bu(\tau)\mathrm{d}\tau\ , \quad u \in L_-^{2,m}, \tag{2.86}$$

and the *causal observability operator* $\Theta : \mathbb{R}^n \to L_+^{2,p}$ as

$$(\Theta x)(t) := Ce^{At}x\ ,\ t \geq 0, \ \ x \in \mathbb{R}^n. \tag{2.87}$$

With these definitions, (2.85) may be written as

$$\mathbb{H}_{\mathcal{G}} = \Theta\Psi. \tag{2.88}$$

Similarly, if the system (2.81) is anticausal, then we can write

$$\widehat{\mathbb{H}}_{\mathcal{G}} = \widehat{\Theta}\widehat{\Psi}, \tag{2.89}$$

where $\widehat{\Psi} : L_+^{2,m} \to \mathbb{R}^n$ is the *anticausal controllability operator*

$$\widehat{\Psi}u := -\int_0^{\infty} e^{-A\tau}Bu(\tau)\mathrm{d}\tau \ , \quad u \in L_+^{2,m}, \tag{2.90}$$

and $\widehat{\Theta} : \mathbb{R}^n \to L_-^{2,p}$ is the *anticausal observability operator*

$$(\widehat{\Theta}x)(t) := Ce^{At}x \ , \ t \leq 0, \quad x \in \mathbb{R}^n. \tag{2.91}$$

It is obvious that the operators Ψ, Θ, $\widehat{\Psi}$, $\widehat{\Theta}$ are finite rank operators. From formulas (2.88) and (2.89), it follows that $\mathbb{H}_{\mathcal{G}}$ and $\widehat{\mathbb{H}}_{\mathcal{G}}$ are also finite rank operators.

The adjoints of these operators are

$$(\Psi^*x)(t) = B^Te^{-A^Tt} \ , \ t \leq 0 \ , \quad x \in \mathbb{R}^n, \tag{2.92}$$

$$\Theta^*y = \int_0^{\infty} e^{A^Tt}C^Ty(t)\mathrm{d}t \ , \quad y \in L_+^{2,p}, \tag{2.93}$$

and

$$(\widehat{\Psi}^*x)(t) = -B^Te^{-A^Tt} \ , \ t \geq 0, \quad x \in \mathbb{R}^n, \tag{2.94}$$

$$\widehat{\Theta}^*y = \int_{-\infty}^{0} e^{A^Tt}C^Ty(t)\mathrm{d}t \ , \ y \in L_-^{2,p}. \tag{2.95}$$

With these expressions we can compute

$$\Psi\Psi^* = \int_{-\infty}^{0} e^{-A\tau}BB^Te^{-A^T\tau}\mathrm{d}\tau = \int_0^{\infty} e^{At}BB^Te^{A^Tt}\mathrm{d}t = P, \tag{2.96}$$

$$\Theta^*\Theta = \int_0^{\infty} e^{A^Tt}C^TCe^{At}\mathrm{d}t = Q, \tag{2.97}$$

and

$$\widehat{\Psi}\widehat{\Psi}^* = \int_0^{\infty} e^{-At}BB^Te^{-A^Tt}\mathrm{d}t = \widehat{P}, \tag{2.98}$$

$$\widehat{\Theta}^*\widehat{\Theta} = \int_0^{\infty} e^{-A^Tt}C^TCe^{-At}\mathrm{d}t = \widehat{Q}, \tag{2.99}$$

where P, $\widehat{P}$, Q, and $\widehat{Q}$ are the causal, anticausal, controllability, and observability Gramians, respectively, as can be seen from (2.19) and (2.23).

Proposition 2.4.8 *Let $\mathcal{G}$ be the input–output operator of a causal linear system. Then $\sigma > 0$ is a Hankel singular value of the system if and only if it is a singular value of $\mathbb{H}_{\mathcal{G}}$.*

Proof. If σ is a non–zero singular value of $\mathbb{H}_{\mathcal{G}}$, then it is a non–zero eigenvalue of $\mathbb{H}_{\mathcal{G}}^*\mathbb{H}_{\mathcal{G}} = \Psi^*\Theta^*\Theta\Psi$. Hence it is also a non–zero eigenvalue of $\Psi\Psi^*\Theta^*\Theta = PQ$ (see (2.88)).

The converse can be established similarly. ■

A simple consequence of the previous assertion is given in the following corollary.

Corollary 2.4.9 *If $\mathcal{G}$ is the L^2 input–output operator of a causal system then*

$$||\mathbb{H}_{\mathcal{G}}|| = \overline{\sigma}(PQ).$$

Corollary 2.4.9 justifies the following definition.

Definition 2.4.10 *If $\mathcal{G}$ is the L^2 input–output operator of a causal system then $||\mathbb{H}_{\mathcal{G}}||$ is called the* causal Hankel norm *and it is usually denoted by $||\mathcal{G}||_H$.*

Both Proposition 2.4.8 and Corollary 2.4.9 admit obvious anticausal duals. Accordingly, we can define as in Definition 2.4.10, the *anticausal Hankel norm* and we shall denote this by $||\mathcal{G}||_{\widehat{H}}$.

Finally, we present a few auxiliary results on general Hankel and Toeplitz operators. The first one, gives the adjoints of these operators.

Proposition 2.4.11 *Let $\mathcal{G}$ be a bounded linear operator from $L^{2,m}$ to $L^{2,p}$. Then the adjoints of the associated Hankel and Toeplitz operators are:*

$$\mathbb{T}_{\mathcal{G}}^* = \mathbb{T}_{\mathcal{G}^*}\ , \quad \widehat{\mathbb{T}}_{\mathcal{G}}^* = \widehat{\mathbb{T}}_{\mathcal{G}^*}\ ,$$

$$\mathbb{H}_{\mathcal{G}}^* = \widehat{\mathbb{H}}_{\mathcal{G}^*}\ , \quad \widehat{\mathbb{H}}_{\mathcal{G}}^* = \mathbb{H}_{\mathcal{G}^*}\ .$$

Proof. We give the proof of one of these relations as the others can be similarly derived. For example, we shall compute the adjoint of $\widehat{\mathbb{H}}_{\mathcal{G}}$. We have, for each $u \in L_-^{2,p}$ and $v \in L_+^{2,m}$,

$$\begin{aligned}\left\langle \widehat{\mathbb{H}}_{\mathcal{G}}v, u\right\rangle_{L_-^2} &= \langle P_-^p\mathcal{G}v, u\rangle_{L_-^2} = \langle \mathcal{G}v, u\rangle_{L^2} = \langle v, \mathcal{G}^*u\rangle_{L^2} = \langle v, P_+^m\mathcal{G}^*u\rangle_{L_+^2} \\ &= \langle v, \mathbb{H}_{\mathcal{G}^*}u\rangle_{L_+^2}\ .\end{aligned}$$

■

A simple, but important result gives the Toeplitz operator of a composition of two operators.

Proposition 2.4.12 *Let $\mathcal{G}$ be a bounded linear operator from $L^{2,m}$ to $L^{2,p}$ and $\mathcal{H}$ be a bounded linear operator from $L^{2,n}$ to $L^{2,m}$. Then*

$$\mathbb{T}_{\mathcal{G}\mathcal{H}} = \mathbb{T}_{\mathcal{G}}\mathbb{T}_{\mathcal{H}} + \mathbb{H}_{\mathcal{G}}\widehat{\mathbb{H}}_{\mathcal{H}}. \tag{2.100}$$

Proof. We have, successively,

$$\begin{aligned}\mathbb{T}_{\mathcal{G}\mathcal{H}} &= P_+^p\mathcal{G}\mathcal{H}P_+^n = P_+^p\mathcal{G}(P_+^m + P_-^m)\mathcal{H}P_+^n = P_+^p\mathcal{G}P_+^m\mathcal{H}P_+^n + P_+^p\mathcal{G}P_-^m\mathcal{H}P_+^n \\ &= \mathbb{T}_{\mathcal{G}}\mathbb{T}_{\mathcal{H}} + \mathbb{H}_{\mathcal{G}}\widehat{\mathbb{H}}_{\mathcal{H}}.\end{aligned}$$

■

Two immediate consequences of Proposition 2.4.12 will be repeatedly used.

Corollary 2.4.13 *Let $\mathcal{G}$ be a bounded linear operator from $L^{2,m}$ to $L^{2,p}$ and $\mathcal{H}$ be a bounded linear operator from $L^{2,n}$ to $L^{2,m}$. If $\mathcal{G}$ is anticausal, or $\mathcal{H}$ is causal, then $\mathbb{T}_{\mathcal{G}\mathcal{H}} = \mathbb{T}_{\mathcal{G}}\mathbb{T}_{\mathcal{H}}$.*

Corollary 2.4.14 *Let $\mathcal{G}$ be a bounded linear operator on $L^{2,m}$ that is boundedly invertible. If $\mathcal{G}$ is causal, or anticausal, then $\mathbb{T}_{\mathcal{G}}$ is invertible and $\mathbb{T}_{\mathcal{G}}^{-1} = \mathbb{T}_{\mathcal{G}^{-1}}$.*

B. The discrete–time case

The next sections are devoted to the discrete–time correspondents of the notions and results presented in the first part of this chapter. The main difficulty posed by discrete–time systems is that they are, in general, not time–reversible. Consequently, unlike the continuous–time case, we need to consider both direct–time and inverse–time notions.

2.5 Linear discrete–time systems

We shall work throughout the next sections with discrete–time state–space systems having two types of representations: direct–time and inverse–time. Since discrete–time systems are not always time–reversible, these two representations are in general not equivalent.

2.5.1 Direct and inverse–time systems

A *discrete–time state–space system with direct–time representation* is given by the equations

$$\begin{aligned} x_{k+1} &= Ax_k + Bu_k, \\ y_k &= Cx_k + Du_k, \end{aligned} \tag{2.101}$$

whereas a *discrete–time state–space system with inverse–time representation* is given by the equations

$$\begin{aligned} x_k &= Ax_{k+1} + Bu_k, \\ y_k &= Cx_{k+1} + Du_k. \end{aligned} \tag{2.102}$$

In both representations, $x = (x_k)_{k\in\mathbb{Z}}$, $u = (u_k)_{k\in\mathbb{Z}}$, and $y = (y_k)_{k\in\mathbb{Z}}$ are sequences in $\mathbb{R}^n$, $\mathbb{R}^m$ and $\mathbb{R}^p$, respectively, and A, B, C and D are real matrices of appropriate dimensions.

We introduce two operators defined on the space of vector valued sequences $w = (w_k)_{k\in\mathbb{Z}}$. The *shift operator* σ is defined by

$$(\sigma w)_k := w_{k+1}, \tag{2.103}$$

and the *time–reversal operator* Ω is defined as

$$(\Omega w)_k := w_{-k}. \tag{2.104}$$

Clearly, both operators are invertible and $\Omega^2 = I$.

If multiplication of a matrix with a vector valued sequence is defined pointwise, then it is clear that (2.101) can be rewritten as

$$\begin{aligned} \sigma x &= Ax + Bu, \\ y &= Cx + Du, \end{aligned} \tag{2.105}$$

whereas (2.102) can be rewritten as

$$\begin{aligned} x &= A\sigma x + Bu, \\ y &= C\sigma x + Du. \end{aligned} \tag{2.106}$$

It is easy to see that

$$(\Omega\sigma x)_k = (\sigma x)_{-k} = x_{-(k+1)} = x_{-k-1},$$

that is,

$$\Omega\sigma x = \sigma^{-1}\Omega x. \tag{2.107}$$

The time–reversal operator can be used to transform an inverse–time representation into a corresponding direct–time representation, as we explain in the sequel. Applying $\sigma\Omega$ in equations (2.106), and using (2.107), we obtain

$$\begin{aligned} \sigma\Omega x &= A\Omega x + B\sigma\Omega u, \\ \sigma\Omega y &= C\Omega x + D\sigma\Omega u. \end{aligned}$$

If we take

$$\widehat{x} := \Omega x \ , \quad \widehat{u} := \sigma\Omega u \ , \quad \widehat{y} := \sigma\Omega y, \tag{2.108}$$

then we obtain that $\widehat{x}$, $\widehat{u}$ and $\widehat{y}$ satisfy the direct–time linear system

$$\begin{aligned} \sigma\widehat{x} &= A\widehat{x} + B\widehat{u}, \\ \widehat{y} &= C\widehat{x} + D\widehat{u}. \end{aligned} \tag{2.109}$$

Relation (2.109) defines the direct–time correspondent of (2.102). We shall use the transformation (2.108) to deduce inverse–time counterparts of established direct–time results.

If A is invertible, the discrete–time system is said to be *time–reversible* and it admits *equivalent* direct–time and inverse–time representations. Indeed, in this case (2.106) is equivalent to the direct–time representation

$$\begin{aligned} \sigma x &= A^{-1}x - A^{-1}Bu, \\ y &= CA^{-1}x + (D - CA^{-1}B)u. \end{aligned} \tag{2.110}$$

Conversely, if we start from the direct–time system (2.105), and A is invertible, the equivalent inverse–time representation is

$$\begin{aligned} x &= A^{-1}\sigma x - A^{-1}Bu, \\ y &= CA^{-1}\sigma x + (D - CA^{-1}B)u. \end{aligned} \tag{2.111}$$

It is obvious, and we shall use this fact later, that if a system is antistable then A is invertible and the time–reversed equivalent is stable.

Although the state–space representations (2.101) or (2.102) are best–suited to describe the dynamical behavior of a system, its input–output properties are better revealed by the TFM. The TFM of the system with direct–time representation (2.101) is the rational $p \times m$ matrix function

$$\mathbf{G}(z) := C(zI - A)^{-1}B + D. \tag{2.112}$$

For a system with inverse–time representation (2.102), the TFM is

$$\widehat{\mathbf{G}}(z) := C(z^{-1}I - A)^{-1}B + D. \tag{2.113}$$

We shall use for the TFM $\mathbf{G}$ a similar notation as in the continuous–time case, that also emphasizes the state–space realization that generated it,

$$\mathbf{G} = \left[\begin{array}{c|c} A & B \\ \hline C & D \end{array}\right]. \tag{2.114}$$

To indicate the time–inverse case we shall use the superscript a:

$$\widehat{\mathbf{G}} = \left[\begin{array}{c|c} A & B \\ \hline C & D \end{array}\right]^{a}. \tag{2.115}$$

As in the continuous–time case, whenever we refer to a direct–time system (or inverse–time system) we understand by this the state–space representation (2.101) (or (2.102)), or the input–output representation as $y(z) = \mathbf{G}(z)u(z)$ (or $y(z) = \widehat{\mathbf{G}}(z)u(z)$). Here $y(z)$ and $u(z)$ stand for the Z–transforms of $y(t)$ and $u(t)$, respectively. However, there are important differences with respect to the continuous–time case. For example, the TFM of a system with a direct–time state–space representation is always proper (it has no poles at infinity), while the TFM of a system with an inverse–time state–space representation may be improper, but it never has poles at zero. Thus, the non–equivalence between direct and inverse–time state–space representations of a system is inherited by the TFM representation. Again, the equivalence is recaptured if A is assumed invertible, which ensures that the corresponding TFM (2.112) (or (2.113)) has no poles at infinity (or at zero). More precisely, a rational matrix $\mathbf{G}$ without poles at infinity and at zero can be seen alternatively as the TFM of a system with a direct or an inverse–time state–space representation, and we can write for $\mathbf{G}$ a realization of the form (2.114) or of the form (2.115). Indeed, (2.114) is the standard realization of the proper rational matrix $\mathbf{G}(z)$, while (2.115) can be obtained by writing a standard realization of the proper rational matrix $\mathbf{G}(\frac{1}{z})$ (see also Section 1.13). In turn, if $\mathbf{G}$ is an arbitrary rational matrix, we can always decompose it as

$$\mathbf{G}(z) = \mathbf{G}_p(z) + \mathbf{G}_i(z)$$

where $\mathbf{G}_p$ has no poles at infinity and $\mathbf{G}_i$ has no poles at zero. Thus, $\mathbf{G}_p$ is the TFM of a system with a direct–time state–space representation, while $\mathbf{G}_i$ is the TFM of a system with an inverse–time representation, and the state–space representations of both systems can be obtained by writing down realizations of the rational matrices $\mathbf{G}_p$ and $\mathbf{G}_i$ in the forms (2.112) and (2.113), respectively. Finally, the above discussion shows that an arbitrary rational matrix $\mathbf{G}$ can be seen as the TFM of a discrete–time state–space system with mixed direct and inverse–time evolutions.

2.5.2 *Stable, antistable, and dichotomic systems*

The system (2.105) is called *stable* if all the eigenvalues of A are in $\mathbb{D}$. The system is called *antistable* if all the eigenvalues of A are in $\mathbb{C}\setminus\overline{\mathbb{D}}$. The system is called *dichotomic* if A has no eigenvalues on $\partial\mathbb{D}$. By a slight terminology abuse, we say that A is stable, antistable, or dichotomic, if it is the state matrix of a stable, antistable, or dichotomic system, respectively. We shall also call the inverse–time system (2.106) *stable*, *antistable*, respectively *dichotomic*, if its state matrix is stable, antistable, or dichotomic, respectively.

As in the continuous–time case, if the system is dichotomic, there exists a similarity transformation T such that

$$\widetilde{A} := TAT^{-1} = \begin{bmatrix} A_- & O \\ O & A_+ \end{bmatrix} \begin{matrix} \}n_- \\ \}n_+ \end{matrix} \tag{2.116}$$

where A_- is square and stable and A_+ is square and antistable. The other system matrices become

$$\widetilde{B} := TB = \begin{bmatrix} B_- \\ B_+ \end{bmatrix} \begin{matrix} \}n_- \\ \}n_+ \end{matrix}, \quad \widetilde{C} := CT^{-1} = [\, \underbrace{C_-}_{n_-} \; \underbrace{C_+}_{n_+} \,]. \tag{2.117}$$

Such a transformation is not unique. However, the subspaces

$$\mathcal{X}_-(A) := \operatorname{Im} T^{-1} \begin{bmatrix} I_{n_-} \\ O \end{bmatrix}, \quad \mathcal{X}_+(A) := \operatorname{Im} T^{-1} \begin{bmatrix} O \\ I_{n_+} \end{bmatrix} \tag{2.118}$$

are independent of T and are called, respectively, the *stable* and the *antistable subspace* of A. It is easy to see that they are invariant subspaces for A that are complementary,

$$\mathcal{X}_-(A) \oplus \mathcal{X}_+(A) = \mathbb{R}^n,$$

and that the projection on these subspaces are given by the formulas

$$\Pi_- = T^{-1} \begin{bmatrix} I_{n_-} & O \\ O & O \end{bmatrix} T, \quad \Pi_+ = T^{-1} \begin{bmatrix} O & O \\ O & I_{n_+} \end{bmatrix} T. \tag{2.119}$$

2.5.3 Controllability, stabilizability, and duals

Controllability and observability are defined as in the continuous–time case. A discrete–time system is called *stabilizable* if the pair (A, B) is d–controllable, and is called *antistabilizable* if the pair (A, B) is $\mathbb{D} \cup \partial\mathbb{D}$–controllable. A discrete–time system is called *detectable* (or *antidetectable*) if the pair (C, A) is d–observable (or $\mathbb{D} \cup \partial\mathbb{D}$–observable). Whenever it is clear from the context that we are in a continuous or a discrete–time setting, we shall also talk about stabilizable, antistabilizable, detectable, and antidetectable pairs of matrices.

2.5.4 Spaces of rational matrices, norms, and continuity

Let $L^\infty(\partial\mathbb{D})$ denote the Banach space of matrix–valued complex functions $G(t)$ that are (essentially) bounded on $[0,\ 2\pi]$, with the so–called $L^\infty(\partial\mathbb{D})$*–norm*,

$$||G||_\infty := \sup_{t \in \partial\mathbb{D}} \overline{\sigma}(G(t)). \tag{2.120}$$

In case of rational matrices $G(t) \equiv \mathbf{G}$, we rewrite (2.120) in the form

$$\|\mathbf{G}\|_\infty := \sup_{\theta\in[0,2\pi)} \overline{\sigma}(\mathbf{G}(e^{j\theta})) = \sup_{\theta\in[0,2\pi]} \|\mathbf{G}(e^{j\theta})\|. \tag{2.121}$$

If the set over which we take the supremum in (2.121) results from the context, we shall talk about L^∞ space and norm. Keeping the notation from the continuous–time case, we denote by $RL^\infty_{p\times m}$ the space of all $p \times m$ rational matrices (not necessarily proper) with entries bounded on the unit circle $\partial\mathbb{D}$, and $RH^\infty_{+,p\times m}$ will denote the space of (proper) rational matrices with entries analytic on $\overline{\mathbb{C}} \setminus \mathbb{D}$. By $RH^\infty_{-,p\times m}$ we denote the set of all $p\times m$ rational functions (not necessarily proper) with elements analytic in $\overline{\mathbb{D}}$. Notice the differences with respect to the continuous–time case in which these spaces are defined with reference to the imaginary axis and the left–half plane instead of the unit circle and unit disk, respectively, and in which the elements belonging to RL^∞ and RH^∞_- must be proper. The $L^\infty(\mathbb{C}_0)$–norm is also called H^∞*–norm* for rational matrices $\mathbf{G} \in RH^\infty_+$.

The adjoint (or conjugate) of the TFM $\mathbf{G} = C(zI - A)^{-1}B + D$ of a discrete–time system is defined as

$$\mathbf{G}^*(z) := \mathbf{G}^T(\frac{1}{z}),$$

such that, for each real θ,

$$\mathbf{G}^*(e^{j\theta}) = \mathbf{G}^T(-e^{j\theta}) = [\mathbf{G}(e^{j\theta})]^*.$$

The realization of $\mathbf{G}^*$ is given in two different notations by

$$\mathbf{G}^* = \left[\begin{array}{c|c} A^T & C^T \\ \hline B^T & D^T \end{array}\right]^a = \left[\begin{array}{c|c} I - zA^T & -C^T \\ \hline zB^T & D^T \end{array}\right]$$

(see (2.115) and (1.76)). It is easy to see now that the conjugate of a proper TFM of a discrete–time system may be improper, provided A is singular.

Just as in the continuous–time case, if A is dichotomic, then the TFM $\mathbf{G}$ given by (2.112) is in RL^∞. If A is stable, then $\mathbf{G}$ is in RH^∞_+, and if A is antistable, then $\mathbf{G}$ is in RH^∞_-.

Let $L^2(\partial\mathbb{D})$ be the Hilbert space of matrix (or vector or scalar)–valued complex functions $G(t)$ for which the so–called $L^2(\partial\mathbb{D})$*–norm*,

$$\|G\|_2 := \{\frac{1}{2\pi}\int_0^{2\pi} \text{Trace}\,[G^*(e^{j\theta})G(e^{j\theta})]\text{d}t\}^{\frac{1}{2}}, \tag{2.122}$$

is finite. If $G \equiv \mathbf{G}$ is a rational matrix then its $L^2(\partial\mathbb{D})$–norm may be finite not only if $\mathbf{G}$ is strictly proper but also if it is proper and even improper. The H^2*–norm* is defined for rational matrices $\mathbf{G} \in RH^\infty_+$ and has the same expression (2.122).

From Theorem 2.1.1, it follows that $\overline{\sigma}(\mathbf{G}(e^{j\theta}))$ is continuous on the compact interval $[0, 2\pi]$ and attains its maximum for some θ in this interval.

2.5.5 Gramians and Hankel singular values

Assume that A is stable. Then we have from Theorem 1.5.2 that the following symmetric Stein equations

$$APA^T - P + BB^T = 0 \tag{2.123}$$

and

$$A^TQA - Q + C^TC = 0 \tag{2.124}$$

have unique positive semidefinite solutions P and Q in $\mathbb{R}^{n\times n}$. P is called the *causal controllability Gramian* associated with the matrix pair (A, B), and Q is called the *causal observability Gramian* associated with the matrix pair (C, A). This is justified by the fact that $P > 0$ if and only if the system is controllable, and similarly $Q > 0$ if and only if the system is observable, as Theorems 1.5.6 and 1.5.4 show. It is well–known that they can also be expressed by

$$P = \sum_{k=0}^{\infty} A^k BB^T (A^T)^k \ , \quad Q = \sum_{k=0}^{\infty} (A^T)^k C^T C A^k. \tag{2.125}$$

Let $\Lambda(PQ) = \Lambda(QP) = \{\sigma_1^2, \ldots, \sigma_n^2\}$. As both P and Q are positive semidefinite, we have that σ_i $(i = 1, \ldots, n)$ are non–negative. Those values σ_i which are non–zero, depend only on the TFM $\mathbf{G}(z)$ and are called the *Hankel singular values* of the system (2.105) or (2.106).

Theorem 2.1.4 has an obvious discrete–time corresponding result.

In the case that A is antistable, we can introduce the *anticausal controllability Gramian* $\widehat{P}$ associated with the matrix pair (A, B) and the *anticausal observability Gramian* $\widehat{Q}$ associated with the matrix pair (C, A), as the unique solutions (see Theorem 1.5.2) of the symmetric Stein equations

$$A\widehat{P}A^T - \widehat{P} - BB^T = 0 \tag{2.126}$$

and

$$A^T\widehat{Q}A - \widehat{Q} - C^TC = 0. \tag{2.127}$$

They can also be expressed as

$$\widehat{P} = \sum_{k=1}^{\infty} (A^{-1})^k BB^T (A^{-T})^k \ , \quad \widehat{Q} = \sum_{k=1}^{\infty} (A^{-T})^k C^T C (A^{-1})^k, \tag{2.128}$$

and enjoy similar properties as their causal counterparts. In fact, it is easy to see that $\widehat{P}$ is the causal controllability Gramian of the pair $(A^{-1}, A^{-1}B)$, whereas $\widehat{Q}$ is the observability Gramian of the pair (CA^{-1}, A^{-1}).

Similar to the continuous case, if $D = 0$ and A is stable,

$$\|\mathbf{G}\|_2 = [\text{Trace}\,(B^TQB)]^{\frac{1}{2}} = [\text{Trace}\,(CPC^T)]^{\frac{1}{2}},$$

and, if A is antistable,

$$\|\mathbf{G}\|_2 = [\text{Trace}\,(B^T\widehat{Q}B)]^{\frac{1}{2}} = [\text{Trace}\,(C\widehat{P}C^T)]^{\frac{1}{2}},$$

where the norms are taken in $L^2(\partial\mathbb{D})$. The formulas for connections of systems obtained in Section 2.1.5 are valid irrespective of the continuous or discrete–time setting.

2.5.6 Bilinear transformation between a continuous and a discrete–time system

There exists a standard way of transforming a discrete–time system into a continuous one that preserves some important properties. This transformation sometimes allows the establishment of discrete–time results, using their continuous–time counterparts or vice versa. The essential ingredient of this transformation is the following bilinear transformation in the complex plane:

$$z = \frac{1+s}{1-s} \Leftrightarrow s = \frac{z-1}{z+1}. \tag{2.129}$$

This is a conformal mapping of the open left–half plane $\mathbb{C}_-$ onto the open unit disk $\mathbb{D}$. It maps the point $s = 1$ into $z = \infty$ and $s = \infty$ into $z = -1$.

We assume that -1 is not an eigenvalue of A. This is particularly true if the discrete–time system is dichotomic. Substituting (2.129) in the expression for the TFM

$$\mathbf{G} = \left[\begin{array}{c|c} A & B \\ \hline C & D \end{array}\right] = C(zI - A)^{-1}B + D,$$

and denoting

$$\mathbf{G}_c(s) := \mathbf{G}(\frac{1+s}{1-s}), \tag{2.130}$$

after some straightforward manipulations this yields

$$\mathbf{G}_c(s) = \left[\begin{array}{c|c} A_c & B_c \\ \hline C_c & D_c \end{array}\right](s) = C_c(sI - A_c)^{-1}B_c + D_c, \tag{2.131}$$

where

$$\begin{array}{rcl} A_c & = & (A+I)^{-1}(A-I), \\ B_c & = & \sqrt{2}(A+I)^{-1}B, \\ C_c & = & \sqrt{2}C(A+I)^{-1}, \\ D_c & = & D - C(A+I)^{-1}B. \end{array} \tag{2.132}$$

Notice that $A + I$ is invertible since we assumed that $-1 \notin \Lambda(A)$.

It is easy to show that A is stable (antistable, or dichotomic) if and only if A_c is stable (antistable, or dichotomic) in the continuous–time sense. This is a mere consequence of the fact that the bilinear mapping (2.129) transforms the left–half plane into the unit disk, the right–half plane into the exterior of the unit disk, and the imaginary axis into the unit circle. This implies that if the TFM $\mathbf{G}$ is in RL^∞ (in a discrete–time sense), then $\mathbf{G}_c$ is in RL^∞ (in a continuous–time sense) and, in this case, their respective norms are equal:

$$||\mathbf{G}||_\infty = ||\mathbf{G}_c||_\infty. \tag{2.133}$$

Here the norms are taken in $L^\infty(\partial\mathbb{D})$ and $L^\infty(\mathbb{C}_0)$, respectively. A remarkable property of this transformation is that it conserves the Gramians. Indeed, let A be stable. Then, as explained, A_c is stable as well (in the continuous–time sense). Let P_c be the controllability Gramian of the pair (A_c, B_c). We then have

$$\begin{array}{rcl} 0 & = & A_cP_c + P_cA_c^T + B_cB_c^T \\ & = & (A+I)^{-1}(A-I)P_c + P_c(A^T - I)(A^T + I)^{-1} \end{array}$$

$$\begin{aligned}
&+ 2(A+I)^{-1}BB^T(A^T+I)^{-1} \\
= \;& (A+I)^{-1}[(A-I)P_c(A^T+I) + (A+I)P_c(A^T-I) \\
&+ 2BB^T](A^T+I)^{-1} \\
= \;& 2(A+I)^{-1}[AP_cA^T - P_c + BB^T](A^T+I)^{-1},
\end{aligned}$$

which implies

$$AP_cA^T - P_c + BB^T = 0.$$

Since the solution of the symmetric Stein equation is unique, this means that P_c is the controllability Gramian of (A, B). In the same way it can be shown that the observability Gramian remains unchanged after performing the bilinear transformation (2.132). In particular, this means that (A, B, C, D) and (A_c, B_c, C_c, D_c) have the same Hankel singular values.

A similar transformation can be defined for the inverse–time system (2.106) by starting from the TFM

$$\widehat{\mathbf{G}}(z) = \left[\begin{array}{c|c} A & B \\ \hline C & D \end{array}\right]^a (z) = C(z^{-1}I - A)^{-1}B + D.$$

Conversely, starting with a continuous–time system, we can use the bilinear transformation (2.129) to convert it into a discrete–time system.

2.5.7 Continuity of the H^∞–norm

We can now state the discrete–time counterpart of Theorem 2.1.3. As explained before, Theorem 2.1.3 can be easily deduced from this result.

Theorem 2.5.1 *Assume that the sequence of linear discrete–time systems*

$$\mathbf{G}_k = \left[\begin{array}{c|c} A_k & B_k \\ \hline C_k & D_k \end{array}\right], \quad k \in \mathbb{N}, \tag{2.134}$$

is such that

$$A_k \to A, B_k \to B, C_k \to C \text{ and } D_k \to D \text{ as } k \to \infty,$$

and A and A_k are all stable, and $\mathbf{G}(z) = C(zI - A)^{-1}B + D \equiv 0$. Then

$$\lim_{k\to\infty} \|\mathbf{G}_k\|_\infty = 0.$$

The proof of this theorem makes use of the following result.

Lemma 2.5.2 *Let $\nu > 0$ and $0 \le \mu < q < 1$ be arbitrary constants. Then there exists a positive $\rho(\nu, \mu, q)$ such that*

$$\|A^k\| \le \rho(\nu, \mu, q)q^k , \tag{2.135}$$

for each $k \in \mathbb{N}$ and for each $A \in \mathbb{R}^{n\times n}$ having all its eigenvalues placed into the closed disk of radius μ and for which $\|A\| \le \nu$.

Proof. We assume the conclusion is not true. That is, for some positive ν and $0 \leq \mu < q < 1$, and for arbitrary $\rho > 0$, there exists an $A \in \mathbb{R}^{n\times n}$ and a $k \in \mathbb{N}$ such that $||A|| \leq \nu$, $\Lambda(A) \subset \{z : \ ||z|| \leq \mu\}$ and $||A^k|| > \rho q^k$. Take $\rho = i \in \mathbb{N}\setminus\{0\}$ and let A_i and k_i be such that $||A_i|| \leq \nu$, $\Lambda(A_i) \subset \{z : ||z|| \leq \mu\}$ and

$$||A_i^{k_i}|| > iq^{k_i}. \tag{2.136}$$

As the sequence $(A_i)_{i\geq 1}$ is bounded, it has a convergent subsequence $(A_{i_j})_{j\geq 0}$ such that $A_{i_j} \to \widetilde{A}$ as $j \to \infty$ and $\Lambda(\widetilde{A}) \subset \{z : ||z|| \leq \mu\}$. This implies that for every $\widetilde{q}$ with $\mu < \widetilde{q} < q$ there exists $\widetilde{\rho}$ such that $||\widetilde{A}^k|| \leq \widetilde{\rho} q^k, k \in \mathbb{N}$. Let j_0 be large enough so that

$$||A_{i_j} - \widetilde{A}|| < \frac{\frac{q}{q} - 1}{\widetilde{\rho}}\widetilde{q},$$

for each $j > j_0$. Using the identity

$$A_{i_j}^k = \widetilde{A}^k + \sum_{s=0}^{k-1} \widetilde{A}^{k-s-1}(A_{i_j} - \widetilde{A})A_{i_j}^s,$$

we can evaluate

$$\begin{aligned} ||A_{i_j}^k|| &\leq ||\widetilde{A}^k|| + \sum_{s=0}^{k-1} ||\widetilde{A}^{k-s-1}||||(A_{i_j} - \widetilde{A})||||A_{i_j}^s|| \\ &\leq \widetilde{\rho}\widetilde{q}^k + \sum_{s=0}^{k-1} \widetilde{\rho}\widetilde{q}^{k-s-1}\frac{\frac{q}{q} - 1}{\widetilde{\rho}}\widetilde{q}||A_{i_j}^s||, \end{aligned}$$

or, equivalently,

$$\widetilde{q}^{-k}||A_{i_j}^k|| \leq \widetilde{\rho} + (\frac{q}{\widetilde{q}} - 1)\sum_{s=0}^{k-1} \widetilde{q}^{-s}||A_{i_j}^s||.$$

Let $u_k := \widetilde{q}^{-k}||A_{i_j}^k||$ for $j > j_0$. With this notation we can readily write the previous inequality as

$$u_k \leq \widetilde{\rho} + (\frac{q}{\widetilde{q}} - 1)\sum_{s=0}^{k-1} u_s$$

for all $k \geq 0$. From this inequality, we can prove by induction that

$$u_k \leq \widetilde{\rho}\left(\frac{q}{\widetilde{q}}\right)^k,$$

which in turn implies that $||A_{i_j}^k|| \leq \widetilde{\rho} q^k$ for all $k \geq 0$ and $j > j_0$. Letting $k = k_{i_j}$ and using inequality (2.136), we get $i_j q^{k_{i_j}} \leq \widetilde{\rho} q^{k_{i_j}}$, that is, $i_j \leq \widetilde{\rho}$ for all $j > j_0$. However, this is clearly impossible as it contradicts the unboundedness of $(i_j)_{j\geq 0}$. Hence, the conclusion of the theorem is true. ■

Proof of Theorem 2.5.1. Convergent sequences are bounded, so we can find a positive ν such that $||A_k|| + ||B_k|| + ||C_k|| + ||D_k|| < \nu$ for each $k \in \mathbb{N}$. As all A_k and

A are stable, there exists a positive $\mu < 1$ such that all the eigenvalues of A_k and of A have modulus smaller than μ. We also choose q such that $\mu < q < 1$. According to Lemma 2.5.2 there exists a ρ depending on ν, μ, and q, such that

$$||A_k^i|| \leq \rho q^i,$$

for each $k, i \in \mathbb{N}$. As a consequence

$$||C_k A_k^i B_k|| \leq \nu^2 \rho q^i \ , \quad k, i \in \mathbb{N}. \tag{2.137}$$

However, the TFM $\mathbf{G}_k$ can be written as

$$\mathbf{G}_k(z) = D_k + C_k(zI - A_k)^{-1}B_k = D_k + \sum_{i=0}^{\infty} C_k A_k^i B_k z^{-(i+1)}$$

for every z on the unit circle. Now we can estimate

$$\begin{aligned} ||\mathbf{G}_k(z)|| &\leq ||D_k|| + \sum_{i=0}^{\infty} ||C_k A_k^i B_k|| \leq ||D_k|| + \sum_{i=0}^{i_0} ||C_k A_k^i B_k|| + \sum_{i=i_0+1}^{\infty} \nu^2 \rho q^i \\ &= ||D_k|| + \sum_{i=0}^{i_0} ||C_k A_k^i B_k|| + \nu^2 \rho \frac{q^{i_0+1}}{1-q}. \end{aligned} \tag{2.138}$$

We choose an arbitrary $\varepsilon > 0$ and an i_0 such that

$$q^{i_0+1} < \frac{1-q}{\nu^2 \rho} \frac{\varepsilon}{3}. \tag{2.139}$$

As $\mathbf{G}_k(z) \to 0$, it follows that $C_k A_k^i B_k \to 0$ as $k \to \infty$ for each $i \in \mathbb{N}$. This means we can choose N large enough such that

$$||D_k|| < \frac{\varepsilon}{3} \quad \text{and} \quad ||C_k A_k^i B_k|| \frac{1}{i_0+1} \frac{\varepsilon}{3}, \tag{2.140}$$

for each $i = 0, 1, \ldots, i_0$ and all $k > N$. Introducing these in estimate (2.138), it follows that, for each z on the unit circle and each $k > N$, we have

$$||\mathbf{G}_k(z)|| < \varepsilon.$$

Taking the supremum after z, we obtain

$$||\mathbf{G}_k||_\infty < \varepsilon,$$

for all $k > N$, and therefore $||\mathbf{G}_k||_\infty \to 0$. ∎

We return now to the proof of Theorem 2.1.3.

Proof of Theorem 2.1.3. Using the bilinear transformation (2.129), we transform the sequence of systems (2.13) into

$$\mathbf{G}_{c,k} = \left[\begin{array}{c|c} A_{c,k} & B_{c,k} \\ \hline C_{c,k} & D_{c,k} \end{array} \right]$$

and

$$\mathbf{G} = \left[\begin{array}{c|c} A & B \\ \hline C & D \end{array}\right]$$

into

$$\mathbf{G}_c = \left[\begin{array}{c|c} A_c & B_c \\ \hline C_c & D_c \end{array}\right],$$

where $\mathbf{G}_{c,k}$ and $\mathbf{G}_c$ are related to $\mathbf{G}_k$ and $\mathbf{G}$ by (2.130), respectively. Clearly, $\mathbf{G}_k$ and $\mathbf{G}$ satisfy all the hypotheses of Theorem 2.5.1. From (2.133) we have

$$||\mathbf{G}_{c,k}||_\infty = ||\mathbf{G}_k||_\infty,$$

from which the conclusion follows immediately.

2.6 State evolutions corresponding to ℓ^2 inputs and associated operators

The purpose of this section is to introduce some important particular solutions of the linear recursion

$$\sigma x = Ax + Bu, \tag{2.141}$$

and of its inverse–time counterpart

$$x = A\sigma x + Bu, \tag{2.142}$$

where $A \in \mathbb{R}^{n \times n}, B \in \mathbb{R}^{n \times m}$, in the case in which u is an $L^2(\mathbb{Z})$ function. We shall use further the notation ℓ^2 as a shorthand for $L^2(\mathbb{Z})$. Thus $\ell^{2,m}$ stands for the Hilbert space of doubly infinite norm–square summable $\mathbb{C}^m$–valued sequences. If $u = (u_k)_{k \in \mathbb{Z}} \in \ell^{2,m}$, $u_k \in \mathbb{C}^m$, then

$$||u||_2 := (\sum_{-\infty}^{\infty} ||u_k||^2)^{\frac{1}{2}} < \infty$$

is the appropriate ℓ^2*–norm* of u. By $\ell_+^{2,m}$ and $\ell_-^{2,m}$ we denote the closed subspaces of $\ell^{2,m}$ consisting of the sequences of positive and (strictly) negative support, respectively. Clearly, $\ell^{2,m} = \ell_+^{2,m} \oplus \ell_-^{2,m}$. Usually we shall work with real–valued sequences, i.e., $u_k \in \mathbb{R}$.

We consider first the direct–time case, and deduce the inverse–time one using the variable transformations (2.108). The most common solution to (2.141) is the *initial-value solution* . For an arbitrary ξ in $\mathbb{R}$ and u in $\ell_+^{2,m}$ we denote by $x^{\xi,u}$ the solution to (2.141) that satisfies $x_0 = \xi$ and that is given by the variations of constants formula

$$x_k^{\xi,u} = A^k\xi + \sum_{i=0}^{k-1} A^{k-i-1}Bu_i \ , \quad k \geq 1. \tag{2.143}$$

For the inverse–time recursion (2.142), the solution $\widehat{x}^{\xi,u}$ that satisfies $x_0 = \xi$ is

$$\widehat{x}_k^{\xi,u} = A^{-k}\xi + \sum_{i=k}^{-1} A^{-k+i}Bu_i \ , \quad k \leq -1. \tag{2.144}$$

This follows directly from the fact that $\widehat{x}^{\xi,u} = \Omega x^{\xi,\sigma\Omega u}$.

We shall also study other solutions of (2.141) and (2.142). We introduce them in this section together with their main properties. As in the continuous–time case, we give whenever possible an operator expression for these solutions.

2.6.1 *The direct–time case: A stable*

If A is stable, the initial value solution $x^{\xi,u}$ is in $\ell_+^{2,n}$ for each $\xi \in \mathbb{R}^n$ and $u \in \ell_+^{2,m}$. We introduce the linear operators

$$\Phi : \mathbb{R}^n \to \ell_+^{2,n}\ , \quad (\Phi\xi)_k := A^k\xi, \quad k \geq 0, \tag{2.145}$$

and

$$\mathcal{L} : \ell_+^{2,m} \to \ell_+^{2,n}\ , \quad (\mathcal{L}u)_k := \sum_{i=0}^{k-1} A^{k-i-1}Bu_i\ , \quad k \geq 1. \tag{2.146}$$

It is easy to show that both are bounded operators. Obviously we can write

$$x^{\xi,u} = \Phi\xi + \mathcal{L}u. \tag{2.147}$$

Let us now consider the case of a doubly infinite input sequence u in $\ell^{2,m}$.

Theorem 2.6.1 *Let A be stable. Then, for each $u = (u_k)_{k\in\mathbb{Z}}$ in $\ell^{2,m}$, there exists a unique sequence $x_e^u = (x_{e,k}^u)_{k\in\mathbb{Z}}$ in $\ell^{2,n}$ that satisfies the recursive equation (2.141) and is given by*

$$x_{e,k}^u = \sum_{i=-\infty}^{k-1} A^{k-i-1}Bu_i\ , \quad k \in \mathbb{Z}. \tag{2.148}$$

Proof. As A is stable, we can find $\rho > 1$ and $0 < q < 1$ such that

$$||A^k|| \leq \rho q^k, \tag{2.149}$$

for every $k \in \mathbb{Z}$. From this estimate, it is easy to show that the series in (2.148) is convergent. Now we prove that the sequence x_e^u given by (2.148) is in $\ell^{2,n}$.

From (2.148) we successively deduce

$$\begin{aligned}
||x_{e,k}^u||^2 &= ||\sum_{i=-\infty}^{k-1} A^{k-i-1}Bu_i||^2 \leq (\sum_{i=-\infty}^{k-1} ||A^{k-i-1}||||B||||u_i||)^2 \\
&\leq (\sum_{i=-\infty}^{k-1} \rho q^{k-i-1}||B||||u_i||)^2 = \rho^2||B||^2(\sum_{i=-\infty}^{k-1} q^{\frac{k-i-1}{2}} q^{\frac{k-i-1}{2}} ||u_i||)^2 \\
&\quad \text{(using the Cauchy – Schwartz inequality)} \\
&\leq \rho^2||B||^2 \sum_{i=-\infty}^{k-1} q^{k-i-1} \sum_{i=-\infty}^{k-1} q^{k-i-1}||u_i||^2 \\
&= \rho^2||B||^2 \sum_{i=0}^{\infty} q^i \sum_{i=-\infty}^{k-1} q^{k-i-1}||u_i||^2 \\
&= \frac{\rho^2||B||^2}{1-q} \sum_{i=-\infty}^{k-1} q^{k-i-1}||u_i||^2.
\end{aligned}$$

Further, we have for arbitrary integers M and N,

$$\begin{aligned}\sum_{k=-N}^{M} ||x_{e,k}^{u}||^2 &\leq \sum_{k=-N}^{M} \frac{\rho^2||B||^2}{1-q} \sum_{i=-\infty}^{k-1} q^{k-i-1}||u_i||^2 \\ &= \frac{\rho^2||B||^2}{1-q} \sum_{i=-\infty}^{M-1} \sum_{k=\max(i+1,-N)}^{M} q^{k-i-1}||u_i||^2 \\ &= \frac{\rho^2||B||^2}{1-q} \sum_{i=-\infty}^{M-1} ||u_i||^2 \sum_{k=\max(0,-N-i-1)}^{M-i-1} q^k \\ &\leq \frac{\rho^2||B||^2}{1-q} \sum_{i=-\infty}^{M-1} ||u_i||^2 \sum_{k=0}^{\infty} q^k = \frac{\rho^2||B||^2}{(1-q)^2} \sum_{i=-\infty}^{M-1} ||u_i||^2.\end{aligned}$$

Taking in this estimate the limit as N and M go to ∞, we conclude that

$$||x_e^u||_2 \leq \frac{\rho||B||}{1-q}||u||_2 \tag{2.150}$$

for each $u \in \ell^{2,m}$, and thus $x_e^u \in \ell^{2,n}$.

Next it is easy to check that (2.148) satisfies (2.141). To prove uniqueness, it is enough to show that $x \equiv 0$ is the unique solution in $\ell^{2,n}$ of $\sigma x = Ax$. Suppose this is not true. Then we can write $x_k = A^{k-i}x_i$ for every k and $i < k$ in $\mathbb{Z}$. Take a k in $\mathbb{Z}$ such that $x_k \neq 0$. Then, for every $i < k$ $||x_k|| \leq \rho q^{k-i}||x_i||$, or, equivalently,

$$\frac{1}{\rho^2}q^{2(i-k)}||x_k||^2 \leq ||x_i||^2.$$

If we sum in this inequality for all $i < k$ we get a divergent series in the left–hand side because q is subunitary. Therefore x cannot be in ℓ_2 as assumed. ■

From the proof of Theorem 2.6.1 it follows that we can define the bounded linear operator $\mathcal{L}_e$ from $\ell^{2,m}$ to $\ell^{2,n}$, given by

$$(\mathcal{L}_e u)_k := \sum_{i=-\infty}^{k-1} A^{k-i-1}Bu_i\ , \quad k \in \mathbb{Z}. \tag{2.151}$$

Accordingly, we have

$$x_e^u = \mathcal{L}_e u \tag{2.152}$$

and

$$||\mathcal{L}_e|| \leq \left(\frac{\rho||B||}{1-q}\right)$$

as (2.150) shows. If we denote by $P_{\pm}^r$, the usual orthogonal projections of $\ell^{2,r}$ onto $\ell_{\pm}^{2,r}$, it is obvious that

$$\mathcal{L} = P_+^n \mathcal{L}_e P_+^m, \tag{2.153}$$

where $\mathcal{L}$ and $\mathcal{L}_e$ are defined in (2.146) and (2.151), respectively.

2.6.2 The inverse–time case: A stable

The inverse–time counterpart of Theorem 2.6.1 can be easily deduced using the time–reversal operator Ω (as in (2.108) and (2.109)).

Theorem 2.6.2 *Let A be stable. Then, for each $u = (u_k)_{k\in\mathbb{Z}}$ in $\ell^{2,m}$, there exists a unique sequence $\widehat{x}_e^u = (\widehat{x}_{ek}^u)_{k\in\mathbb{Z}}$ in $\ell^{2,n}$ that satisfies the inverse–time recursive equation (2.142) and is given by*

$$\widehat{x}_{e,k}^u = \sum_{i=k}^{\infty} A^{i-k} B u_i \,, \quad k \in \mathbb{Z}. \tag{2.154}$$

Let us introduce the bounded linear operator $\widehat{\mathcal{L}}_e$ from $\ell^{2,m}$ to $\ell^{2,n}$, given by

$$(\widehat{\mathcal{L}}_e u)_k := \sum_{i=k+1}^{\infty} A^{i-k-1} B u_i \,, \quad k \in \mathbb{Z}. \tag{2.155}$$

Accordingly, we have

$$\sigma \widehat{x}_e^u = \widehat{\mathcal{L}}_e u. \tag{2.156}$$

Notice the slight difference with respect to the direct–time case. The reason why the operator $\mathcal{L}_e$ is introduced to express σx_e^u in place of x_e^u is simply a matter of convenience for later use.

It is also convenient to introduce the linear bounded operator from $\ell_+^{2,m}$ to $\ell_+^{2,n}$

$$\widehat{\mathcal{L}} := P_+^n \widehat{\mathcal{L}}_e P_+^m , \tag{2.157}$$

that is

$$(\widehat{\mathcal{L}} u)_k = \sum_{i=k+1}^{\infty} A^{i-k-1} B u_i \,, \quad k \geq 0. \tag{2.158}$$

Defining

$$\widehat{x}^u := P_+^n \widehat{x}_e^u , \tag{2.159}$$

we have

$$\sigma \widehat{x}^u = \widehat{\mathcal{L}} u. \tag{2.160}$$

2.6.3 The direct–time case: A antistable

Returning to (2.141), but considering A to be antistable, we have the following result.

Theorem 2.6.3 *Let A be antistable. Then, for each $u = (u_k)_{k\in\mathbb{Z}}$ in $\ell^{2,m}$, there exists a unique sequence $x_e^u = (x_{e\,k}^u)_{k\in\mathbb{Z}}$ in $\ell^{2,n}$ that satisfies the recursive equation (2.141) and is given by*

$$x_{e,k}^u = -\sum_{i=k}^{\infty} (A^{-1})^{i-k+1} B u_i \,, \quad k \in \mathbb{Z}. \tag{2.161}$$

(Notice that, being antistable, A is invertible).

Proof. Rewrite (2.141) as

$$x = A^{-1}\sigma x - A^{-1}Bu$$

and apply Theorem 2.6.2 to this inverse–time evolution, using the fact that A^{-1} is stable. ■

We shall also need the restriction of x_e^u to $\mathbb{N}$. Therefore, for $u \in \ell_+^{2,m}$ we introduce

$$x_k^u := x_{e,k}^u, \quad k \geq 0. \tag{2.162}$$

If we introduce the operator $\mathcal{L}$ in this case through the relation (2.153), we can write

$$x^u = \mathcal{L}u. \tag{2.163}$$

It is obviously possible to give the inverse–time version of Theorem 2.6.3. In fact, as the direct–time antistable case can always be deduced from the inverse–time stable case, in the same way in which we established Theorem 2.6.3 from Theorem 2.6.2, we shall only rarely consider direct–time antistable systems.

2.6.4 The direct–time case: A dichotomic

By combining Theorem 2.6.1 and Theorem 2.6.3, we obtain the following result.

Theorem 2.6.4 *Let A be dichotomic. Then, for each u in $\ell^{2,m}$, the equation (2.141) has a unique solution x_e^u in $\ell^{2,n}$, which is given by*

$$x_{e,k}^u = \sum_{i=-\infty}^{k-1} A^{k-i-1}\Pi_- Bu_i \; - \sum_{i=k}^{\infty}(A^\dagger)^{i-k+1}\Pi_+ Bu_i\,, \quad k \in \mathbb{Z}, \tag{2.164}$$

where Π_- and Π_+ are the projections (2.119) of $\mathbb{R}^n$ on the stable and antistable subspace of A, and $A^\dagger$ is a matrix that leaves $\mathcal{X}_+(A)$ invariant and is inverting the action of A on $\mathcal{X}_+(A)$ (i.e., $A^\dagger Ax = AA^\dagger x = x$, for $x \in \mathcal{X}_+(A)$).

Proof. Notice first that in (2.164), $A^\dagger$ acts on the elements of $\mathcal{X}_+(A)$ exclusively, where its image is well defined. The proof is identical to the one for the continuous–time counterpart and is left to the reader. Note only that if A is invertible then $A^\dagger = A^{-1}$ and (2.164) becomes

$$x_{e,k}^u = \sum_{i=-\infty}^{k-1} A^{k-i-1}\Pi_- Bu_i \; - \sum_{i=k}^{\infty}(A)^{k-i-1}\Pi_+ Bu_i\,, \quad k \in \mathbb{Z}.$$

■

Relation (2.164) defines a bounded linear $\mathcal{L}_e$ operator from $\ell^{2,m}$ to $\ell^{2,n}$ and

$$x_e^u = \mathcal{L}_e u. \tag{2.165}$$

We shall also make use of the operator

$$\mathcal{L} := P_+^n \mathcal{L}_e P_+^m. \tag{2.166}$$

In particular, in the case that A is antistable (and therefore invertible),

$$(\mathcal{L}_e u)_k = -\sum_{i=k}^{\infty} A^{k-i-1} B u_i \ , \quad k \in \mathbb{Z}, \tag{2.167}$$

and

$$(\mathcal{L} u)_k = -\sum_{i=k}^{\infty} A^{k-i-1} B u_i \ , \quad k \geq 0. \tag{2.168}$$

2.6.5 *The inverse–time case: A dichotomic*

Theorem 2.6.4 admits an inverse–time dual.

Theorem 2.6.5 *Let A be dichotomic. Then, for each u in $\ell^{2,m}$, the inverse–time recursion (2.142) has a unique solution $\widehat{x}_e^u$ in $\ell^{2,n}$, which is given, in the same notation as in Theorem 2.6.4, by*

$$\widehat{x}_{e,k}^u = -\sum_{i=-\infty}^{k-1} (A^\dagger)^{k-i} \Pi_+ B u_i + \sum_{i=k}^{\infty} A^{i-k} \Pi_- B u_i \ , \quad k \in \mathbb{Z}. \tag{2.169}$$

Accordingly, we introduce the bounded linear operator $\widehat{\mathcal{L}}_e$ from $\ell^{2,m}$ to $\ell^{2,n}$ such that

$$\sigma \widehat{x}_e^u = \widehat{\mathcal{L}}_e u. \tag{2.170}$$

We shall also make use of the operator

$$\widehat{\mathcal{L}} := P_+^n \widehat{\mathcal{L}}_e P_+^m. \tag{2.171}$$

If we define, as in the stable case,

$$\widehat{x}^u := P_+^n \widehat{x}_e^u, \tag{2.172}$$

we have

$$\sigma \widehat{x}^u = \widehat{\mathcal{L}} u. \tag{2.173}$$

In particular, in the case that A is antistable,

$$(\widehat{\mathcal{L}}_e u)_k = -\sum_{i=-\infty}^{k-1} A^{i-k} B u_i \ , \quad k \in \mathbb{Z}, \tag{2.174}$$

and

$$(\widehat{\mathcal{L}} u)_k = -\sum_{i=0}^{k-1} A^{i-k} B u_i \ , \quad k \geq 0. \tag{2.175}$$

2.6.6 The case: A arbitrary

As we have seen, in this case we can define the initial value solution $x^{\xi,u}$ through (2.143). In general, $x^{\xi,u}$ is not in $\ell_+^{2,n}$. It is interesting to consider the set of those input functions u that make $x^{\xi,u}$ norm–square summable. These inputs can be regarded as stabilizing the system in open–loop. We introduce

$$\mathcal{U}_\xi := \{u : u \in \ell_+^{2,m} \text{ such that } x^{\xi,u} \in \ell_+^{2,n}\}. \tag{2.176}$$

If A is stable then $\mathcal{U}_\xi = \ell_+^{2,m}$. In general, the following result holds.

Lemma 2.6.6 *Assume that (A, B) is stabilizable and F is a stabilizing feedback. Let $\xi \in \mathbb{R}^n$ and $u \in \ell_+^{2,m}$. Then $u \in \mathcal{U}_\xi$ if and only if $u = F x_F^{\xi,v} + v$, where v is arbitrary in $\ell_+^{2,m}$ and $x_F^{\xi,v}$ is the unique solution in $\ell_+^{2,n}$ of*

$$\sigma x_F = (A + BF)x_F + Bv\ , \quad x_F(0) = \xi.$$

In the same vein, we may consider $\widehat{x}^u$ defined by (2.159), that is

$$\widehat{x}_k^u = \sum_{i=k}^{\infty} A^{i-k} B u_i\ , \quad k \in \mathbb{N},$$

for those u for which the series converges. Again, we may consider the set of those input sequences that make $\widehat{x}^u$ a norm–square summable sequence:

$$\widehat{\mathcal{U}} := \{u : u \in \ell_+^{2,m} \text{ such that } \widehat{x}^u \in \ell_+^{2,n}\}. \tag{2.177}$$

This set is completely characterized in the following lemma.

Lemma 2.6.7 *Assume that (A, B) is stabilizable and F is a stabilizing feedback. Then $u \in \widehat{\mathcal{U}}$ if and only if $u = F\sigma\widehat{x}_F^v + v$, where v is arbitrary in $\ell_+^{2,m}$ and $\widehat{x}_F^v$ is the unique $\ell_+^{2,n}$ solution of*

$$x_F = (A + BF)\sigma x_F + Bv.$$

2.7 ℓ^2 input–output operators

2.7.1 The direct–time case

Let us consider again the linear discrete direct–time system

$$\begin{aligned} \sigma x &= Ax + Bu, \\ y &= Cx + Du. \end{aligned} \tag{2.178}$$

We assume throughout this section that A is dichotomic. Then, by (2.165), we can write $x_e^u = \mathcal{L}_e u$, for $u \in \ell^{2,m}$. Accordingly, we have $y = (C\mathcal{L}_e + D)u$ and the output y is in $\ell^{2,p}$. Thus we deduce that the system (2.178) defines a linear bounded input–output operator from $\ell^{2,m}$ to $\ell^{2,p}$, whose expression follows easily from (2.164) as

$$(\mathcal{G}u)_k = \sum_{i=-\infty}^{k-1} CA^{k-i-1}\Pi_- Bu_i - \sum_{i=k}^{\infty} C(A^\dagger)^{i-k+1}\Pi_+ Bu_i + Du_k. \tag{2.179}$$

An alternative expression for $\mathcal{G}$ can be given by applying a state coordinate transformation T that realizes the decomposition (2.116), (2.117), into a stable and an antistable part. As a coordinate change does not affect the relation between the input and the output of the system, we can write

$$(\mathcal{G}u)_k = \sum_{i=-\infty}^{k-1} C_- A_-^{k-i-1} B_- u_i - \sum_{i=k}^{\infty} C_+ (A_+^{-1})^{i-k+1} B_+ u_i + D u_k. \tag{2.180}$$

The following is the discrete direct–time counterpart of Theorem 2.3.1.

Theorem 2.7.1 *Assume that D is invertible and $A - BD^{-1}C$ is dichotomic. Then the input–output operator $\mathcal{G}$ is invertible and its inverse is the input–output operator of the direct–time system*

$$\begin{array}{rcll} x_{k+1} & = & (A - BD^{-1}C)x_k & + \; BD^{-1}y_k, \\ u_k & = & -D^{-1}Cx_k & + \; D^{-1}y_k. \end{array} \tag{2.181}$$

Since A is dichotomic, the TFM $\mathbf{G}(z) = C(zI - A)^{-1}B + D$ of (2.178) belongs to $RL^{\infty}_{p\times m}$. If $y \in \ell^{2,p}$ and $u \in \ell^{2,m}$ are such that

$$y = \mathcal{G}u,$$

then their Fourier series $\widehat{y}$ and $\widehat{u}$ are elements in $L^2(\partial\mathbb{D})$, and

$$\widehat{y}(e^{j\theta}) = \mathbf{G}(e^{j\theta})\widehat{u}(e^{j\theta})$$

for almost every θ.

The following is the discrete–time analogue of Theorem 2.3.2.

Theorem 2.7.2 *If the system (2.178) is dichotomic, then the norm of its input–output operator is equal to the $L^{\infty}(\partial\mathbb{D})$–norm of its TFM, i.e.,*

$$||\mathcal{G}|| = ||\mathbf{G}||_{\infty}. \tag{2.182}$$

2.7.2 The inverse–time case

All the concepts and results introduced in the previous subsection can be given for the inverse–time discrete system

$$\begin{array}{rcl} x & = & A\sigma x + Bu, \\ y & = & C\sigma x + Du. \end{array} \tag{2.183}$$

We assume as before that A is dichotomic. From (2.169) we deduce that the ℓ^2 input–output operator of the inverse–time discrete system is

$$(\widehat{\mathcal{G}}u)_k = -\sum_{i=-\infty}^{k} C(A^{\dagger})^{k+1-i}\Pi_+ B u_i + \sum_{i=k+1}^{\infty} C A^{i-k-1}\Pi_- B u_i + D u_k \tag{2.184}$$

or, in terms of the decomposition (2.116), (2.117),

$$(\widehat{\mathcal{G}}u)_k = -\sum_{i=-\infty}^{k} C_+(A_+^{-1})^{k+1-i} B_+ u_i + \sum_{i=k+1}^{\infty} C_- A_-^{i-k-1} B_- u_i + D u_k. \tag{2.185}$$

Using the transformation (2.108), we can prove that the input–output operator of the inverse–time system (2.183) can be expressed in terms of the input–output operator of the direct–time system (2.178) as

$$\widehat{\mathcal{G}} = \Omega \mathcal{G} \Omega. \tag{2.186}$$

This can also be checked directly using expressions (2.184) and (2.179).

The following is the discrete inverse–time counterpart of Theorem 2.3.1.

Theorem 2.7.3 *Assume that D is invertible and $A - BD^{-1}C$ is dichotomic. Then the input–output operator $\widehat{\mathcal{G}}$ is invertible and its inverse is the input–output operator of the inverse–time system*

$$\begin{array}{rclcl} x_k & = & (A - BD^{-1}C)x_{k+1} & + & BD^{-1}y_k, \\ u_k & = & -D^{-1}Cx_{k+1} & + & D^{-1}y_k. \end{array} \tag{2.187}$$

Since A is dichotomic, the TFM of (2.183), $\widehat{\mathbf{G}}(z) = C(z^{-1}I - A)^{-1}B + D$, belongs to $RL^{\infty}_{p\times m}$ and relation (2.182) holds as well, i.e., $\|\widehat{\mathcal{G}}\| = \|\widehat{\mathbf{G}}\|_{\infty}$.

Obviously, $\widehat{\mathbf{G}}(z) = \mathbf{G}(z^{-1})$, and therefore $\|\widehat{\mathbf{G}}\|_{\infty} = \|\mathbf{G}\|_{\infty}$.

2.8 Hankel and Toeplitz operators

In the previous section we introduced the discrete–time analogue of the input–output operator. Now we can state and prove the discrete–time analogues of the results in Section 2.4 for the associated Hankel and Toeplitz operators.

We denote by $P^r_{\pm}$, the orthogonal projections of $\ell^{2,r}$ onto $\ell^{2,r}_{\pm}$.

Definition 2.8.1 *Let $\mathcal{G}$ be a bounded linear operator from $\ell^{2,m}$ to $\ell^{2,p}$. We define:*

1. *The* causal Hankel operator *associated with $\mathcal{G}$,*

$$\mathbb{H}_{\mathcal{G}} := P^p_+ \mathcal{G}|_{\ell^{2,m}_-}.$$

2. *The* anticausal Hankel operator *associated with $\mathcal{G}$,*

$$\widehat{\mathbb{H}}_{\mathcal{G}} := P^p_- \mathcal{G}|_{\ell^{2,m}_+}.$$

3. *The* causal Toeplitz operator *associated with $\mathcal{G}$,*

$$\mathbb{T}_{\mathcal{G}} := P^p_+ \mathcal{G}|_{\ell^{2,m}_+}.$$

4. *The* anticausal Toeplitz operator *associated with $\mathcal{G}$,*

$$\widehat{\mathbb{T}}_{\mathcal{G}} := P^p_- \mathcal{G}|_{\ell^{2,m}_-}.$$

Similarly to the continuous case, we use further the following modified definitions:

$$\mathbb{H}_{\mathcal{G}} := P^p_+ \mathcal{G} P^m_-, \quad \widehat{\mathbb{H}}_{\mathcal{G}} := P^p_- \mathcal{G} P^m_+, \quad \mathbb{T}_{\mathcal{G}} := P^p_+ \mathcal{G} P^m_+, \quad \widehat{\mathbb{T}}_{\mathcal{G}} := P^p_- \mathcal{G} P^m_-.$$

The discrete–time counterpart of Proposition 2.4.4 is immediate.

Remark 2.8.2 Let $\mathcal{L}_e$, $\mathcal{L}$, $\widehat{\mathcal{L}}_e$ and $\widehat{\mathcal{L}}$ be the operators introduced in Section 2.6. Relations (2.166) and (2.171) express the fact that $\mathcal{L}$ and $\widehat{\mathcal{L}}$ are the causal Toeplitz operators associated with $\mathcal{L}_e$ and $\widehat{\mathcal{L}}_e$, i.e., $\mathcal{L} = \mathbb{T}_{\mathcal{L}_e}$ and $\widehat{\mathcal{L}}_e = \mathbb{T}_{\widehat{\mathcal{L}}_e}$.

In the case in which $\mathcal{G}$ is the input–output operator of a linear system, the corresponding TFM $\mathbf{G}(z)$ will be called the *symbol* of the Toeplitz (Hankel) operator. We shall sometimes indicate a Toeplitz operator by its symbol $\mathbb{T}_{\mathbf{G}}$, instead of by the corresponding input–output operator $\mathbb{T}_{\mathcal{G}}$.

The discrete–time counterpart of Theorem 2.4.5 can also be readily proved.

Theorem 2.8.3 *Assume the direct–time system (2.178) is dichotomic and let $\mathcal{G}$ be its ℓ^2 input–output operator.*

If A is stable, then $\widehat{\mathbb{H}}_{\mathcal{G}} = 0$. Conversely, if $\widehat{\mathbb{H}}_{\mathcal{G}} = 0$ and, additionally, (A, B) is controllable and (C, A) is observable, then A is stable.

Dually, if A is antistable, then $\mathbb{H}_{\mathcal{G}} = 0$. Conversely, if $\mathbb{H}_{\mathcal{G}} = 0$ and (A, B) is controllable and (C, A) is observable, then A is antistable.

Proof. Assume A is stable. Then the contribution of the antistable part of the system in (2.180) vanishes and the input–output operator becomes

$$(\mathcal{G}u)_k = \sum_{i=-\infty}^{k-1} CA^{k-i-1}Bu_i + Du_k. \tag{2.188}$$

If we use in this expression $u_i = 0$, for $i \leq -1$, then it is clear that $(\mathcal{G}u)_k = 0$ for $k \leq -1$. This means that $\widehat{\mathbb{H}}_{\mathcal{G}} := P_-^p \mathcal{G} P_+^m = 0$.

Conversely, assume now that $\widehat{\mathbb{H}}_{\mathcal{G}} = 0$ and that the system is controllable and observable. Assuming that the system is not stable, we can perform a change of coordinates that realizes the decomposition (2.116), (2.117). Then the realization of the system can be written as

$$\mathbf{G} = \left[\begin{array}{cc|c} A_- & O & B_- \\ O & A_+ & B_+ \\ \hline C_- & C_+ & D \end{array}\right],$$

with $A_- \in \mathbb{R}^{n_- \times n_-}$ stable and $A_+ \in \mathbb{R}^{n_+ \times n_+}$ antistable. Moreover, the pairs $(A_\pm, B_\pm)$ and $(C_\pm, A_\pm)$ are controllable and observable, respectively. If we apply (2.180) for the input $u_+ = P_+^m u$, then it is easy to see that

$$(\widehat{\mathbb{H}}_{\mathcal{G}}u)_k = -C_+(A_+^{-1})^{-k} \sum_{i=0}^{\infty} (A_+^{-1})^{i+1} B_+ u_i \,, \quad k \leq -1,$$

or, equivalently,

$$(\widehat{\mathbb{H}}_{\mathcal{G}}u)_k = C_+(A_+^{-1})^{-k} \xi_{u_+}, \tag{2.189}$$

where

$$\xi_{u_+} := -\sum_{i=0}^{\infty} (A_+^{-1})^{i+1} B_+ u_i \,. \tag{2.190}$$

If $(\widehat{\mathbb{H}}_{\mathcal{G}}u)_k = 0$ for every $k \leq -1$, it follows from the observability of (C_+, A_+) that $\xi_{u_+} = 0$ for every $u_+ \in \ell_+^{2,m}$. If we use now $u_i = -B_+^T(A_+^{-T})^{i+1}z$ in (2.190) for an arbitrary $z \in \mathbb{R}^{n+}$, we obtain

$$0 = \sum_{i=0}^{\infty}(A_+^{-1})^{i+1}B_+B_+^T(A_+^{-T})^{i+1}z = \widehat{P}z$$

where, according to (2.128), $\widehat{P}$ is the anticausal controllability Gramian. However, as (A_+, B_+) is controllable, $\widehat{P}$ must be nonsingular. This means that $n_+ = 0$, and thus the system is stable.

The dual part of the theorem follows analogously. ■

Remark 2.8.4 Theorem 2.8.3 has an inverse–time correspondent, that follows easily if we notice that the Hankel operators associated with the direct–time input–output operator $\mathcal{G}$ of (2.178) and the inverse–time input–output operator $\widehat{\mathcal{G}}$ of (2.183) are related through

$$\mathbb{H}_{\widehat{\mathcal{G}}} = \sigma\Omega\widehat{\mathbb{H}}_{\mathcal{G}}\Omega\sigma^{-1}. \tag{2.191}$$

A similar relation is valid for the corresponding Toeplitz operators:

$$\mathbb{T}_{\widehat{\mathcal{G}}} = \sigma\Omega\widehat{\mathbb{T}}_{\mathcal{G}}\Omega\sigma^{-1}. \tag{2.192}$$

The proof of these relations is based on (2.186) combined with the fact that $\Omega\sigma\ell_+^{2,m} = \ell_-^{2,m}$.

Definition 2.8.5 *A bounded linear operator $\mathcal{G}$ from $\ell^{2,m}$ to $\ell^{2,p}$ is called* causal *if $\widehat{\mathbb{H}}_{\mathcal{G}} = 0$ and* anticausal *if $\mathbb{H}_{\mathcal{G}} = 0$.*

We shall say that the linear system (2.101) is (anti)causal if its input–output operator is (anti)causal.

According to the discrete–time counterpart of Proposition 2.4.4, causality of $\mathcal{G}$ is equivalent to

$$\mathbb{T}_{\mathcal{G}} = \mathcal{G}P_+^m,$$

whereas anticausality is equivalent to

$$\widehat{\mathbb{T}}_{\mathcal{G}} = \mathcal{G}P_-^m.$$

According to Theorem 2.8.3, a stable direct–time system is causal, whereas a causal direct–time system that is controllable and observable, is stable. For an inverse–time system, stability corresponds to anticausality.

If system (2.178) is causal, then from (2.188) we can deduce

$$(\mathbb{H}_{\mathcal{G}}u)_k = (P_+^m\mathcal{G}u)_k = CA^k\sum_{i=-\infty}^{-1}A^{-i-1}Bu_i\ , \quad k \geq 0. \tag{2.193}$$

Let us define the *causal direct–time controllability operator* $\Psi : \ell_-^{2,m} \to \mathbb{R}^n$ as

$$\Psi u := \sum_{i=-\infty}^{-1}A^{-i-1}Bu_i\ , \quad u \in \ell_-^{2,m}, \tag{2.194}$$

and the *causal direct–time observability operator* $\Theta : \mathbb{R}^n \to \ell_+^{2,p}$ as

$$(\Theta\xi)_k := CA^k\xi \ , \ k \geq 0, \quad \xi \in \mathbb{R}^n. \tag{2.195}$$

With these notations (2.193) may be written as

$$\mathbb{H}_{\mathcal{G}} = \Theta\Psi. \tag{2.196}$$

Similarly, if the inverse–time system (2.183) is anticausal, then we can write

$$\widehat{\mathbb{H}}_{\widehat{\mathcal{G}}} = \widehat{\Theta}\widehat{\Psi}, \tag{2.197}$$

where $\widehat{\Psi} : \ell_+^{2,m} \to \mathbb{R}^n$ is the *anticausal inverse–time controllability operator*

$$\widehat{\Psi}u := \sum_{i=0}^{\infty} A^i B u_i \ , \quad u \in \ell_+^{2,m}, \tag{2.198}$$

and $\widehat{\Theta} : \mathbb{R}^n \to \ell_-^{2,p}$ is the *anticausal inverse–time observability operator*

$$(\widehat{\Theta}\xi)_k := CA^{-k-1}\xi \ , \ k \leq -1, \quad \xi \in \mathbb{R}^n. \tag{2.199}$$

It is obvious that the operators Ψ, Θ, $\widehat{\Psi}$, $\widehat{\Theta}$ are finite rank operators. From (2.196) and (2.197), it follows that $\mathbb{H}_{\mathcal{G}}$ and $\widehat{\mathbb{H}}_{\widehat{\mathcal{G}}}$ are also finite rank operators.

The adjoints of these operators are

$$(\Psi^*\xi)_k = B^T(A^T)^{-k-1}\xi \ , \quad k \leq -1 \ , \quad \xi \in \mathbb{R}^n, \tag{2.200}$$

$$\Theta^* y = \sum_{i=0}^{\infty}(A^T)^i C^T y_i \ , \quad y \in \ell_+^{2,p}, \tag{2.201}$$

and

$$(\widehat{\Psi}^* x)_k = B^T(A^T)^k x \ , \quad k \geq 0, \ , \quad x \in \mathbb{R}^n, \tag{2.202}$$

$$\widehat{\Theta}^* y = \sum_{k=-\infty}^{-1} (A^T)^{-k-1} C^T y_k \ , \quad y \in \ell_-^{2,p}. \tag{2.203}$$

With these expressions we can compute

$$\Psi\Psi^* = \sum_{i=-\infty}^{-1} A^{-i-1}BB^T(A^T)^{-i-1} = \sum_{k=0}^{\infty} A^k BB^T(A^T)^k = P, \tag{2.204}$$

$$\Theta^*\Theta = \sum_{i=0}^{\infty}(A^T)^i C^T C A^i = Q, \tag{2.205}$$

where P and Q are, respectively, the causal controllability and observability Gramians defined in (2.125). On the other hand,

$$\widehat{\Psi}\widehat{\Psi}^* = \sum_{i=0}^{\infty} A^i BB^T(A^T)^i = P, \tag{2.206}$$

$$\widehat{\Theta}^*\widehat{\Theta} = \sum_{k=-\infty}^{-1} (A^T)^{-k-1}C^TCA^{-k-1} = \sum_{i=0}^{\infty}(A^T)^iC^TCA^i = Q. \tag{2.207}$$

We could also consider the dual cases in which the direct–time system is anticausal and the inverse–time system is causal, and then obtain similar relations to those above involving the anticausal controllability and observability Gramians defined in (2.128).

Now we can state the discrete–time counterpart of Proposition 2.4.8. The proof is omitted since it is very similar to that in the continuous–time case.

Proposition 2.8.6 *Let $\mathcal{G}$ be the input–output operator of a linear causal direct–time discrete system. Then $\sigma > 0$ is a Hankel singular value of the system if and only if it is a singular value of $\mathbb{H}_{\mathcal{G}}$.*

A simple consequence of the previous proposition is given below.

Corollary 2.8.7 *If $\mathcal{G}$ is the ℓ^2 input–output operator of a causal direct–time discrete system then*

$$||\mathbb{H}_{\mathcal{G}}|| = \overline{\sigma}(PQ).$$

Corollary 2.8.7 justifies the following definition.

Definition 2.8.8 *If $\mathcal{G}$ is the ℓ^2 input–output operator of a causal system then $||\mathbb{H}_{\mathcal{G}}||$ is called the* causal Hankel norm *and it is usually denoted by $||\mathcal{G}||_H$.*

Both Proposition 2.8.6 and Corollary 2.8.7 admit obvious inverse–time duals. Accordingly, we can define as in Definition 2.8.8, the *anticausal Hankel norm* and we shall denote this by $||\mathcal{G}||_{\widehat{H}}$.

Proposition 2.4.8, Proposition 2.4.12, Corollary 2.4.13 and Corollary 2.4.14 remain valid for operators on ℓ^2–spaces.

Notes and references

The most comprehensive treatment of linear systems from a theoretical viewpoint is given by Kailath [Kai80]. Standard references for Gramians and Hankel singular values are Glover [Glo84], and McFarlane and Glover [MG90]. Basic results on L^2 (ℓ^2) solutions to differential (difference) equations are found in Tadmor and Verma [TV92] and in Halanay and Ionescu [HI94d]. For Hardy spaces (H^2 and H^∞) and Toeplitz operator theory we refer to Douglas [Dou72]. Hankel operators are studied in detail in Partington [Par88]. The Rellich's results on perturbation theory of eigenvalue problems are found in Kato [Kat66]. General results on Fourier transforms and L^p spaces are given in Rudin [Rud66]. A book that introduces the operator viewpoint in systems theory is Francis [Fra87].

Part II

Generalized Riccati Theory

3

Popov Triplets

The notion of a Popov triplet and the various associated objects provide the general framework for all the theoretical developments in this book. In particular, the equivalence relation for Popov triplets enables an elegant exposition of the general Riccati equation theory.

We introduce the notion of a Popov triplet together with an associated equivalence relation, and discuss their significance in Section 3.1. In Section 3.2 we associate with a Popov triplet several mathematical objects: various quadratic indices; the continuous–time algebraic Riccati system, which is an extension of the continuous–time algebraic Riccati equation; the Kalman–Popov–Yakubovich system, which is a very useful alternative way of writing a Riccati equation; the continuous–time Hamiltonian system; the transmission pencil and the input–output transfer function matrix of the Hamiltonian system, also known as the extended Hamiltonian pencil and the Popov function, respectively. We study the relationships between these objects and between pairs of objects associated with equivalent Popov triplets. In Section 3.3 we introduce a set of operators associated with the Hamiltonian system that are the cornerstone of the operator–based approach to the Riccati theory presented in Chapter 4. We give in Section 3.4 several examples of important Popov triplets.

The discrete–time counterpart theory needs individual presentation, and this is given in Sections 3.5–3.8.

A. The continuous–time case

3.1 Popov triplets: definition, significance, and equivalence

Definition 3.1.1 (Popov triplet) *A triplet of matrices*

$$\Sigma := (A, B; P), \quad P := \begin{bmatrix} Q & L \\ L^T & R \end{bmatrix} = P^T,$$

where $A \in \mathbb{R}^{n\times n}$, $B \in \mathbb{R}^{n\times m}$, $Q = Q^T \in \mathbb{R}^{n\times n}$, $L \in \mathbb{R}^{n\times m}$, $R = R^T \in \mathbb{R}^{m\times m}$, *and* $P \in \mathbb{R}^{(n+m)\times(n+m)}$, *is called a* Popov triplet. *Alternatively, and most frequently, we shall use the explicit representation of a Popov triplet as*

$$\Sigma = (A, B; Q, L, R).$$

A Popov triplet can be seen as a synthetic (or abbreviated) representation of a couple consisting of a controlled dynamical system

$$\dot{x} = Ax + Bu \tag{3.1}$$

and an associated quadratic performance index

$$J = \int_{t_0}^{t_1} w^T(t)Pw(t)\mathrm{d}t, \quad w(t) := \begin{bmatrix} x(t) \\ u(t) \end{bmatrix}, \tag{3.2}$$

where $x(t)$ and $u(t)$ satisfy (3.1).

We introduce now an equivalence transformation for Popov triplets.

Definition 3.1.2 (Equivalence of Popov triplets) *Two Popov triplets*

$$\begin{aligned} \Sigma &= (A, B; Q, L, R), \\ \widetilde{\Sigma} &= (\widetilde{A}, \widetilde{B}; \widetilde{Q}, \widetilde{L}, \widetilde{R}), \end{aligned}$$

are called (X, F)–equivalent *(or simply,* equivalent*) if there exist* $F \in \mathbb{R}^{m\times n}$ *and* $X = X^T \in \mathbb{R}^{n\times n}$ *such that*

$$\begin{aligned} \widetilde{A} &= A + BF, \\ \widetilde{B} &= B, \\ \widetilde{Q} &= Q + LF + F^T L^T + F^T RF + \widetilde{A}^T X + X\widetilde{A}, \\ \widetilde{L} &= L + F^T R + XB, \\ \widetilde{R} &= R. \end{aligned} \tag{3.3}$$

We say that $\widetilde{\Sigma}$ *is a* feedback equivalent *(or F–equivalent) of* Σ *if it is* $(0, F)$*–equivalent for some matrix* F.

We say that $\widetilde{\Sigma}$ *is a* reduced equivalent *of* Σ *if it is* $(X, 0)$*–equivalent and, in addition,* $\widetilde{Q} = 0$.

It can be easily checked that the relation defined above is indeed an equivalence relation satisfying the corresponding axioms. Straightforward computations show that if Σ and $\widetilde{\Sigma}$ are (X, F)–equivalent, and $\widetilde{\Sigma}$ and $\widetilde{\widetilde{\Sigma}}$ are $(\widetilde{X}, \widetilde{F})$–equivalent, then Σ and $\widetilde{\widetilde{\Sigma}}$ are $(\widetilde{\widetilde{X}}, \widetilde{\widetilde{F}})$–equivalent, where $\widetilde{\widetilde{X}} := X + \widetilde{X}$ and $\widetilde{\widetilde{F}} := F + \widetilde{F}$.

Remark 3.1.3 From (3.3) we see that if $\widetilde{\Sigma}$ is a reduced equivalent of Σ, then X must satisfy the Lyapunov equation

$$A^T X + XA + Q = 0. \tag{3.4}$$

If A is stable, then Theorem 1.5.2 shows that the Lyapunov equation (3.4) has a unique solution. If the pair (A, B) is stabilizable, it is possible to determine first a feedback such that $\widetilde{A} = A + BF$ is stable and then solve (3.4) for the transformed F–equivalent triplet to obtain a reduced equivalent.

In the sequel, we shall call the Lyapunov equation (3.4), the *reducing Lyapunov equation* of Σ. In those cases in which this equation has a unique solution we shall denote this by X_r, and we shall denote the corresponding $(X_r, 0)$–equivalent triplet, by $\Sigma_r = (A, B; 0, L_r, R)$, where $L_r = L + X_r B$.

Remark 3.1.4 The equivalence relation introduced in Definition 3.1.2 is a particular case of a similar relation introduced in Popov's book [Pop66] and this reference gives substantial motivation for its introduction. Essentially, it is a quite general transformation that leaves unchanged the value of the quadratic index associated with the Popov triplet. We shall make extensive use of this and other properties of the equivalence transformation. It is easier to give at this point a motivation for the feedback equivalence. If in the controlled system (3.1), we perform the change of input

$$u = Fx + v \tag{3.5}$$

– called a *feedback transformation* – and we rewrite the controlled system (3.1) and the quadratic index (3.2) in terms of the new input, it is clear that the new controlled system and the new quadratic index correspond to the feedback equivalent of Σ.

3.2 Objects associated with a Popov triplet

Throughout this section we fix a Popov triplet $\Sigma = (A, B; Q, L, R)$, with $A \in \mathbb{R}^{n\times n}$, $B \in \mathbb{R}^{n\times m}$. Following our plan we associate with Σ several mathematical objects and we study the effect that the equivalence transformation has on these objects.

3.2.1 Quadratic indices

The *quadratic index* associated with Σ is the functional

$$J_\Sigma(\xi, u) := \left\langle \begin{bmatrix} x^{\xi,u} \\ u \end{bmatrix}, \begin{bmatrix} Q & L \\ L^T & R \end{bmatrix} \begin{bmatrix} x^{\xi,u} \\ u \end{bmatrix} \right\rangle_{L_+^{2,n} \times L_+^{2,m}}, \tag{3.6}$$

where $x^{\xi,u}$ is the solution of (3.1) that satisfies $x(0) = \xi$ as defined in (2.54) and $u \in L_+^{2,m}$ is such that $x^{\xi,u} \in L_+^{2,n}$, i.e., u ranges in the set $\mathcal{U}_\xi$ introduced in (2.74). If A is stable, $J_\Sigma(\xi, u)$ is defined for all $\xi \in \mathbb{R}^n$ and $u \in L_+^{2,m}$.

In the case that A is dichotomic, we introduce the *extended quadratic index* associated with Σ as

$$J_{e,\Sigma}(u) := \left\langle \begin{bmatrix} x_e^u \\ u \end{bmatrix}, \begin{bmatrix} Q & L \\ L^T & R \end{bmatrix} \begin{bmatrix} x_e^u \\ u \end{bmatrix} \right\rangle_{L^{2,n} \times L^{2,m}}, \tag{3.7}$$

where x_e^u is the L^2–solution on the entire axis of (3.1) defined by (2.69).

Limiting the time–horizon in the extended quadratic index (3.7) to the positive real axis, we introduce the *restricted quadratic index* associated with Σ as the functional

$$J_\Sigma(u) := \left\langle \begin{bmatrix} x^u \\ u \end{bmatrix}, \begin{bmatrix} Q & L \\ L^T & R \end{bmatrix} \begin{bmatrix} x^u \\ u \end{bmatrix} \right\rangle_{L_+^{2,n} \times L_+^{2,m}}, \tag{3.8}$$

where x^u is the solution of (3.1) defined by (2.66) and u is such that x^u is norm–square integrable, i.e., u ranges in the set $\widehat{\mathcal{U}}$ introduced in (2.75). If A is antistable, $J_\Sigma(u)$ is defined for each $u \in L_+^{2,m}$.

The following result shows how the quadratic indices just introduced transform under the equivalence transformation of the underlying Popov triplets.

Proposition 3.2.1 *Let $\Sigma = (A, B; Q, L, R)$ and $\widetilde{\Sigma} = (\widetilde{A}, \widetilde{B}; \widetilde{Q}, \widetilde{L}, \widetilde{R})$ be two (X, F)-equivalent Popov triplets. Then we have:*

1. $J_\Sigma(\xi, u) = J_{\widetilde{\Sigma}}(\xi, \widetilde{u}) + \xi^T X \xi, \quad (u \in \mathcal{U}_\xi)$, *where* $\widetilde{u} = -Fx^{\xi,u} + u$.
2. $J_\Sigma(u) = J_{\widetilde{\Sigma}}(\widetilde{u}) + (x^u(0))^T X x^u(0), \quad (u \in \widehat{\mathcal{U}})$, *where* $\widetilde{u} = -Fx^u + u$.
3. *If* $A + BF$ *is dichotomic, then* $J_{e,\Sigma}(u) = J_{e,\widetilde{\Sigma}}(\widetilde{u})$, *where* $\widetilde{u} = -Fx_e^u + u$.

Proof. 1. Let us denote by $\widetilde{x}$ the solution of

$$\dot{\widetilde{x}} = \widetilde{A}\widetilde{x} + B\widetilde{u}, \tag{3.9}$$

such that $\widetilde{x}(0) = \xi$. It is easy to check that $\widetilde{x} = x^{\xi,u}$. In particular, $\widetilde{x} \in L_+^{2,n}$. Therefore we can write using (3.3)

$$J_{\widetilde{\Sigma}}(\xi, \widetilde{u}) = \left\langle \begin{bmatrix} \widetilde{x} \\ \widetilde{u} \end{bmatrix}, \begin{bmatrix} \widetilde{Q} & \widetilde{L} \\ \widetilde{L}^T & \widetilde{R} \end{bmatrix} \begin{bmatrix} \widetilde{x} \\ \widetilde{u} \end{bmatrix} \right\rangle$$

$$= \left\langle \begin{bmatrix} x^{\xi,u} \\ u - Fx^{\xi,u} \end{bmatrix}, \begin{bmatrix} Q + LF + F^T L^T + F^T RF & L + F^T R \\ L^T + RF & R \end{bmatrix} \begin{bmatrix} x^{\xi,u} \\ u - Fx^{\xi,u} \end{bmatrix} \right\rangle$$

$$+ \left\langle \begin{bmatrix} \widetilde{x} \\ \widetilde{u} \end{bmatrix}, \begin{bmatrix} \widetilde{A}^T X + X\widetilde{A} & XB \\ B^T X & O \end{bmatrix} \begin{bmatrix} \widetilde{x} \\ \widetilde{u} \end{bmatrix} \right\rangle$$

where the scalar products have been taken over $L_+^{2,n} \times L_+^{2,m}$. After straightforward manipulations the first term in the last relation equals $J_\Sigma(\xi, u)$, whereas the second term can be rewritten using (3.9) as

$$\int_0^\infty \frac{\mathrm{d}}{\mathrm{d}t}(\widetilde{x}(t)^T X \widetilde{x}(t))\mathrm{d}t = -\xi^T X \xi,$$

since $\lim_{t\to\infty} \widetilde{x}(t)^T X \widetilde{x}(t) = 0$.

The proof of point 2 is similar, with the difference that $\widetilde{x}$ is defined to satisfy $\widetilde{x}(0) = x^u(0)$, which in turn implies that $\widetilde{x} = x^u$.

For the case of point 3, $\widetilde{x}$ is defined as the L^2 solution of (3.9), and it is checked that $\widetilde{x} = x_e^u$. The rest of the proof is entirely similar. ■

3.2.2 The continuous-time algebraic Riccati system (CTARS)

For a symmetric $X \in \mathbb{R}^{n\times n}$, we define the *dissipation matrix* associated with Σ as

$$D_\Sigma(X) := \begin{bmatrix} A^T X + XA + Q & L + XB \\ B^T X + L^T & R \end{bmatrix}. \tag{3.10}$$

The system of equations

$$D_\Sigma(X) \begin{bmatrix} I \\ F \end{bmatrix} = 0 \tag{3.11}$$

in the unknowns $F \in \mathbb{R}^{m\times n}$, $X \in \mathbb{R}^{n\times n}$, is called *the continuous-time algebraic Riccati system (CTARS)* associated with Σ, and is denoted as CTARS(Σ).

Definition 3.2.2 *A pair (X, F) satisfying (3.11), with $X = X^T$ and $A + BF$ stable, is called a stabilizing solution to the CTARS(Σ).*

For the solution to the CTARS we can prove the following uniqueness result.

Proposition 3.2.3 *Let (X, F) and $(\widehat{X}, \widehat{F})$ be two stabilizing solutions to the CTARS(Σ) given in (3.11). Then $X = \widehat{X}$.*

Proof. From (3.11) and (3.10) we can write

$$\begin{aligned} A^T X + XA + Q + (L + XB)F &= 0, &(3.12)\\ B^T X + L^T + RF &= 0. &(3.13)\end{aligned}$$

Let $S := A + BF$. Then we rewrite (3.12) as

$$A^T X + XS + LF + Q = 0. \tag{3.14}$$

Let $\widehat{S} := A + B\widehat{F}$ and add to (3.14), (3.13) premultiplied by $\widehat{F}^T$ to obtain

$$\widehat{S}^T X + XS + \widehat{F}^T L^T + LF + \widehat{F}^T RF + Q = 0. \tag{3.15}$$

Similarly, if we start with

$$D_\Sigma(\widehat{X}) \begin{bmatrix} I \\ \widehat{F} \end{bmatrix} = 0,$$

and interchange the role of X and F with the role of $\widehat{X}$ and $\widehat{F}$ in the above reasoning, we arrive at

$$S^T \widehat{X} + \widehat{X}\widehat{S} + F^T L^T + L\widehat{F} + F^T R\widehat{F} + Q = 0. \tag{3.16}$$

Take the transpose of (3.16) and then subtract it from (3.15). As X and $\widehat{X}$ are both symmetric,

$$\widehat{S}^T (X - \widehat{X}) + (X - \widehat{X})S = 0. \tag{3.17}$$

Since both S and $\widehat{S}$ are stable, Theorem 1.5.1 shows that the unique solution to the Sylvester equation (3.17) is $X - \widehat{X} = 0$ and the conclusion follows. ■

The next result gives the relation between solutions to CTARSs associated with equivalent Popov triplets.

Proposition 3.2.4 *Let $\Sigma = (A, B; Q, L, R)$ and $\widetilde{\Sigma} = (\widetilde{A}, \widetilde{B}; \widetilde{Q}, \widetilde{L}, \widetilde{R})$ be two (X, F)-equivalent Popov triplets. Then (X_s, F_s) is a (stabilizing) solution to the CTARS(Σ) if and only if $(X_s - X, F_s - F)$ is a (stabilizing) solution to the CTARS($\widetilde{\Sigma}$).*

Proof. Let $X_\Delta := X_s - X = X_\Delta^T$, $F_\Delta := F_s - F$. We show first that

$$D_\Sigma(X_s) \begin{bmatrix} I \\ F_s \end{bmatrix} = 0 \quad \Leftrightarrow \quad D_{\widetilde{\Sigma}}(X_\Delta) \begin{bmatrix} I \\ F_\Delta \end{bmatrix} = 0$$

which will prove the relation between the solutions of the CTARSs. Indeed, we have

$$\begin{bmatrix} A^T X_s + X_s A + Q & X_s B + L \\ B^T X_s + L^T & R \end{bmatrix} \begin{bmatrix} I \\ F_s \end{bmatrix} = 0$$

$$\Leftrightarrow \begin{bmatrix} A^TX_\Delta + X_\Delta A + Q + A^TX + XA & L + XB + X_\Delta B \\ B^TX_\Delta + L^T + B^TX & R \end{bmatrix} \begin{bmatrix} I \\ F_s \end{bmatrix} = 0$$

$$\Leftrightarrow \begin{bmatrix} I & F^T \\ O & I \end{bmatrix} \begin{bmatrix} A^TX_\Delta + X_\Delta A + Q + A^TX + XA & L + XB + X_\Delta B \\ B^TX_\Delta + L^T + B^TX & R \end{bmatrix}$$

$$\times \begin{bmatrix} I & O \\ F & I \end{bmatrix} \begin{bmatrix} I & O \\ -F & I \end{bmatrix} \begin{bmatrix} I \\ F_s \end{bmatrix} = 0$$

$$\Leftrightarrow \begin{bmatrix} \widetilde{A}^TX_\Delta + X_\Delta \widetilde{A} + \widetilde{Q} & \widetilde{L} + X_\Delta \widetilde{B} \\ \widetilde{B}^TX_\Delta + \widetilde{L}^T & \widetilde{R} \end{bmatrix} \begin{bmatrix} I \\ F_\Delta \end{bmatrix} = 0.$$

The fact that (X_s, F_s) is stabilizing if and only if (X_Δ, F_Δ) is stabilizing is trivial. ■

3.2.3 The continuous–time algebraic Riccati equation (CTARE)

Let $\Sigma = (A, B; Q, L, R)$ be a Popov triplet with R nonsingular. Then from (3.13) we get

$$F = -R^{-1}(B^TX + L^T). \tag{3.18}$$

Substituting (3.18) in (3.12) one obtains, for $X = X^T$,

$$A^TX + XA - (XB + L)R^{-1}(B^TX + L^T) + Q = 0 \tag{3.19}$$

which is the well–known *continuous–time algebraic Riccati equation (CTARE)* associated with Σ, denoted CTARE(Σ).

Definition 3.2.5 *A symmetric solution X to (3.19) is called* a stabilizing (antistabilizing) solution *to CTARE(Σ) if $A+BF$ is stable (antistable) for F defined by (3.18). The matrix F in (3.18) is called a* stabilizing (antistabilizing) Riccati feedback *and will be occasionally denoted by F_{Ricc}.*

From Proposition 3.2.3 we have directly the following result.

Corollary 3.2.6 *If the CTARE(Σ) has a stabilizing solution then it is unique.*

Accordingly, we can speak about *the* stabilizing solution and *the* stabilizing Riccati feedback.

An immediate consequence of Proposition 3.2.4 is stated in the following corollary.

Corollary 3.2.7 *Let $\Sigma = (A, B; Q, L, R)$ and $\widetilde{\Sigma} = (\widetilde{A}, \widetilde{B}; \widetilde{Q}, \widetilde{L}, \widetilde{R})$ be two (X, F)–equivalent Popov triplets. Then X_s is the stabilizing solution to the CTARE(Σ) if and only if $X_s - X$ is the stabilizing solution to the CTARE($\widetilde{\Sigma}$). The stabilizing Riccati feedbacks are related by the formula*

$$\widetilde{F}_{Ricc} = F_{Ricc} - F. \tag{3.20}$$

3.2.4 *The Kalman–Popov–Yakubovich system (KPYS)*

Let $\Sigma = (A, B; Q, L, R)$ be a Popov triplet with R invertible. Denote by J a sign matrix of dimension $m \times m$. The nonlinear system in the unknowns $X = X^T$, V and W, given by

$$\begin{aligned} R &= V^T J V, \\ L + XB &= W^T J V, \\ Q + A^T X + XA &= W^T J W, \end{aligned} \tag{3.21}$$

is called *the Kalman–Popov–Yakubovich system (KPYS) in J form* associated with Σ and J, denoted KPYS(Σ, J). Since R is assumed invertible, it follows from the first equation (3.21) that two necessary solvability condition for the KPYS are the invertibility of J, i.e., J should take the form

$$J = \begin{bmatrix} -I_{m_1} & \\ & I_{m_2} \end{bmatrix} \tag{3.22}$$

for some m_1 and m_2, with $m_1 + m_2 = m$, and $\operatorname{sgn}(R) = J$. Throughout the book we shall use KPYSs defined for nonsingular sign matrices J.

Definition 3.2.8 *A triplet of matrices $(X = X^T, V, W)$ satisfying (3.21) such that $A + BF$ is stable, for*

$$F := -V^{-1} W, \tag{3.23}$$

is called a stabilizing solution *to the KPYS(Σ, J) and F is called a* stabilizing feedback.

Remark 3.2.9 The relation between the CTARE in (3.19) and the KPYS in (3.21) associated with the same Popov triplet Σ is the following. If $(X = X^T, V, W)$ satisfies the KPYS(Σ, J), then X is a symmetric solution to the CTARE(Σ). Indeed, from the first equation (3.21) it follows that V can be eliminated together with W in (3.21), and we arrive at the CTARE(Σ). Conversely, if X is a symmetric solution to (3.19), then we can choose a V such that the first equation in (3.21) is satisfied, provided $\operatorname{sgn}(R) = J$. Then W can be defined such that it satisfies the second equation in (3.21) and it follows that it satisfies also the third equation in (3.21). Thus the triplet (X, V, W) is a solution to the KPYS(Σ, J). It is also easy to check that the Riccati feedback in (3.18) coincides in this case with the feedback matrix given by (3.23). This construction establishes a relation between the solutions to the CTARE and the solutions to the KPYS as follows. The KPYS(Σ, J) has a (stabilizing) solution $(X = X^T, V, W)$ if and only if the CTARE(Σ) has a (stabilizing) solution X and $\operatorname{sgn}(R) = J$.

As an easy consequence we have the following result.

Corollary 3.2.10 *If (X, V, W) and $(\widehat{X}, \widehat{V}, \widehat{W})$ are two stabilizing solutions to the KPYS(Σ, J), then $X = \widehat{X}$.*

Remark 3.2.11 If the sign matrix J is either $J = I_m$ or $J = -I_m$, then (3.21) reduces to either

$$\begin{aligned} R &= V^T V, \\ L + XB &= W^T V, \\ Q + A^T X + XA &= W^T W, \end{aligned} \tag{3.24}$$

or

$$\begin{aligned} R &= -V^T V, \\ L + XB &= -W^T V, \\ Q + A^T X + XA &= -W^T W. \end{aligned} \tag{3.25}$$

The KPYSs (3.24) and (3.25) are known as the KPYS in *positivity form* KPYS(Σ, I) and the KPYS in *negativity form* KPYS$(\Sigma, -I)$, respectively.

The KPYS in (3.21) can be used to write convenient representations for the quadratic indices associated with Σ as explained in the following result. We make use of the notation introduced in Section 3.2.1.

Proposition 3.2.12 *Let Σ be a Popov triplet and let (X, V, W) be a solution to the KPYS(Σ, J) in (3.21) (not necessarily a stabilizing one).*

1. For each $u \in \mathcal{U}_\xi$ we have

$$J_\Sigma(\xi, u) = \langle Vu + Wx^{\xi,u}, J(Vu + Wx^{\xi,u}) \rangle_{L_+^{2,m}} + \xi^T X \xi. \tag{3.26}$$

2. For each $u \in \widehat{U}$ we have

$$J_\Sigma(u) = \langle Vu + Wx^u, J(Vu + Wx^u) \rangle_{L_+^{2,m}} + (x^u(0))^T X x^u(0). \tag{3.27}$$

3. If A is dichotomic, we have

$$J_{e,\Sigma}(u) = \langle Vu + Wx_e^u, J(Vu + Wx_e^u) \rangle_{L^{2,m}}, \quad u \in L^{2,m}. \tag{3.28}$$

Proof. We use Proposition 3.2.1 for the $(X, 0)$–equivalent $\widetilde{\Sigma}$ of Σ. We have $\widetilde{u} = u$ and, using the KPYS(Σ, J) given in (3.21), we have

$$\begin{aligned} J_{\widetilde{\Sigma}}(\xi, u) &= \left\langle \begin{bmatrix} x^{\xi,u} \\ u \end{bmatrix}, \begin{bmatrix} \widetilde{Q} & \widetilde{L} \\ \widetilde{L}^T & R \end{bmatrix} \begin{bmatrix} x^{\xi,u} \\ u \end{bmatrix} \right\rangle_{L_+^{2,n} \times L_+^{2,m}} \\ &= \left\langle \begin{bmatrix} x^{\xi,u} \\ u \end{bmatrix}, \begin{bmatrix} A^T X + XA + Q & XB + L \\ B^T X + L^T & R \end{bmatrix} \begin{bmatrix} x^{\xi,u} \\ u \end{bmatrix} \right\rangle_{L_+^{2,n} \times L_+^{2,m}} \\ &= \left\langle \begin{bmatrix} x^{\xi,u} \\ u \end{bmatrix}, \begin{bmatrix} W^T \\ V^T \end{bmatrix} J \begin{bmatrix} W & V \end{bmatrix} \begin{bmatrix} x^{\xi,u} \\ u \end{bmatrix} \right\rangle_{L_+^{2,n} \times L_+^{2,m}} \\ &= \langle Vu + Wx^{\xi,u}, J(Vu + Wx^{\xi,u}) \rangle_{L_+^{2,m}} \end{aligned}$$

which establishes relation (3.26). The other relations are obtained in a perfectly similar manner. ∎

The next result gives the relation between the solutions to the KPYSs associated with equivalent Popov triplets.

Proposition 3.2.13 *Let $\Sigma = (A, B; Q, L, R)$ and $\widetilde{\Sigma} = (\widetilde{A}, \widetilde{B}; \widetilde{Q}, \widetilde{L}, \widetilde{R})$ be two (X, F)-equivalent Popov triplets. Then (X_s, V_s, W_s) is a (stabilizing) solution to the KPYS(Σ, J) if and only if $(X_s - X, V_s, W_s + V_s F)$ is a (stabilizing) solution to the KPYS($\widetilde{\Sigma}, J$).*

Proof. We have

$$\begin{aligned} W_s^T J V_s &= L + X_s B = L + F^T R + XB + (X_s - X)B - F^T R \\ &= \widetilde{L} + (X_s - X)B - F^T V_s^T J V_s \end{aligned}$$

which gives

$$\widetilde{L} + (X_s - X)B = (W_s + V_s F)^T J V_s, \tag{3.29}$$

and

$$\begin{aligned} &(W_s + V_s F)^T J (W_s + V_s F) \\ &= W_s^T J W_s + F^T V_s^T J W_s + W_s^T J V_s F + F^T V_s^T J V_s F \\ &= A^T X_s + X_s A + Q + F^T (L^T + B^T X_s) + (L + X_s B) F + F^T R F \\ &= A^T (X_s - X) + (X_s - X) A + F^T B^T (X_s - X) + (X_s - X) B F \\ &\quad + F^T B^T X + XBF + Q + F^T L^T + LF + F^T R F + A^T X + XA \\ &= \widetilde{A}^T (X_s - X) + (X_s - X) \widetilde{A} + \widetilde{Q}. \end{aligned} \tag{3.30}$$

From (3.29) and (3.30) we get that if (X_s, V_s, W_s) is a solution to the KPYS(Σ, J) then $(X_s - X, V_s, W_s + V_s F)$ is a solution to the KPYS$(\widetilde{\Sigma}, J)$. The converse follows by simply reversing the above line of reasoning. The result for the stabilizing solutions follows form the equality $A - BV_s^{-1} W_s = \widetilde{A} - \widetilde{B} V_s^{-1} (W_s + V_s F)$. ■

3.2.5 The continuous-time Hamiltonian system (CTHS)

Let $\Sigma = (A, B; Q, L, R)$ be a Popov triplet. The linear differential system

$$\begin{aligned} \dot{x} &= Ax & &+ Bu, \\ \dot{\lambda} &= -Qx & - A^T \lambda \quad &- Lu, \\ v &= L^T x & + B^T \lambda \quad &+ Ru, \end{aligned} \tag{3.31}$$

is called the *continuous-time Hamiltonian system (CTHS)* associated with Σ, denoted CTHS(Σ).

We shall be interested in those triplets (u, x, λ) that annihilate the output v of the Hamiltonian system. Later, we shall impose further conditions on these triplets. So far, we can state the following result

Proposition 3.2.14 *Let $\Sigma = (A, B; Q, L, R)$ and $\widetilde{\Sigma} = (\widetilde{A}, \widetilde{B}; \widetilde{Q}, \widetilde{L}, \widetilde{R})$ be two (X, F)-equivalent Popov triplets. Then (u, x, λ) annihilates the output of the CTHS(Σ) on a certain time interval if and only if $(u - Fx, x, \lambda - Xx)$ annihilates the output of the CTHS($\widetilde{\Sigma}$) on the same time interval.* ■

Proof. The proof is a matter of direct substitutions. ■

3.2.6 The extended Hamiltonian pencil (EHP)

Let $\Sigma = (A, B; Q, L, R)$ be a Popov triplet. The matrix pencil $sM_\Sigma - N_\Sigma$, with

$$M_\Sigma := \begin{bmatrix} I & O & O \\ O & I & O \\ O & O & O \end{bmatrix}, \quad N_\Sigma := \begin{bmatrix} A & O & B \\ -Q & -A^T & -L \\ L^T & B^T & R \end{bmatrix}, \tag{3.32}$$

is called the *extended Hamiltonian pencil (EHP)* associated with Σ, denoted EHP(Σ).

Notice that the EHP(Σ) can be regarded as the transmission pencil of the CTHS(Σ) in (3.31). As we are interested in output–annihilating solutions of the CTHS it is not surprising that its transmission pencil is going to play a significant role as well. If we take $v = 0$ in (3.31) and we denote

$$z(t) := \begin{bmatrix} x(t) \\ \lambda(t) \\ u(t) \end{bmatrix},$$

then we can rewrite (3.31) in the descriptor form $M_\Sigma \dot{z} = N_\Sigma z$. The result that follows relates EHPs associated with equivalent Popov triplets.

Proposition 3.2.15 *Let $\Sigma = (A, B; Q, L, R)$ and $\widetilde{\Sigma} = (\widetilde{A}, \widetilde{B}; \widetilde{Q}, \widetilde{L}, \widetilde{R})$ be two (X, F)–equivalent Popov triplets. Then the EHP(Σ) and EHP($\widetilde{\Sigma}$) are related by*

$$sM_\Sigma - N_\Sigma = U(sM_{\widetilde{\Sigma}} - N_{\widetilde{\Sigma}})V, \tag{3.33}$$

where

$$U := \begin{bmatrix} I & O & O \\ X & I & F^T \\ O & O & I \end{bmatrix}, \quad V := \begin{bmatrix} I & O & O \\ -X & I & O \\ -F & O & I \end{bmatrix}. \tag{3.34}$$

Proof. The proof is again a matter of straightforward substitutions. ∎

3.2.7 The continuous–time Popov function

Let $\Sigma = (A, B; Q, L, R)$ be a Popov triplet. For $s \in \mathbb{C} \setminus \{\Lambda(A) \cup \Lambda(-A^T)\}$ we define the rational matrix function

$$\mathbf{\Pi}_\Sigma(s) := [B^T(-sI - A^T)^{-1} \;\; I] \begin{bmatrix} Q & L \\ L^T & R \end{bmatrix} \begin{bmatrix} (sI - A)^{-1}B \\ I \end{bmatrix}, \tag{3.35}$$

that is called the *Popov function* associated with Σ.

It can be easily checked that

$$\mathbf{\Pi}_\Sigma = \left[\begin{array}{cc|c} A & O & B \\ -Q & -A^T & -L \\ \hline L^T & B^T & R \end{array}\right], \tag{3.36}$$

that is, $\mathbf{\Pi}_\Sigma$ is exactly the TFM from u to v of the CTHS(Σ) in (3.31). Moreover, the EHP(Σ) is exactly the transmission pencil associated with the realization (3.36) of the Popov function. The next result relates the Popov functions associated with two equivalent Popov triplets.

Proposition 3.2.16 *Let* $\Sigma = (A, B; Q, L, R)$ *and* $\widetilde{\Sigma} = (\widetilde{A}, \widetilde{B}; \widetilde{Q}, \widetilde{L}, \widetilde{R})$ *be two* (X, F)*–equivalent Popov triplets. Then for* $s \in \mathbb{C} \setminus \{\Lambda(A) \cup \Lambda(-A^T) \cup \Lambda(\widetilde{A}) \cup \Lambda(-\widetilde{A}^T)\}$ *we have*

$$\mathbf{\Pi}_{\Sigma}(s) = \mathbf{S}_F^*(s)\mathbf{\Pi}_{\widetilde{\Sigma}}(s)\mathbf{S}_F(s), \tag{3.37}$$

where

$$\mathbf{S}_F := \left[\begin{array}{c|c} A & B \\ \hline -F & I \end{array}\right]. \tag{3.38}$$

Moreover, if $(X_s = X_s^T, F_s)$ *is a solution to the CTARS(*Σ*) in (3.11), then*

$$\mathbf{\Pi}_{\Sigma}(s) = \mathbf{S}_{F_s}^*(s)R\mathbf{S}_{F_s}(s) \tag{3.39}$$

where $\mathbf{S}_{F_s}$ *is (3.38) written for* $F = F_s$*. (The identity (3.39) is known as the* spectral factorisation identity *and (3.38) is called a* spectral factor*).*

Proof. We have successively

$$\begin{aligned} &(sI - \widetilde{A})^{-1}B\mathbf{S}_F(s) \\ &= (sI - A - BF)^{-1}B[I - F(sI - A)^{-1}B] \\ &= (sI - A - BF)^{-1}[B - BF(sI - A)^{-1}B] \\ &= (sI - A - BF)^{-1}[I - BF(sI - A)^{-1}]B \\ &= (sI - A - BF)^{-1}(sI - A - BF)(sI - A)^{-1}B \\ &= (sI - A)^{-1}B. \end{aligned}$$

In conclusion, $(sI - \widetilde{A})^{-1}B\mathbf{S}_F(s) = (sI - A)^{-1}B$. Consequently,

$$\begin{bmatrix} (sI - \widetilde{A})^{-1}B \\ I \end{bmatrix} \mathbf{S}_F(s) = \begin{bmatrix} (sI - A)^{-1}B \\ I - F(sI - A)^{-1}B \end{bmatrix} = \begin{bmatrix} I & O \\ -F & I \end{bmatrix} \begin{bmatrix} (sI - A)^{-1}B \\ I \end{bmatrix}.$$

Using this identity, we readily obtain

$$\begin{aligned} \mathbf{S}_F^*(s)\mathbf{\Pi}_{\widetilde{\Sigma}}(s)\mathbf{S}_F(s) &= [B^T(-sI - A^T)^{-1} \; I] \begin{bmatrix} I & -F^T \\ O & I \end{bmatrix} \begin{bmatrix} \widetilde{Q} & \widetilde{L} \\ \widetilde{L}^T & \widetilde{R} \end{bmatrix} \\ &\quad \times \begin{bmatrix} I & O \\ -F & I \end{bmatrix} \begin{bmatrix} (sI - A)^{-1}B \\ I \end{bmatrix}. \end{aligned}$$

A straightforward computation gives

$$\begin{bmatrix} I & -F^T \\ O & I \end{bmatrix} \begin{bmatrix} \widetilde{Q} & \widetilde{L} \\ \widetilde{L}^T & \widetilde{R} \end{bmatrix} \begin{bmatrix} I & O \\ -F & I \end{bmatrix} = \begin{bmatrix} Q & L \\ L^T & R \end{bmatrix} + \begin{bmatrix} A^TX + XA & XB \\ B^TX & O \end{bmatrix}.$$

Combining these identities, we conclude that (3.37) is proved, if we show that

$$[\, B^T(-sI - A^T)^{-1} \quad I \,] \begin{bmatrix} \widetilde{A}^TX + X\widetilde{A} & X\widetilde{B} \\ \widetilde{B}^TX & O \end{bmatrix} \begin{bmatrix} (sI - A)^{-1}B \\ I \end{bmatrix} = 0. \tag{3.40}$$

To this end, let

$$\begin{bmatrix} A^T\widetilde{X}+\widetilde{X}A & \widetilde{X}B \\ B^T\widetilde{X} & O \end{bmatrix} = \begin{bmatrix} A^T \\ B^T \end{bmatrix} \begin{bmatrix} \widetilde{X} & O \end{bmatrix} + \begin{bmatrix} \widetilde{X} \\ O \end{bmatrix} \begin{bmatrix} A & B \end{bmatrix}. \tag{3.41}$$

We have

$$\begin{bmatrix} A & B \end{bmatrix} \begin{bmatrix} (sI-A)^{-1}B \\ I \end{bmatrix} = [A(sI-A)^{-1}+I]B = s(sI-A)^{-1}B \tag{3.42}$$

and

$$\begin{bmatrix} B^T(-sI-A^T)^{-1} & I \end{bmatrix} \begin{bmatrix} A^T \\ B^T \end{bmatrix} = -sB^T(-sI-A^T)^{-1}. \tag{3.43}$$

With (3.41), (3.42) and (3.43) the left–hand side of (3.40) becomes

$$-sB^T(-sI-A^T)^{-1}\widetilde{X}(sI-A)^{-1}B + B^T(-sI-A^T)^{-1}\widetilde{X}s(sI-A)^{-1}B = 0$$

and thus (3.40) is proved.

If (X_s, F_s) is a solution of the CTARS (3.11) associated with Σ, then it is easily checked that the (X_s, F_s)–equivalent of Σ has $\widetilde{Q} = 0$ and $\widetilde{L} = 0$. Consequently, the Popov function associated with $\widetilde{\Sigma}$ is in this case $\mathbf{\Pi}_{\widetilde{\Sigma}}(s) = R$, and (3.39) follows as a consequence of (3.37). ■

3.3 Associated operators

Throughout this section we fix a Popov triplet $\Sigma = (A, B; Q, L, R)$, with $A \in \mathbb{R}^{n\times n}$, $B \in \mathbb{R}^{n\times m}$. In Section 2.2 we have given various solutions of the controlled linear differential equation

$$\dot{x} = Ax + Bu \tag{3.44}$$

in operator form (as, for example, in (2.57) and (2.67)). By substituting these expressions into the quadratic indices associated with Σ, we shall obtain in this section operatorial expressions for the quadratic indices. The operators that will turn up in these expressions will play a major role in the theory presented in Chapter 4 and further in this book. For the convenience of notation we start with the case in which A is dichotomic.

3.3.1 The case: A dichotomic

Assume that A is *dichotomic* and $u \in L^{2,m}$. Substituting the $L^{2,n}$ solution $x_e^u = \mathcal{L}_e u$ of equation (3.44) (see Section 2.2.3) in the expression of the extended quadratic index (3.7), we have

$$\begin{aligned} J_{e,\Sigma}(u) &= \left\langle \begin{bmatrix} x_e^u \\ u \end{bmatrix}, \begin{bmatrix} Q & L \\ L^T & R \end{bmatrix} \begin{bmatrix} x_e^u \\ u \end{bmatrix} \right\rangle \\ &= \left\langle u, \begin{bmatrix} \mathcal{L}_e^* & I \end{bmatrix} \begin{bmatrix} Q & L \\ L^T & R \end{bmatrix} \begin{bmatrix} \mathcal{L}_e \\ I \end{bmatrix} u \right\rangle = \langle u, \mathcal{R}_{e,\Sigma} u \rangle, \end{aligned} \tag{3.45}$$

where

$$\mathcal{R}_{e,\Sigma} := R + L^T\mathcal{L}_e + \mathcal{L}_e^*L + \mathcal{L}_e^*Q\mathcal{L}_e = \mathcal{R}_{e,\Sigma}^* \tag{3.46}$$

is a linear bounded operator from $L^{2,m}$ into $L^{2,m}$. The next result relates the operators $\mathcal{R}_e$ associated with two equivalent Popov triplets.

Proposition 3.3.1 *Let $\Sigma = (A, B; Q, L, R)$ and $\widetilde{\Sigma} = (\widetilde{A}, \widetilde{B}; \widetilde{Q}, \widetilde{L}, \widetilde{R})$ be two (X, F)–equivalent Popov triplets. If A and $\widetilde{A} = A + BF$ are dichotomic, then*

$$\mathcal{R}_{e,\Sigma} = \mathcal{N}_F^*\mathcal{R}_{e,\widetilde{\Sigma}}\mathcal{N}_F, \tag{3.47}$$

where $\mathcal{N}_F$ is the input–output operator of the system

$$\begin{aligned} \dot{x} &= Ax + Bu, \\ v &= -Fx + u. \end{aligned} \tag{3.48}$$

In particular, if $\widetilde{\Sigma}$ is $(X, 0)$–equivalent to Σ, then $\mathcal{R}_{e,\Sigma} = \mathcal{R}_{e,\widetilde{\Sigma}}$.

Proof. From (3.45) and point 3 of Proposition 3.2.1 we have

$$J_{e,\Sigma}(u) = \langle u, \mathcal{R}_{e,\Sigma}u\rangle = J_{e,\widetilde{\Sigma}}(\widetilde{u}) = \left\langle \widetilde{u}, \mathcal{R}_{e,\widetilde{\Sigma}}\widetilde{u}\right\rangle = \left\langle \mathcal{N}_F u, \mathcal{R}_{e,\widetilde{\Sigma}}\mathcal{N}_F u\right\rangle$$

where $\widetilde{u} = -Fx_e^u + u = \mathcal{N}_F u$. Alternatively, we may compare (3.47) with (3.37) and see that the two expressions are related by the Fourier transform. ■

3.3.2 *The case: A stable*

Assume now that A is *stable*. Then the quadratic index $J_\Sigma(\xi, u)$ in (3.6) is defined for each $u \in L_+^{2,m}$. Substituting in (3.6) the solution $x^{\xi,u} = \Phi\xi + \mathcal{L}u$ to (3.44), for $x(0) = \xi$ and $u \in L_+^{2,m}$ (see Section 2.2.1), we have

$$\begin{aligned} J_\Sigma(\xi, u) &= \left\langle \begin{bmatrix} x^{\xi,u} \\ u \end{bmatrix}, \begin{bmatrix} Q & L \\ L^T & R \end{bmatrix} \begin{bmatrix} x^{\xi,u} \\ u \end{bmatrix} \right\rangle \\ &= \left\langle \begin{bmatrix} \Phi & \mathcal{L} \\ O & I \end{bmatrix} \begin{bmatrix} \xi \\ u \end{bmatrix}, \begin{bmatrix} Q & L \\ L^T & R \end{bmatrix} \begin{bmatrix} \Phi & \mathcal{L} \\ O & I \end{bmatrix} \begin{bmatrix} \xi \\ u \end{bmatrix} \right\rangle \\ &= \left\langle \begin{bmatrix} \xi \\ u \end{bmatrix}, \begin{bmatrix} \Phi^* & O \\ \mathcal{L}^* & I \end{bmatrix} \begin{bmatrix} Q & L \\ L^T & R \end{bmatrix} \begin{bmatrix} \Phi & \mathcal{L} \\ O & I \end{bmatrix} \begin{bmatrix} \xi \\ u \end{bmatrix} \right\rangle \\ &= \left\langle \begin{bmatrix} \xi \\ u \end{bmatrix}, \begin{bmatrix} \mathcal{P}_{o,\Sigma} & \mathcal{P}_\Sigma \\ \mathcal{P}_\Sigma^* & \mathcal{R}_\Sigma \end{bmatrix} \begin{bmatrix} \xi \\ u \end{bmatrix} \right\rangle, \end{aligned} \tag{3.49}$$

where

$$\mathcal{P}_{o,\Sigma} : \mathbb{R}^n \to \mathbb{R}^n, \quad \mathcal{P}_{o,\Sigma} := \Phi^*Q\Phi, \tag{3.50}$$

$$\mathcal{P}_\Sigma : L_+^{2,m} \to \mathbb{R}^n, \quad \mathcal{P}_\Sigma := \Phi^*(Q\mathcal{L} + L), \tag{3.51}$$

$$\mathcal{R}_\Sigma : L_+^{2,m} \to L_+^{2,m}, \quad \mathcal{R}_\Sigma := R + L^T\mathcal{L} + \mathcal{L}^*L + \mathcal{L}^*Q\mathcal{L} = \mathcal{R}_\Sigma^* \tag{3.52}$$

are all linear bounded operators.

The adjoint of Φ is given by

$$\Phi^* x = \int_0^\infty e^{A^T t} x(t) \mathrm{d}t, \quad x \in L_+^{2,n}, \tag{3.53}$$

as can be derived from formula (2.93) if we notice that Φ is the causal observability operator in the particular case that $C = I_n$.

Writing (3.50) explicitly, we have

$$\mathcal{P}_{o,\Sigma} = \int_0^\infty e^{A^T t} Q e^{At} \mathrm{d}t. \tag{3.54}$$

From the last expression, it is easy to check that $\mathcal{P}_{o,\Sigma}$ is the unique solution to the reducing Lyapunov equation of Σ,

$$A^T X + XA + Q = 0. \tag{3.55}$$

This means that $\mathcal{P}_{o,\Sigma} = X_r$.

The adjoint of $\mathcal{L}$ turns out to be

$$(\mathcal{L}^* x)(t) = \int_t^\infty B^T e^{-A^T(t-\tau)} x(\tau) \mathrm{d}\tau, \ \forall x \in L_+^{2,n}, \quad t \geq 0. \tag{3.56}$$

Since A is stable the operator $\mathcal{L}_e$ is causal as follows from Theorem 2.4.5, by taking $C = I_n$, $D = 0$. Since $\mathcal{L}$ is the Toeplitz operator corresponding to $\mathcal{L}_e$, and using Corollary 2.4.13 in relation (3.46), we deduce that

$$\mathcal{R}_\Sigma = \mathbb{T}_{\mathcal{R}_{e,\Sigma}}, \tag{3.57}$$

that is, the operator $\mathcal{R}_\Sigma$ is a Toeplitz operator.

The following proposition relates pairwise the operators introduced so far and associated with two equivalent Popov triplets.

Proposition 3.3.2 *Let $\Sigma = (A, B; Q, L, R)$ and $\widetilde{\Sigma} = (\widetilde{A}, \widetilde{B}; \widetilde{Q}, \widetilde{L}, \widetilde{R})$ be two (X, F)-equivalent Popov triplets. If A and $\widetilde{A} = A + BF$ are stable, then*

$$\mathcal{P}_{o,\Sigma} = X + \mathcal{P}_{o,\widetilde{\Sigma}} - \mathcal{P}_{\widetilde{\Sigma}} F \Phi - \Phi^* F^T \mathcal{P}_{\widetilde{\Sigma}}^* + \Phi^* F^T \mathcal{R}_{\widetilde{\Sigma}} F \Phi\ , \tag{3.58}$$

$$\mathcal{P}_\Sigma = (\mathcal{P}_{\widetilde{\Sigma}} - \Phi^* F^T \mathcal{R}_{\widetilde{\Sigma}}) \mathbb{T}_{\mathcal{N}_F}\ , \tag{3.59}$$

$$\mathcal{R}_\Sigma = \mathbb{T}_{\mathcal{N}_F}^* \mathcal{R}_{\widetilde{\Sigma}} \mathbb{T}_{\mathcal{N}_F}, \tag{3.60}$$

where $\mathcal{N}_F$ is the input-output operator of the system (3.48).

In particular, if $\widetilde{\Sigma}$ is $(X, 0)$-equivalent with Σ, then $\mathcal{P}_{o,\Sigma} = X + \mathcal{P}_{o,\widetilde{\Sigma}}$, $\mathcal{P}_\Sigma = \mathcal{P}_{\widetilde{\Sigma}}$, $\mathcal{R}_\Sigma = \mathcal{R}_{\widetilde{\Sigma}}$.

Proof. From point 1 of Proposition 3.2.1, we have

$$\left\langle \begin{bmatrix} \xi \\ u \end{bmatrix}, \begin{bmatrix} \mathcal{P}_{o,\Sigma} & \mathcal{P}_\Sigma \\ \mathcal{P}_\Sigma^* & \mathcal{R}_\Sigma \end{bmatrix} \begin{bmatrix} \xi \\ u \end{bmatrix} \right\rangle = \left\langle \begin{bmatrix} \xi \\ \widetilde{u} \end{bmatrix}, \begin{bmatrix} \mathcal{P}_{o,\widetilde{\Sigma}} & \mathcal{P}_{\widetilde{\Sigma}} \\ \mathcal{P}_{\widetilde{\Sigma}}^* & \mathcal{R}_{\widetilde{\Sigma}} \end{bmatrix} \begin{bmatrix} \xi \\ \widetilde{u} \end{bmatrix} \right\rangle + \xi^T X \xi,$$

where $\widetilde{u} = -F x^{\xi,u} + u = -F\Phi\xi + \mathcal{N}_F u$, for each $\xi \in \mathbb{R}^n$ and $u \in L_+^{2,m}$. The proof follows now by straightforward substitution and identification of corresponding terms in the last formula. Notice that since A is stable, the system (3.48) is causal, that is, $\mathbb{T}_{\mathcal{N}_F} = \mathcal{N}_F P_+^m$. ∎

3.3.3 The case: A antistable

We assume now that A is *antistable*. Then the quadratic index $J_\Sigma(u)$ in (3.8) is defined for each $u \in L_+^{2,m}$. Substituting in (3.8) the solution $x^u = \mathcal{L}u$ to (3.44), for $u \in L_+^{2,m}$ (see Section 2.2.2), we have

$$J_\Sigma(u) = \langle u, \mathcal{R}_\Sigma u \rangle , \tag{3.61}$$

where $\mathcal{R}_\Sigma : L_+^{2,m} \to L_+^{2,m}$ is the linear bounded operator

$$\mathcal{R}_\Sigma = R + L^T \mathcal{L} + \mathcal{L}^* L + \mathcal{L}^* Q \mathcal{L} = \mathcal{R}_\Sigma{}^*. \tag{3.62}$$

Although $\mathcal{L}$ is in this case also the Toeplitz operator corresponding to $\mathcal{L}_e$, the latter is not causal and thus $\mathcal{R}_\Sigma$ is not, in general, a Toeplitz operator. The Toeplitz operator of $\mathcal{R}_{e,\Sigma}$ can still be computed from (3.46) using Proposition 2.4.12. We obtain, making use also of Proposition 2.4.11, that $\mathbb{T}_{\mathcal{R}_{e,\Sigma}} = \mathcal{R}_\Sigma + \widehat{\mathbb{H}}_{\mathcal{L}_e^*} Q \widehat{\mathbb{H}}_{\mathcal{L}_e}$, that is,

$$\mathcal{R}_\Sigma = \mathbb{T}_{\mathcal{R}_{e,\Sigma}} - \widehat{\mathbb{H}}^*_{\mathcal{L}_e} Q \widehat{\mathbb{H}}_{\mathcal{L}_e}. \tag{3.63}$$

This relation is sometimes called the Toeplitz–Hankel representation of $\mathcal{R}_\Sigma$. An important consequence of this relation is that $\mathcal{R}_\Sigma$ is a Toeplitz operator if $Q = 0$. In particular, if Σ_r is the reduced equivalent of Σ, then $\mathcal{R}_{\Sigma_r}$ is a Toeplitz operator; more precisely,

$$\mathcal{R}_{\Sigma_r} = \mathbb{T}_{\mathcal{R}_{e,\Sigma_r}} = \mathbb{T}_{\mathcal{R}_{e,\Sigma}}. \tag{3.64}$$

Applying (2.89) for the particular case $C = I_n$, $D = 0$, we have $\widehat{\mathbb{H}}_{\mathcal{L}_e} = \widehat{\Phi}\widehat{\Psi}$, where $\widehat{\Psi} : L_+^{2,m} \to \mathbb{R}^n$ is the anticausal controllability operator (2.90) of the pair (A, B), and $\widehat{\Phi} : \mathbb{R}^n \to L_-^{2,n}$ is the anticausal observability operator (2.91) of the pair (I_n, A), i.e., $(\widehat{\Phi}\xi)(t) = e^{At}\xi$, $t \leq 0$, $\xi \in \mathbb{R}^n$. Substituting in (3.63), we have

$$\mathcal{R}_\Sigma = \mathbb{T}_{\mathcal{R}_{e,\Sigma}} - \widehat{\Psi}^* \widehat{\Phi}^* Q \widehat{\Phi} \widehat{\Psi}. \tag{3.65}$$

But

$$-\widehat{\Phi}^* Q \widehat{\Phi} = - \int_{-\infty}^{0} e^{A^T t} Q e^{At} \mathrm{d}t \tag{3.66}$$

is the unique solution X_r of the reducing Lyapunov equation of Σ given in (3.55). The operator $\mathcal{R}_\Sigma$ can again be rewritten as

$$\mathcal{R}_\Sigma = \mathbb{T}_{\mathcal{R}_{e,\Sigma}} + \widehat{\Psi}^* X_r \widehat{\Psi}. \tag{3.67}$$

For the operators $\mathcal{R}$ associated with two equivalent Popov triplets we obtain the following result.

Proposition 3.3.3 *Let $\Sigma = (A, B; Q, L, R)$ and $\widetilde{\Sigma} = (\widetilde{A}, \widetilde{B}; \widetilde{Q}, \widetilde{L}, \widetilde{R})$ be two (X, F)–equivalent Popov triplets. If A and $A + BF$ are antistable, then*

$$\mathcal{R}_\Sigma = \mathbb{T}^*_{\mathcal{N}_F} \mathcal{R}_{\widetilde{\Sigma}} \mathbb{T}_{\mathcal{N}_F} + \widehat{\Psi}^* X \widehat{\Psi}, \tag{3.68}$$

where $\mathcal{N}_F$ is the input–output operator of the system (3.48).

In particular, if $\widetilde{\Sigma}$ is $(X, 0)$–equivalent with Σ, then $\mathcal{R}_\Sigma = \mathcal{R}_{\widetilde{\Sigma}} + \widehat{\Psi}^* X \widehat{\Psi}$.

Proof. From point 2 of Proposition 3.2.1 we have

$$J_\Sigma(u) = \langle u, \mathcal{R}_\Sigma u\rangle = J_{\widetilde{\Sigma}}(\widetilde{u}) + (x^u(0))^T X x^u(0) = \left\langle \widetilde{u}, \mathcal{R}_{\widetilde{\Sigma}}\widetilde{u}\right\rangle + (x^u(0))^T X x^u(0),$$

where $\widetilde{u} = -Fx^u + u = \mathbb{T}_{\mathcal{N}_F}u$. Substituting here $x^u(0) = \widehat{\Psi}u$, we immediately reach the conclusion. ■

3.3.4 Reduction of the case A antistable to the case A stable

We show now that the operator $\mathcal{R}_{\Sigma_r}$ in (3.64) is an operator of the type (3.52) associated with an appropriate Popov triplet with a stable A–matrix. We introduce first the following definition.

Definition 3.3.4 (Dual of a Popov triplet) *Let $\Sigma = (A, B; Q, L, R)$ be a Popov triplet with A antistable and let $\Sigma_r = (A, B; O, L_r = L + X_r B, R)$ be its reduced equivalent, where X_r is the unique solution to the reducing Lyapunov equation $A^T X + XA + Q = 0$. Then the Popov triplet $\Sigma_d = (A_d, B_d; 0, L_d, R_d)$, where*

$$A_d = -A^T, \quad B_d = -L_r = -L - X_r B, \quad L_d = B, \quad R_d = R, \tag{3.69}$$

is called the dual Popov triplet *of Σ.*

Notice that A_d in (3.69) is stable, as A is antistable. The following result reveals some interesting relationships among certain operators associated with the original triplet Σ and its dual Σ_d and will be a basic tool for several proofs in Chapter 4.

Lemma 3.3.5 *Let $\Sigma = (A, B; Q, L, R)$ be a Popov triplet with A antistable, and let $\Sigma_r = (A, B; O, L_r = L + X_r B, R)$ and $\Sigma_d = (A_d, B_d; 0, L_d, R_d)$ be the reduced and the dual Popov triplets of Σ, respectively. Then*

$$\mathcal{R}_{\Sigma_d} = \mathcal{R}_{\Sigma_r}, \tag{3.70}$$

$$\mathcal{P}_{\Sigma_d} = -\widehat{\Psi}, \tag{3.71}$$

where $\widehat{\Psi}$ is the anticausal controllability operator (2.90) associated with the pair (A, B), and the Popov functions are related through

$$\mathbf{\Pi}_\Sigma = \mathbf{\Pi}_{\Sigma_d}. \tag{3.72}$$

Proof. Let Φ_d and $\mathcal{L}_d$ be the operators (2.55) and (2.56) corresponding to Σ_d, i.e.,

$$(\Phi_d \mu)(t) = e^{-A^T t}\mu, \ t \geq 0, \quad \forall \mu \in \mathbb{R}^n, \tag{3.73}$$

and

$$(\mathcal{L}_d u)(t) = -\int_0^t e^{-A^T(t-\tau)} L_r u(\tau) \mathrm{d}\tau. \tag{3.74}$$

The adjoints of these operators are, according to (3.53) and (3.56),

$$\Phi_d^* \lambda = \int_0^\infty e^{-At}\lambda(t)\mathrm{d}t, \quad \forall \lambda \in L_+^{2,n}, \tag{3.75}$$

$$(\mathcal{L}_d^*\lambda)(t) = -\int_t^\infty L_r^T e^{A(t-\tau)}\lambda(\tau)\mathrm{d}\tau, \quad \forall\lambda \in L_+^{2,n}. \tag{3.76}$$

Then for all $u \in L_+^{2,m}$ we have successively

$$(\mathcal{L}_d^* L_d u)(t) = -\int_t^\infty {L_r}^T e^{A(t-\tau)} Bu(\tau)\mathrm{d}\tau = ({L_r}^T \mathcal{L}u)(t), \quad (t \geq 0), \tag{3.77}$$

$$(L_d^T \mathcal{L}_d u)(t) = -\int_0^t B^T e^{-A^T(t-\tau)} L_r u(\tau)\mathrm{d}\tau = (\mathcal{L}^* L_r u)(t), \quad (t \geq 0). \tag{3.78}$$

Combining (3.52), (3.69), (3.77) and (3.78) we get

$$\mathcal{R}_{\Sigma_d} = R_d + L_d^T \mathcal{L}_d + \mathcal{L}_d^* L_d = R + \mathcal{L}^* L_r + L_r^T \mathcal{L} = \mathcal{R}_{\Sigma_r}.$$

For (3.71) we have from (3.51)

$$\mathcal{P}_{\Sigma_d} u = \Phi_d^* L_d u = \int_0^\infty e^{-At} Bu(t)\mathrm{d}t = -\widehat{\Psi}u.$$

Equality (3.72) follows by direct checking. ■

3.3.5 $\mathcal{R}_e$ as the input-output operator of the CTHS

There is an intimate relationship between the operators introduced in this section and the mathematical objects introduced in Section 3.2. This relationship will be further revealed in Chapter 4. For now we prove the following key result.

Theorem 3.3.6 *Let $\Sigma = (A, B; Q, L, R)$ be a Popov triplet. If A is dichotomic, then the operator $\mathcal{R}_{e,\Sigma}$ in (3.46) is the L^2 input-output operator of the CTHS(Σ) in (3.31).*

Proof. Let A be dichotomic and take an arbitrary u in $L^{2,m}$. Then the unique x in $L^{2,n}$ satisfying the first equation of (3.31), $\dot{x} = Ax + Bu$, is $x_e^u = \mathcal{L}_e u$. Using now formula (2.69) and the second equation in (3.31), $\dot{\lambda} = -Qx - A^T\lambda - Lu$, one gets

$$\begin{aligned} B^T \lambda_e^u(t) &= -\int_{-\infty}^t B^T e^{-A^T(t-\tau)} \Pi_-(-Lu(\tau) - Qx_e^u(\tau))\mathrm{d}\tau \\ &\quad -\int_t^\infty B^T e^{-A^T(t-\tau)} \Pi_+(-Lu(\tau) - Qx_e^u(\tau))\mathrm{d}\tau \\ &= ((\mathcal{L}_e^* L + \mathcal{L}_e^* Q \mathcal{L}_e)u)(t). \end{aligned}$$

Using the last equation in (3.31) yields

$$\begin{aligned} v_e^u(t) &= L^T x_e^u(t) + B^T \lambda_e^u(t) + Ru(t) \\ &= (L^T \mathcal{L}_e u)(t) + ((\mathcal{L}_e^* L + \mathcal{L}_e^* Q \mathcal{L}_e)u)(t) + Ru(t) \end{aligned}$$

or, eventually,

$$v_e^u = L^T \mathcal{L}_e u + \mathcal{L}_e^* Q \mathcal{L}_e u + \mathcal{L}_e^* L u + Ru.$$

Thus, with (3.46), we have $v_e^u = \mathcal{R}_{e,\Sigma} u$, and the proof is complete. ■

An immediate consequence of Theorem 3.3.6 is the fact that if A is stable then $\mathcal{R}_\Sigma$ is the Toeplitz operator with the symbol the Popov function $\mathbf{\Pi}_\Sigma$. Indeed, the Popov function is the TFM of the Hamiltonian system and $\mathcal{R}_\Sigma$ is the Toeplitz operator associated with the input–output operator of the Hamiltonian system.

3.4 Particular relevant cases

Several particular Popov triplets play a prominent role in the applications and deserve to be singled out.

Consider the system

$$\mathbf{G} = \left[\begin{array}{c|c} A & B \\ \hline C & D \end{array}\right]. \tag{3.79}$$

We associate with it the following Popov triplets:

$$\begin{aligned} \Sigma_p &= (A, B; C^T C, C^T D, D^T D), & (3.80) \\ \Sigma_n &= (A, B; C^T C, C^T D, I + D^T D), & (3.81) \\ \Sigma_c &= (A, B; C^T C, C^T D, -\gamma^2 I + D^T D), & (3.82) \end{aligned}$$

for any $\gamma > 0$, called, respectively, the *positiveness, normalized* and γ*-contractiveness* Popov triplets associated with the system (3.79). If $D = 0$, Σ_n becomes

$$\Sigma_s = (A, B; C^T C, 0, I), \tag{3.83}$$

usually termed the *standard Popov triplet (for control)* associated with the strictly proper system

$$\mathbf{G} = \left[\begin{array}{c|c} A & B \\ \hline C & O \end{array}\right]. \tag{3.84}$$

Notice that Σ_s could be obtained from Σ_p if the following orthogonality conditions hold:

$$C^T D = 0, \quad D^T D = I. \tag{3.85}$$

This is the case for example if (3.79) has the particular form

$$\mathbf{G} = \left[\begin{array}{c|c} A & B \\ \hline C & O \\ O & I \end{array}\right]. \tag{3.86}$$

The Popov functions associated with (3.80), (3.81) and (3.82) are

$$\begin{aligned} \mathbf{\Pi}_{\Sigma_p}(s) &= \mathbf{G}^*(s)\mathbf{G}(s), & (3.87) \\ \mathbf{\Pi}_{\Sigma_n}(s) &= I + \mathbf{G}^*(s)\mathbf{G}(s), & (3.88) \\ \mathbf{\Pi}_{\Sigma_c}(s) &= -\gamma^2 I + \mathbf{G}^*(s)\mathbf{G}(s), & (3.89) \end{aligned}$$

as can be easily checked. If A is dichotomic then $\mathbf{G}(j\omega)$ is well defined for all $\omega \in \mathbb{R}$. Hence $\mathbf{G}^*(j\omega)\mathbf{G}(j\omega) \geq 0$ for $\omega \in \mathbb{R}$. This fact justifies the attribute of positiveness given to Σ_p. If $-\gamma^2 I + \mathbf{G}^*(j\omega)\mathbf{G}(j\omega) \leq 0$ for $\omega \in \mathbb{R}$, then $||\mathbf{G}||_\infty \leq \gamma$ and the attribute of contractiveness for Σ_c arises also in a natural way. The terminology adopted for Σ_n will be justified later in connection with what is called the normalized (coprime) factorization.

The quadratic index (3.6) associated with Σ_s is

$$J_{\Sigma_s}(\xi, u) = \int_0^\infty (||y||^2 + ||u||^2)\mathrm{d}t = ||y||_2^2 + ||u||_2^2, \tag{3.90}$$

(here $y = Cx$), and the CTARE (3.19) associated with Σ_s is

$$A^T X + XA - XBB^T X + C^T C = 0 \tag{3.91}$$

which is precisely the *standard CTARE (for control)* associated with the usual linear quadratic problem

$$\min_{u \in \mathcal{U}_\xi} J_{\Sigma_s}(\xi, u).$$

This problem has been thoroughly investigated since the early 1960s. Of equal importance is the dual form of (3.91),

$$AY + YA^T - YC^T CY + BB^T = 0, \tag{3.92}$$

known as the *standard CTARE for estimation* and associated with the *standard Popov triplet (for estimation)* $\Sigma_e = (A^T, C^T; BB^T, 0, I)$.

In Chapter 7 we shall investigate several properties of the system $\mathbf{G}$ in terms of the CTAREs and KPYSs associated with the various Popov triplets introduced in this section.

B. The discrete–time case

3.5 Popov triplets: the double interpretation, significance, and equivalence

In the discrete–time case a Popov triplet has the same form as in Definition 3.1.1 for the continuous–time case, i.e.,

$$\Sigma := (A, B; P), \quad P := \begin{bmatrix} Q & L \\ L^T & R \end{bmatrix} = P^T, \tag{3.93}$$

where $A \in \mathbb{R}^{n \times n}$, $B \in \mathbb{R}^{n \times m}$, $Q = Q^T \in \mathbb{R}^{n \times n}$, $L \in \mathbb{R}^{n \times m}$, $R = R^T \in \mathbb{R}^{m \times m}$, and $P \in \mathbb{R}^{(n+m)\times(n+m)}$. Alternatively, we shall use the explicit notation $\Sigma = (A, B; Q, L, R)$. However, an important difference is that in the discrete–time case the Popov triplet (3.93) can be regarded as corresponding to the direct–time controlled recursion

$$\sigma x = Ax + Bu \tag{3.94}$$

and a quadratic cost of the form

$$J := \sum_{k=r}^{s} \begin{bmatrix} x_k \\ u_k \end{bmatrix}^T P \begin{bmatrix} x_k \\ u_k \end{bmatrix}, \tag{3.95}$$

or, alternatively, to the inverse–time controlled recursion

$$x = A\sigma x + Bu \tag{3.96}$$

and a quadratic cost of the form

$$\widehat{J} := \sum_{k=r}^{s} \begin{bmatrix} x_{k+1} \\ u_k \end{bmatrix}^T P \begin{bmatrix} x_{k+1} \\ u_k \end{bmatrix}. \tag{3.97}$$

In the first case we say that Σ has a *direct–time* significance, while in the latter one we say that Σ has an *inverse–time* significance.

Notice in (3.96), that if we introduce the change of variables $x = \Omega\widehat{x}$, $u = \sigma\Omega\widehat{u}$, similar to (2.108), then $\widehat{x}$ and $\widehat{u}$ will satisfy the direct–time control recursion (3.94), whereas the quadratic cost becomes

$$\widehat{J} = \sum_{k=r}^{s} \begin{bmatrix} \widehat{x}_{-k-1} \\ \widehat{u}_{-k-1} \end{bmatrix}^T P \begin{bmatrix} \widehat{x}_{-k-1} \\ \widehat{u}_{-k-1} \end{bmatrix} = \sum_{k=-s-1}^{-r-1} \begin{bmatrix} \widehat{x}_k \\ \widehat{u}_k \end{bmatrix}^T P \begin{bmatrix} \widehat{x}_k \\ \widehat{u}_k \end{bmatrix}.$$

We conclude that, with this change of variables, the quadratic index (3.97) takes the direct–time form (3.95) with different limits of summation. We shall make use of this transformation to establish inverse–time counterparts for various objects and results related to the Popov triplet (3.93). In particular, a feedback transformation for (3.94) of the form

$$u = Fx + v,$$

that yields

$$\sigma x = (A + BF)x + Bv,$$

becomes in the inverse–time case (3.96)

$$u = F\sigma x + v, \tag{3.98}$$

and yields

$$x = (A + BF)\sigma x + Bv. \tag{3.99}$$

We introduce now an equivalence transformation for Popov triplets in discrete–time.

The equivalence transformation

Definition 3.5.1 (Equivalence of Popov triplets in discrete–time)
Two Popov triplets

$$\begin{aligned} \Sigma &= (A, B; Q, L, R), \\ \widetilde{\Sigma} &= (\widetilde{A}, \widetilde{B}; \widetilde{Q}, \widetilde{L}, \widetilde{R}), \end{aligned}$$

are said to be (X, F)–equivalent *(or simply,* equivalent*) if there exist* $F \in \mathbb{R}^{m\times n}$ *and* $X = X^T \in \mathbb{R}^{n\times n}$ *such that*

$$\begin{aligned} \widetilde{A} &= A + BF, \\ \widetilde{B} &= B, \\ \widetilde{Q} &= Q + LF + F^TL^T + F^TRF + \widetilde{A}^TX\widetilde{A} - X, \\ \widetilde{L} &= L + F^TR + \widetilde{A}^TXB, \\ \widetilde{R} &= R + B^TXB. \end{aligned} \tag{3.100}$$

We say that $\widetilde{\Sigma}$ *is a* feedback equivalent *(or is* F*–equivalent) with* Σ *if it is* $(0, F)$*–equivalent for some matrix* F.

We say that $\widetilde{\Sigma}$ *is a* reduced equivalent *of* Σ *if it is* $(X, 0)$*–equivalent and, in addition,* $\widetilde{Q} = 0$.

It can be easily checked that the relation defined above is indeed an equivalence relation satisfying the corresponding axioms. Straightforward computations show that if Σ and $\widetilde{\Sigma}$ are (X,F)–equivalent, and $\widetilde{\Sigma}$ and $\widetilde{\widetilde{\Sigma}}$ are $(\widetilde{X},\widetilde{F})$–equivalent, then Σ and $\widetilde{\widetilde{\Sigma}}$ are $(\widetilde{\widetilde{X}},\widetilde{\widetilde{F}})$–equivalent where $\widetilde{\widetilde{X}} = X + \widetilde{X}$ and $\widetilde{\widetilde{F}} = F + \widetilde{F}$.

Remark 3.5.2 It is easy to see that if $\widetilde{\Sigma}$ is a reduced equivalent of Σ, then X satisfies the symmetric Stein equation

$$A^T X A - X + Q = 0. \tag{3.101}$$

If A is stable, then Theorem 1.5.2 shows that the symmetric Stein equation (3.101) has a unique solution. If the pair (A, B) is stabilizable, it is possible to take first a feedback transformation such that $\widetilde{A} = A + BF$ is stable and then solve (3.101) for the transformed triplet to obtain a reduced equivalent.

In the sequel, we shall call the symmetric Stein equation (3.101), the *reducing Stein equation* of Σ. In those cases that this equation has a unique solution we shall denote this by X_r, and we shall denote the corresponding $(X_r, 0)$–equivalent triplet by $\Sigma_r = (A, B; 0, L_r, R_r)$, where $L_r = L + A^T X_r B$ and $R_r = R + B^T X_r B$.

Remark 3.1.4 is appropriate in the discrete–time case as well.

3.6 Objects associated with a Popov triplet in discrete–time

Throughout this section we fix a Popov triplet $\Sigma = (A, B; Q, L, R)$, with $A \in \mathbb{R}^{n\times n}$, $B \in \mathbb{R}^{n\times m}$. We associate with Σ several mathematical objects that are the analogues of those introduced in Section 3.2 and we study the effect that the equivalence transformation has on these objects. As a rule, we consider first the direct–time interpretation for Σ and then the inverse–time interpretation. Most of the proofs will be omitted as they run similar to those for the continuous–time correspondents in Section 3.2.

3.6.1 Quadratic indices

The *direct–time quadratic index* associated with Σ is the functional

$$J_\Sigma(\xi, u) := \left\langle \begin{bmatrix} x^{\xi,u} \\ u \end{bmatrix}, \begin{bmatrix} Q & L \\ L^T & R \end{bmatrix} \begin{bmatrix} x^{\xi,u} \\ u \end{bmatrix} \right\rangle_{\ell_+^{2,n} \times \ell_+^{2,m}}, \tag{3.102}$$

where $x^{\xi,u}$ is the solution of (3.94) that satisfies $x_0 = \xi$ as defined in (2.143) and $u \in \ell_+^{2,m}$ is such that $x^{\xi,u} \in \ell_+^{2,n}$, i.e., u ranges in the set $\mathcal{U}_\xi$ introduced in (2.176). If A in (3.94) is stable then $J_\Sigma(\xi, u)$ is defined for each $\xi \in \mathbb{R}^n$ and $u \in \ell_+^{2,m}$.

The *inverse–time quadratic index* associated with Σ is the functional

$$\widehat{J}_\Sigma(u) := \left\langle \begin{bmatrix} \sigma\widehat{x}^u \\ u \end{bmatrix}, \begin{bmatrix} Q & L \\ L^T & R \end{bmatrix} \begin{bmatrix} \sigma\widehat{x}^u \\ u \end{bmatrix} \right\rangle_{\ell_+^{2,n} \times \ell_+^{2,m}}, \tag{3.103}$$

where $\widehat{x}^u$ is the solution of the inverse–time recursion (3.96) defined by (2.159) and $u \in \ell_+^{2,m}$ is such that $\widehat{x}^u \in \ell_+^{2,m}$, i.e., u ranges in the set $\widehat{\mathcal{U}}$ introduced in (2.177). If A is stable, $\widehat{J}_\Sigma(u)$ is defined for each $u \in \ell_+^{2,m}$.

In the case that A is dichotomic, we introduce the *extended direct–time quadratic index* associated with Σ as

$$J_{e,\Sigma}(u) := \left\langle \begin{bmatrix} x_e^u \\ u \end{bmatrix}, \begin{bmatrix} Q & L \\ L^T & R \end{bmatrix} \begin{bmatrix} x_e^u \\ u \end{bmatrix} \right\rangle_{\ell^{2,n} \times \ell^{2,m}}, \tag{3.104}$$

where x_e^u is the ℓ^2 solution on the entire axis of (3.94) defined by (2.164). Similarly, we define the *extended inverse–time quadratic index* associated with Σ as

$$\widehat{J}_{e,\Sigma}(u) := \left\langle \begin{bmatrix} \sigma\widehat{x}_e^u \\ u \end{bmatrix}, \begin{bmatrix} Q & L \\ L^T & R \end{bmatrix} \begin{bmatrix} \sigma\widehat{x}_e^u \\ u \end{bmatrix} \right\rangle_{\ell^{2,n} \times \ell^{2,m}}, \tag{3.105}$$

where $\widehat{x}_e^u$ is the ℓ^2 solution on the entire axis of the inverse–time recursion (3.96) defined by (2.169).

Limiting the time–horizon in the extended direct–time quadratic index to the positive integers, we introduce the *restricted quadratic index* associated with Σ as the functional

$$J_\Sigma(u) := \left\langle \begin{bmatrix} x^u \\ u \end{bmatrix}, \begin{bmatrix} Q & L \\ L^T & R \end{bmatrix} \begin{bmatrix} x^u \\ u \end{bmatrix} \right\rangle_{\ell_+^{2,n} \times \ell_+^{2,m}}, \tag{3.106}$$

where x^u is the solution of (3.94) defined by (2.162) and u is such that x^u is norm–square summable. If A is antistable, $J_\Sigma(u)$ is defined for each $u \in \ell_+^{2,m}$. The restricted quadratic index has similar properties to the inverse–time quadratic index. In fact, we shall see later that the two are closely linked in the case that A is antistable.

The effect of the equivalence transformation on these indices is given by the following result.

Proposition 3.6.1 *Let $\Sigma = (A, B; Q, L, R)$ and $\widetilde{\Sigma} = (\widetilde{A}, \widetilde{B}; \widetilde{Q}, \widetilde{L}, \widetilde{R})$ be two (X, F)–equivalent Popov triplets. Then we have:*

1. $J_\Sigma(\xi, u) = J_{\widetilde{\Sigma}}(\xi, \widetilde{u}) + \xi^T X \xi$, $\quad (u \in \mathcal{U}_\xi)$, *where* $\widetilde{u} = -Fx^{\xi,u} + u$.
2. $\widehat{J}_\Sigma(u) = \widehat{J}_{\widetilde{\Sigma}}(\widetilde{u}) + (\widehat{x}_0^u)^T X \widehat{x}_0^u$, $\quad (u \in \widehat{\mathcal{U}})$, *where* $\widetilde{u} = -F\sigma\widehat{x}^u + u$.
3. *If $A + BF$ is dichotomic, then* $J_{e,\Sigma}(u) = J_{e,\widetilde{\Sigma}}(\widetilde{u})$, *where* $\widetilde{u} = -Fx_e^u + u$.
4. *If $A+BF$ is dichotomic, then* $\widehat{J}_{e,\Sigma}(u) = \widehat{J}_{e,\widetilde{\Sigma}}(\widetilde{u})$, *where* $\widetilde{u} = -F\sigma x_e^u + u$.

3.6.2 The discrete–time algebraic Riccati system (DTARS)

For a symmetric $X \in \mathbb{R}^{n\times n}$, we define the *dissipation matrix* associated with Σ as

$$D_\Sigma(X) := \begin{bmatrix} A^TXA - X + Q & L + A^TXB \\ B^TXA + L^T & R + B^TXB \end{bmatrix}. \tag{3.107}$$

The system of equations

$$D_\Sigma(X) \begin{bmatrix} I \\ F \end{bmatrix} = 0 \tag{3.108}$$

in the unknowns $F \in \mathbb{R}^{m\times n}, X \in \mathbb{R}^{n\times n}$, is called *the discrete–time algebraic Riccati system (DTARS)* associated with Σ, and is denoted DTARS(Σ).

Definition 3.6.2 *A pair (X, F) satisfying (3.108), with $X = X^T$, and $A + BF$ stable is called a stabilizing solution to the DTARS(Σ).*

For the solution to DTARS we can prove the following uniqueness result.

Proposition 3.6.3 *Let (X, F) and $(\widehat{X}, \widehat{F})$ be two stabilizing solutions to the DTARS(Σ) given in (3.108). Then $X = \widehat{X}$.*

Proof. From (3.108) with (3.107) we get

$$A^T XA - X + Q + (L + A^T XB)F = 0, \quad (3.109)$$
$$B^T XA + L^T + (R + B^T XB)F = 0. \quad (3.110)$$

Let $S := A + BF$ and rewrite (3.109) and (3.110) as

$$A^T XS - X + Q + LF = 0, \quad (3.111)$$
$$B^T XS + L^T + RF = 0. \quad (3.112)$$

Let $\widehat{S} := A + B\widehat{F}$ and add to (3.111), (3.112) premultiplied by $\widehat{F}^T$ and obtain

$$\widehat{S}^T XS - X + \widehat{F}^T L^T + LF + \widehat{F}^T RF + Q = 0. \quad (3.113)$$

Similarly, if we start with

$$D_\Sigma(\widehat{X}) \begin{bmatrix} I \\ \widehat{F} \end{bmatrix} = 0$$

and further we interchange the role of X and F with the role of $\widehat{X}$ and $\widehat{F}$ in the above reasoning, we arrive at

$$S^T \widehat{X}\widehat{S} - \widehat{X} + F^T L^T + L\widehat{F} + F^T R\widehat{F} + Q = 0. \quad (3.114)$$

Take the transpose of (3.114) and then subtract it from (3.113) and obtain

$$\widehat{S}^T (X - \widehat{X})S - (X - \widehat{X}) = 0. \quad (3.115)$$

As both S and $\widehat{S}$ are stable, Theorem 1.5.1 shows that the Stein equation (3.115) has the unique solution $X - \widehat{X} = 0$, i.e., $X = \widehat{X}$. ■

The next result gives the relation between the solutions to the DTARSs associated with equivalent Popov triplets.

Proposition 3.6.4 *Let Σ and $\widetilde{\Sigma}$ be two (X, F)-equivalent Popov triplets. Then (X_s, F_s) is a (stabilizing) solution to the DTARS(Σ) if and only if $(X_s - X, F_s - F)$ is a (stabilizing) solution to the DTARS($\widetilde{\Sigma}$).*

3.6.3 The discrete-time algebraic Riccati equation (DTARE)

Assume that in (3.108) $R + B^T XB$ is nonsingular. Then from (3.110) we get

$$F = -(R + B^T XB)^{-1}(B^T XA + L^T). \quad (3.116)$$

Substituting (3.116) in (3.109) one obtains for $X = X^T$,

$$A^T XA - X + Q - (L + A^T XB)(R + B^T XB)^{-1}(B^T XA + L^T) = 0 \qquad (3.117)$$

which is the well–known *discrete–time algebraic Riccati equation (DTARE)* associated with Σ, denoted DTARE(Σ).

Definition 3.6.5 *A symmetric solution X to (3.117) is called a* stabilizing (antistabilizing) solution *to DTARE(Σ) if $A + BF$ is stable (antistable) for F defined by (3.116). The matrix F in (3.116) is called a* stabilizing (antistabilizing) Riccati feedback *and will be occasionally denoted by F_{Ricc}.*

From Proposition 3.6.3 we have the following result.

Corollary 3.6.6 *If the DTARE(Σ) has a stabilizing solution then it is unique.*

Accordingly, we can speak about *the* stabilizing solution and *the* stabilizing Riccati feedback. An immediate consequence of Proposition 3.6.4 stated below relates the solutions to the DTAREs associated with two equivalent Popov triplets.

Corollary 3.6.7 *Let $\Sigma = (A, B; Q, L, R)$ and $\widetilde{\Sigma} = (\widetilde{A}, \widetilde{B}; \widetilde{Q}, \widetilde{L}, \widetilde{R})$ be two (X, F)–equivalent Popov triplets. Then X_s is the stabilizing solution to the DTARE(Σ) if and only if $X_s - X$ is the stabilizing solution to the DTARE($\widetilde{\Sigma}$). The stabilizing Riccati feedbacks are related by the formula*

$$\widetilde{F}_{Ricc} = F_{Ricc} - F. \qquad (3.118)$$

3.6.4 The Kalman–Szego–Popov–Yakubovich system (KSPYS)

Let $\Sigma = (A, B; Q, L, R)$ be a Popov triplet and J be an $m \times m$ sign matrix. The nonlinear system in the unknowns $X = X^T$, V and W, given by

$$\begin{aligned} R + B^T XB &= V^T JV, \\ L + A^T XB &= W^T JV, \\ Q + A^T XA - X &= W^T JW, \end{aligned} \qquad (3.119)$$

where $R + B^T XB$ is nonsingular, is called *the Kalman–Szego–Popov–Yakubovich system (KSPYS) in J form* associated with Σ and J, denoted KSPYS(Σ, J). Since $R + B^T XB$ is assumed invertible, it follows from the first equation in (3.119) that a necessary solvability condition for the KSPYS is the invertibility of J, i.e.,

$$J = \begin{bmatrix} -I_{m_1} & \\ & I_{m_2} \end{bmatrix}, \qquad (3.120)$$

for some m_1 and m_2, with $m_1 + m_2 = m$. Moreover, X from any solution (X, V, W) to the KSPYS(Σ, J) should satisfy $\mathrm{sgn}\,(R + B^T XB) = J$. Throughout the book we shall use KSPYSs defined for nonsingular sign matrices J.

Definition 3.6.8 *A triplet of matrices $(X = X^T, V, W)$ satisfying (3.119) such that $A + BF$ is stable, for*

$$F := -V^{-1}W, \qquad (3.121)$$

is called a stabilizing solution *to the KSPYS(Σ, J) and F is called a* stabilizing feedback.

Remark 3.6.9 The relation between the DTARE in (3.117) and the KSPYS in (3.119) associated with the same Popov triplet is the following. If $(X = X^T, V, W)$ satisfies the KSPYS(Σ, J), then X is a symmetric solution to the DTARE(Σ). This follows by eliminating V and W in (3.119). Conversely, if X is a symmetric solution to the DTARE(Σ), we can choose a V such that the first equation in (3.119) is satisfied, provided sgn $(R+B^T XB) = J$. Then W can be defined such that it satisfies the second equation in (3.119) and it follows that it satisfies also the third equation in (3.119). Thus the triplet $(X = X^T, V, W)$ is a solution to the KSPYS(Σ, J). It is also easy to check that the Riccati feedback in (3.116) coincides in this case with the feedback matrix given by (3.121). This construction establishes a relation between the solutions to the DTARE and the solutions to the KSPYS as follows. The KSPYS(Σ, J) has a (stabilizing) solution $(X = X^T, V, W)$ if and only if the DTARE(Σ) has a (stabilizing) solution X and sgn $(R + B^T XB) = J$.

As an easy consequence we have the following result.

Corollary 3.6.10 *If (X, V, W) and $(\widehat{X}, \widehat{V}, \widehat{W})$ are two stabilizing solutions to the KSPYS(Σ, J), then $X = \widehat{X}$.*

Remark 3.6.11 If the sign matrix J is either $J = I_m$ or $J = -I_m$, then (3.119) reduces either to

$$\begin{aligned} R + B^T XB &= V^T V, \\ L + A^T XB &= W^T V, \\ Q + A^T XA - X &= W^T W, \end{aligned} \tag{3.122}$$

or to

$$\begin{aligned} R + B^T XB &= -V^T V, \\ L + A^T XB &= -W^T V, \\ Q + A^T XA - X &= -W^T W, \end{aligned} \tag{3.123}$$

respectively. The KSPYSs (3.122) and (3.123) are usually known as the KSPYS in *positivity form* KSPYS(Σ, I) and the KSPYS in *negativity form* KSPYS$(\Sigma, -I)$, respectively.

The KSPYS in (3.119) can be used to write convenient representations for the quadratic indices associated with Σ as explained further.

Proposition 3.6.12 *Let Σ be a Popov triplet and let (X, V, W) be a solution to the KSPYS(Σ, J) in (3.119) (not necessarily a stabilizing one).*

1. *For each $u \in \mathcal{U}_\xi$ we have*

$$J_\Sigma(\xi, u) = \langle Vu + Wx^{\xi,u}, J(Vu + Wx^{\xi,u})\rangle_{\ell_+^{2,m}} + \xi^T X\xi. \tag{3.124}$$

2. *For each $u \in \widehat{\mathcal{U}}$ we have*

$$\widehat{J}_\Sigma(u) = \langle Vu + W\sigma\widehat{x}^u, J(Vu + W\sigma\widehat{x}^u)\rangle_{\ell_+^{2,m}} + (\widehat{x}_0^u)^T X\widehat{x}_0^u. \tag{3.125}$$

3. *If A is dichotomic, and $u \in \ell^{2,m}$, we have*

$$J_{e,\Sigma}(u) = \langle Vu + Wx_e^u, J(Vu + Wx_e^u)\rangle_{\ell^{2,m}}, \tag{3.126}$$

and

$$\widehat{J}_{e,\Sigma}(u) = \langle Vu + W\sigma\widehat{x}_e^u, J(Vu + W\sigma\widehat{x}_e^u)\rangle_{\ell^{2,m}}. \tag{3.127}$$

Proof. We use Proposition 3.6.1 for the $(X,0)$–equivalent of Σ. We have $\widetilde{u}=u$ and, using the KSPYS(Σ, J) in (3.119), we have

$$
\begin{aligned}
J_{\widetilde{\Sigma}}(\xi,u) &= \left\langle \begin{bmatrix} x^{\xi,u} \\ u \end{bmatrix}, \begin{bmatrix} \widetilde{Q} & \widetilde{L} \\ \widetilde{L}^T & R \end{bmatrix} \begin{bmatrix} x^{\xi,u} \\ u \end{bmatrix} \right\rangle_{\ell_+^{2,n}\times \ell_+^{2,m}} \\
&= \left\langle \begin{bmatrix} x^{\xi,u} \\ u \end{bmatrix}, \begin{bmatrix} A^TXA - X + Q & A^TXB + L \\ B^TXA + L^T & R + B^TXB \end{bmatrix} \begin{bmatrix} x^{\xi,u} \\ u \end{bmatrix} \right\rangle_{\ell_+^{2,n}\times \ell_+^{2,m}} \\
&= \left\langle \begin{bmatrix} x^{\xi,u} \\ u \end{bmatrix}, \begin{bmatrix} W^T \\ V^T \end{bmatrix} J \begin{bmatrix} W & V \end{bmatrix} \begin{bmatrix} x^{\xi,u} \\ u \end{bmatrix} \right\rangle_{\ell_+^{2,n}\times \ell_+^{2,m}} \\
&= \langle Vu + Wx^{\xi,u}, J(Vu + Wx^{\xi,u})\rangle_{\ell_+^{2,m}}
\end{aligned}
$$

which establishes relation (3.124). The other relations are obtained similarly. ■

The next result gives the relation between the solutions to the KSPYSs associated with equivalent Popov triplets.

Proposition 3.6.13 *Let $\Sigma = (A,B;Q,L,R)$ and $\widetilde{\Sigma} = (\widetilde{A},\widetilde{B};\widetilde{Q},\widetilde{L},\widetilde{R})$ be two (X,F)–equivalent Popov triplets. Then (X_s,V_s,W_s) is a (stabilizing) solution to the KSPYS(Σ,J) if and only if $(X_s - X, V_s, W_s + V_sF)$ is a (stabilizing) solution to the KSPYS$(\widetilde{\Sigma},J)$.*

3.6.5 *The discrete–time Hamiltonian system (DTHS)*

Let $\Sigma = (A,B;Q,L,R)$ be a Popov triplet. The linear system

$$
\begin{aligned}
\sigma x &= Ax & &+ Bu, \\
\lambda &= Qx &+ A^T\sigma\lambda &+ Lu, \\
v &= L^Tx &+ B^T\sigma\lambda &+ Ru,
\end{aligned}
\tag{3.128}
$$

is called the *direct–time (discrete–time) Hamiltonian system* (DTHS) associated with Σ, denoted DTHS(Σ).

We shall be interested in those triplets (u,x,λ) that annihilate the output v of the DTHS. We have the following result.

Proposition 3.6.14 *Let Σ and $\widetilde{\Sigma}$ be two (X,F)–equivalent Popov triplets. Then (u,x,λ) annihilates the output of the DTHS(Σ) on a certain (discrete–) time interval if and only if $(u - Fx, x, \lambda - Xx)$ annihilates the output of the DTHS$(\widetilde{\Sigma})$ on the same time interval.*

The system

$$
\begin{aligned}
x &= A\sigma x & &+ Bu, \\
\sigma\lambda &= Q\sigma x &+ A^T\lambda &+ Lu, \\
v &= L^T\sigma x &+ B^T\lambda &+ Ru,
\end{aligned}
\tag{3.129}
$$

is called the *inverse–time (discrete–time) Hamiltonian system* (ITHS) associated with Σ, denoted ITHS(Σ).

We give now the correspondent of Proposition 3.6.14.

Proposition 3.6.15 *Let Σ and $\widetilde{\Sigma}$ be two (X,F)–equivalent Popov triplets. Then (u,x,λ) annihilates the output of the ITHS(Σ) on a certain (discrete–) time interval if and only if $(u-F\sigma x, x, \lambda - Xx)$ annihilates the output of the ITHS($\widetilde{\Sigma}$) on the same time interval.*

Proof. It follows immediately from Proposition 3.6.14 by noticing that the two Hamiltonian systems can be related to one another through a variable transformation like (2.108), more precisely,

$$\widehat{x} = \Omega x\ ,\quad \widehat{\lambda} = \Omega\lambda\ ,\quad \widehat{u} = \sigma\Omega u\ ,\quad \widehat{v} = \sigma\Omega v. \tag{3.130}$$

■

3.6.6 *The extended symplectic pencils (ESPs)*

Let $\Sigma = (A, B; Q, L, R)$ be a Popov triplet. The matrix pencil $zM_\Sigma - N_\Sigma$, with

$$M_\Sigma := \begin{bmatrix} I & O & O \\ O & -A^T & O \\ O & -B^T & O \end{bmatrix},\quad N_\Sigma := \begin{bmatrix} A & O & B \\ Q & -I & L \\ L^T & O & R \end{bmatrix}, \tag{3.131}$$

is called the *(direct–time) extended symplectic pencil (ESP)* associated with Σ, denoted ESP(Σ).

If we take $v = 0$ in (3.128) and we denote

$$p := \begin{bmatrix} x \\ \lambda \\ u \end{bmatrix},$$

then we can rewrite the DTHS (3.128) in the descriptor form

$$M_\Sigma \sigma p = N_\Sigma p.$$

The matrix pencil $z\widehat{M}_\Sigma - \widehat{N}_\Sigma$, with

$$\widehat{M}_\Sigma := \begin{bmatrix} -A & O & O \\ -Q & I & O \\ -L^T & O & O \end{bmatrix},\quad \widehat{N}_\Sigma := \begin{bmatrix} -I & O & B \\ O & A^T & L \\ O & B^T & R \end{bmatrix}, \tag{3.132}$$

is called the *inverse–time extended symplectic pencil* associated with Σ.

The ITHS (3.129) can be written in the descriptor form

$$\widehat{M}_\Sigma \sigma p = \widehat{N}_\Sigma p.$$

The relation between the direct–time and the inverse–time symplectic pencils is

$$z\widehat{M}_\Sigma - \widehat{N}_\Sigma = (M_\Sigma - zN_\Sigma)\begin{bmatrix} I & O & O \\ O & I & O \\ O & O & z^{-1}I \end{bmatrix}.$$

Proposition 3.6.16 *Let $\Sigma = (A, B; Q, L, R)$ and $\widetilde{\Sigma} = (\widetilde{A}, \widetilde{B}; \widetilde{Q}, \widetilde{L}, \widetilde{R})$ be two (X, F)–equivalent Popov triplets. Then the ESP(Σ) and the ESP($\widetilde{\Sigma}$) are related by*

$$zM_\Sigma - N_\Sigma = U(zM_{\widetilde{\Sigma}} - N_{\widetilde{\Sigma}})V \tag{3.133}$$

where

$$U := \begin{bmatrix} I & O & O \\ -\widetilde{A}^T X & I & -F^T \\ -B^T X & O & I \end{bmatrix}, \quad V := \begin{bmatrix} I & O & O \\ -X & I & O \\ -F & O & I \end{bmatrix}. \tag{3.134}$$

3.6.7 *The discrete-time Popov functions*

Let $\Sigma = (A, B; Q, L, R)$ be a Popov triplet. For $z \in \mathbb{C} \setminus \{\Lambda(A) \cup 1/\Lambda(A)\}$ we define the rational matrix function

$$\mathbf{\Pi}_\Sigma(z) := [B^T(z^{-1}I - A^T)^{-1} \;\; I] \begin{bmatrix} Q & L \\ L^T & R \end{bmatrix} \begin{bmatrix} (zI - A)^{-1}B \\ I \end{bmatrix} \tag{3.135}$$

that is called the *discrete direct–time Popov function* associated with Σ. For $z \in \mathbb{C} \setminus \{\Lambda(A) \cup 1/\Lambda(A)\}$, we define the rational matrix function

$$\widehat{\mathbf{\Pi}}_\Sigma(z) := [B^T(zI - A^T)^{-1} \;\; I] \begin{bmatrix} Q & L \\ L^T & R \end{bmatrix} \begin{bmatrix} (z^{-1}I - A)^{-1}B \\ I \end{bmatrix} \tag{3.136}$$

that is called the *discrete inverse–time Popov function* associated with Σ. Notice that

$$\widehat{\mathbf{\Pi}}_\Sigma(z) = \mathbf{\Pi}_\Sigma(\frac{1}{z}) = \mathbf{\Pi}_\Sigma^T(z). \tag{3.137}$$

It can be easily checked that a descriptor realization of $\mathbf{\Pi}_\Sigma(z)$ is given by

$$\mathbf{\Pi}_\Sigma = \left[\begin{array}{cc|c} zI - A & O & -B \\ -Q & I - zA^T & -L \\ \hline L^T & B^T z & R \end{array}\right] \tag{3.138}$$

and a descriptor realization of $\widehat{\mathbf{\Pi}}_\Sigma(z)$ is given by

$$\widehat{\mathbf{\Pi}}_\Sigma = \left[\begin{array}{cc|c} I - zA & O & -B \\ -Qz & zI - A^T & -L \\ \hline L^T z & B^T & R \end{array}\right]. \tag{3.139}$$

As in the continuous case, the direct–time Popov function is the TFM from u to v of the DTHS(Σ) given in (3.128), whereas the inverse–time Popov function is the TFM from u to v of the ITHS(Σ) given in (3.129). Moreover, the ESP(Σ) is exactly the transmission pencil associated with the realization (3.138) of the discrete direct–time Popov function, and the inverse–time ESP(Σ) is exactly the transmission pencil associated with the realization (3.139) of the discrete inverse–time Popov function.

In view of relation (3.137) it is enough to look at the effect of the equivalence transformation on the direct–time Popov function.

Proposition 3.6.17 *Let $\Sigma = (A, B; Q, L, R)$ and $\widetilde{\Sigma} = (\widetilde{A}, \widetilde{B}; \widetilde{Q}, \widetilde{L}, \widetilde{R})$ be two (X, F)-equivalent Popov triplets. Then for $z \in \mathbb{C} \setminus \{\Lambda(A) \cup 1/\Lambda(A) \cup \Lambda(\widetilde{A}) \cup 1/\Lambda(\widetilde{A})\}$ we have*

$$\mathbf{\Pi}_\Sigma(z) = \mathbf{S}_F^*(z)\mathbf{\Pi}_{\widetilde{\Sigma}}(z)\mathbf{S}_F(z) \tag{3.140}$$

where

$$\mathbf{S}_F := \left[\begin{array}{c|c} A & B \\ \hline -F & I \end{array}\right]. \tag{3.141}$$

Moreover, if $(X_s = X_s^T, F_s)$ is a solution to the DTARS(Σ) in (3.108), then

$$\mathbf{\Pi}_\Sigma(z) = \mathbf{S}_{F_s}^*(z)(R + B^T X_s B)\mathbf{S}_{F_s}(z), \tag{3.142}$$

where $\mathbf{S}_{F_s}$ is (3.141) written for $F = F_s$. (The identity (3.142) is known as the spectral factorization identity *and (3.141) is called a* spectral factor*).*

3.7 The time–reversed Popov triplet

Let $\Sigma = (A, B; Q, L, R)$ be a Popov triplet with $A \in \mathbb{R}^{n\times n}$, $B \in \mathbb{R}^{n\times m}$. In the case in which A is invertible, the inverse–time controlled recursion (3.96) can be written in the direct–time form

$$\sigma x = A^{-1}x - A^{-1}Bu = \widehat{A}x + \widehat{B}u. \tag{3.143}$$

We may substitute σx as given by (3.143) into the expression of the inverse–time quadratic index

$$\widehat{J} = \left\langle \begin{bmatrix} \sigma x \\ u \end{bmatrix}, \begin{bmatrix} Q & L \\ L^T & R \end{bmatrix} \begin{bmatrix} \sigma x \\ u \end{bmatrix} \right\rangle,$$

and we rewrite the index in terms of x and u. We are led to consider a new Popov triplet $\widehat{\Sigma}$ that we shall call *the time-reversed Popov triplet* associated with Σ and that is given by $\widehat{\Sigma} = (\widehat{A}, \widehat{B}; \widehat{Q}, \widehat{L}, \widehat{R})$, where

$$\begin{aligned} \widehat{A} &:= A^{-1}, \\ \widehat{B} &:= -A^{-1}B, \\ \widehat{Q} &:= A^{-T}QA^{-1}, \\ \widehat{L} &:= A^{-T}L - A^{-T}QA^{-1}B, \\ \widehat{R} &:= R - L^TA^{-1}B - B^TA^{-T}L + B^TA^{-T}QA^{-1}B. \end{aligned} \tag{3.144}$$

We shall refer to the direct–time mathematical objects associated with $\widehat{\Sigma}$ as the *time-reversed* objects associated with Σ. For example, the direct–time quadratic index $J_{\widehat{\Sigma}}$ associated with $\widehat{\Sigma}$ will be referred as the time–reversed quadratic index associated with Σ. The transformation from Σ to $\widehat{\Sigma}$ is involutive, that is, the time–reversed form of $\widehat{\Sigma}$ is Σ. A Popov triplet with A invertible, and that admits therefore a time–reversed correspondent, will be called *reversible.*

The rest of this section is devoted to the study of the relations between inverse–time objects and time–reversed objects associated with Σ.

Proposition 3.7.1 *Let $\Sigma = (A, B; Q, L, R)$ be a Popov triplet with A dichotomic and invertible, and let $\widehat{\Sigma}$ be the time–reversed Popov triplet of Σ. Then the inverse–time* *quadratic index (3.103) associated with* Σ *equals the restricted quadratic index (3.106) associated with* $\widehat{\Sigma}$,

$$\widehat{J}_{\Sigma}(u) = J_{\widehat{\Sigma}}(u),$$

for each $u \in \widehat{\mathcal{U}}$, *where* $\widehat{\mathcal{U}}$ *is defined in (2.177).*

Proof. Assume that u is in $\ell_+^{2,m}$ such that $\widehat{x}^u$ is in $\ell_+^{2,n}$. It is clear that $\widehat{x}^u$ satisfies (3.143), that is, it coincides with the solution x^u for this direct–time equation. In particular, the two indices are defined for the same input functions. Their equality is a matter of simple substitutions. ■

It is also easy to prove that the direct–time Popov function associated with $\widehat{\Sigma}$ coincides with the inverse–time Popov function associated with Σ, i.e.,

$$\mathbf{\Pi}_{\widehat{\Sigma}} = \widehat{\mathbf{\Pi}}_{\Sigma}.$$

Other relations are less obvious as we shall see immediately. The reason for this is essentially related to the way in which an inverse–time feedback transformation $u = F\sigma x + v$ on the inverse–time recursion (3.96) reflects in the time–reversed form (3.143). Indeed, if we substitute σx from (3.143) into the feedback transformation, we obtain

$$u = FA^{-1}x - FA^{-1}Bu + v.$$

If $I + FA^{-1}B$ is invertible, this can be written as a direct–time feedback transformation

$$u = (I + FA^{-1}B)^{-1}FA^{-1}x + v = \widehat{F}x + v,$$

(for a new v). But $I + FA^{-1}B$ is invertible if and only if $A + BF$ is invertible. We call a feedback transformation that satisfies this condition *admissible*. Such a feedback transformation preserves the reversibility of recursions. Moreover,

$$\widehat{A} + \widehat{B}\widehat{F} = (A + BF)^{-1}, \tag{3.145}$$

which means that the closed–loop recursions remain related through a time–reversal transformation. Notice also that

$$\widehat{F} = (I + FA^{-1}B)^{-1}FA^{-1} = F(A + BF)^{-1}. \tag{3.146}$$

Also, from (3.145), we have

$$F = \widehat{F}(\widehat{A} + \widehat{B}\widehat{F})^{-1}. \tag{3.147}$$

The restriction to admissible feedback transformations and the non–trivial relation between F and $\widehat{F}$ explain why the relation between a Popov triplet and its time–reversed correspondent is not always transparent. As an example, we have the next proposition.

Proposition 3.7.2 *Let* $\Sigma = (A, B; Q, L, R)$ *be a reversible Popov triplet with* $A \in \mathbb{R}^{n\times n}$, $B \in \mathbb{R}^{n\times m}$.

1. *Let* X *be a symmetric solution to the DTARE(*Σ*). Then* $-X$ *is a solution to the time-reversed DTARE(*Σ*) if and only if* $R - L^TA^{-1}B$ *is nonsingular.*

2. *Let* $\widehat{X}$ *be a symmetric solution to the time-reversed DTARE(*Σ*). Then* $-\widehat{X}$ *is a solution to the DTARE(*Σ*) if and only if* $R - L^TA^{-1}B$ *is nonsingular.*

3. *Assume that* X *is the stabilizing solution to the DTARE(*Σ*). Then the stabilizing Riccati feedback is admissible, if and only if* $R - L^TA^{-1}B$ *is nonsingular.*

Proof. We make first some preparations. Let $D_\Sigma(X)$ and $D_{\widehat{\Sigma}}(\widehat{X})$ be the dissipation matrices associated with Σ and $\widehat{\Sigma}$, respectively (see (3.107)). Here $\widehat{\Sigma}$ stands for the time–reversed version of Σ given in (3.144). Introduce the following notation:

$$\begin{array}{rcl} G(X) & := & R + B^T X B, \\ H(X) & := & L + A^T X B, \\ K(X) & := & A^T X A - X + Q, \\ L(X) & := & G(X) - H^T(X) A^{-1} B, \end{array} \tag{3.148}$$

and let $\widehat{G}(\widehat{X}), \widehat{H}(\widehat{X}), \widehat{K}(\widehat{X})$, and $\widehat{L}(\widehat{X})$ be their time–reversed correspondents. Thus

$$D_\Sigma(X) = \begin{bmatrix} K(X) & H(X) \\ H^T(X) & G(X) \end{bmatrix} \tag{3.149}$$

and

$$D_{\widehat{\Sigma}}(\widehat{X}) = \begin{bmatrix} \widehat{K}(\widehat{X}) & \widehat{H}(\widehat{X}) \\ \widehat{H}^T(\widehat{X}) & \widehat{G}(\widehat{X}) \end{bmatrix}. \tag{3.150}$$

By using (3.144) it is now easy to check that between the dissipation matrices (3.149), (3.150), the following equality holds:

$$D_{\widehat{\Sigma}}(-X) = \begin{bmatrix} A^{-T} & O \\ -B^T A^{-T} & I \end{bmatrix} D_\Sigma(X) \begin{bmatrix} A^{-1} & -A^{-1}B \\ O & I \end{bmatrix}. \tag{3.151}$$

1. Assume now that X is a symmetric solution to the DTARE(Σ) in (3.117). Then $G(X)$ is nonsingular and its Schur complement in $D_\Sigma(X)$ vanishes because $K(X) - H(X)G^{-1}(X)H(X)^T = 0$ is exactly the DTARE(Σ). Thus we may write from (3.151) with (3.149) and (3.150), that

$$\begin{aligned} \begin{bmatrix} \widehat{K}(-X) & \widehat{H}(-X) \\ \widehat{H}^T(-X) & \widehat{G}(-X) \end{bmatrix} &= \begin{bmatrix} A^{-T} & O \\ -B^T A^{-T} & I \end{bmatrix} \begin{bmatrix} I & H(X)G^{-1}(X) \\ O & I \end{bmatrix} \begin{bmatrix} O & O \\ O & G(X) \end{bmatrix} \\ &\quad \times \begin{bmatrix} I & O \\ G^{-1}(X)H^T(X) & I \end{bmatrix} \begin{bmatrix} A^{-1} & -A^{-1}B \\ O & I \end{bmatrix} \end{aligned}$$

$$= \begin{bmatrix} A^{-T}H(X)G^{-1}(X)H^T(X)A^{-1} & A^{-T}H(X)G^{-1}(X)L(X) \\ L^T(X)G^{-1}(X)H^T(X)A^{-1} & L^T(X)G^{-1}(X)L(X) \end{bmatrix}. \tag{3.152}$$

But from (3.148) we get

$$L(X) = G(X) - H^T(X)A^{-1}B = R - L^T A^{-1} B. \tag{3.153}$$

If $R - L^T A^{-1} B$ is nonsingular then (3.152) shows, via (3.153), that $\widehat{G}(-X)$ is also nonsingular. On the other hand, in the light of the above considerations regarding the Schur complement of $G(X)$ in $D_\Sigma(X)$, X is a solution to the DTARE(Σ) if and only if

$$\operatorname{rank} D_\Sigma(X) = \operatorname{rank} G(X) = m. \tag{3.154}$$

According to (3.151) combined with the invertibility of $\widehat{G}(-X)$, just proved above, it follows that

$$\operatorname{rank} D_{\widehat{\Sigma}}(-X) = \operatorname{rank} \widehat{G}(-X) = m. \tag{3.155}$$

Hence (3.155) shows that $-X$ is a solution to the time–reversed DTARE(Σ).

Conversely, if $-X$ is a solution to the time–reversed DTARE(Σ) then $\widehat{G}(-X)$ is nonsingular and we deduce from (3.151) that $L(X) = G(X) - H^T(X)A^{-1}B$ must be nonsingular as well. From here, it follows via (3.153) that $R - L^T A^{-1} B$ is nonsingular.

2. The proof runs similarly to 1.

3. The stabilizing Riccati feedback F_{Ricc} is admissible if and only if $A + BF_{Ricc}$ is invertible or, equivalently, if and only if $I + F_{Ricc}A^{-1}B$ is invertible. We compute successively

$$\begin{aligned} I + F_{Ricc}A^{-1}B &= I - (R + B^T XB)^{-1}(L^T + B^T XA)A^{-1}B \\ &= (R + B^T XB)^{-1}(R - L^T A^{-1}B), \end{aligned}$$

which demonstrates the assertion. ■

Remark 3.7.3 Let $\Sigma = (A, B; Q, L, R)$ be a reversible Popov triplet. Then the invertibility of $R - L^T A^{-1}B$ is equivalent to the invertibility of the matrix N_Σ of the ESP(Σ) in (3.131) because of

$$\begin{aligned} \operatorname{rank} N_\Sigma &= n + \operatorname{rank} \begin{bmatrix} A & B \\ L^T & R \end{bmatrix} = n + \operatorname{rank} \begin{bmatrix} A & B \\ O & R - L^T A^{-1}B \end{bmatrix} \\ &= 2n + \operatorname{rank}(R - L^T A^{-1}B). \end{aligned}$$

In turn, the invertibility of N_Σ is equivalent to the regularity of the ESP(Σ) together with the absence of generalized eigenvalues in zero for the ESP(Σ).

3.8 Associated operators

Throughout this section we fix a Popov triplet $\Sigma = (A, B; Q, L, R)$, with $A \in \mathbb{R}^{n\times n}$, $B \in \mathbb{R}^{n\times m}$. In Section 2.6 we have written various solutions of the direct–time controlled recursion

$$\sigma x = Ax + Bu \tag{3.156}$$

and of the inverse–time controlled recursion

$$x = A\sigma x + Bu \tag{3.157}$$

in operator form (as, for example, in (2.147), or in (2.160)). By substituting these expressions into the quadratic indices associated with Σ, we shall obtain in this section operatorial expressions for the quadratic indices. The operators that will turn up in these expressions will play a major role in the theory presented in Chapter 4 and further in this book.

3.8.1 The direct–time case: A dichotomic

Assume that A is dichotomic and $u \in \ell^{2,m}$. Substituting the $\ell^{2,n}$ solution $x_e^u = \mathcal{L}_e u$ of (3.156) (see Section 2.6.4) in the expression of the extended quadratic index (3.104), we have

$$
\begin{aligned}
J_{e,\Sigma}(u) &= \left\langle \begin{bmatrix} x_e^u \\ u \end{bmatrix}, \begin{bmatrix} Q & L \\ L^T & R \end{bmatrix} \begin{bmatrix} x_e^u \\ u \end{bmatrix} \right\rangle \\
&= \left\langle u, \begin{bmatrix} \mathcal{L}_e^* & I \end{bmatrix} \begin{bmatrix} Q & L \\ L^T & R \end{bmatrix} \begin{bmatrix} \mathcal{L}_e \\ I \end{bmatrix} u \right\rangle \\
&= \langle u, \mathcal{R}_{e,\Sigma} u \rangle , \qquad (3.158)
\end{aligned}
$$

where

$$\mathcal{R}_{e,\Sigma} := R + L^T \mathcal{L}_e + \mathcal{L}_e^* L + \mathcal{L}_e^* Q \mathcal{L}_e = \mathcal{R}_{e,\Sigma}^* \qquad (3.159)$$

is a linear bounded operator from $\ell^{2,m}$ into $\ell^{2,m}$. The following result relates the operators $\mathcal{R}_e$ associated with two equivalent Popov triplets.

Proposition 3.8.1 *Let $\Sigma = (A, B; Q, L, R)$ and $\widetilde{\Sigma} = (\widetilde{A}, \widetilde{B}; \widetilde{Q}, \widetilde{L}, \widetilde{R})$ be two (X, F)–equivalent Popov triplets. If A and $\widetilde{A} = A + BF$ are dichotomic, then*

$$\mathcal{R}_{e,\Sigma} = \mathcal{N}_F^* \mathcal{R}_{e,\widetilde{\Sigma}} \mathcal{N}_F, \qquad (3.160)$$

where $\mathcal{N}_F$ is the input–output operator of the system

$$
\begin{aligned}
\sigma x &= Ax + Bu, \\
v &= -Fx + u.
\end{aligned} \qquad (3.161)
$$

In particular, if $\widetilde{\Sigma}$ is $(X, 0)$–equivalent to Σ, then $\mathcal{R}_{e,\Sigma} = \mathcal{R}_{e,\widetilde{\Sigma}}$.

The proof is identical to the one for the continuous counterpart and it is omitted.

3.8.2 The direct–time case: A stable

Assume now that A is *stable.* Then the quadratic index $J_\Sigma(\xi, u)$ is defined for each $u \in \ell_+^{2,m}$ as in (3.102). Substituting in (3.102) the solution $x^{\xi,u} = \Phi\xi + \mathcal{L}u$ to (3.156) for $x_0 = \xi$ and $u \in \ell_+^{2,m}$ (see Section 2.6.1) we can write, just as in the continuous–time case,

$$J_\Sigma(\xi, u) = \left\langle \begin{bmatrix} x^{\xi,u} \\ u \end{bmatrix}, \begin{bmatrix} Q & L \\ L^T & R \end{bmatrix} \begin{bmatrix} x^{\xi,u} \\ u \end{bmatrix} \right\rangle = \left\langle \begin{bmatrix} \xi \\ u \end{bmatrix}, \begin{bmatrix} \mathcal{P}_{o,\Sigma} & \mathcal{P}_\Sigma \\ \mathcal{P}_\Sigma^* & \mathcal{R}_\Sigma \end{bmatrix} \begin{bmatrix} \xi \\ u \end{bmatrix} \right\rangle, \qquad (3.162)$$

where

$$\mathcal{P}_{o,\Sigma} : \mathbb{R}^n \to \mathbb{R}^n, \quad \mathcal{P}_{o,\Sigma} := \Phi^* Q \Phi, \qquad (3.163)$$

$$\mathcal{P}_\Sigma : \ell_+^{2,m} \to \mathbb{R}^n, \quad \mathcal{P}_\Sigma := \Phi^*(Q\mathcal{L} + L), \qquad (3.164)$$

$$\mathcal{R}_\Sigma : \ell_+^{2,m} \to \ell_+^{2,m}, \quad \mathcal{R}_\Sigma := R + L^T \mathcal{L} + \mathcal{L}^* L + \mathcal{L}^* Q \mathcal{L} = \mathcal{R}_\Sigma^* \qquad (3.165)$$

are all linear bounded operators.

The adjoint of Φ is given by

$$\Phi^* x = \sum_{k=0}^{\infty} (A^T)^k x_k, \quad x = (x_k)_{k \geq 0} \in \ell_+^{2,n}, \qquad (3.166)$$

as can be derived from formula (2.201), if we notice that Φ is the causal observability operator in the particular case that $C = I_n$.

Writing (3.163) explicitly, we have

$$\mathcal{P}_{o,\Sigma} = \sum_{k=0}^{\infty} (A^T)^k Q A^k. \tag{3.167}$$

From the last expression, it is easy to check that $\mathcal{P}_{o,\Sigma}$ is the unique solution to the reducing Stein equation of Σ,

$$A^T X A - X + Q = 0. \tag{3.168}$$

This means that $\mathcal{P}_{o,\Sigma} = X_r$.

The adjoint of $\mathcal{L}$ turns out to be

$$(\mathcal{L}^* x)_k = \sum_{i=k+1}^{\infty} B^T (A^T)^{i-k-1} x_i, \quad k \geq 0, \quad x = (x_i)_{i \geq 0} \in \ell_+^{2,n}. \tag{3.169}$$

If A is stable the operator $\mathcal{L}_e$ is causal as follows from Theorem 2.8.3, taking $C = I_n$, $D = 0$. Taking into account that $\mathcal{L}$ is the Toeplitz operator corresponding to $\mathcal{L}_e$, and using the discrete version of Corollary 2.4.13 in relation (3.159), we deduce that

$$\mathcal{R}_\Sigma = \mathbb{T}_{\mathcal{R}_{e,\Sigma}}, \tag{3.170}$$

that is the operator $\mathcal{R}_\Sigma$ is a Toeplitz operator.

Proposition 3.8.2 *Let $\Sigma = (A, B; Q, L, R)$ and $\widetilde{\Sigma} = (\widetilde{A}, \widetilde{B}; \widetilde{Q}, \widetilde{L}, \widetilde{R})$ be two (X, F)–equivalent Popov triplets. If A and $\widetilde{A} = A + BF$ are stable, then*

$$\mathcal{P}_{o,\Sigma} = X + \mathcal{P}_{o,\widetilde{\Sigma}} - \mathcal{P}_{\widetilde{\Sigma}} F \Phi - \Phi^* F^T \mathcal{P}^*_{\widetilde{\Sigma}} + \Phi^* F^T \mathcal{R}_{\widetilde{\Sigma}} F \Phi \ , \tag{3.171}$$

$$\mathcal{P}_\Sigma = (\mathcal{P}_{\widetilde{\Sigma}} - \Phi^* F^T \mathcal{R}_{\widetilde{\Sigma}}) \mathbb{T}_{\mathcal{N}_F} \ , \tag{3.172}$$

$$\mathcal{R}_\Sigma = \mathbb{T}^*_{\mathcal{N}_F} \mathcal{R}_{\widetilde{\Sigma}} \mathbb{T}_{\mathcal{N}_F}, \tag{3.173}$$

where $\mathcal{N}_F$ is the input–output operator of the system (3.161).

In particular, if $\widetilde{\Sigma}$ is $(X, 0)$–equivalent with Σ, then $\mathcal{P}_{o,\Sigma} = X + \mathcal{P}_{o,\widetilde{\Sigma}}$, $\mathcal{P}_\Sigma = \mathcal{P}_{\widetilde{\Sigma}}$, $\mathcal{R}_\Sigma = \mathcal{R}_{\widetilde{\Sigma}}$.

The proof is identical to that of its continuous–time counterpart Proposition 3.3.2.

We consider now the inverse–time interpretation of Σ.

3.8.3 The inverse–time case: A dichotomic

Assume again that A is *dichotomic*. Introducing the solution $\sigma \widehat{x_e^u} = \widehat{\mathcal{L}}_e u$ to (3.157) (see (2.170)) in the expression of the extended quadratic index (3.105), we have for each $u \in \ell^{2,m}$,

$$\begin{aligned} \widehat{J}_{e,\Sigma}(u) &= \left\langle \begin{bmatrix} \sigma \widehat{x}_e^u \\ u \end{bmatrix}, \begin{bmatrix} Q & L \\ L^T & R \end{bmatrix} \begin{bmatrix} \sigma \widehat{x}_e^u \\ u \end{bmatrix} \right\rangle \\ &= \left\langle u, \begin{bmatrix} \widehat{\mathcal{L}}_e^* & I \end{bmatrix} \begin{bmatrix} Q & L \\ L^T & R \end{bmatrix} \begin{bmatrix} \widehat{\mathcal{L}}_e \\ I \end{bmatrix} u \right\rangle = \left\langle u, \widehat{\mathcal{R}}_{e,\Sigma} u \right\rangle, \end{aligned} \tag{3.174}$$

where

$$\widehat{\mathcal{R}}_{e,\Sigma} := R + L^T \widehat{\mathcal{L}}_e + \widehat{\mathcal{L}}_e^* L + \widehat{\mathcal{L}}_e^* Q \widehat{\mathcal{L}}_e = \mathcal{R}_{e,\Sigma}^* \tag{3.175}$$

is a linear bounded operator from $\ell^{2,m}$ into $\ell^{2,m}$.

Proposition 3.8.3 *Let $\Sigma = (A, B; Q, L, R)$ and $\widetilde{\Sigma} = (\widetilde{A}, \widetilde{B}; \widetilde{Q}, \widetilde{L}, \widetilde{R})$ be two (X, F)–equivalent Popov triplets. If $\widetilde{A} = A + BF$ is dichotomic, then*

$$\widehat{\mathcal{R}}_{e,\Sigma} = \widehat{\mathcal{N}}_F^* \widehat{\mathcal{R}}_{e,\widetilde{\Sigma}} \widehat{\mathcal{N}}_F, \tag{3.176}$$

where $\widehat{\mathcal{N}}_F$ is the input–output operator of the system

$$\begin{aligned} x &= A\sigma x + Bu, \\ v &= -F\sigma x + u. \end{aligned} \tag{3.177}$$

In particular, if $\widetilde{\Sigma}$ is $(X, 0)$–equivalent to Σ, then $\widehat{\mathcal{R}}_{e,\Sigma} = \widehat{\mathcal{R}}_{e,\widetilde{\Sigma}}$.

3.8.4 The inverse–time case: A stable

Assume now that A is stable. Then the inverse–time quadratic index $\widehat{J}_\Sigma(u)$ is defined for each $u \in \ell_+^{2,m}$ as in (3.103). Substituting in this relation $\sigma \widehat{x}^u = \widehat{\mathcal{L}} u$ from (2.160), we have, for each $u \in \ell_+^{2,m}$,

$$\widehat{J}_\Sigma(u) = \left\langle u, \widehat{\mathcal{R}}_\Sigma u \right\rangle, \tag{3.178}$$

where $\widehat{\mathcal{R}}_\Sigma : \ell_+^{2,m} \to \ell_+^{2,m}$ is the linear bounded operator

$$\widehat{\mathcal{R}}_\Sigma := R + L^T \widehat{\mathcal{L}} + \widehat{\mathcal{L}}^* L + \widehat{\mathcal{L}}^* Q \widehat{\mathcal{L}} = \widehat{\mathcal{R}}_\Sigma^*. \tag{3.179}$$

Although $\widehat{\mathcal{L}}$ is the Toeplitz operator corresponding to $\widehat{\mathcal{L}}_e$, the latter is not causal and thus $\widehat{\mathcal{R}}_\Sigma$ is not, in general, a Toeplitz operator. The Toeplitz operator of $\widehat{\mathcal{R}}_{e,\Sigma}$ can still be computed from (3.175) using the discrete–time version of Proposition 2.4.12. We obtain, making use also of the discrete–time version of Proposition 2.4.11, that

$$\mathbb{T}_{\widehat{\mathcal{R}}_{e,\Sigma}} = \widehat{\mathcal{R}}_\Sigma + \widehat{\mathbb{H}}_{\widehat{\mathcal{L}}_e^*} Q \widehat{\mathbb{H}}_{\widehat{\mathcal{L}}_e},$$

that is,

$$\widehat{\mathcal{R}}_\Sigma = \mathbb{T}_{\widehat{\mathcal{R}}_{e,\Sigma}} - \widehat{\mathbb{H}}_{\widehat{\mathcal{L}}_e}^* Q \widehat{\mathbb{H}}_{\widehat{\mathcal{L}}_e}. \tag{3.180}$$

This is a Toeplitz–Hankel representation of $\widehat{\mathcal{R}}_\Sigma$. An important consequence of this relation is that $\widehat{\mathcal{R}}_\Sigma$ is a Toeplitz operator if $Q = 0$. In particular, if Σ_r is the reduced equivalent of Σ, then $\widehat{\mathcal{R}}_{\Sigma_r}$ is a Toeplitz operator or, more precisely,

$$\widehat{\mathcal{R}}_{\Sigma_r} = \mathbb{T}_{\widehat{\mathcal{R}}_{e,\Sigma_r}} = \mathbb{T}_{\widehat{\mathcal{R}}_{e,\Sigma}}. \tag{3.181}$$

Applying (2.197) for the particular case $C = I_n$, $D = 0$, we have $\widehat{\mathbb{H}}_{\mathcal{L}_e} = \widehat{\Phi}\widehat{\Psi}$, where $\widehat{\Psi}$ is the anticausal inverse–time controllability operator (2.198) associated with the pair (A, B), and $\widehat{\Phi} : \mathbb{R}^n \to \ell_-^{2,n}$ is the anticausal inverse–time observability operator

(2.199) associated with the pair (I_n, A), i.e., $(\widehat{\Phi}\xi)_k = A^{-k-1}\xi$, $k \leq -1$, $\xi \in \mathbb{R}^n$. Substituting in (3.180), we have

$$\widehat{\mathcal{R}}_\Sigma = \mathbb{T}_{\widehat{\mathcal{R}}_{e,\Sigma}} - \widehat{\Psi}^* \widehat{\Phi}^* Q \widehat{\Phi} \widehat{\Psi}. \tag{3.182}$$

However, the operator

$$\widehat{\Phi}^* Q \widehat{\Phi} = \sum_{k=-\infty}^{-1} (A^T)^{-k-1} Q A^{-k-1} = \sum_{i=0}^{\infty} (A^T)^i Q A^i \tag{3.183}$$

is equal to the unique solution X_r of the reducing Stein equation of Σ given in (3.168). The operator $\widehat{\mathcal{R}}_\Sigma$ can again be rewritten as

$$\widehat{\mathcal{R}}_\Sigma = \mathbb{T}_{\widehat{\mathcal{R}}_{e,\Sigma}} - \widehat{\Psi}^* X_r \widehat{\Psi}. \tag{3.184}$$

Proposition 3.8.4 *Let $\Sigma = (A, B; Q, L, R)$ and $\widetilde{\Sigma} = (\widetilde{A}, \widetilde{B}; \widetilde{Q}, \widetilde{L}, \widetilde{R})$ be two (X, F)–equivalent Popov triplets. If $\widetilde{A} = A + BF$ is stable, then*

$$\widehat{\mathcal{R}}_\Sigma = \mathbb{T}^*_{\widehat{\mathcal{N}}_F} \widehat{\mathcal{R}}_{\widetilde{\Sigma}} \mathbb{T}_{\widehat{\mathcal{N}}_F} + \widehat{\Psi}^* X \widehat{\Psi}, \tag{3.185}$$

where $\widehat{\mathcal{N}}_F$ is the input–output operator of the system (3.177).

In particular, if $\widetilde{\Sigma}$ is $(X, 0)$–equivalent to Σ, then $\widehat{\mathcal{R}}_\Sigma = \widehat{\mathcal{R}}_{\widetilde{\Sigma}} + \widehat{\Psi}^* X \widehat{\Psi}$.

Proof. From point 2 of Proposition 3.6.1 we have

$$J_{\widehat{\Sigma}}(u) = \left\langle u, \widehat{\mathcal{R}}_\Sigma u \right\rangle = J_{\widetilde{\Sigma}}(\widetilde{u}) + (\widehat{x}_0^u)^T X \widehat{x}_0^u = \left\langle \widetilde{u}, \widehat{\mathcal{R}}_{\widetilde{\Sigma}} \widetilde{u} \right\rangle + (\widehat{x}_0^u)^T X \widehat{x}_0^u,$$

where $\widetilde{u} = -F\sigma\widehat{x}^u + u = \widehat{\mathcal{N}}_F u$. Substituting here $\widehat{x}_0^u = \widehat{\Psi} u$, we immediately reach the conclusion. ∎

3.8.5 Reduction of the inverse–time case to the direct–time case

As in the continuous–time case, we show now that the operator $\widehat{\mathcal{R}}_{\Sigma_r}$ in (3.181) is an operator of the type (3.165) associated with a direct time recursion (with an appropriate Popov triplet seen with direct–time significance and with the A–matrix stable). We introduce first the following definition.

Definition 3.8.5 (Dual of a Popov triplet in discrete–time)
Consider a Popov triplet $\Sigma = (A, B; Q, L, R)$ with A stable and seen with inverse-time significance. Let $\Sigma_r = (A, B; 0, L_r := L + A^T X_r B, R_r := R + B^T X_r B)$ be its reduced equivalent, where X_r is the unique solution to the reducing Stein equation $A^T X A - X + Q = 0$. Then the Popov triplet $\Sigma_d = (A_d, B_d; 0, L_d, R_d)$ where

$$A_d := A^T, \quad B_d := L_r = L + A^T X_r B, \quad L_d := B, \quad R_d := R_r = R + B^T X_r B \tag{3.186}$$

is called the dual Popov triplet *of Σ.*

Lemma 3.8.6 *Let $\Sigma = (A, B; Q, L, R)$ be a Popov triplet with A stable and seen with inverse–time significance, and let $\Sigma_r = (A, B; 0, L_r, R_r)$ and $\Sigma_d = (A_d, B_d; 0, L_d, R_d)$ be the reduced and the dual Popov triplets associated with Σ. Then*

$$\mathcal{R}_{\Sigma_d} = \widehat{\mathcal{R}}_{\Sigma_r}, \tag{3.187}$$

$$\mathcal{P}_{\Sigma_d} = \widehat{\Psi}, \tag{3.188}$$

where $\widehat{\Psi}$ is the anticausal inverse–time controllability operator (2.198) associated with the pair (A, B), and the Popov functions are related through

$$\widehat{\mathbf{\Pi}}_{\Sigma} = \mathbf{\Pi}_{\Sigma_d}, \tag{3.189}$$

where $\widehat{\mathbf{\Pi}}_{\Sigma}$ is the inverse–time Popov function (3.136) associated with Σ and $\mathbf{\Pi}_{\Sigma_d}$ is the direct–time Popov function (3.135) associated with Σ_d.

Proof. Let Φ_d and $\mathcal{L}_d$ be the operators (2.145) and (2.146) corresponding to Σ_d, i.e.,

$$(\Phi_d \mu)_k = (A^T)^k \mu, \quad k \geq 0, \quad \forall \mu \in \mathbb{R}, \tag{3.190}$$

and

$$(\mathcal{L}_d u)_k = \sum_{i=0}^{k-1} (A^T)^{k-i-1} L_r u_i\ , \quad k \geq 0. \tag{3.191}$$

The adjoints of these operators are

$$\Phi_d^* \lambda = \sum_{k=0}^{\infty} A^k \lambda_k, \quad \lambda \in \ell_+^{2,n}, \tag{3.192}$$

and

$$(\mathcal{L}_d^* \lambda)_k = \sum_{i=k+1}^{\infty} L_r^T A^{i-k-1} \lambda_i, \quad k \geq 0, \quad \lambda \in \ell_+^{2,n}. \tag{3.193}$$

Then for all $u \in \ell_+^{2,m}$ we have

$$(\mathcal{L}_d^* L_d u)_k = \sum_{i=k+1}^{\infty} L_r^T A^{i-k-1} B u_i = (L_r^T \widehat{\mathcal{L}} u)_k \tag{3.194}$$

and

$$(L_d^T \mathcal{L}_d u)_k = \sum_{i=0}^{k-1} B^T (A^T)^{k-i-1} L_r u_i = (\widehat{\mathcal{L}}^* L_r u)_k. \tag{3.195}$$

Combining (3.165), (3.186), (3.194), and (3.195) we get

$$\mathcal{R}_{\Sigma_d} = R_d + \mathcal{L}_d^* L_d + L_d^T \mathcal{L}_d = R_r + L_r^T \widehat{\mathcal{L}} + \widehat{\mathcal{L}}^* L_r = \widehat{\mathcal{R}}_{\Sigma_r}$$

and (3.187) follows.

Further, from (3.164) we have $\mathcal{P}_{\Sigma_d} u = \Phi_d^* L_d u = \sum_{k=0}^{\infty} A^k B u_k = \widehat{\Psi} u$. Equality (3.189) follows by direct checking. ■

3.8.6 The direct–time case: A antistable

If A is *antistable*, we can still define the operator $\mathcal{R}_\Sigma : \ell_+^{2,m} \to \ell_+^{2,m}$ as the linear bounded operator

$$\mathcal{R}_\Sigma := R + L^T\mathcal{L} + \mathcal{L}^*L + \mathcal{L}^*Q\mathcal{L} = \mathcal{R}_\Sigma{}^*, \tag{3.196}$$

which is not a Toeplitz operator as $\mathcal{L}$ is not causal in this case. There is no difficulty in showing that, in this case, the restricted quadratic index associated with Σ is

$$J_\Sigma(u) = \langle u, \mathcal{R}_\Sigma u\rangle\,. \tag{3.197}$$

Passing to the time–reversed Popov triplet and using the theory for the stable, inverse–time case we can deduce successively that

$$\mathcal{R}_\Sigma = \mathbb{T}_{\mathcal{R}_{e,\Sigma}} - \widehat{\mathbb{H}}_{\mathcal{L}_e}^* Q \widehat{\mathbb{H}}_{\mathcal{L}_e}, \tag{3.198}$$

and, for the reduced Popov triplet,

$$\mathcal{R}_{\Sigma_r} = \mathbb{T}_{\mathcal{R}_{e,\Sigma_r}} = \mathbb{T}_{\mathcal{R}_{e,\Sigma}}. \tag{3.199}$$

Proposition 3.8.7 *Assume that A is antistable. If $\widetilde{\Sigma}$ is $(X,0)$–equivalent to Σ, then*

$$\mathcal{R}_\Sigma = \mathcal{R}_{\widetilde{\Sigma}} + \widehat{\Psi}^* X \widehat{\Psi}. \tag{3.200}$$

Since A is antistable it is nonsingular and the Popov triplet $\Sigma = (A, B; Q, L, R)$, seen with direct–time significance is reversible, and so is its reduced equivalent $\Sigma_r = (A, B; 0, L_r, R_r)$. Let $\widehat{\Sigma}_r$ be the time–reversed Popov triplet of Σ_r and let also $\widehat{\Sigma}_d = (A_d, B_d; 0, L_d, R_d)$ be the dual of $\widehat{\Sigma}_r$ with the entries given explicitly by (3.144), that is,

$$A_d = A^{-T}, B_d = -A^{-T}L_r, L_d = -A^{-1}B, R_d = R - B^TA^{-T}L_r - L_r^TA^{-1}B. \tag{3.201}$$

Notice that A_d is stable, since A is antistable. The following result reveals some interesting relationships between the newly introduced Popov triplet $\widehat{\Sigma}_d$ and the original triplet.

Lemma 3.8.8 *Let $\widehat{\Sigma}_d = (A_d, B_d; 0, L_d, R_d)$ be the Popov triplet defined through (3.201). Then*

$$\mathcal{R}_{\widehat{\Sigma}_d} = \mathcal{R}_{\Sigma_r}, \tag{3.202}$$

$$\mathcal{P}_{\widehat{\Sigma}_d} = \widehat{\Psi}, \tag{3.203}$$

where $\widehat{\Psi}$ is the anticausal controllability operator (2.198) associated with the pair (A, B), and the Popov functions of the two triplets are related through

$$\mathbf{\Pi}_\Sigma = \mathbf{\Pi}_{\widehat{\Sigma}_d}. \tag{3.204}$$

Proof. Let $\mathcal{L}_d$ be the input–to–the–state trajectory operator of the pair (A_d, B_d) given by formula (2.146), that is

$$(\mathcal{L}_d u)_k = \sum_{i=0}^{k-1} A_d^{k-i-1} B_d u_i = -\sum_{i=0}^{k-1} (A^{-T})^{k-i} L_r u_i, \tag{3.205}$$

and Φ_d be the free evolution operator of A_d given by formula (2.145), i.e.,

$$(\Phi_d \mu)_k = A_d^k \mu = (A^{-T})^k \mu. \tag{3.206}$$

According to (3.166) and (3.169), the adjoints of these operators are

$$\Phi_d^* x = \sum_{k=0}^{\infty} (A^{-1})^k x_k, \quad x \in \ell_+^{2,n} \tag{3.207}$$

and

$$(\mathcal{L}_d^* u)_k = -\sum_{i=k+1}^{\infty} (L_r^T A^{-1})(A^{-1})^{i-k-1} u_i, \quad u \in \ell_+^{2,n}. \tag{3.208}$$

We compute

$$\mathcal{R}_{\widehat{\Sigma}_d} = R_d + L_d^T \mathcal{L}_d + \mathcal{L}_d^* L_d = R - B^T A^{-T} L_r + L_d^T \mathcal{L}_d - L_r^T A^{-1} B + \mathcal{L}_d^* L_d.$$

With (3.205), we have for all $u \in \ell_+^{2,m}$

$$\begin{aligned} -B^T A^{-T} L_r u_k + (L_d^T \mathcal{L}_d u)_k &= -B^T A^{-T} L_r u_k - \sum_{i=0}^{k-1} B^T (A^{-T})^{k-i+1} L_r u_i \\ &= -\sum_{i=0}^{k} B^T (A^{-T})^{k-i+1} L_r u_i = (\mathcal{L}_r^* L_r u)_k. \end{aligned} \tag{3.209}$$

Further, with (3.208), we have for all $u \in \ell_+^{2,m}$

$$-L_r^T A^{-1} B u_k + (\mathcal{L}_d^* L_d u)_k = -L_r^T A^{-1} B u_k - \sum_{i=k+1}^{\infty} L_r^T (A^{-1})^{i-k+1} B u_i = ({L_r}^T \mathcal{L} u)_k. \tag{3.210}$$

We conclude that $\mathcal{R}_{\widehat{\Sigma}_d} = R + \mathcal{L}_r^* L_r + L_r^T \mathcal{L} = \mathcal{R}_{\Sigma_r}$ and (3.202) is established.

For (3.203) we have by (3.164)

$$\mathcal{P}_{\widehat{\Sigma}_d} u = \Phi_d^* L_d u = \sum_{k=0}^{\infty} (A^{-1})^k A^{-1} B u_k = \sum_{k=0}^{\infty} (A^{-1})^{k+1} B u_k = \widehat{\Psi} u.$$

Equality (3.204) follows by direct checking. ∎

3.8.7 $\mathcal{R}_e$ and $\widehat{\mathcal{R}}_e$ as input-output operators

Here follows the discrete–time counterpart of Theorem 3.3.6.

Theorem 3.8.9 *Let $\Sigma = (A, B; Q, L, R)$ be a Popov triplet. If A is dichotomic, then the operator $\mathcal{R}_{e,\Sigma}$ is the ℓ^2 input-output operator of the DTHS(Σ) in (3.128), and $\widehat{\mathcal{R}}_{e,\Sigma}$ is the ℓ^2 input-output operator of the ITHS(Σ) in (3.129).*

Proof. Let A be dichotomic, and take an arbitrary u in $\ell^{2,m}$. Then the unique x in $\ell^{2,n}$ satisfying the first equation of (3.128),

$$\sigma x = Ax + Bu,$$

is $x_e^u = \mathcal{L}_e u$. As A^T is also dichotomic, by using formula (2.169) for the second equation in (3.128), that is an inverse–time recursion,

$$\lambda = Qx + A^T\sigma\lambda + Lu,$$

one gets

$$\begin{aligned}(B^T\sigma\lambda_e^u)_k &= -\sum_{i=-\infty}^{k} B^T((A^T)^\dagger)^{k+1-i}\Pi_+(Lu_i + Qx_{e,i}^u)\\ &\quad + \sum_{i=k+1}^{\infty} B^T(A^T)^{i-k-1}(Lu_i + Qx_{e,i}^u)\\ &= ((\mathcal{L}_e^*Q\mathcal{L}_e + \mathcal{L}_e^*L)u)_k.\end{aligned}$$

Using this formula in the last equation in (3.128) gives $v_e^u = L^T\mathcal{L}_e u + \mathcal{L}_e^*Q\mathcal{L}_e u + \mathcal{L}_e^*Lu + Ru$, or, with (3.159), $v_e^u = \mathcal{R}_{e,\Sigma}u$, and the proof of the first part is completed.
The second part of the theorem follows identically. ■

An immediate consequence of Theorem 3.8.9 is the fact that $\mathcal{R}_\Sigma$ is the Toeplitz operator with the symbol the Popov function $\mathbf{\Pi}_\Sigma$. Indeed, the Popov function is the TFM of the Hamiltonian system and $\mathcal{R}_\Sigma$ is the Toeplitz operator associated with the input–output operator of the Hamiltonian system.

Notes and references

The notion of a Popov triplet originates in Popov's monograph [Pop66] although the name was only recently adopted in a series of works by the authors. The equivalence relation as defined in this chapter is a particular case of a more general equivalence relation studied in detail in Chapter 2 of Popov [Pop66]. Various objects associated with a Popov triplet are already present in Popov [Pop66]. Others, as for example the extended Hamiltonian and symplectic pencils, were introduced later by Van Dooren [Van81]. In both Popov [Pop66] and Van Dooren [Van81] the attention was restricted to the the positivity case. The extended Hamiltonian and symplectic pencils, the Kalman–Popov–Yakubovich systems in J form, and the algebraic Riccati systems appeared in their full generality in several papers of Ionescu, Oară, and Weiss, including [IW93b, IO96d, IO96e, IOW97]. Discrete inverse–time objects and Popov triplets with inverse–time significance were studied in connection with the time–reversed linear quadratic problem and the Nehari problem by Ionescu [Ion96], and Oară [Oar96b]. The Popov triplets and their equivalence are the underlying algebraic concepts throughout the theory of linear quadratic control and robust control, although these were not recognized and studied as such.

4

Riccati Theory: An Operator–based Approach

We present in this chapter existence criteria for the stabilizing solution to the Riccati equation, as well as some related results and formulas in terms of the special operators introduced in Chapter 3. The main result given in Section 4.1 relates the existence of the stabilizing solution to the invertibility of a Toeplitz operator in the case in which the state matrix A is stable. This result contains most of the classical results in linear quadratic control which we present as particular cases. We relax the hypothesis on the stability of the state matrix A and replace it with the stabilizability of the pair (A, B) in Section 4.2. Again, we pay a particular attention to the classical positivity results. In Section 4.3 we discuss the case in which the matrix A is antistable and obtain an existence result for the stabilizing solution to the Riccati equation that we apply further to obtain the solution to the time–reversed linear quadratic control problem. The characterization of the existence of the stabilizing solution to the algebraic Riccati equation in terms of the invertibility of an associated Toeplitz operator obtained in Section 4.1, although very general, is difficult to check. A remedy to this is given in Section 4.4 where we present the so–called signature condition which is a sufficient condition for the invertibility of the Toeplitz operator. The signature condition plays an important role in game–theoretic situations, as we show by solving the maxmin quadratic problem. It is also key to the Nehari problem, the H^∞–control problem, and to the robust stabilization problem, all discussed in Part III. We return to an engineering viewpoint in the spirit of the Popov's celebrated monograph [Pop66] in Section 4.5 and translate all sufficient (and necessary) existence conditions obtained so far in frequency domain terms, i.e., in terms of an evaluation of the Popov function over the imaginary axis.

The main technical difficulty encountered when extending the continuous–time results given in Sections 4.1–4.6 to a discrete–time setting is that a discrete system is in general not time–reversible. We overcome this difficulty and present the same level of generality for the results on discrete–time Riccati equations in Sections 4.7–4.13.

A. The continuous–time case

4.1 Existence of the stabilizing solution: the stable case

Given a Popov triplet $\Sigma = (A, B; Q, L, R)$ with $A \in \mathbb{R}^{n\times n}$, $B \in \mathbb{R}^{n\times m}$, A stable, the aim of this section is to derive necessary and sufficient conditions for the existence of the stabilizing solution to the CTARE(Σ)

$$A^T X + XA - (XB + L)R^{-1}(B^T X + L^T) + Q = 0, \tag{4.1}$$

in terms of the invertibility of the Toeplitz operator $\mathcal{R}_\Sigma$ given by

$$\mathcal{R}_\Sigma : L_+^{2,m} \to L_+^{2,m}, \quad \mathcal{R}_\Sigma := R + L^T\mathcal{L} + \mathcal{L}^*L + \mathcal{L}^*Q\mathcal{L} = \mathcal{R}_\Sigma^*, \tag{4.2}$$

where

$$\mathcal{L} : L_+^{2,m} \to L_+^{2,n}, \quad (\mathcal{L}u)(t) := \int_0^t e^{A(t-\tau)}Bu(\tau)\mathrm{d}\tau, \quad t \geq 0. \tag{4.3}$$

Recall from Chapter 3 that $\mathcal{R}_\Sigma$ is the Toeplitz operator associated with the input–output operator

$$\mathcal{R}_{e,\Sigma} : L^{2,m} \to L^{2,m}, \quad \mathcal{R}_{e,\Sigma} := R + L^T\mathcal{L}_e + \mathcal{L}_e^*L + \mathcal{L}_e^*Q\mathcal{L}_e = \mathcal{R}_{e,\Sigma}^* \tag{4.4}$$

of the CTHS(Σ)

$$\begin{array}{rcllll} \dot{x} & = & Ax & & + Bu, \\ \dot{\lambda} & = & -Qx & - A^T\lambda & - Lu, \\ v & = & L^T x & + B^T\lambda & + Ru, \end{array} \tag{4.5}$$

where

$$\mathcal{L}_e : L^{2,m} \to L^{2,n}, \quad (\mathcal{L}_e u)(t) := \int_{-\infty}^t e^{A(t-\tau)}Bu(\tau)\mathrm{d}\tau, \quad t \in \mathbb{R}. \tag{4.6}$$

The following operators, defined in Chapters 2 and 3,

$$\Phi : \mathbb{R}^n \to L_+^{2,n}, \quad (\Phi\xi)(t) := e^{At}\xi, \quad t \geq 0, \tag{4.7}$$

$$\mathcal{P}_{o,\Sigma} : \mathbb{R}^n \to \mathbb{R}^n, \quad \mathcal{P}_{o,\Sigma} := \Phi^*Q\Phi, \tag{4.8}$$

$$\mathcal{P}_\Sigma : L_+^{2,m} \to \mathbb{R}^n, \quad \mathcal{P}_\Sigma := \Phi^*(Q\mathcal{L} + L), \tag{4.9}$$

will also play an important role in the derivations in this chapter.

4.1.1 The main result

Here follows the continuous–time version of the main result of the chapter.

Theorem 4.1.1 *Let $\Sigma = (A, B; Q, L, R)$ be a Popov triplet with A stable. Then the following assertions are equivalent:*

1. R is nonsingular and the CTARE(Σ) given in (4.1) has a (unique) stabilizing solution X.

2. The Toeplitz operator $\mathcal{R}_\Sigma$ given in (4.2) has a bounded inverse.

The difficult part in proving this theorem is the implication 2. $\Rightarrow$ 1. We shall prove first the other implication.

Proof. 1. $\Rightarrow$ 2. Assume that the CTARE(Σ) has a stabilizing solution X and let F_{Ricc} be the associated stabilizing feedback. If $\widetilde{\Sigma}$ is the (X, F_{Ricc})–equivalent of Σ, then $\widetilde{Q} = 0$ and $\widetilde{L} = 0$ as can be seen from (3.3). Consequently, $\mathcal{R}_{e,\widetilde{\Sigma}} = R$. By (3.60), we have

$$\mathcal{R}_\Sigma = \mathbb{T}^*_{\mathcal{N}_{F_{Ricc}}} R \mathbb{T}_{\mathcal{N}_{F_{Ricc}}}. \tag{4.10}$$

By hypothesis, R is invertible. Using Corollary 2.4.14, $\mathbb{T}_{\mathcal{N}_{F_{Ricc}}}$ is also invertible. Indeed, $\mathcal{N}_{F_{Ricc}}$ is causal as it is the input–output operator of the system

$$\begin{aligned} \dot{x} = & \quad Ax + Bu, \\ v = & \quad -F_{Ricc}x + u. \end{aligned}$$

According to Theorem 2.3.1, the inverse of $\mathcal{N}_{F_{Ricc}}$ is the input–output operator of

$$\begin{aligned} \dot{x} = & \quad (A + BF_{Ricc})x + Bv, \\ u = & \quad F_{Ricc}x + v. \end{aligned}$$

Thus Corollary 2.4.14 is applicable, and we deduce that $\mathcal{R}_\Sigma$ is invertible. ■

We turn now to the proof of the implication 2. $\Rightarrow$ 1.

4.1.2 Proof of the implication 2. $\Rightarrow$ 1.

We shall state the main steps of the proof as auxiliary results. First of all, if $\mathcal{R}_\Sigma$ is assumed to be invertible and since $\mathcal{R}_\Sigma$ is the Toeplitz operator with the symbol the Popov function, it follows that $\mathbf{\Pi}_\Sigma$ is invertible in RL^∞ and, in particular, $R = \mathbf{\Pi}_\Sigma(\infty)$ is invertible. It remains to show that the CTARE(Σ) has a stabilizing solution.

The first result is actually independent of the invertibility hypothesis on $\mathcal{R}_\Sigma$ and is quite similar to Theorem 3.3.6.

Lemma 4.1.2 *If $\xi \in \mathbb{R}^n$ and $u \in L^{2,m}_+$, then there exists a unique $L^{2,n}_+ \times L^{2,n}_+$ state trajectory (x, λ) of the CTHS(Σ) (4.5) such that $x(0) = \xi$. (We denote this trajectory by $(x^{\xi,u}, \lambda^{\xi,u})$). Moreover, the corresponding output $v^{\xi,u} \in L^{2,m}_+$ of the CTHS(Σ) satisfies*

$$v^{\xi,u} = \mathcal{P}^*_\Sigma \xi + \mathcal{R}_\Sigma u. \tag{4.11}$$

Proof. The proof follows closely that of Theorem 3.3.6, so we shall be brief. According to (2.57), $x^{\xi,u} = \Phi\xi + \mathcal{L}u$. By Theorem 2.2.2, we have

$$\lambda^{\xi,u}(t) = -\int_t^\infty e^{-A^T(t-\tau)}(-Lu(\tau) - Qx^{\xi,u}(\tau))\mathrm{d}\tau. \tag{4.12}$$

Formula (4.11) follows by straightforward substitutions similar to those used in the proof of Theorem 3.3.6. ■

We assume now that $\mathcal{R}_\Sigma$ is invertible. It is then immediate that for each ξ there exists a unique u such that $v^{\xi,u} = 0$. If we denote such a u by u^ξ, we have from (4.11) that

$$u^\xi = -\mathcal{R}_\Sigma^{-1}\mathcal{P}^*_\Sigma \xi. \tag{4.13}$$

Let us denote $x^\xi := x^{\xi,u^\xi}$, $\lambda^\xi := \lambda^{\xi,u^\xi}$, the corresponding solutions of the CTHS(Σ) (4.5).

A simple observation is that, because of the time–invariance of the CTHS, if $\xi' = x^\xi(\tau)$, then $u^{\xi'}(t) = u^\xi(t+\tau)$ and $\lambda^{\xi'}(t) = \lambda^\xi(t+\tau)$. We use this fact in the proof of our next two results.

Lemma 4.1.3 *We have*

$$\lambda^\xi = Xx^\xi, \tag{4.14}$$

where X is the symmetric $n \times n$ matrix

$$X := \mathcal{P}_{o,\Sigma} - \mathcal{P}_\Sigma \mathcal{R}_\Sigma^{-1} \mathcal{P}_\Sigma^*. \tag{4.15}$$

Proof. Obviously, λ^ξ is absolutely continuous as a part of the state trajectory of the CTHS. From (4.12), we get

$$\lambda^\xi(0) = \int_0^\infty e^{A^T\tau}(Lu^\xi(\tau) + Qx^\xi(\tau))\mathrm{d}\tau.$$

From (4.13) we have

$$x^\xi = \Phi\xi + \mathcal{L}u^\xi = (\Phi - \mathcal{L}\mathcal{R}_\Sigma^{-1}\mathcal{P}_\Sigma^*)\xi. \tag{4.16}$$

Using expression (3.53) for the adjoint of Φ, with (4.13) and (4.16),

$$\lambda^\xi(0) \quad = \quad \Phi^*(Lu^\xi + Qx^\xi) = \Phi^* L(-\mathcal{R}_\Sigma^{-1}\mathcal{P}_\Sigma^*\xi) + \Phi^* Q(\Phi - \mathcal{L}\mathcal{R}_\Sigma^{-1}\mathcal{P}_\Sigma^*)\xi.$$

Using now the definitions of $\mathcal{P}_{o,\Sigma}$ and $\mathcal{P}_\Sigma$ in (4.8) and (4.9), respectively, we derive easily that $\lambda^\xi(0) = X\xi$, with X given by (4.15). Applying this formula for $\xi' = x^\xi(\tau)$ we have $\lambda^\xi(\tau) = \lambda^{\xi'}(0) = X\xi' = Xx^\xi(\tau)$, and formula (4.14) is proved. ■

It will turn out that X introduced in (4.15) – called the *representation formula* – is the stabilizing solution of the CTARE(Σ).

Lemma 4.1.4 *The function u^ξ is absolutely continuous. Moreover, the matrix*

$$F := -R^{-1}(B^TX + L^T), \tag{4.17}$$

where X is given by (4.15) satisfies

$$u^\xi = Fx^\xi. \tag{4.18}$$

Proof. As u^ξ is, by definition, such that $v^{\xi,u^\xi} = 0$, from the last equation of the CTHS (4.5) we deduce $u^\xi = -R^{-1}(L^Tx^\xi + B^T\lambda^\xi)$. Since both x^ξ and λ^ξ are absolutely continuous, u^ξ is absolutely continuous as well. Substituting λ^ξ from (4.14), we obtain directly (4.18). ■

The proof of Theorem 4.1.1 is completed with the proof of the following result.

Lemma 4.1.5 *If F is given by (4.17) and X is given by (4.15), then $A+BF$ is stable, and X satisfies the CTARE(Σ) (4.1). In other words, X is the stabilizing solution of the CTARE(Σ) and $F = F_{Ricc}$ is the stabilizing Riccati feedback.*

Proof. Introducing (4.18) in the first equation of the CTHS (4.5), we deduce that x^ξ satisfies for each ξ

$$\dot{x}^\xi = (A + BF)x^\xi. \tag{4.19}$$

This means that all the initial–value solutions of the differential equation $\dot{x} = (A + BF)x$ are in $L_+^{2,n}$. Moreover, $\|x^\xi\|_2 \leq \rho_0\|\xi\|$ for some $\rho_0 > 0$ and all $\xi \in \mathbb{R}^n$, as directly follows from (4.16). Hence

$$Z := \int_0^\infty e^{(A+BF)^T t} e^{(A+BF)t} \mathrm{d}t \leq \rho_0^2 I$$

is well defined and $Z > 0$. Furthermore, it is a matter of simple computation to check that Z satisfies the Lyapunov equation $(A + BF)^T Z + Z(A + BF) + I = 0$. From this we conclude with Theorem 1.5.4 that $A + BF$ is stable.

Substituting (4.14) and (4.18) in the second equation of the CTHS (4.5) we obtain $X\dot{x}^\xi = -Qx^\xi - A^T X x^\xi - LFx^\xi$. Using now (4.19) and (4.17), we deduce that

$$\{X[A - BR^{-1}(B^T X + LT)] + Q + A^T X - LR^{-1}(B^T X + L^T)\}\xi = 0$$

for x^ξ evaluated at $t = 0$. Since ξ is arbitrary in $\mathbb{R}^n$, it follows that X satisfies the CTARE(Σ) (4.1).

Comparing (4.17) with (3.18), and taking into account that $A + BF$ is stable, we get that X is the stabilizing solution and F is the stabilizing Riccati feedback. ■

We can now incidentally prove the following result.

Corollary 4.1.6 *Let $\Sigma = (A, B; Q, L, R)$ be a Popov triplet with A stable. Assume that the Toeplitz operator $\mathcal{R}_\Sigma$ in (4.2) has a bounded inverse and let X be the stabilizing solution to the CTARE(Σ). Then we have*

$$J_\Sigma(\xi, u^\xi) = \left\langle \begin{bmatrix} x^{\xi,u^\xi} \\ u \end{bmatrix}, \begin{bmatrix} Q & L \\ L^T & R \end{bmatrix} \begin{bmatrix} x^{\xi,u^\xi} \\ u \end{bmatrix} \right\rangle_{L_+^{2,n} \times L_+^{2,m}} = \xi^T X \xi \tag{4.20}$$

for every $\xi \in \mathbb{R}^n$ and $u^\xi = -\mathcal{R}_\Sigma^{-1}\mathcal{P}_\Sigma^*\xi$, with $\mathcal{P}_\Sigma$ given in (4.9).

Proof. We simply introduce $u^\xi = -\mathcal{R}_\Sigma^{-1}\mathcal{P}_\Sigma^*\xi$ in the operator form (3.49) of the quadratic index and then (4.20) is a consequence of the representation formula (4.15) of the stabilizing solution X. ■

4.1.3 The positiveness case

Based on the obvious fact that a coercive operator is invertible, Theorem 4.1.1 leads directly to the following result that is related to the so–called *linear quadratic problem (LQP)*. The underlying basic assumption here is that the R matrix (the coefficient of the quadratic term) is positive definite.

Corollary 4.1.7 *Let $\Sigma = (A, B; Q, L, R)$ be a Popov triplet with A stable.*

1. The operator $\mathcal{R}_{e,\Sigma}$ given in (4.4) is coercive if and only if $R > 0$ and the CTARE(Σ) has a stabilizing solution X.

2. If the operator $\mathcal{R}_{e,\Sigma}$ is coercive then the minimum of the quadratic index (3.6) is

$$\min_{u \in L_+^{2,m}} J_\Sigma(\xi, u) = \xi^T X \xi, \quad \forall \xi \in \mathbb{R}^n, \tag{4.21}$$

and the minimum is attained for $u = u^\xi$ given in (4.13). In particular, u^ξ satisfies the feedback relation (4.18).

3. If, in addition,

$$\begin{bmatrix} Q & L \\ L^T & R \end{bmatrix} \geq 0, \tag{4.22}$$

and the operator $\mathcal{R}_{e,\Sigma}$ is coercive then $X \geq 0$.

Proof. 1. If $\mathcal{R}_{e,\Sigma}$ is coercive, then $\mathcal{R}_\Sigma$ is coercive hence is invertible and, according to Theorem 4.1.1, the CTARE(Σ) has a stabilizing solution. The positiveness of R follows now from (4.10). Conversely, if the CTARE has a stabilizing solution X and F_{Ricc} is the stabilizing Riccati feedback, then for $\widetilde{\Sigma}$, the (X, F_{Ricc})–equivalent of Σ, we have $\mathcal{R}_{e,\widetilde{\Sigma}} = R$. With (3.47), we can write

$$\mathcal{R}_{e,\Sigma} = \mathcal{N}^*_{F_{Ricc}} R \mathcal{N}_{F_{Ricc}}.$$

As R is positive definite and $\mathcal{N}_{F_{Ricc}}$ is invertible, it follows that $\mathcal{R}_{e,\Sigma}$ is coercive.

2. Based on Remarks 3.2.9 and 3.2.11 it follows that the KPYS(Σ) in positivity form (3.24) has a stabilizing solution (X, V, W) where X is exactly the stabilizing solution to the CTARE(Σ). According to point 1 of Proposition 3.2.12, we have

$$J_\Sigma(\xi, u) = \|Vu + Wx^{\xi,u}\|_2^2 + \xi^T X \xi, \quad u \in L_+^{2,m}. \tag{4.23}$$

It follows that the minimum of $J_\Sigma(\xi, u)$ is attained for the feedback law

$$u = -V^{-1} W x = F_{Ricc} x \tag{4.24}$$

and it equals $\xi^T X \xi$.

3. If (4.22) holds, then $J_\Sigma(\xi, u) \geq 0$, for each $u \in L_+^{2,m}$, and the conclusion follows trivially from (4.21). ■

We may relax the coerciveness condition in point 1 of Corollary 4.1.7 as follows.

Corollary 4.1.8 *Let $\Sigma = (A, B; Q, L, R)$ be a Popov triplet with A stable. Assume that $R > 0$ and that the pair (A, B) is controllable. If*

$$\mathcal{R}_{e,\Sigma} \geq 0, \tag{4.25}$$

then there exists a symmetric solution X to the CTARE(Σ), such that for F defined by (4.17) we have $\Lambda(A + BF) \subset \mathbb{C}_- \cup \mathbb{C}_0$. (We call X a semi–stabilizing solution *to the CTARE).*

Proof. Let $\varepsilon > 0$ and let $\Sigma_\varepsilon = (A, B; Q, L, R + \varepsilon I)$. Then $\mathcal{R}_{e,\Sigma_\varepsilon} := \mathcal{R}_{e,\Sigma} + \varepsilon I >> 0$. According to Corollary 4.1.7, the CTARE(Σ_ε)

$$A^T X_\varepsilon + X_\varepsilon A - (X_\varepsilon B + L)(R + \varepsilon I)^{-1}(B^T X_\varepsilon + L^T) + Q = 0 \tag{4.26}$$

has a stabilizing solution X_ε, i.e., $A+BF_\varepsilon$ is stable, for $F_\varepsilon = -(R+\varepsilon I)^{-1}(B^T X_\varepsilon + L^T)$.

Rewrite (4.26) as

$$A^T X_\varepsilon + X_\varepsilon A - F_\varepsilon^T (R+\varepsilon I) F_\varepsilon + Q = 0. \tag{4.27}$$

Since (A,B) is controllable, we may choose $\widetilde{F}$ such that $\widetilde{A} := A + B\widetilde{F}$ is antistable. By Corollary 3.2.7, (4.27) is equivalent to

$$\widetilde{A}^T X_\varepsilon + X_\varepsilon \widetilde{A} - (F_\varepsilon - \widetilde{F})^T (R+\varepsilon I)(F_\varepsilon - \widetilde{F}) + \widetilde{Q}_\varepsilon = 0, \tag{4.28}$$

where

$$\widetilde{Q}_\varepsilon := Q + L\widetilde{F} + \widetilde{F}^T L^T + \widetilde{F}^T (R+\varepsilon I)\widetilde{F}.$$

On the other hand, as $\widetilde{A}$ is antistable, Theorem 1.5.2 implies that the Lyapunov equation

$$\widetilde{A}^T Z + Z\widetilde{A} + Q + L\widetilde{F} + \widetilde{F}^T L^T + \widetilde{F}^T R_1 \widetilde{F} = 0 \tag{4.29}$$

has a unique solution Z for every symmetric matrix R_1. Choose $R_1 \geq R + \varepsilon I$ and subtract (4.29) from (4.28). Then

$$\widetilde{A}^T (X_\varepsilon - Z) + (X_\varepsilon - Z)\widetilde{A} - (F_\varepsilon - \widetilde{F})^T (R+\varepsilon I)(F_\varepsilon - \widetilde{F}) - \widetilde{F}^T (R_1 - R - \varepsilon I)\widetilde{F} = 0. \tag{4.30}$$

As $\widetilde{A}$ is antistable and $(F_\varepsilon - \widetilde{F})^T (R+\varepsilon I)(F_\varepsilon - \widetilde{F}) + \widetilde{F}^T (R_1 - R - \varepsilon I)\widetilde{F} \geq 0$, relation (4.30) implies that $X_\varepsilon - Z \geq 0$. Hence,

$$X_\varepsilon \geq Z \tag{4.31}$$

for every $\varepsilon > 0$. Using (4.15) we deduce that $X_\epsilon = \mathcal{P}_{o,\Sigma} - \mathcal{P}_\Sigma (\mathcal{R}_{e,\Sigma} + \epsilon I)^{-1} \mathcal{P}_\Sigma^*$ which shows that X_ε decreases as ε decreases to 0. Hence with (4.31) we conclude that there exists $X = X^T$ such that

$$\lim_{\varepsilon \downarrow 0} X_\varepsilon = X. \tag{4.32}$$

Taking the limit $\varepsilon \downarrow 0$ in (4.26) it follows that X satisfies the CTARE(Σ). In addition, $\lim_{\varepsilon\downarrow 0} F_\varepsilon = F = -R^{-1}(B^T X + L)$ and $A + BF_\varepsilon$ is stable for each $\varepsilon > 0$. Therefore $A + BF$ will have all the eigenvalues in the closed left–half plane. ■

4.2 Removing the stability assumption

Using a feedback transformation it is trivial to relax the stability assumption in Theorem 4.1.1 to the stabilizability assumption on the pair (A, B). Of course, if the CTARE(Σ) were to have a stabilizing solution, the pair (A, B) has to be stabilizable. We obtain the following result.

Theorem 4.2.1 *Let $\Sigma = (A, B; Q, L, R)$ be a Popov triplet with (A, B) stabilizable. Then the following two assertions are equivalent.*

1. R is nonsingular and the CTARE(Σ) (4.1) has a (unique) stabilizing solution.

2. There exists $F \in \mathbb{R}^{m\times n}$ such that $A+BF$ is stable and such that the F–equivalent of Σ, $\widetilde{\Sigma} = (\widetilde{A}, \widetilde{B}; \widetilde{Q}, \widetilde{L}, \widetilde{R})$, has an invertible associated Toeplitz operator $\mathcal{R}_{\widetilde{\Sigma}}$.

Proof. Indeed, the CTARE(Σ) has a stabilizing solution if and only if CTARE($\widetilde{\Sigma}$) has one, and this, in turn, is equivalent to the invertibility of $\mathcal{R}_{\widetilde{\Sigma}}$ according to Theorem 4.1.1. ■

Let us also notice that if point 2 of Theorem 4.2.1 holds for an arbitrary stabilizing F, then it holds for *every* stabilizing F. This is easy to prove using Proposition 3.3.2.

The fact that the existence of the stabilizing solution, in the absence of stability, is assessed by means of a transformed triplet instead of the original one is not very satisfactory. We shall see in Section 4.3 that this situation can be partially corrected in the case that A is antistable. There is still an important case in which we can express the existence of the stabilizing solution exclusively in terms of the original Popov triplet: the positiveness case.

4.2.1 The positiveness case: the standard Riccati equation for control

The following result is well–known in classical positivity theory.

Corollary 4.2.2 *Let $\Sigma_s = (A, B; C^TC, 0, I)$ be the standard Popov triplet for control and assume that the pairs (A, B) and (C, A) are stabilizable and detectable, respectively. Then the CTARE(Σ_s)*

$$A^TX + XA - XBB^TX + C^TC = 0 \tag{4.33}$$

has a positive semidefinite stabilizing solution.

Proof. Let F be such that $A + BF$ is stable and consider the F–equivalent $\widetilde{\Sigma}_s = (\widetilde{A}, \widetilde{B}; \widetilde{Q}, \widetilde{L}, \widetilde{R})$ of Σ_s, where

$$\widetilde{A} := A + BF, \widetilde{B} := B, \widetilde{Q} := C^TC + F^TF, \widetilde{L} := F^T, \widetilde{R} := I.$$

In this case

$$\begin{aligned} \mathcal{R}_{e,\widetilde{\Sigma}_s} &= I + F\widetilde{\mathcal{L}}_e + \widetilde{\mathcal{L}}_e^* F^T + \widetilde{\mathcal{L}}_e^*(C^TC + F^TF)\widetilde{\mathcal{L}}_e \\ &= (I + F\widetilde{\mathcal{L}}_e)^*(I + F\widetilde{\mathcal{L}}_e) + \widetilde{\mathcal{L}}_e^* C^TC\widetilde{\mathcal{L}}_e \geq 0. \end{aligned}$$

For $\varepsilon > 0$ let $\widetilde{\Sigma}_{s,\varepsilon} = (\widetilde{A}, \widetilde{B}; \widetilde{Q}, \widetilde{L}, \widetilde{R} + \varepsilon I)$. Then

$$\mathcal{R}_{e,\widetilde{\Sigma}_{s,\varepsilon}} := \mathcal{R}_{e,\widetilde{\Sigma}_s} + \varepsilon I >> 0. \tag{4.34}$$

According to point 1 of Corollary 4.1.7, the CTARE($\widetilde{\Sigma}_{s,\varepsilon}$)

$$\widetilde{A}^TX_\varepsilon + X_\varepsilon\widetilde{A} - (X_\varepsilon\widetilde{B} + \widetilde{L})(\widetilde{R} + \varepsilon I)^{-1}(\widetilde{B}^TX_\varepsilon + \widetilde{L}^T) + \widetilde{Q} = 0 \tag{4.35}$$

has a stabilizing solution X_ε and, according to point 2 of the same corollary,

$$\min_{u \in L_+^{2,m}} J_{\widetilde{\Sigma}_{s,\varepsilon}}(\xi, u) = \xi^T X_\varepsilon \xi. \tag{4.36}$$

But

$$\begin{bmatrix} \widetilde{Q} & \widetilde{L} \\ \widetilde{L}^T & \widetilde{R} + \varepsilon I \end{bmatrix} \geq 0. \tag{4.37}$$

Hence, following point 3 of Corollary 4.1.7 we get $X_\varepsilon \geq 0$. Using formula (4.15) it is easy to check that X_ε decreases as ε decreases to zero. Since $X_\varepsilon \geq 0$ it follows that

$$\lim_{\varepsilon \downarrow 0} X_\varepsilon = X \geq 0$$

exists. Taking the limit in (4.35), it follows that X satisfies the CTARE($\widetilde{\Sigma}_s$) and, hence, the original CTARE(Σ_s) in (4.33). We rewrite now the CTARE(Σ_s) in the equivalent form

$$(A - BB^T X)^T X + X(A - BB^T X) + C^T C + XBB^T X = 0. \qquad (4.38)$$

As $X \geq 0$ and the pair $\left(\begin{bmatrix} C \\ B^T X \end{bmatrix}, A - BB^T X\right)$ is clearly detectable it follows by Theorem 1.5.5 that $A - BB^T X$ is stable and consequently X is the stabilizing solution of the CTARE(Σ_s) in (4.33). ■

4.3 Existence of the stabilizing solution: the antistable case

Clearly, the technical tools developed in the previous section work in the case in which A is antistable as well. However, if we concentrate on the case in which A is antistable we obtain an interesting existence result that will come in handy for the solution to the Nehari problem in Chapter 8.

4.3.1 The main result

The approach we take is to start with a Popov triplet $\Sigma = (A, B; Q, L, R)$, with A antistable, and to reduce this case to the one in which the matrix A is stable. In doing this we shall need the dual Popov triplet introduced in Definition 3.3.4. We have the following result.

Theorem 4.3.1 *Let $\Sigma = (A, B; Q, L, R)$ be a Popov triplet with A antistable. Let $\Sigma_r = (A, B; 0, L + X_r B, R)$ and $\Sigma_d = (-A^T, -L - X_r B; 0, B, R)$ be the reduced equivalent and the dual of Σ, respectively, where X_r is the unique solution to the reducing Lyapunov equation $A^T X + XA + Q = 0$. Then the following assertions are equivalent.*

1. R is nonsingular and the CTARE(Σ_d) has a stabilizing solution Z.

2. The Toeplitz operator $\mathcal{R}_{\Sigma_r}$ is invertible, where $\mathcal{R}_{\Sigma_r}$ is the Toeplitz operator (4.2) written for the Popov triplet Σ_r.

If point 1 holds, then Z is given by the representation formula

$$Z = -\widehat{\Psi} \mathcal{R}_{\Sigma_r}^{-1} \widehat{\Psi}^*,$$

where $\widehat{\Psi}$ is the anticausal controllability operator (2.90) associated with the pair (A, B).

Proof. The proof follows by combining Lemma 3.3.5 with Theorem 4.1.1 and (4.15). ■

We now focus on a particular type of stabilizing solution that satisfies an additional condition. The next theorem contains the main result of the section.

Theorem 4.3.2 *Let $\Sigma = (A, B; Q, L, R)$ be a Popov triplet with A antistable. Let $\Sigma_r = (A, B; 0, L_r, R)$ be the reduced equivalent of Σ where X_r is the unique solution to the reducing Lyapunov equation $A^T X + XA + Q = 0$. Then the following assertions are equivalent:*

1. R is invertible and the CTARE(Σ) has a stabilizing solution X such that $X - X_r$ is nonsingular.

2. The Toeplitz operator $\mathcal{R}_{\Sigma_r}$ is invertible, and $\widehat{\Psi}\mathcal{R}_{\Sigma_r}^{-1}\widehat{\Psi}^*$ is invertible.

Before proceeding with the proof we need an auxiliary result stated as a lemma below.

Lemma 4.3.3 *Let $\Sigma = (A, B; Q, L, R)$ be a Popov triplet with A antistable and R invertible and let $\Sigma_r = (A, B; 0, L_r = L + X_r B, R)$ and $\Sigma_d = (A_d = -A^T, B_d = -L_r = -L - X_r B; 0, B, R)$ be the reduced equivalent and the dual Popov triplets associated with Σ, respectively, where X_r is the unique solution to the reducing Lyapunov equation $A^T X + XA + Q = 0$. Then the CTARE(Σ_r) has a nonsingular stabilizing solution X if and only if the CTARE(Σ_d) has a nonsingular stabilizing solution Z, and in this case*

$$X = -Z^{-1}.$$

Proof. Assume that the CTARE(Σ_d) has a nonsingular solution Z. Then

$$-AZ - ZA^T - (-ZL_r + B)R^{-1}(B^T - L_r^T Z) = 0 \tag{4.39}$$

and

$$\widetilde{A}_d := A_d + B_d F_{d,Ricc} = -A^T + L_r R^{-1}(B^T - L_r^T Z) \tag{4.40}$$

is stable. Let $X := -Z^{-1}$. From definition X is nonsingular. We show that X is the stabilizing solution of the CTARE(Σ_r). An immediate check using (4.39) shows that X satisfies

$$A^T X + XA - (XB + L_r)R^{-1}(B^T X + L_r^T) = 0 \tag{4.41}$$

which is exactly the CTARE(Σ_r). By (4.39) and (4.40) we also have successively

$$\begin{aligned} Z\widetilde{A}_d Z^{-1} &= Z[-A^T + L_r R^{-1}(B^T - L_r^T Z)]Z^{-1} \\ &= -ZA^T Z^{-1} + ZL_r R^{-1}(B^T Z^{-1} - L_r^T) \\ &= A + (-ZL_r + B)R^{-1}(B^T - L_r^T Z)Z^{-1} + ZL_r R^{-1}(B^T Z^{-1} - L_r^T) \\ &= A + BR^{-1}(B^T Z^{-1} - L_r^T) = A - BR^{-1}(B^T X + L_r^T). \end{aligned} \tag{4.42}$$

As $\widetilde{A}_d$ is stable, the rightmost term in (4.42) is stable too, which means that X is the stabilizing solution to the CTARE(Σ_r). The inverse implication follows by reversing the arguments. ■

Now we can proceed with the proof of Theorem 4.3.2.

Proof of Theorem 4.3.2 1. $\Rightarrow$ 2. Assume that the CTARE(Σ) has a stabilizing solution X such that $X - X_r$ is nonsingular. This means that the CTARE(Σ_r) has a nonsingular solution. According to Lemma 4.3.3, this implies that the CTARE(Σ_d) has a nonsingular stabilizing solution Z. By Theorem 4.1.1, the Toeplitz operator $\mathcal{R}_{\Sigma_d}$

is invertible. Now, by Lemma 3.3.5, we conclude that $\mathcal{R}_{\Sigma_r}$ is invertible. Furthermore, from Theorem 4.3.1 we have

$$Z = -\widehat{\Psi}\mathcal{R}_{\Sigma_r}^{-1}\widehat{\Psi}^*. \tag{4.43}$$

Since Z is nonsingular, assertion 2 follows. In particular, combining with Lemma 4.3.3, we obtain the following operator representation:

$$X = X_r + (\widehat{\Psi}\mathcal{R}_{\Sigma_r}^{-1}\widehat{\Psi}^*)^{-1}. \tag{4.44}$$

2. $\Rightarrow$ 1. Assume now that $\mathcal{R}_{\Sigma_r}$ and $\widehat{\Psi}\mathcal{R}_{\Sigma_r}^{-1}\widehat{\Psi}^*$ are invertible. We have from Lemma 3.3.5 that $\mathcal{R}_{\Sigma_d}$ is invertible. By Theorem 4.1.1, R is invertible and the CTARE(Σ_d) has a stabilizing solution, that can be represented as in (4.43). Since we assumed that $\widehat{\Psi}\mathcal{R}_{\Sigma_r}^{-1}\widehat{\Psi}^*$ is invertible, we deduce that the CTARE(Σ_d) has a nonsingular stabilizing solution. According to Lemma 4.3.3, the CTARE(Σ_r) has a stabilizing solution, thus the CTARE(Σ) has a stabilizing solution X with $X - X_r$ nonsingular. ■

4.3.2 *The positiveness case*

We explore now the *positiveness case* and show that the stabilizing solution that we have studied in Section 4.3.1 is closely related to a type of time–reversed linear quadratic optimal problem.

Corollary 4.3.4 *Let $\Sigma = (A, B; Q, L, R)$ be a Popov triplet with A antistable. We assume that:*

a. *$\mathcal{R}_{e,\Sigma}$ is coercive.*

b. *(A, B) is controllable.*

Then:

1. $R > 0$ and the CTARE(Σ) has a stabilizing solution X such that $X - X_r$ is positive definite.

2. We have

$$\min_{\substack{\widehat{\Psi}u = \xi \\ u \in L_+^{2,m}}} J_\Sigma(u) = \xi^T X \xi, \tag{4.45}$$

where

$$J_\Sigma(u) = \left\langle \begin{bmatrix} x^u \\ u \end{bmatrix}, \begin{bmatrix} Q & L \\ L^T & R \end{bmatrix} \begin{bmatrix} x^u \\ u \end{bmatrix} \right\rangle_{L_+^{2,n} \times L_+^{2,m}}.$$

Proof. 1. We check that point 2 of Theorem 4.3.2 holds. By Proposition 3.3.1, $\mathcal{R}_{e,\Sigma} = \mathcal{R}_{e,\Sigma_r}$. From (3.64), we deduce that $\mathcal{R}_{\Sigma_r}$ is coercive, and thus invertible. Of course, its inverse is also coercive. On the other hand, by the controllability assumption, the anticausal controllability Gramian $\widehat{P} = \widehat{\Psi}\widehat{\Psi}^*$ is positive definite. Take a $\nu > 0$ such that $\xi^T\widehat{P}\xi \geq \nu||\xi||^2$ and a $\delta > 0$ such that $\langle u, \mathcal{R}_{\Sigma_r}^{-1}u\rangle \geq \delta||u||^2$. Then we have

$$\begin{aligned} \xi^T\widehat{\Psi}\mathcal{R}_{\Sigma_r}^{-1}\widehat{\Psi}^*\xi &= \left\langle \widehat{\Psi}^*\xi, \mathcal{R}_{\Sigma_r}^{-1}\widehat{\Psi}^*\xi \right\rangle \geq \delta||\widehat{\Psi}^*\xi||^2 \\ &= \delta\left\langle \xi, \widehat{\Psi}\widehat{\Psi}^*\xi \right\rangle = \delta\xi^T\widehat{P}\xi \geq \delta\nu||\xi||^2, \ \forall \xi \in \mathbb{R}^n. \end{aligned} \tag{4.46}$$

We deduce that $\widehat{\Psi}\mathcal{R}_{\Sigma_r}^{-1}\widehat{\Psi}^*$ is positive definite, and therefore invertible. The existence of the stabilizing solution follows from Theorem 4.3.2 and the sign of $X - X_r$ follows from the operator representation (4.44), and assertion 1 is proved.

2. By Proposition 3.2.1, for each $u \in L_+^{2,m}$ such that $x^u(0) = \widehat{\Psi}u = \xi$, we have

$$J_\Sigma(u) = J_{\Sigma_r}(u) + \xi^T X_r \xi, \tag{4.47}$$

and by (3.61),

$$J_{\Sigma_r}(u) = \langle u, \mathcal{R}_{\Sigma_r} u\rangle$$

where $\mathcal{R}_{\Sigma_r} >> 0$. Therefore the problem reduces to compute

$$\min_{\widehat{\Psi}u=\xi} J_{\Sigma_r}(u).$$

To this end, we zero the Frechet derivative of the Lagrange functional

$$F(u) = J_{\Sigma_r}(u) + 2\lambda^T(\widehat{\Psi}u - \xi)$$

where λ is the Lagrange multiplier, and get

$$\mathcal{R}_{\Sigma_r} u + \widehat{\Psi}^*\lambda = 0$$

from where

$$u = -\mathcal{R}_{\Sigma_r}^{-1}\widehat{\Psi}^*\lambda. \tag{4.48}$$

As u must also satisfy $\Psi u = \xi$, one gets with (4.48) that

$$\lambda = -(\widehat{\Psi}\mathcal{R}_{\Sigma_r}^{-1}\widehat{\Psi}^*)^{-1}\xi$$

and (4.48) becomes

$$u^\xi = \mathcal{R}_{\Sigma_r}^{-1}\widehat{\Psi}^*(\widehat{\Psi}\mathcal{R}_{\Sigma_r}^{-1}\widehat{\Psi}^*)^{-1}\xi \tag{4.49}$$

which is exactly the optimal control. Thus we eventually obtain

$$J_{\Sigma_r}(u^\xi) = \xi^T(\widehat{\Psi}\mathcal{R}_{\Sigma_r}^{-1}\widehat{\Psi}^*)^{-1}\xi = \xi^T(X - X_r)\xi$$

and (4.45) follows immediately with (4.47). ■

4.4 The signature condition

As we have seen in Theorem 4.1.1, the existence of the stabilizing solution to the CTARE is intimately related to the existence of the bounded inverse of the Toeplitz operator $\mathcal{R}_\Sigma$. A simple case in which the invertibility of this operator is guaranteed is the so-called *positiveness case* that is the object of Corollary 4.1.7.

In this section a more elaborate sufficient condition for the invertibility of the Toeplitz operator $\mathcal{R}_\Sigma$ will be explored. This condition, called the *signature condition*, is relevant for game theoretic situations and will play a central role in solving several important problems in robust control as we shall see in Part III.

Let $\Sigma = (A, B; Q, L, R)$ be a Popov triplet with $A \in \mathbb{R}^{n\times n}$, $B \in \mathbb{R}^{n\times m}$, and A stable. Let the matrix B be partitioned as

$$B = [B_1 \;\; B_2], \;\; B_i \in \mathbb{R}^{n\times m_i}, \;\; i = 1, 2, \;\; m_1 + m_2 = m, \tag{4.50}$$

and the matrices L and R be partitioned in accordance with B, i.e.,

$$L = [L_1 \;\; L_2], \quad R = \begin{bmatrix} R_{11} & R_{12} \\ R_{12}^T & R_{22} \end{bmatrix}. \tag{4.51}$$

The solution to the initial value problem

$$\dot{x} = Ax + B_1u_1 + B_2u_2 \;, \quad x(0) = \xi, \tag{4.52}$$

will be written as

$$x^{\xi,u_1,u_2} = \Phi\xi + \mathcal{L}_1u_1 + \mathcal{L}_2u_2,$$

where $\mathcal{L}_i$, $i = 1, 2$, are the $\mathcal{L}$ operators (4.3) corresponding to the pairs (A, B_i). Following (4.2), the operator $\mathcal{R}_\Sigma$ can be partitioned conformably as

$$\mathcal{R}_\Sigma = \begin{bmatrix} \mathcal{R}_{11} & \mathcal{R}_{12} \\ \mathcal{R}_{12}^* & \mathcal{R}_{22} \end{bmatrix}. \tag{4.53}$$

It is easy to see that $\mathcal{R}_{22}$ coincides with the Toeplitz operator $\mathcal{R}_{\Sigma_2}$ associated with the Popov triplet

$$\Sigma_2 = (A, B_2; Q, L_2, R_{22}). \tag{4.54}$$

Now we are able to state the main result of this section.

Theorem 4.4.1 *Let $\Sigma = (A, B; Q, L, R)$ be a Popov triplet with its entries partitioned as in (4.50)–(4.51) and A stable. If $\mathcal{R}_{22} >> 0$ and*

$$\mathcal{R}_{11}^\times = \mathcal{R}_{11} - \mathcal{R}_{12}\mathcal{R}_{22}^{-1}\mathcal{R}_{12}^* << 0, \tag{4.55}$$

(see (4.53)) then we have:

1. *The CTARE(Σ_2) has a stabilizing solution X_2, where $\Sigma_2 = (A, B_2; Q, L_2, R_{22})$.*
2. *The CTARE(Σ) has a stabilizing solution X such that*

$$X \geq X_2. \tag{4.56}$$

Proof. 1. follows directly from Theorem 4.1.1 applied to the Popov triplet Σ_2.

2. We show that $\mathcal{R}_\Sigma$ is invertible to deduce from Theorem 4.1.1 the existence of the stabilizing solution X. We have

$$\mathcal{R}_\Sigma = \begin{bmatrix} I_1 & \mathcal{R}_{12}\mathcal{R}_{22}^{-1} \\ O & I_2 \end{bmatrix} \begin{bmatrix} \mathcal{R}_{11}^\times & O \\ O & \mathcal{R}_{22} \end{bmatrix} \begin{bmatrix} I_1 & O \\ \mathcal{R}_{22}^{-1}\mathcal{R}_{12}^* & I_2 \end{bmatrix},$$

where $I_i = I_{L_+^{2,m_i}}$, $i = 1, 2$. Accordingly, $\mathcal{R}_\Sigma$ is invertible and

$$\begin{aligned} \mathcal{R}_\Sigma^{-1} &= \begin{bmatrix} I_1 & O \\ -\mathcal{R}_{22}^{-1}\mathcal{R}_{12}^* & I_2 \end{bmatrix} \begin{bmatrix} (\mathcal{R}_{11}^\times)^{-1} & O \\ O & \mathcal{R}_{22}^{-1} \end{bmatrix} \begin{bmatrix} I_1 & -\mathcal{R}_{12}\mathcal{R}_{22}^{-1} \\ O & I_2 \end{bmatrix} \\ &= \begin{bmatrix} (\mathcal{R}_{11}^\times)^{-1} & -(\mathcal{R}_{11}^\times)^{-1}\mathcal{R}_{12}\mathcal{R}_{22}^{-1} \\ -\mathcal{R}_{22}^{-1}\mathcal{R}_{12}^*(\mathcal{R}_{11}^\times)^{-1} & \mathcal{R}_{22}^{-1} + \mathcal{R}_{22}^{-1}\mathcal{R}_{12}^*(\mathcal{R}_{11}^\times)^{-1}\mathcal{R}_{12}\mathcal{R}_{22}^{-1} \end{bmatrix}. \end{aligned} \tag{4.57}$$

We use now the representation formula (4.15) of the stabilizing solution to write

$$\begin{aligned} X &= \mathcal{P}_{o,\Sigma} - \mathcal{P}_\Sigma \mathcal{R}_\Sigma^{-1} \mathcal{P}_\Sigma^* \\ &= \mathcal{P}_{o,\Sigma} - \begin{bmatrix} \mathcal{P}_1 & \mathcal{P}_2 \end{bmatrix} \mathcal{R}_\Sigma^{-1} \begin{bmatrix} \mathcal{P}_1^* \\ \mathcal{P}_2^* \end{bmatrix}, \end{aligned}$$

where, according to (4.9),

$$\mathcal{P}_\Sigma = \Phi^*(Q\mathcal{L} + L) = \begin{bmatrix} \Phi^*(Q\mathcal{L}_1 + L_1) & \Phi^*(Q\mathcal{L}_2 + L_2) \end{bmatrix} =: \begin{bmatrix} \mathcal{P}_1 & \mathcal{P}_2 \end{bmatrix}.$$

Obviously, $\mathcal{P}_2 = \mathcal{P}_{\Sigma_2}$. Straightforward computations lead to

$$X = X_2 - (\mathcal{P}_1 + \mathcal{P}_2 \mathcal{R}_{22}^{-1} \mathcal{R}_{12}^*)(\mathcal{R}_{11}^\times)^{-1}(\mathcal{P}_1 + \mathcal{P}_2 \mathcal{R}_{22}^{-1} \mathcal{R}_{12}^*)^*, \tag{4.58}$$

where X_2 is the stabilizing solution of the CTARE(Σ_2) and is given by the same formula (4.15),

$$X_2 = \mathcal{P}_{o,\Sigma_2} - \mathcal{P}_{\Sigma_2} \mathcal{R}_{\Sigma_2}^{-1} \mathcal{P}_{\Sigma_2}^* = \mathcal{P}_{o,\Sigma} - \mathcal{P}_2 \mathcal{R}_{22}^{-1} \mathcal{P}_2^*.$$

The inequality (4.56) follows directly from (4.58) and the fact that $\mathcal{R}_{11}^\times << 0$. ■

The following result shows that the situation considered in Theorem 4.4.1 is related to a maxmin quadratic problem, or in another interpretation, an open–loop quadratic game optimization problem.

Theorem 4.4.2 *With the same notation and under the same hypotheses as in Theorem 4.4.1, let*

$$J_\Sigma(\xi, u_1, u_2) = \left\langle \begin{bmatrix} x^{\xi,u_1,u_2} \\ u_1 \\ u_2 \end{bmatrix}, \begin{bmatrix} Q & L_1 & L_2 \\ L_1^T & R_{11} & R_{12} \\ L_2^T & R_{12}^T & R_{22} \end{bmatrix} \begin{bmatrix} x^{\xi,u_1,u_2} \\ u_1 \\ u_2 \end{bmatrix} \right\rangle$$

be the quadratic index (3.6) for all $(\xi, u_1, u_2) \in \mathbb{R}^n \times L_+^{2,m_1} \times L_+^{2,m_2}$. *Then:*

1. For each $\xi \in \mathbb{R}^n$ *and each* $u_1 \in L_+^{2,m_1}$, *there exists a unique* u_2 *in* L_+^{2,m_2}, *that we denote by* u_2^{ξ,u_1}, *that minimizes* $J_\Sigma(\xi, u_1, u_2)$, *i.e.,*

$$J_\Sigma(\xi, u_1, u_2) \geq J_\Sigma(\xi, u_1, u_2^{\xi,u_1}) \tag{4.59}$$

for all $u_2 \in L_+^{2,m_2}$.

2. For each ξ *in* $\mathbb{R}^n$, *there exists a unique* u_1, *that we denote* u_1^ξ, *that maximizes* $J_\Sigma(\xi, u_1, u_2^{\xi,u_1})$.

3. If we denote $u_2^\xi = u_2^{\xi,u_1^\xi}$, *then*

$$J_\Sigma(\xi, u_1^\xi, u_2^\xi) = \xi^T X \xi \tag{4.60}$$

for every $\xi \in \mathbb{R}^n$, *where* X *is the stabilizing solution to the CTARE(Σ).*

Moreover, if x^ξ *is the solution of (4.52) corresponding to* $u_1 = u_1^\xi$ *and* $u_2 = u_2^\xi$, *then*

$$u^\xi = \begin{bmatrix} u_1^\xi \\ u_2^\xi \end{bmatrix} = F_{Ricc}\, x^\xi, \tag{4.61}$$

where F_{Ricc} *is the stabilizing Riccati feedback.*

Remark 4.4.3 The conclusion of Theorem 4.4.2 can be synthetically expressed as

$$\max_{u_1} \min_{u_2} J_\Sigma(\xi, u_1, u_2) = \xi^T X \xi. \tag{4.62}$$

Also the maxmin solution is available in the "feedback" form (4.61). As it will appear from the proof, the notations u^ξ and x^ξ introduced in Theorem 4.4.2 are consistent with the notation introduced in Section 4.1. It will appear therefore that (4.61) is simply a repetition of relation (4.18), whereas (4.62) is a consequence of (4.20).

Proof. 1. In the operator form of the quadratic index (3.49), we perform a "square separation",

$$\begin{aligned}
&J_\Sigma(\xi, u_1, u_2)\\
&= \left\langle \begin{bmatrix} \xi \\ u_1 \\ u_2 \end{bmatrix} \begin{bmatrix} \mathcal{P}_{o,\Sigma} & \mathcal{P}_1 & \mathcal{P}_2 \\ \mathcal{P}_1^* & \mathcal{R}_{11} & \mathcal{R}_{12} \\ \mathcal{P}_2^* & \mathcal{R}_{12}^* & \mathcal{R}_{22} \end{bmatrix} \begin{bmatrix} \xi \\ u_1 \\ u_2 \end{bmatrix} \right\rangle \\
&= \left\langle \begin{bmatrix} \xi \\ u_1 \end{bmatrix}, \begin{bmatrix} \mathcal{P}_{o,\Sigma} & \mathcal{P}_1 \\ \mathcal{P}_1^* & \mathcal{R}_{11} \end{bmatrix} \begin{bmatrix} \xi \\ u_1 \end{bmatrix} \right\rangle + \left\langle u_2, \begin{bmatrix} \mathcal{P}_2^* & \mathcal{R}_{12}^* \end{bmatrix} \begin{bmatrix} \xi \\ u_1 \end{bmatrix} \right\rangle \\
&\quad + \left\langle \begin{bmatrix} \xi \\ u_1 \end{bmatrix}, \begin{bmatrix} \mathcal{P}_2 \\ \mathcal{R}_{12} \end{bmatrix} u_2 \right\rangle + \langle u_2, \mathcal{R}_{22} u_2 \rangle \\
&= \left\langle \begin{bmatrix} \xi \\ u_1 \end{bmatrix}, \left(\begin{bmatrix} \mathcal{P}_{o,\Sigma} & \mathcal{P}_1 \\ \mathcal{P}_1^* & \mathcal{R}_{11} \end{bmatrix} - \begin{bmatrix} \mathcal{P}_2 \\ \mathcal{R}_{12} \end{bmatrix} \mathcal{R}_{22}^{-1} \begin{bmatrix} \mathcal{P}_2^* & \mathcal{R}_{12}^* \end{bmatrix} \right) \begin{bmatrix} \xi \\ u_1 \end{bmatrix} \right\rangle \\
&\quad + \left\langle u_2 + \mathcal{R}_{22}^{-1} \begin{bmatrix} \mathcal{P}_2^* & \mathcal{R}_{12}^* \end{bmatrix} \begin{bmatrix} \xi \\ u_1 \end{bmatrix}, \mathcal{R}_{22} \left(u_2 + \mathcal{R}_{22}^{-1} \begin{bmatrix} \mathcal{P}_2^* & \mathcal{R}_{12}^* \end{bmatrix} \begin{bmatrix} \xi \\ u_1 \end{bmatrix} \right) \right\rangle.
\end{aligned}$$

Since $\mathcal{R}_{22} >> 0$, the minimum of $J_\Sigma(\xi, u_1, u_2)$ is unique and is attained for

$$u_2^{\xi, u_1} = -\mathcal{R}_{22}^{-1} \begin{bmatrix} \mathcal{P}_2^* & \mathcal{R}_{12}^* \end{bmatrix} \begin{bmatrix} \xi \\ u_1 \end{bmatrix}. \tag{4.63}$$

2. We substitute (4.63) into the expression of $J_\Sigma(\xi, u_1, u_2)$ to obtain

$$J_\Sigma(\xi, u_1, u_2^{\xi,u_1}) = \left\langle \begin{bmatrix} \xi \\ u_1 \end{bmatrix}, \begin{bmatrix} \mathcal{P}_{o,\Sigma} - \mathcal{P}_2 \mathcal{R}_{22}^{-1} \mathcal{P}_2^* & \mathcal{P}_1 - \mathcal{P}_2 \mathcal{R}_{22}^{-1} \mathcal{R}_{12}^* \\ \mathcal{P}_1^* - \mathcal{R}_{12} \mathcal{R}_{22}^{-1} \mathcal{P}_2^* & \mathcal{R}_{11}^\times \end{bmatrix} \begin{bmatrix} \xi \\ u_1 \end{bmatrix} \right\rangle. \tag{4.64}$$

Performing again a "square separation" in which we also make use of relation (4.58), we have

$$\begin{aligned}
J_\Sigma(\xi, u_1, u_2^{\xi,u_1}) &= \langle \xi, X\xi \rangle \\
+ \langle u_1 + (\mathcal{R}_{11}^\times)^{-1}(\mathcal{P}_1^* - \mathcal{R}_{12}\mathcal{R}_{22}^{-1}\mathcal{P}_2^*)\xi&, \mathcal{R}_{11}^\times [u_1 + (\mathcal{R}_{11}^\times)^{-1}(\mathcal{P}_1^* - \mathcal{R}_{12}\mathcal{R}_{22}^{-1}\mathcal{P}_2^*)\xi] \rangle .
\end{aligned}$$

Since $\mathcal{R}_{11}^\times << 0$, the previous expression takes its unique maximum value in

$$u_1^\xi = -(\mathcal{R}_{11}^\times)^{-1}(\mathcal{P}_1^* - \mathcal{R}_{12}\mathcal{R}_{22}^{-1}\mathcal{P}_2^*)\xi. \tag{4.65}$$

3. Substituting (4.65) into (4.63), we obtain

$$\begin{aligned}
u_2^\xi &= -\mathcal{R}_{22}^{-1} \begin{bmatrix} \mathcal{P}_2^* & \mathcal{R}_{12}^* \end{bmatrix} \begin{bmatrix} \xi \\ u_1^\xi \end{bmatrix} \\
&= -\mathcal{R}_{22}^{-1}[\mathcal{P}_2^* - \mathcal{R}_{12}^*(\mathcal{R}_{11}^\times)^{-1}(\mathcal{P}_1^* - \mathcal{R}_{12}\mathcal{R}_{22}^{-1}\mathcal{P}_2^*)]\xi.
\end{aligned}$$

Comparing the expressions of u_1^ξ and u_2^ξ with (4.57), we realize that

$$\begin{bmatrix} u_1^\xi \\ u_2^\xi \end{bmatrix} = -\mathcal{R}_\Sigma^{-1}\mathcal{P}_\Sigma^*\xi, \tag{4.66}$$

that is (4.66) coincides with the input u^ξ given by (4.13). Now (4.60) and (4.61) are simple consequences of (4.20) and (4.18). ■

Remark 4.4.4 The quadratic index associated with the Popov triplet Σ_2 is

$$J_{\Sigma_2}(\xi, u_2) = J_\Sigma(\xi, 0, u_2).$$

Since Corollary 4.1.7 is applicable to the Popov triplet Σ_2 we deduce that

$$\min_{u_2} J_\Sigma(\xi, 0, u_2) = \min_{u_2} J_{\Sigma_2}(\xi, u_2) = \xi^T X_2 \xi. \tag{4.67}$$

Combining this with (4.62), we obtain yet another proof of the fact that $X \geq X_2$. Further, we see that $\xi^T X \xi = \xi^T X_2 \xi$ if and only if $u_1^\xi = 0$.

4.5 Frequency domain conditions

We have given so far a number of (necessary and) sufficient conditions for the existence of the stabilizing solution to the CTARE. These conditions were expressed in terms of various operators that we associated with a Popov triplet. In this section we are expressing these conditions in frequency domain terms. Traditionally, this has been a very important issue and one of the points that made Positivity Theory so attractive. Frequency domain conditions were "easy" to check. It was highly convenient therefore to have a link between such conditions and state–space properties. With modern control techniques, the accent has been somehow shifted. The frequency domain H^∞ constraints are being achieved using state–space manipulations that are much more convenient for multivariable systems. Nonetheless, the interest in connections between state–space domain properties and frequency domain properties remains high.

4.5.1 The main result

As we have seen in Section 4.1, the existence of the stabilizing solution to the CTARE (4.1) was related to the existence of the bounded inverse of a Toeplitz operator $\mathcal{R}_\Sigma$. There is a well–known group of operator–theoretic results linking the invertibility properties of a Toeplitz operator to certain factorization properties of the symbol. We use such a result to give a frequency domain complement of Theorems 4.1.1 and 4.2.1.

First, some concepts that allow a simpler formulation of our result.

Definition 4.5.1 *Let* $\Sigma = (A, B; Q, L, R)$ *be a Popov triplet and let* $\mathbf{\Pi}_\Sigma$ *be the associated Popov function, that is*

$$\mathbf{\Pi}_\Sigma(s) = \begin{bmatrix} B^T(-sI - A^T)^{-1} & I \end{bmatrix} \begin{bmatrix} Q & L \\ L^T & R \end{bmatrix} \begin{bmatrix} (sI - A)^{-1}B \\ I \end{bmatrix}. \tag{4.68}$$

If A is stable and there exist $\mathbf{\Xi}$ and $\mathbf{\Omega}$, two units in $RH^{\infty}_{+,m\times m}$ such that

$$\mathbf{\Pi}_{\Sigma}(s) = [\mathbf{\Xi}(s)]^{*}\mathbf{\Omega}(s), \tag{4.69}$$

then we shall say that $\mathbf{\Pi}_{\Sigma}$ is antianalytic factorizable.

If the pair (A,B) is stabilizable, and there exists F such that $A+BF$ is stable, and the Popov function $\mathbf{\Pi}_{\widetilde{\Sigma}}$ is antianalytic factorizable, where $\widetilde{\Sigma}$ is the F–equivalent of Σ, then we say that $\mathbf{\Pi}_{\Sigma}$ is antianalytic prefactorizable.

An easy consequence of Proposition 3.2.16 is the following.

Proposition 4.5.2 *Let $\Sigma = (A,B;Q,L,R)$ be a Popov triplet with the pair (A,B) stabilizable. If there exists an F such that $A+BF$ is stable and for which $\mathbf{\Pi}_{\widetilde{\Sigma}}$ is antianalytic factorizable, where $\widetilde{\Sigma}$ is the F–equivalent of Σ, then $\mathbf{\Pi}_{\widetilde{\Sigma}}$ is antianalytic factorizable for* every *F–equivalent triplet $\widetilde{\Sigma}$ for which $A+BF$ is stable.*

With these preparations we can state our main result in frequency domain setting.

Theorem 4.5.3 *Let $\Sigma = (A,B;P)$ be a Popov triplet with A stable. Then the following two assertions are equivalent.*

1. R is nonsingular and the CTARE(Σ) has a stabilizing solution.

2. The Popov function $\mathbf{\Pi}_{\Sigma}$ is antianalytic factorizable.

Proof. 1. $\Rightarrow$ 2. Let X be the stabilizing solution and F_{Ricc} the stabilizing Riccati feedback. If $\widetilde{\Sigma}$ is the (X,F_{Ricc})–equivalent of Σ, then $\widetilde{Q}=0$, $\widetilde{L}=0$ and consequently $\mathbf{\Pi}_{\widetilde{\Sigma}}(s) = R$. By Proposition 3.2.16,

$$\mathbf{\Pi}_{\Sigma}(s) = (I - F_{Ricc}(sI-A)^{-1}B)^{*}R(I - F_{Ricc}(sI-A)^{-1}B).$$

For an antianalytic factorization, we may take for example

$$\mathbf{\Omega}(s) := R(I - F_{Ricc}(sI-A)^{-1}B)$$

and

$$\mathbf{\Xi}(s) := I - F_{Ricc}(sI-A)^{-1}B.$$

Indeed, they are both in $RH^{\infty}_{+,m\times m}$ since A is stable and their inverses are in $RH^{\infty}_{+,m\times m}$ since $A+BF_{Ricc}$ is stable.

2. $\Rightarrow$ 1. Let $\mathbf{\Pi}_{\Sigma} = \mathbf{\Xi}^{*}\mathbf{\Omega}$, where $\mathbf{\Xi}$ and $\mathbf{\Omega}$ are units in $RH^{\infty}_{+,m\times m}$. First of all, this implies that $R = \mathbf{\Pi}_{\Sigma}(\infty) = \mathbf{\Xi}^{*}(\infty)\mathbf{\Omega}(\infty)$ is invertible.

According to Proposition 2.4.12 we have

$$\mathcal{R}_{\Sigma} = \mathbb{T}_{\mathbf{\Pi}_{\Sigma}} = \mathbb{T}^{*}_{\mathbf{\Xi}}\mathbb{T}_{\mathbf{\Omega}}.$$

By Corollary 2.4.14, both $\mathbb{T}^{*}_{\mathbf{\Xi}}$ and $\mathbb{T}_{\mathbf{\Omega}}$ are invertible. Thus $\mathcal{R}_{\Sigma}$ is invertible and assertion 1 follows from Theorem 4.1.1. ∎

We remove now the stability assumption in a similar way as we did in the case of Theorem 4.2.1 and get the following result.

Theorem 4.5.4 *Let $\Sigma = (A,B;P)$ be a Popov triplet with (A,B) stabilizable. Then the following two assertions are equivalent.*

1. R is nonsingular and the CTARE(Σ) has a stabilizing solution.

2. The Popov function $\mathbf{\Pi}_{\Sigma}$ is antianalytic prefactorizable.

4.5.2 The positiveness case: A stable

For the positiveness case we have the following result that allows for a frequency domain formulation of Corollary 4.1.7 and Corollary 4.1.8.

Corollary 4.5.5 *Let $\Sigma = (A, B; P)$ be a Popov triplet with A stable. Then we have:*
1. There exists $\rho > 0$ such that $\mathbf{\Pi}_\Sigma(j\omega) \geq \rho I$ for each $\omega \in \mathbb{R}$, if and only if $\mathcal{R}_{e,\Sigma}$ is coercive.
2. $\mathbf{\Pi}_\Sigma(j\omega) \geq 0$ for each $\omega \in \mathbb{R}$, if and only if $\mathcal{R}_{e,\Sigma} \geq 0$.

Proof. For $u \in L^{2,m}$, we denote by $\widehat{u}$ its Fourier transform. By Parseval's identity, we have

$$\langle u, \mathcal{R}_{e,\Sigma} u \rangle_{L^{2,m}} = \frac{1}{2\pi} \int_{-\infty}^{\infty} \langle \widehat{u}(j\omega), \mathbf{\Pi}_\Sigma(j\omega)\widehat{u}(j\omega) \rangle \, d\omega. \tag{4.70}$$

1. If for some $\rho > 0$ we have $\mathbf{\Pi}_\Sigma(j\omega) \geq \rho I$, then from (4.70), it results

$$\langle u, \mathcal{R}_{e,\Sigma} u \rangle_{L^{2,m}} \geq \frac{1}{2\pi} \rho ||\widehat{u}||_2^2 = \rho ||u||_2^2$$

and $\mathcal{R}_{e,\Sigma}$ is coercive.

Conversely, assume that $\mathcal{R}_{e,\Sigma}$ is coercive. Then there exists $\mu > 0$ such that $\mathcal{R}_{e,\Sigma} \geq \mu I$. If it does not exist $\rho > 0$ such that $\mathbf{\Pi}_\Sigma(j\omega) \geq \rho I, \forall \omega \in \mathbb{R}$, then for each $\rho > 0$ there exists $\omega_0 \in \mathbb{R}$ such that $\mathbf{\Pi}_\Sigma(j\omega_o) - \rho I$ is not positive semidefinite. Take $0 < \rho < \mu$. Then there exists a nonzero $u_o \in \mathbb{C}^m$ such that

$$\langle u_o, \mathbf{\Pi}_\Sigma(j\omega_o) u_o \rangle_{\mathbb{C}^m} \leq \rho ||u_o||^2.$$

By continuity, there exists an $\varepsilon > 0$ such that

$$\langle u_o, \mathbf{\Pi}_\Sigma(j\omega) u_o \rangle_{\mathbb{C}^m} < \mu ||u_o||^2$$

for all ω such that $|\omega - \omega_o| < \frac{\varepsilon}{2}$. Let us define

$$\widehat{u}_\varepsilon(j\omega) = \begin{cases} u_o, & |\omega - \omega_o| < \frac{\varepsilon}{2} \\ 0, & \text{otherwise.} \end{cases}$$

Then $||\widehat{u}_\varepsilon||_2 = ||u_o||\varepsilon$. This implies

$$\langle u_\varepsilon, \mathcal{R}_{e,\Sigma} u_\varepsilon \rangle_{L^{2,m}} \geq \frac{1}{2\pi} ||u_o||^2 \mu\varepsilon,$$

where u_ε is the inverse Fourier transform of $\widehat{u}_\varepsilon$. However, by construction,

$$\langle \widehat{u}_\varepsilon, \mathbf{\Pi}_\Sigma \widehat{u}_\varepsilon \rangle_{L^{2,m}(\mathbb{C}_0)} < \mu ||u_o||^2 \varepsilon$$

from which

$$\frac{1}{2\pi} \langle \widehat{u}_\varepsilon, \mathbf{\Pi}_\Sigma \widehat{u}_\varepsilon \rangle_{L^{2,m}(\mathbb{C}_0)} < \langle u_\epsilon, \mathcal{R}_{e,\Sigma} u_\epsilon \rangle_{L^{2,m}}$$

which contradicts the identity (4.70).

The proof of assertion 2 is perfectly similar. ■

Corollary 4.5.6 *Corollaries 4.1.7 and 4.1.8 hold if conditions $\mathcal{R}_{e,\Sigma} >> 0$ and $\mathcal{R}_{e,\Sigma} \geq 0$ are replaced by their frequency domain counterparts $\mathbf{\Pi}_\Sigma > 0$ on $\overline{\mathbb{C}}_0$ and $\mathbf{\Pi}_\Sigma \geq 0$ on $\mathbb{C}_0$, respectively.*

4.5.3 The positiveness case: A arbitrary

We remove now the stability condition and get the following result.

Corollary 4.5.7 *Let $\Sigma = (A, B; Q, L, R)$ be a Popov triplet. Assume that:*
a. *The pair (A, B) is stabilizable.*
b. *The realization of $\mathbf{\Pi}_\Sigma$,*

$$\mathbf{\Pi}_\Sigma = \left[\begin{array}{cc|c} A & O & B \\ -Q & -A^T & -L \\ \hline L^T & B^T & R \end{array}\right], \tag{4.71}$$

is $\mathbb{C}_0$–controllable and $\mathbb{C}_0$–observable.
c. $\mathbf{\Pi}_\Sigma(j\omega) > 0, \quad \forall \omega \in \overline{\mathbb{R}}$.
Then:
1. The CTARE(Σ) has a stabilizing solution X.
2. We have

$$\min_{u \in \mathcal{U}^\xi} J_\Sigma(\xi, u) = \xi^T X \xi$$

where $\mathcal{U}_\xi := \{u : u \in L_+^{2,m}$ such that $x^{\xi,u} \in L_+^{2,n}\}$.
3. If, in addition,

$$\begin{bmatrix} Q & L \\ L^T & R \end{bmatrix} \geq 0, \tag{4.72}$$

then $X \geq 0$.

Proof. Let F be such that $\widetilde{A} := A + BF$ is stable and let $\widetilde{\Sigma} = (\widetilde{A}, \widetilde{B}; \widetilde{Q}, \widetilde{L}, \widetilde{R})$ be the F–equivalent of Σ. According to (3.37) we have

$$\mathbf{\Pi}_{\widetilde{\Sigma}}(j\omega) \geq 0, \quad \forall \omega \in \overline{\mathbb{R}}. \tag{4.73}$$

Since both realizations of $\mathbf{\Pi}_\Sigma$ and

$$\mathbf{\Pi}_{\widetilde{\Sigma}} = \left[\begin{array}{cc|c} \widetilde{A} & O & \widetilde{B} \\ -\widetilde{Q} & -\widetilde{A}^T & -\widetilde{L} \\ \hline \widetilde{L}^T & \widetilde{B}^T & \widetilde{R} \end{array}\right]$$

share the same zeros (as their transmission pencils, which are the EHP(Σ) and the EHP($\widetilde{\Sigma}$), are related through (3.33)), hypotheses b and c imply that the inequality (4.73) is strict. By Corollary 4.5.5, we see that we can apply Corollary 4.1.7 to $\widetilde{\Sigma}$. Assertions 1 and 2 follow now if we relate back to the original Popov triplet with the help of Corollary 3.2.7 and Proposition 3.2.1.

Assertion 3 is a simple consequence of the fact that (4.72) implies that $J_\Sigma(\xi, u) \geq 0$ for each ξ and $u \in \mathcal{U}^\xi$. ∎

4.5.4 The signature condition in the frequency domain

We give now a frequency domain version of Theorem 4.4.1. In fact we shall prove a stronger result for the corresponding KPYS(Σ). Consider again a Popov triplet

$\Sigma = (A, B; Q, L, R)$ with B, L and R partitioned as

$$B = \begin{bmatrix} B_1 & B_2 \end{bmatrix}, L = \begin{bmatrix} L_1 & L_2 \end{bmatrix}, R = \begin{bmatrix} R_{11} & R_{12} \\ R_{12}^T & R_{22} \end{bmatrix}, \tag{4.74}$$

($B_i \in \mathbb{R}^{n \times m_i}, i = 1, 2, m = m_1 + m_2$). Let the Popov function $\mathbf{\Pi}_\Sigma$ be partitioned in accordance with R in (4.74), i.e.,

$$\mathbf{\Pi}_\Sigma = \begin{bmatrix} \mathbf{\Pi}_{11} & \mathbf{\Pi}_{12} \\ \mathbf{\Pi}_{12}^* & \mathbf{\Pi}_{22} \end{bmatrix}. \tag{4.75}$$

Introduce also the sign matrix

$$J := \begin{bmatrix} -I_{m_1} & \\ & I_{m_2} \end{bmatrix}. \tag{4.76}$$

Remember also the Popov triplet $\Sigma_2 = (A, B_2; Q, L_2, R_{22})$. It is easy to see that $\mathbf{\Pi}_{22} = \mathbf{\Pi}_{\Sigma_2}$.

A sufficient condition

Theorem 4.5.8 *Let $\Sigma = (A, B; Q, L, R)$ be a Popov triplet with A stable and its entries partitioned as in (4.74). Assume $\mathbf{\Pi}_{22} > 0$ on $\overline{\mathbb{C}}_0$ (see (4.75)). If there exists an $\mathbf{S} \in RH^{\infty}_{+, m_2 \times m_1}$ such that*

$$\begin{bmatrix} I & \mathbf{S}^* \end{bmatrix} \mathbf{\Pi}_\Sigma \begin{bmatrix} I \\ \mathbf{S} \end{bmatrix} < 0 \text{ on } \overline{\mathbb{C}}_0, \tag{4.77}$$

then the KPYS(Σ, J) has a stabilizing solution

$$(X, V, W) = (X, \begin{bmatrix} V_{11} & O \\ V_{21} & V_{22} \end{bmatrix}, \begin{bmatrix} W_1 \\ W_2 \end{bmatrix})$$

where the partitions of V and W are in accordance with J given in (4.76).

Proof. We prove first that the CTARE(Σ) has a stabilizing solution by showing that the hypotheses of Theorem 4.4.1 are satisfied.

The fact $\mathcal{R}_{22} >> 0$ follows from the positiveness assumption on $\mathbf{\Pi}_{22}$ using Corollary 4.5.5 for the triplet Σ_2. To prove that $\mathcal{R}_{11}^{\times} = \mathcal{R}_{11} - \mathcal{R}_{12}\mathcal{R}_{22}^{-1}\mathcal{R}_{12}^* << 0$, let us denote the input–output operator corresponding to $\mathbf{S}$ with $\mathcal{S}_e$. Since $\mathbf{S} \in RH^{\infty}_{+, m_2 \times m_1}$, $\mathcal{S}_e$ is causal. Let $\mathcal{S}$ denote the Toeplitz operator corresponding to $\mathcal{S}_e$. We introduce

$$\mathcal{T}_e := \begin{bmatrix} I_1 & \mathcal{S}_e^* \\ O & I_2 \end{bmatrix} \begin{bmatrix} \mathcal{R}_{e11} & \mathcal{R}_{e12} \\ \mathcal{R}_{e12}^* & \mathcal{R}_{e22} \end{bmatrix} \begin{bmatrix} I_1 & O \\ \mathcal{S}_e & I_2 \end{bmatrix} = \begin{bmatrix} \mathcal{T}_{e11} & \mathcal{T}_{e12} \\ \mathcal{T}_{e12}^* & \mathcal{T}_{e22} \end{bmatrix}, \tag{4.78}$$

where $I_1 := I_{L^{2,m_1}}$, $I_2 := I_{L^{2,m_2}}$. Notice that the transfer function of $\mathcal{T}_{e11}$ is the transfer function of the left–hand side of (4.77). Therefore $\mathcal{T}_{e11} << 0$. Also $\mathcal{T}_{e22} = \mathcal{R}_{e22} >> 0$. Since $\mathcal{S}_e$ is causal, we can apply Corollary 2.4.13 to write

$$\mathcal{T} := \mathbb{T}_{\mathcal{T}_e} = \begin{bmatrix} I_1^+ & \mathcal{S}^* \\ O & I_2^+ \end{bmatrix} \begin{bmatrix} \mathcal{R}_{11} & \mathcal{R}_{12} \\ \mathcal{R}_{12}^* & \mathcal{R}_{22} \end{bmatrix} \begin{bmatrix} I_1^+ & O \\ \mathcal{S} & I_2^+ \end{bmatrix} = \begin{bmatrix} \mathcal{T}_{11} & \mathcal{T}_{12} \\ \mathcal{T}_{12}^* & \mathcal{T}_{22} \end{bmatrix}$$

where $I_1^+ := I_{L_+^{2,m_1}}$, $I_2^+ := I_{L_+^{2,m_2}}$. This means that

$$\begin{bmatrix} \mathcal{R}_{11} & \mathcal{R}_{12} \\ \mathcal{R}_{12}^* & \mathcal{R}_{22} \end{bmatrix} = \begin{bmatrix} I_1^+ & -\mathcal{S}^* \\ O & I_2^+ \end{bmatrix} \begin{bmatrix} \mathcal{T}_{11} & \mathcal{T}_{12} \\ \mathcal{T}_{12}^* & \mathcal{T}_{22} \end{bmatrix} \begin{bmatrix} I_1^+ & O \\ -\mathcal{S} & I_2^+ \end{bmatrix}.$$

Using this identity straightforward computations lead to $\mathcal{R}_{11}^\times = \mathcal{T}_{11} - \mathcal{T}_{12}\mathcal{T}_{22}^{-1}\mathcal{T}_{12}^*$. Since $\mathcal{T}_{11} << 0$ and $\mathcal{T}_{22} = \mathcal{R}_{22} >> 0$ the conclusion follows.

In order to show that the KPYS(Σ, J) has a stabilizing solution (X, V, W), with V of lower–left block triangular form, notice that $R_{22} = \mathbf{\Pi}_{22}(j\infty) > 0$. Further, if we denote $D_S = \mathbf{S}(\infty)$, then we have from (4.77) that

$$\begin{bmatrix} I & D_S^T \end{bmatrix} \begin{bmatrix} R_{11} & R_{12} \\ R_{12}^T & R_{22} \end{bmatrix} \begin{bmatrix} I \\ D_S \end{bmatrix} < 0$$

which can be rewritten as

$$R_{11} - R_{12}R_{22}^{-1}R_{12}^T + (R_{12} + D_S^T R_{22})R_{22}^{-1}(R_{12}^T + R_{22}D_S) < 0.$$

Since $R_{22} > 0$ it follows that $R_{11}^\times := R_{11} - R_{12}R_{22}^{-1}R_{12}^T < 0$. Hence we can write

$$R = \begin{bmatrix} R_{11} & R_{12} \\ R_{12}^T & R_{22} \end{bmatrix} = \begin{bmatrix} I & R_{12}R_{22}^{-1} \\ O & I \end{bmatrix} \begin{bmatrix} R_{11}^\times & O \\ O & R_{22} \end{bmatrix} \begin{bmatrix} I & O \\ R_{22}^{-1}R_{12}^T & I \end{bmatrix} = V^T J V$$

for

$$V = \begin{bmatrix} (-R_{11}^\times)^{\frac{1}{2}} & O \\ R_{22}^{-\frac{1}{2}}R_{12}^T & R_{22}^{\frac{1}{2}} \end{bmatrix} = \begin{bmatrix} V_{11} & O \\ V_{21} & V_{22} \end{bmatrix}.$$

The conclusion follows by invoking Remark 3.2.9. ■

A necessary and sufficient condition

As we see from the proof of Theorem 4.5.8, condition (4.77) is merely a sufficient condition in order to satisfy (4.55), which in turn is a sufficient condition for the invertibility of $\mathcal{R}_\Sigma$ (as $\mathcal{R}_{22} >> 0$). Since we know that the existence of the stabilizing solution is equivalent to the invertibility of $\mathcal{R}_\Sigma$, it is quite interesting to find a situation in which the existence of the stabilizing solution implies condition (4.77). This is the aim of our next result which is the cornerstone of our solutions to several robust control problems in Part III: the Nehari problem, the H^∞–control problem, and robust stabilization.

Theorem 4.5.9 *Let $\Sigma = (A, B; Q, L, R)$ be a Popov triplet with its entries partitioned as in (4.74). Assume that:*

a. *A is stable.*

b. *$\mathbf{\Pi}_{22} > 0$ on $\overline{\mathbb{C}}_0$.*

c.

$$\begin{bmatrix} Q & L_2 \\ L_2^T & R_{22} \end{bmatrix} \geq 0.$$

Then we have:

I. *The following assertions are equivalent:*

1. There exists $\mathbf{S} \in RH^{\infty}_{+,m_2 \times m_1}$ *which satisfies the signature inequality*

$$\begin{bmatrix} I & \mathbf{S}^* \end{bmatrix} \mathbf{\Pi}_{\Sigma} \begin{bmatrix} I \\ \mathbf{S} \end{bmatrix} < 0 \text{ on } \overline{\mathbb{C}}_0. \tag{4.79}$$

2. The KPYS(Σ, J) has a stabilizing solution

$$(X, V, W) = (X, \begin{bmatrix} V_{11} & O \\ V_{21} & V_{22} \end{bmatrix}, \begin{bmatrix} W_1 \\ W_2 \end{bmatrix})$$

with $X \geq 0$ and where the partitions of V and W are in accordance with J in (4.76).

II. *If assertion 1 or 2 holds, then there exists a rational matrix function*

$$\mathbf{G} = \begin{bmatrix} \mathbf{G}_1 & \mathbf{G}_2 \\ \mathbf{G}_3 & \mathbf{G}_4 \end{bmatrix}$$

partitioned in accordance with $\mathbf{\Pi}_{\Sigma}$ in (4.75) such that $\mathbf{G}$ and $\mathbf{G}_4$ are units in RH^{∞}_{+} and

$$\mathbf{\Pi}_{\Sigma} = \mathbf{G}^* J \mathbf{G}, \tag{4.80}$$

(i.e., $\mathbf{G}$ is a J–spectral factor of $\mathbf{\Pi}_{\Sigma}$). The class of all $\mathbf{S}$ for which (4.79) holds is given by

$$\mathbf{S} = \mathbf{S}_2 \mathbf{S}_1^{-1}, \quad \begin{bmatrix} \mathbf{S}_1 \\ \mathbf{S}_2 \end{bmatrix} = \mathbf{G}^{-1} \begin{bmatrix} I \\ \boldsymbol{\theta} \end{bmatrix}, \tag{4.81}$$

where $\boldsymbol{\theta}$ is a free parameter in $RH^{\infty}_{+,m_2 \times m_1}$ with $\|\boldsymbol{\theta}\|_{\infty} < 1$.

Proof. I. For 1. $\Rightarrow$ 2., we only need to prove that the stabilizing solution (X, V, W) to the KPYS(Σ) that exists according to Theorem 4.5.8 satisfies $X \geq 0$. This follows from the fact that $X \geq X_2$ where X_2 is the stabilizing solution to the CTARE(Σ_2) and, since

$$\begin{bmatrix} Q & L_2 \\ L_2^T & R_{22} \end{bmatrix} \geq 0,$$

we can apply Corollary 4.1.7 point 3, to conclude that X_2 is positive semidefinite.

I. 2. $\Rightarrow$ 1. Let us assume now that the KPYS(Σ, J) has a stabilizing solution

$$(X, V, W) = (X, \begin{bmatrix} V_{11} & O \\ V_{21} & V_{22} \end{bmatrix}, \begin{bmatrix} W_1 \\ W_2 \end{bmatrix})$$

with $X \geq 0$. We prove the existence of $\mathbf{S}$ satisfying (4.79), by actually proving part II of the theorem. That is, we do not obtain only one $\mathbf{S}$ satisfying (4.79), but the whole class.

By Proposition 3.2.16, we can write

$$\mathbf{\Pi}_{\Sigma} = \mathbf{G}^* J \mathbf{G},$$

where

$$\mathbf{G} = V\mathbf{S}_F = \left[\begin{array}{c|c} A & B \\ \hline W & V \end{array}\right],$$

$$\mathbf{S}_F = \left[\begin{array}{c|c} A & B \\ \hline -F_{Ricc} & I \end{array}\right],$$

$$V = \begin{bmatrix} V_{11} & O \\ V_{21} & V_{22} \end{bmatrix}, \tag{4.82}$$

and F_{Ricc} is the stabilizing Riccati feedback.

Clearly $\mathbf{G}$ is a unit in RH_+^∞ with the realization

$$\mathbf{G} = \left[\begin{array}{c|cc} A & B_1 & B_2 \\ \hline W_1 & V_{11} & O \\ W_2 & V_{21} & V_{22} \end{array}\right] = \begin{bmatrix} \mathbf{G}_1 & \mathbf{G}_2 \\ \mathbf{G}_3 & \mathbf{G}_4 \end{bmatrix} \tag{4.83}$$

and

$$\mathbf{G}_4 = \left[\begin{array}{c|c} A & B_2 \\ \hline W_2 & V_{22} \end{array}\right]. \tag{4.84}$$

In order to prove that $\mathbf{G}_4$ is a unit in RH_+^∞ we shall prove that $A - B_2V_{22}^{-1}W_2$ is stable. Indeed, in this case

$$\mathbf{G}_4^{-1} = \left[\begin{array}{c|c} A - B_2V_{22}^{-1}W_2 & B_2V_{22}^{-1} \\ \hline -V_{22}^{-1}W_2 & V_{22}^{-1} \end{array}\right] \tag{4.85}$$

is in RH_+^∞. To this end, consider the feedback matrix

$$\widetilde{F} = \begin{bmatrix} O \\ \widetilde{F}_2 \end{bmatrix} = \begin{bmatrix} O \\ -V_{22}^{-1}W_2 \end{bmatrix} \tag{4.86}$$

and denote by $\widetilde{\Sigma} = (\widetilde{A}, \widetilde{B}; \widetilde{Q}, \widetilde{L}, \widetilde{R})$ the $\widetilde{F}$–equivalent of Σ. Thus $\widetilde{A} = A + B\widetilde{F} = A + B_2F_2 = A - B_2V_{22}^{-1}W_2$, $\widetilde{B} = B$. Also, by the special structure of $\widetilde{F}$,

$$\widetilde{Q} = \begin{bmatrix} I & \widetilde{F}_2^T \end{bmatrix} \begin{bmatrix} Q & L_2 \\ L_2^T & R \end{bmatrix} \begin{bmatrix} I \\ \widetilde{F}_2 \end{bmatrix}.$$

This implies that $\widetilde{Q} \geq 0$ and thus we may write $\widetilde{Q} = \widetilde{C}^T\widetilde{C}$, with $\widetilde{C}$ square. By Proposition 3.2.13 the last equation of the KPYS$(\widetilde{\Sigma}, J)$ gives

$$\widetilde{C}^T\widetilde{C} + \widetilde{A}^TX + X\widetilde{A} = (W + V\widetilde{F})^TJ(W + V\widetilde{F}). \tag{4.87}$$

We compute

$$W + V\widetilde{F} = \begin{bmatrix} W_1 \\ W_2 \end{bmatrix} + \begin{bmatrix} V_{11} & O \\ V_{21} & V_{22} \end{bmatrix} \begin{bmatrix} O \\ -V_{22}^{-1}W_2 \end{bmatrix} = \begin{bmatrix} W_1 \\ O \end{bmatrix}. \tag{4.88}$$

In consequence,

$$\widetilde{A}^TX + X\widetilde{A} + W_1^TW_1 + \widetilde{C}^T\widetilde{C} = 0. \tag{4.89}$$

On the other hand,

$$A + BF_{Ricc} = A - \begin{bmatrix} B_1 & B_2 \end{bmatrix} \begin{bmatrix} V_{11} & O \\ V_{21} & V_{22} \end{bmatrix}^{-1} \begin{bmatrix} W_1 \\ W_2 \end{bmatrix}$$

$$= A - B_2V_{22}^{-1}W_2 + (-B_1V_{11}^{-1} + B_2V_{22}^{-1}V_{21}V_{11}^{-1})W_1$$

$$= \widetilde{A} + \begin{bmatrix} K & O \end{bmatrix} \begin{bmatrix} W_1 \\ \widetilde{C} \end{bmatrix} \tag{4.90}$$

is stable for $K := -B_1V_{11}^{-1} + B_2V_{22}^{-1}V_{21}V_{11}^{-1}$ and consequently, as the rightmost term in (4.90) shows, the pair $(\begin{bmatrix} W_1 \\ \widetilde{C} \end{bmatrix}, \widetilde{A})$ is detectable. This combined with the fact that $X \geq 0$ in (4.89) leads via Theorem 1.5.5 to the stability of $\widetilde{A}$.

We have so far proved the required properties for $\mathbf{G}$. Let us complete the proof by showing that (4.81) gives the parameterization of all the $\mathbf{S}$ functions. To this end notice first that by identifying the "2,2" entries of the both sides of (4.80) one gets

$$-\mathbf{G}_2^*\mathbf{G}_2 + \mathbf{G}_4^*\mathbf{G}_4 = \mathbf{\Pi}_{22}.$$

Hence

$$(\mathbf{G}_2\mathbf{G}_4^{-1})^*\mathbf{G}_2\mathbf{G}_4^{-1} = I - \mathbf{G}_4^{-*}\mathbf{\Pi}_{22}\mathbf{G}_4^{-1} \tag{4.91}$$

and by taking into account that $\mathbf{\Pi}_{22} > 0$ on $\overline{\mathbb{C}}_0$, (4.91) provides

$$(\mathbf{G}_2\mathbf{G}_4^{-1})^*\mathbf{G}_2\mathbf{G}_4^{-1} < I \quad \text{on} \quad \overline{\mathbb{C}}_0.$$

Hence

$$\|\mathbf{G}_2\mathbf{G}_4^{-1}\|_\infty < 1, \quad \mathbf{G}_2\mathbf{G}_4^{-1} \in RH_+^\infty. \tag{4.92}$$

Let $\mathbf{G}_1^\times$ be the Schur complement of $\mathbf{G}_4$ in $\mathbf{G}$, i.e.,

$$\mathbf{G}_1^\times := \mathbf{G}_1 - \mathbf{G}_3\mathbf{G}_4^{-1}\mathbf{G}_2. \tag{4.93}$$

By using the standard inversion formula one gets

$$\mathbf{G}^{-1} = \begin{bmatrix} (\mathbf{G}_1^\times)^{-1} & -(\mathbf{G}_1^\times)^{-1}\mathbf{G}_2\mathbf{G}_4^{-1} \\ -\mathbf{G}_4^{-1}\mathbf{G}_3(\mathbf{G}_1^\times)^{-1} & \mathbf{G}_4^{-1} + \mathbf{G}_4^{-1}\mathbf{G}_3(\mathbf{G}_1^\times)^{-1}\mathbf{G}_2\mathbf{G}_4^{-1} \end{bmatrix}. \tag{4.94}$$

Since both $\mathbf{G}$ and $\mathbf{G}_4$ are units in RH_+^∞, (4.93) and (4.94) show that $\mathbf{G}_1^\times$ is a unit in $RH_{+,m_1\times m_1}^\infty$ as well.

Let $\mathbf{S}_1$ and $\mathbf{S}_2$ be defined by

$$\begin{bmatrix} \mathbf{S}_1 \\ \mathbf{S}_2 \end{bmatrix} := \mathbf{G}^{-1} \begin{bmatrix} I \\ \boldsymbol{\theta} \end{bmatrix}, \quad \forall \boldsymbol{\theta} \in RH_+^\infty, \quad \|\boldsymbol{\theta}\|_\infty < 1. \tag{4.95}$$

Hence $\mathbf{S}_1 \in RH_{+,m_1\times m_1}^\infty$, $\mathbf{S}_2 \in RH_{+,m_2\times m_1}^\infty$. Now we show that $\mathbf{S}_1$ is a unit in $RH_{+,m_1\times m_1}^\infty$. With (4.94), (4.95) provides

$$\mathbf{S}_1 = (\mathbf{G}_1^\times)^{-1}(I - \mathbf{G}_2\mathbf{G}_4^{-1}\boldsymbol{\theta}). \tag{4.96}$$

But $\|\mathbf{G}_2\mathbf{G}_4^{-1}\boldsymbol{\theta}\|_\infty \leq \|\mathbf{G}_2\mathbf{G}_4^{-1}\|_\infty\|\boldsymbol{\theta}\|_\infty < 1$ as follows from (4.92) and (4.95). Hence by a variation of argument formula it follows that $I - \mathbf{G}_2\mathbf{G}_4^{-1}\boldsymbol{\theta}$ is a unit in RH_+^∞. Thus $\mathbf{S}_1$ is a unit in RH_+^∞ as (4.96) shows.

Let $\mathbf{S} := \mathbf{S}_2\mathbf{S}_1^{-1} \in RH^{\infty}_{+,m_2\times m_1}$ and we shall show that (4.79) holds for such $\mathbf{S}$. Indeed, by using (4.80) and (4.95) one obtains

$$\begin{bmatrix} I & \mathbf{S}^* \end{bmatrix} \mathbf{\Pi}_\Sigma \begin{bmatrix} I \\ \mathbf{S} \end{bmatrix} = \begin{bmatrix} I & \mathbf{S}_1^{-*}\mathbf{S}_2^* \end{bmatrix} \mathbf{G}^* J \mathbf{G} \begin{bmatrix} I \\ \mathbf{S}_2\mathbf{S}_1^{-1} \end{bmatrix}$$

$$= \mathbf{S}_1^{-*} \begin{bmatrix} \mathbf{S}_1^* & \mathbf{S}_2^* \end{bmatrix} \mathbf{G}^* J \mathbf{G} \begin{bmatrix} \mathbf{S}_1 \\ \mathbf{S}_2 \end{bmatrix} = \mathbf{S}_1^{-*}(\boldsymbol{\theta}^*\boldsymbol{\theta} - I)\mathbf{S}_1^{-1} < 0 \quad \text{on} \quad \overline{\mathbb{C}}_0$$

where the strict inequality holds because of $\|\boldsymbol{\theta}\|_\infty < 1$. Thus the implication 1. ⇒ 2. is completely proved.

To prove part II completely, we must show that any $\mathbf{S} \in RH^{\infty}_{+,m_2\times m_1}$ for which (4.79) holds is given by (4.81). To this end consider any $\mathbf{S} \in RH^{\infty}_{+}$ for which (4.79) holds and let $\boldsymbol{\theta}_1 \in RH^{\infty}_{+,m_1\times m_1}, \boldsymbol{\theta}_2 \in RH^{\infty}_{+,m_2\times m_1}$ be defined by

$$\begin{bmatrix} \boldsymbol{\theta}_1 \\ \boldsymbol{\theta}_2 \end{bmatrix} := \mathbf{G} \begin{bmatrix} I \\ \mathbf{S} \end{bmatrix}. \tag{4.97}$$

Now we shall show that $\boldsymbol{\theta}_1$ is a unit in $RH^{\infty}_{+,m_1\times m_1}$. To this end use first (4.80) and obtain with (4.79) and (4.97)

$$-\boldsymbol{\theta}_1^*\boldsymbol{\theta}_1 + \boldsymbol{\theta}_2^*\boldsymbol{\theta}_2 = \begin{bmatrix} \boldsymbol{\theta}_1^* & \boldsymbol{\theta}_2^* \end{bmatrix} J \begin{bmatrix} \boldsymbol{\theta}_1 \\ \boldsymbol{\theta}_2 \end{bmatrix} = \begin{bmatrix} I & \mathbf{S}^* \end{bmatrix} \mathbf{G}^* J \mathbf{G} \begin{bmatrix} I \\ \mathbf{S} \end{bmatrix}$$

$$= \begin{bmatrix} I & \mathbf{S}^* \end{bmatrix} \mathbf{\Pi}_\Sigma \begin{bmatrix} I \\ \mathbf{S} \end{bmatrix} < 0, \quad \text{on} \quad \overline{\mathbb{C}}_0. \tag{4.98}$$

Hence $\boldsymbol{\theta}_1^*\boldsymbol{\theta}_1 > 0$ on $\overline{\mathbb{C}}_0$ from where it follows that $\boldsymbol{\theta}_1^{-1} \in RL^{\infty}_{m_1\times m_1}$. Let $\boldsymbol{\theta} := \boldsymbol{\theta}_2\boldsymbol{\theta}_1^{-1} \in RL^{\infty}_{m_2\times m_1}$. Then by using again (4.98) one obtains

$$(\boldsymbol{\theta}_2\boldsymbol{\theta}_1^{-1})^*\boldsymbol{\theta}_2\boldsymbol{\theta}_1^{-1} = \boldsymbol{\theta}^*\boldsymbol{\theta} < I \quad \text{on} \quad \overline{\mathbb{C}}_0$$

or, equivalently,

$$\|\boldsymbol{\theta}\|_\infty < 1, \quad \boldsymbol{\theta} \in RL^\infty. \tag{4.99}$$

From (4.97) one gets

$$\begin{bmatrix} I & O \end{bmatrix} \mathbf{G}^{-1} \begin{bmatrix} \boldsymbol{\theta}_1 \\ \boldsymbol{\theta}_2 \end{bmatrix} = I$$

or, explicitly, with (4.94),

$$(\mathbf{G}_1^\times)^{-1}(\boldsymbol{\theta}_1 - \mathbf{G}_2\mathbf{G}_4^{-1}\boldsymbol{\theta}_2) = I. \tag{4.100}$$

Further, for $\boldsymbol{\theta} = \boldsymbol{\theta}_2\boldsymbol{\theta}_1^{-1}$, (4.100) provides

$$(\mathbf{G}_1^\times)^{-1}(I - \mathbf{G}_2\mathbf{G}_4^{-1}\boldsymbol{\theta})\boldsymbol{\theta}_1 = I, \quad \mathbf{G}_2\mathbf{G}_4^{-1}\boldsymbol{\theta} \in RL^\infty. \tag{4.101}$$

From (4.101) we have

$$(I - \mathbf{G}_2\mathbf{G}_4^{-1}\boldsymbol{\theta})^{-1} = \boldsymbol{\theta}_1(\mathbf{G}_1^\times)^{-1} \tag{4.102}$$

which yields

$$(I - \mathbf{G}_2\mathbf{G}_4^{-1}\boldsymbol{\theta})^{-1} \in RH^{\infty}_{+,m_1\times m_1}. \tag{4.103}$$

On the other hand, from (4.92) and (4.99) we get

$$\|\mathbf{G}_2\mathbf{G}_4^{-1}\boldsymbol{\theta}\|_\infty \leq \|\mathbf{G}_2\mathbf{G}_4^{-1}\|_\infty\|\boldsymbol{\theta}\|_\infty < 1.$$

By invoking again the variation of argument formula we get with (4.103) that $I - \mathbf{G}_2\mathbf{G}_4^{-1}\boldsymbol{\theta}$ is a unit in $RH^\infty_{+,m_1\times m_1}$ and consequently $\boldsymbol{\theta}_1^{-1} \in RH^\infty_{+,m_1\times m_1}$ as follows from (4.102). Therefore $\boldsymbol{\theta} = \boldsymbol{\theta}_2\boldsymbol{\theta}_1^{-1} \in RH^\infty_{+,m_2\times m_1}$ and using (4.97) we can express $\mathbf{S} = \mathbf{S}_2\mathbf{S}_1^{-1}$, $\mathbf{S}_1 := \boldsymbol{\theta}_1^{-1}$, $\mathbf{S}_2 := \mathbf{S}\boldsymbol{\theta}_1^{-1}$ and

$$\begin{bmatrix} \mathbf{S}_1 \\ \mathbf{S}_2 \end{bmatrix} = \mathbf{G}^{-1}\begin{bmatrix} I \\ \boldsymbol{\theta} \end{bmatrix}, \qquad \boldsymbol{\theta} \in RH^\infty_{+,m_2\times m_1}, \|\boldsymbol{\theta}\|_\infty < 1,$$

and the theorem is completely proved. ■

The following result is a simple consequence of the signature condition in frequency domain setting, and will play a major role in the state–space feedback solution to the H^∞–control problem in Part III of the book.

Theorem 4.5.10 *Let $\Sigma = (A, B; Q, L, R)$ be a Popov triplet with its entries partitioned as in (4.74) (with A not necessarily stable). Assume that $R_{22} > 0$ and $R_{11} < 0$ and that the KPYS(Σ, J) has a stabilizing solution*

$$(X, V, W) = (X, \begin{bmatrix} V_{11} & O \\ V_{21} & V_{22} \end{bmatrix}, \begin{bmatrix} W_1 \\ W_2 \end{bmatrix}),$$

with $X \geq 0$.

If the stabilizing Riccati feedback is partitioned in accordance with R, i.e.,

$$F_{Ricc} = \begin{bmatrix} F_1 \\ F_2 \end{bmatrix}$$

then we have:

1. $u_2 = F_2x$ stabilizes the system $\dot{x} = Ax + B_1u_1 + B_2u_2$.

2. There exists $\rho > 0$ such that for $u_2 = F_2x$, the following inequality holds

$$J_{e,\Sigma}(u) \leq -\rho^2\|u_1\|_2^2, \quad \forall u_1 \in L^{2,m_1}, \tag{4.104}$$

where $J_{e,\Sigma}$ is the extended quadratic index (3.7), and $u = \begin{bmatrix} u_1 \\ u_2 \end{bmatrix}$.

Proof. Obviously, if R satisfies the hypotheses, then V is of lower–left block triangular form. Then we have

$$\begin{aligned} F_{Ricc} &= -V^{-1}W = -\begin{bmatrix} V_{11} & O \\ V_{21} & V_{22} \end{bmatrix}^{-1}\begin{bmatrix} W_1 \\ W_2 \end{bmatrix} \\ &= \begin{bmatrix} -V_{11}^{-1}W_1 \\ V_{22}^{-1}V_{21}V_{11}^{-1}W_1 - V_{22}^{-1}W_2 \end{bmatrix} = \begin{bmatrix} F_1 \\ F_2 \end{bmatrix}. \end{aligned} \tag{4.105}$$

1. Consider

$$\widetilde{F} := \begin{bmatrix} O \\ F_2 \end{bmatrix} = \begin{bmatrix} O \\ V_{22}^{-1}V_{21}V_{11}^{-1}W_1 - V_{22}^{-1}W_2 \end{bmatrix} \tag{4.106}$$

and $\widetilde{\Sigma} = (\widetilde{A}, \widetilde{B}; \widetilde{Q}, \widetilde{L}, \widetilde{R})$, the $\widetilde{F}$–equivalent of Σ. Since $\widetilde{F}$ has the same structure as in the proof of Theorem 4.5.9 we still have $\widetilde{Q} \geq 0$ and

$$\widetilde{Q} + \widetilde{A}^T X + X\widetilde{A} = (W + V\widetilde{F})^T J (W + V\widetilde{F}). \tag{4.107}$$

Now we have

$$\begin{aligned} W + V\widetilde{F} &= \begin{bmatrix} W_1 \\ W_2 \end{bmatrix} + \begin{bmatrix} V_{11} & O \\ V_{21} & V_{22} \end{bmatrix} \begin{bmatrix} O \\ V_{22}^{-1}V_{21}V_{11}^{-1}W_1 - V_{22}^{-1}W_2 \end{bmatrix} \\ &= \begin{bmatrix} W_1 \\ V_{21}V_{11}^{-1}W_1 \end{bmatrix}. \end{aligned} \tag{4.108}$$

Therefore

$$\begin{aligned} (W + V\widetilde{F})^T J (W + V\widetilde{F}) &= -W_1^T W_1 + W_1^T V_{11}^{-T} V_{21}^T V_{21} V_{11}^{-1} W_1 \\ &= W_1^T V_{11}^{-T}(-V_{11}^T V_{11} + V_{21}^T V_{21}) V_{11}^{-1} W_1. \end{aligned} \tag{4.109}$$

On the other hand,

$$R_{11} = -V_{11}^T V_{11} + V_{21}^T V_{21} \tag{4.110}$$

and therefore we obtain finally

$$\widetilde{A}^T X + X\widetilde{A} - W_1^T V_{11}^{-T} R_{11} V_{11}^{-1} W_1 + \widetilde{C}^T \widetilde{C} = 0 \tag{4.111}$$

where

$$\widetilde{A} = A + B\widetilde{F} = A + B_2F_2 = A + B_2(V_{22}^{-1}V_{21}V_{11}^{-1}W_1 - V_{22}^{-1}W_2), \tag{4.112}$$

and $\widetilde{Q} = \widetilde{C}^T\widetilde{C}$ with $\widetilde{C}$ square.

Since $R_{11} < 0$, let

$$-R_{11} = L^T L \tag{4.113}$$

with L nonsingular. Thus, with (4.113), (4.111) becomes

$$\widetilde{A}^T X + X\widetilde{A} + W_1^T V_{11}^{-T} L^T L V_{11}^{-1} W_1 + \widetilde{C}^T \widetilde{C} = 0 \tag{4.114}$$

and

$$A + BF_{Ricc} = \widetilde{A} - B_1V_{11}^{-1}W_1 = \widetilde{A} - B_1L^{-1}LV_{11}^{-1}W_1 = \widetilde{A} + \begin{bmatrix} K & O \end{bmatrix} \begin{bmatrix} LV_{11}^{-1}W_1 \\ \widetilde{C} \end{bmatrix} \tag{4.115}$$

for $K := -B_1L^{-1}$. Since $A + BF_{Ricc}$ is stable, the rightmost term in (4.115) is stable too and, consequently, the pair $\left(\begin{bmatrix} LV_{11}^{-1}W_1 \\ \widetilde{C} \end{bmatrix}, \widetilde{A}\right)$ is detectable. By combining this conclusion with (4.114) which is fulfilled for $X \geq 0$, the stability of $\widetilde{A} = A + BF_2$ follows from Theorem 1.5.5.

2. The extended quadratic index $J_{e,\Sigma}(u) = J_{e,\Sigma}(u_1, u_2)$ can be expressed as

$$J_{e,\Sigma}(u_1, u_2) = \left\langle \begin{bmatrix} V_{11} & O \\ V_{21} & V_{22} \end{bmatrix} \begin{bmatrix} u_1 \\ u_2 \end{bmatrix} + \begin{bmatrix} W_1 \\ W_2 \end{bmatrix} x_e^u, \begin{bmatrix} -I_{m_1} & O \\ O & I_{m_2} \end{bmatrix} \right.$$

$$\times \left(\begin{bmatrix} V_{11} & O \\ V_{21} & V_{22} \end{bmatrix} \begin{bmatrix} u_1 \\ u_2 \end{bmatrix} + \begin{bmatrix} W_1 \\ W_2 \end{bmatrix} x_e^u) \right\rangle$$

$$= \left\langle \begin{bmatrix} V_{11}u_1 + W_1 x_e^u \\ V_{21}u_1 + V_{22}u_2 + W_2 x_e^u \end{bmatrix}, \begin{bmatrix} -I_{m_1} & O \\ O & I_{m_2} \end{bmatrix} \begin{bmatrix} V_{11}u_1 + W_1 x_e^u \\ V_{21}u_1 + V_{22}u_2 + W_2 x_e^u \end{bmatrix} \right\rangle$$

$$\stackrel{(4.105)}{=} \left\langle \begin{bmatrix} V_{11}(u_1 - F_1 x_e^u) \\ V_{21}(u_1 - F_1 x_e^u) + V_{22}(u_2 - F_2 x_e^u) \end{bmatrix}, \begin{bmatrix} -I_{m_1} & O \\ O & I_{m_2} \end{bmatrix} \right.$$

$$\left. \times \begin{bmatrix} V_{11}(u_1 - F_1 x_e^u) \\ V_{21}(u_1 - F_1 x_e^u) + V_{22}(u_2 - F_2 x_e^u) \end{bmatrix} \right\rangle$$

$$\stackrel{u_2 = F_2 x}{=} \left\langle \begin{bmatrix} V_{11} \\ V_{21} \end{bmatrix} (u_1 - F_1 x_e^u), \begin{bmatrix} -V_{11} \\ V_{21} \end{bmatrix} (u_1 - F_1 x_e^u) \right\rangle$$

$$= \langle (u_1 - F_1 x_e^u), (-V_{11}^T V_{11} + V_{21}^T V_{21})(u_1 - F_1 x_e^u) \rangle$$

$$\stackrel{(4.110)}{=} \langle u_1 - F_1 x_e^u, R_{11}(u_1 - F_1 x_e^u) \rangle \leq -\lambda \| u_1 - F_1 x_e^u \|_2^2 \tag{4.116}$$

where $\lambda > 0$ is the smallest eigenvalue of $-R_{11}$. Using point 1 of the theorem we conclude that the TFM of the system

$$\widetilde{\mathbf{G}}_{11} = \left[\begin{array}{c|c} A + B_2 F_2 & B_1 \\ \hline -F_1 & I \end{array} \right], \quad v_1 = \widetilde{\mathbf{G}}_{11} u_1,$$

is a unit in $RH^{\infty}_{+, m_1 \times m_1}$. Hence there exists $\widetilde{\rho} > 0$ such that $\|\widetilde{\mathbf{G}}_{11} u_1\|_2 \geq \widetilde{\rho} \|u_1\|_2$ and (4.116) yields

$$J_{e,\Sigma}(u_1, u_2)|_{u_2 = F_2 x} \leq -\lambda \|v_1\|_2^2 = -\lambda \|\widetilde{\mathbf{G}}_{11} u_1\|_2^2 \leq -\rho \|u_1\|_2^2$$

for $\rho := \lambda \widetilde{\rho}^2$ and all $u_1 \in L^{2, m_1}$. Thus the theorem is completely proved. ∎

4.6 Algebraic Riccati inequalities

In this section we present a remarkable result on algebraic Riccati inequalities. To this end we introduce first the following notation: for a fixed Popov triplet $\Sigma = (A, B; Q, L, R)$ and for $X = X^T$, denote

$$\mathrm{CRICOP}(\Sigma, X) := A^T X + XA - (XB + L) R^{-1} (B^T X + L^T) + Q.$$

Theorem 4.6.1 *Let $\Sigma = (A, B; Q, L, R)$ be a Popov triplet with $Q \geq 0$. Then the following two assertions are equivalent.*

1. *A is stable and $\mathbf{\Pi}_\Sigma < 0$ on $\overline{\mathbb{C}}_0$.*
2. *$R < 0$ and there exists $X > 0$ such that* $\mathrm{CRICOP}(\Sigma, X) < 0$.

Proof. 1. ⇒ 2. We have $R = \mathbf{\Pi}_\Sigma(\infty) < 0$. On the other hand, there exists $\rho > 0$ such that $\mathbf{\Pi}_\Sigma \leq -\rho I$ on $\overline{\mathbb{C}}_0$. Let

$$\mathbf{G} = \left[\begin{array}{c|c} A & B \\ \hline I & O \end{array} \right]$$

and choose $0 < \epsilon \le \frac{\rho}{2\|\mathbf{G}\|_\infty^2}$. For $Q_\epsilon := Q + \epsilon I$ let $\Sigma_\epsilon = (A, B; Q_\epsilon, L, R)$. Then

$$\mathbf{\Pi}_{\Sigma\epsilon} = \mathbf{\Pi}_\Sigma + \epsilon \mathbf{G}^*\mathbf{G} \le -\rho I + \frac{\rho}{2} = -\frac{\rho}{2} I \quad \text{on} \quad \overline{\mathbb{C}}_0.$$

Hence $\mathbf{\Pi}_{\Sigma_\epsilon} < 0$ on $\overline{\mathbb{C}}_0$ and according to Corollary 4.5.5 the CTARE(Σ_ϵ) has a (stabilizing) solution X, i.e.,

$$A^T X + XA - (L + XB)R^{-1}(B^T X + L^T) + Q + \epsilon I = 0 \tag{4.117}$$

from which it follows that CRICOP(Σ, X) < 0. From (4.117) we get $A^T X + XA < 0$. As A is stable, clearly $X > 0$ and the implication is proved.

2. $\Rightarrow$ 1. Let $S > 0$ be defined by $-S =$ CRICOP(Σ, X). Then

$$A^T X + XA - (L + XB)R^{-1}(B^T X + L^T) + S + Q = 0 \tag{4.118}$$

where $S + Q > 0$. Hence $A^T X + XA < 0$. As $X > 0$ the stability of A follows from Theorem 1.5.4. Consider now $\widetilde{\Sigma} = (A, B; \widetilde{Q}, \widetilde{L}, R)$, the $(X, 0)$–equivalent of Σ, where $\widetilde{L} = L + XB$ and $\widetilde{Q} = Q + A^T X + XA$. According to Proposition 3.2.16, it is enough to prove that $\mathbf{\Pi}_{\widetilde{\Sigma}} < 0$ on $\overline{\mathbb{C}}_0$. Now

$$\mathbf{\Pi}_{\widetilde{\Sigma}}(j\omega) = \begin{bmatrix} (j\omega I - A)^{-1}B \\ I \end{bmatrix}^* \begin{bmatrix} \widetilde{Q} & \widetilde{L} \\ \widetilde{L}^T & R \end{bmatrix} \begin{bmatrix} (j\omega I - A)^{-1}B \\ I \end{bmatrix},$$

and the assertion follows if we prove that

$$\begin{bmatrix} \widetilde{Q} & \widetilde{L} \\ \widetilde{L}^T & R \end{bmatrix} < 0.$$

Since $R < 0$, we prove this inequality by showing that $\widetilde{Q} - \widetilde{L}R^{-1}\widetilde{L}^T < 0$. Indeed, using (4.118), we have $\widetilde{Q} - \widetilde{L}R^{-1}\widetilde{L}^T = -S < 0$ and the proof is complete. ■

Remark 4.6.2 Assertion 2 in Theorem 4.6.1 is equivalent to

$$D_\Sigma(X) = \begin{bmatrix} A^T X + XA + Q & XB + L \\ L^T + B^T X & R \end{bmatrix} < 0$$

for $X > 0$.

B. The discrete–time case

4.7 Existence of the stabilizing solution: the stable case

This section is devoted to the discrete counterpart of the theory given in Section 4.1.

Let $\Sigma = (A, B; Q, L, R)$ be a Popov triplet with $A \in \mathbb{R}^{n\times n}$, $B \in \mathbb{R}^{n\times m}$, and A stable. The main result of this section gives necessary and sufficient conditions for the existence of the stabilizing solution to the DTARE(Σ)

$$A^T XA - X - (A^T XB + L)(R + B^T XB)^{-1}(B^T XA + L^T) + Q = 0 \tag{4.119}$$

in terms of the invertibility of the Toeplitz operator $\mathcal{R}_\Sigma$ given by

$$\mathcal{R}_\Sigma : \ell_+^{2,m} \to \ell_+^{2,m}, \quad \mathcal{R}_\Sigma := R + L^T\mathcal{L} + \mathcal{L}^*L + \mathcal{L}^*Q\mathcal{L} = \mathcal{R}_\Sigma^* \tag{4.120}$$

where

$$\mathcal{L} : \ell_+^{2,m} \to \ell_+^{2,n}, \quad (\mathcal{L}u)_k := \sum_{i=0}^{k-1} A^{k-i-1}Bu_i, \quad k \geq 0. \tag{4.121}$$

Recall from Chapter 3 that $\mathcal{R}_\Sigma$ is the Toeplitz operator associated with the input–output operator

$$\mathcal{R}_{e,\Sigma} : \ell^{2,m} \to \ell^{2,m}, \quad \mathcal{R}_{e,\Sigma} := R + L^T\mathcal{L}_e + \mathcal{L}_e^*L + \mathcal{L}_e^*Q\mathcal{L}_e = \mathcal{R}_{e,\Sigma}^* \tag{4.122}$$

of the DTHS(Σ)

$$\begin{array}{rcllll} \sigma x & = & Ax & & +Bu, \\ \lambda & = & Qx & +A^T\sigma\lambda & +Lu, \\ v & = & L^Tx & +B^T\sigma\lambda & +Ru, \end{array} \tag{4.123}$$

where

$$\mathcal{L}_e : \ell^{2,m} \to \ell^{2,n}, \quad (\mathcal{L}_e u)_k := \sum_{i=-\infty}^{k-1} A^{k-i-1}Bu_i, \quad k \in \mathbb{Z}. \tag{4.124}$$

The following operators defined in Chapters 2 and 3

$$\Phi : \mathbb{R}^n \to \ell_+^{2,n}, \quad (\Phi\xi)_k := A^k\xi, \quad k \geq 0, \tag{4.125}$$

$$\mathcal{P}_{o,\Sigma} : \mathbb{R}^n \to \mathbb{R}^n, \quad \mathcal{P}_{o,\Sigma} := \Phi^*Q\Phi, \tag{4.126}$$

$$\mathcal{P}_\Sigma : \ell_+^{2,m} \to \mathbb{R}^n, \quad \mathcal{P}_\Sigma := \Phi^*(Q\mathcal{L} + L), \tag{4.127}$$

will also play an important role in the derivations in this chapter.

4.7.1 The main result

Here follows the discrete–time version of the main result of the chapter formulated for Popov triplets with *direct–time* significance.

Theorem 4.7.1 *Let $\Sigma = (A, B; Q, L, R)$ be a Popov triplet with direct–time significance and A stable. Then the following two assertions are equivalent:*

1. *The DTARE(Σ) given in (4.119) has a (unique) stabilizing solution X.*
2. *The Toeplitz operator $\mathcal{R}_\Sigma$ given in (4.120) has a bounded inverse.*

As in the continuous–time version, the difficult part in proving this theorem is the implication 2. ⇒ 1. We shall prove first the other implication.

Proof. 1. ⇒ 2. Assume that the DTARE(Σ) has a stabilizing solution X and let F_{Ricc} be the associated stabilizing feedback. If $\widetilde{\Sigma}$ is the (X, F_{Ricc})–equivalent of Σ, then $\widetilde{Q} = 0$ and $\widetilde{L} = 0$ as can be easily seen from (3.100). Consequently, $\mathcal{R}_{e,\widetilde{\Sigma}} = \widetilde{R} = R + B^TXB$. By (3.173), we have

$$\mathcal{R}_\Sigma = \mathbb{T}^*_{\mathcal{N}_{F_{Ricc}}}(R + B^TXB)\mathbb{T}_{\mathcal{N}_{F_{Ricc}}}. \tag{4.128}$$

By hypothesis, $R + B^T XB$ is invertible. Using the discrete–time version of Corollary 2.4.14, $\mathbb{T}_{\mathcal{N}_{F_{Ricc}}}$ is also invertible. Indeed, $\mathcal{N}_{F_{Ricc}}$ is causal as it is the input–output operator of the system

$$\begin{aligned} \sigma x = &\quad Ax + Bu, \\ v = &\quad -F_{Ricc}x + u. \end{aligned}$$

According to Theorem 2.7.1, the inverse of $\mathcal{N}_{F_{Ricc}}$ is the input–output operator of

$$\begin{aligned} \sigma x = &\quad (A + BF_{Ricc})x + Bv, \\ u = &\quad F_{Ricc}x + v. \end{aligned}$$

Thus Corollary 2.4.14 is applicable, and we deduce that $\mathcal{R}_\Sigma$ is invertible. ■

4.7.2 Proof of the implication 2. ⇒ 1.

We turn now to the proof of the implication 2. ⇒ 1. We shall state the main steps of the proof as auxiliary results.

The first result is actually independent of the invertibility hypothesis on $\mathcal{R}_\Sigma$.

Lemma 4.7.2 *If $\xi \in \mathbb{R}^n$ and $u \in \ell_+^{2,m}$, then there exists a unique $\ell_+^{2,n} \times \ell_+^{2,n}$ state trajectory (x, λ) of the DTHS(Σ) (4.123) associated with Σ such that $x(0) = \xi$. (We denote this trajectory by $(x^{\xi,u}, \lambda^{\xi,u})$).*

Moreover, the corresponding output $v^{\xi,u} \in \ell_+^{2,m}$ of the DTHS(Σ) satisfies

$$v^{\xi,u} = \mathcal{P}_\Sigma^* \xi + \mathcal{R}_\Sigma u. \tag{4.129}$$

Proof. The proof follows closely that for Theorem 3.8.9, so we shall be brief. According to (2.147), $x^{\xi,u} = \Phi\xi + \mathcal{L}u$. By Theorem 2.6.2, we have

$$\lambda_k^{\xi,u} = \sum_{i=k}^{\infty} (A^T)^{i-k}(Lu_i + Qx_i^{\xi,u}). \tag{4.130}$$

Formula (4.129) follows by straightforward substitutions similar to the proof of Theorem 3.8.9. ■

We assume now that $\mathcal{R}_\Sigma$ is invertible. It is then immediate that for each ξ there exists a unique u such that $v^{\xi,u} = 0$. If we denote such a u by u^ξ, we have from (4.129) that

$$u^\xi = -\mathcal{R}_\Sigma^{-1}\mathcal{P}_\Sigma^* \xi. \tag{4.131}$$

Let us denote by

$$x^\xi := x^{\xi,u^\xi}, \quad \lambda^\xi := \lambda^{\xi,u^\xi},$$

the corresponding solutions of the DTHS(Σ) (4.123).

Because of the time–invariance of the DTHS, if $\xi' = x_p^\xi$, then $u_k^{\xi'} = u_{k+p}^\xi$ and $\lambda_k^{\xi'} = \lambda_{k+p}^\xi$. We use this fact in the proof of our next two results.

Lemma 4.7.3 *We have*

$$\lambda^\xi = Xx^\xi, \tag{4.132}$$

where X is the symmetric $n \times n$ matrix

$$X := \mathcal{P}_{o,\Sigma} - \mathcal{P}_\Sigma \mathcal{R}_\Sigma^{-1} \mathcal{P}_\Sigma^*. \tag{4.133}$$

Proof. From (4.130), we get

$$\lambda_0^{\xi,u} = \sum_{i=0}^{\infty} (A^T)^i (Lu_i + Qx_i^{\xi,u}). \tag{4.134}$$

Obviously, we have

$$x^\xi = \Phi\xi + \mathcal{L}u^\xi = (\Phi - \mathcal{L}\mathcal{R}_\Sigma^{-1}\mathcal{P}_\Sigma^*)\xi. \tag{4.135}$$

Using in (4.134) the expression for the adjoint of Φ given in (3.166), we compute with (4.135)

$$\begin{aligned} \lambda_0^\xi &= \Phi^*(Lu^\xi + Qx^\xi) \\ &= \Phi^* L(-\mathcal{R}_\Sigma^{-1}\mathcal{P}_\Sigma^*\xi) + \Phi^* Q(\Phi - \mathcal{L}\mathcal{R}_\Sigma^{-1}\mathcal{P}_\Sigma^*)\xi. \end{aligned}$$

Using now the definitions of $\mathcal{P}_{o,\Sigma}$ and $\mathcal{P}_\Sigma$ given in (4.126) and (4.127), respectively, we derive easily that

$$\lambda_0^\xi = X\xi.$$

Applying this formula for $\xi' = x_p^\xi$, $p \in \mathbb{N}$, we have

$$\lambda_p^\xi = \lambda_0^{\xi'} = X\xi' = Xx_p^\xi,$$

and formula (4.132) is proved. ■

It will turn out that X introduced in (4.133) – called the *representation formula* – is the stabilizing solution of the DTARE(Σ).

Lemma 4.7.4 *There exists a matrix F such that*

$$u^\xi = Fx^\xi, \tag{4.136}$$

and which satisfies

$$(R + B^T XB)F + (B^T XA + L^T) = 0, \tag{4.137}$$

where X is given by (4.133).

Proof. From (4.131) it follows that u_0^ξ depends linearly on ξ. Let F be such that

$$u_0^\xi = F\xi.$$

By a time–invariance argument as the one used in the proof of the previous result, F satisfies (4.136). Since u^ξ is, by definition, such that $v^{\xi,u^\xi} = 0$, from the last equation of the DTHS(Σ) (4.123) we deduce that

$$0 = L^T x^\xi + B^T X\sigma x^\xi + Ru^\xi.$$

Using here, $\sigma x^\xi = Ax^\xi + Bu^\xi$ and $u^\xi = Fx^\xi$, we obtain immediately (4.137). ■

The proof of Theorem 4.7.1 is completed with the proof of the following result.

Lemma 4.7.5 *If X is given by (4.133), then $R+B^TXB$ is invertible and X satisfies the DTARE(Σ) in (4.119). Moreover, $A+BF$ is stable, and*

$$F = -(R+B^TXB)^{-1}(B^TXA+L^T). \tag{4.138}$$

In other words, X is the stabilizing solution of the DTARE(Σ) and $F = F_{Ricc}$ is the stabilizing Riccati feedback.

Proof. If we use (4.132) and (4.136) in the second equation of the DTHS(Σ) (4.123) we obtain by a similar reasoning as in the proof of Lemma 4.1.5 that

$$A^TXA - X + Q + (L+A^TXB)F = 0. \tag{4.139}$$

This together with (4.137) imply that (X, F) is a solution to the DTARS(Σ) in (3.108). Furthermore, (X, F) is a stabilizing solution to DTARS(Σ), i.e., $A+BF$ is stable. Indeed, substituting (4.136) in the first equation (4.123) one gets

$$\sigma x^\xi = (A+BF)x^\xi$$

with $x^\xi \in \ell_+^{2,n}$ and $\|x^\xi\|_2 \leq \rho_0\|\xi\|$ for some ρ_0 and all $\xi \in \mathbb{R}^n$ as follows from (4.135). Hence

$$Z := \sum_{k=0}^{\infty}((A+BF)^T)^k(A+BF)^k \leq \rho_0^2 I$$

is well defined and $Z > 0$. Furthermore, it is a matter of simple computation to check that Z satisfies the symmetric Stein equation $(A+BF)^TZ(A+BF) - Z + I = 0$. Invoking Theorem 1.5.4, the stability of $A+BF$ follows.

Let $\widetilde{\Sigma}$ be the (X, F_{Ricc}) equivalent of Σ. Then $\widetilde{R} = R+B^TXB$ and (4.128) holds with $\mathcal{R}_\Sigma$ and $\mathbb{T}_{\mathcal{N}_{F_{Ricc}}}$ boundedly invertible. Hence $(R+B^TXB)$ is invertible. Notice that the bounded invertibility of $\mathbb{T}_{\mathcal{N}_{F_{Ricc}}}$ follows from the stability of $A+BF$ in conjunction with similar arguments used in the proof of implication 2. $\Rightarrow$ 1. of Theorem 4.7.1. Therefore, F can be solved from (4.137) to give (4.138) and introducing it in (4.139) it follows that X satisfies the DTARE(Σ). ∎

We can now incidentally prove the following corollary.

Corollary 4.7.6 *Let $\Sigma = (A, B; Q, L, R)$ be a Popov triplet with A stable. Assume that the Toeplitz operator $\mathcal{R}_\Sigma$ in (4.120) has a bounded inverse and let X be the stabilizing solution to the DTARE(Σ). Then we have*

$$J_\Sigma(\xi, u^\xi) = \left\langle \begin{bmatrix} x^{\xi,u^\xi} \\ u \end{bmatrix}, \begin{bmatrix} Q & L \\ L^T & R \end{bmatrix} \begin{bmatrix} x^{\xi,u^\xi} \\ u \end{bmatrix} \right\rangle = \xi^TX\xi \tag{4.140}$$

for every $\xi \in \mathbb{R}^n$ and $u^\xi = -\mathcal{R}_\Sigma^{-1}\mathcal{P}_\Sigma^*\xi$, with $\mathcal{P}_\Sigma$ given in (4.127).

Proof. We simply introduce $u^\xi = -\mathcal{R}_\Sigma^{-1}\mathcal{P}_\Sigma^*\xi$ in the operator form (3.162) of the quadratic index and then (4.140) is a consequence of the operator expression (4.133) of the stabilizing solution X. ∎

4.7.3 The positiveness case

Based on the obvious fact that a coercive operator is invertible, Theorem 4.7.1 leads directly to the following result that is related to the so–called *linear quadratic problem (LQP)* and positivity theory.

Corollary 4.7.7 *Let $\Sigma = (A, B; Q, L, R)$ be a Popov triplet with A stable. Then we have:*

1. The operator $\mathcal{R}_{e,\Sigma}$ in (4.122) is coercive if and only if the DTARE(Σ) has a stabilizing solution X and $R + B^T XB > 0$.

2. If the operator $\mathcal{R}_{e,\Sigma}$ is coercive, then the minimum of the quadratic index (3.102) is

$$\min_{u \in \ell_+^{2,m}} J_\Sigma(\xi, u) = \xi^T X \xi, \quad \forall \xi \in \mathbb{R}^n, \tag{4.141}$$

and the minimum is attained for $u = u^\xi$ given in (4.131). In particular, u^ξ satisfies the feedback relation (4.136).

3. If, in addition,

$$\begin{bmatrix} Q & L \\ L^T & R \end{bmatrix} \geq 0 \tag{4.142}$$

and the operator $\mathcal{R}_{e,\Sigma}$ is coercive, then $X \geq 0$.

The proof is quite similar to its continuous–time counterpart Corollary 4.1.7, and it is therefore omitted.

We may relax the coerciveness condition in point 1 of Corollary 4.7.7 as we did to obtain Corollary 4.1.8. Here follows the discrete–time version of Corollary 4.1.8. Notice however the important differences with respect to the continuous–time case.

Corollary 4.7.8 *Let $\Sigma = (A, B; Q, L, R)$ be a Popov triplet with $A \in \mathbb{R}^{n\times n}$, $B \in \mathbb{R}^{n\times m}$, A stable. Assume the pair (A, B) is controllable. If*

$$\mathcal{R}_{e,\Sigma} \geq 0, \tag{4.143}$$

then there exists a triple (X, V, W) with $X = X^T$ that satisfies

$$\begin{array}{rcl} R + B^T XB & = & V^T V, \\ L + A^T XB & = & W^T V, \\ Q + A^T XA - X & = & W^T W. \end{array} \tag{4.144}$$

Moreover, if V is nonsingular, then $A - BV^{-1}W$ has all the eigenvalues in the closed unit disk.

Proof. Let $\varepsilon > 0$ and let $\Sigma_\varepsilon = (A, B; Q, L, R + \varepsilon I)$. Then $\mathcal{R}_{e,\Sigma_\varepsilon} := \mathcal{R}_{e,\Sigma} + \varepsilon I >> 0$. According to Corollary 4.7.7, $R + B^T X_\varepsilon B + \varepsilon I > 0$ and the DTARE (Σ_ε)

$$A^T X_\varepsilon A - X_\varepsilon - (A^T X_\varepsilon B + L)(R + B^T X_\varepsilon B + \varepsilon I)^{-1}(B^T X_\varepsilon A + L^T) + Q = 0 \tag{4.145}$$

has a stabilizing solution X_ε, i.e., $A + BF_\varepsilon$ is stable, for

$$F_\varepsilon = -(R + B^T X_\varepsilon B + \varepsilon I)^{-1}(B^T X_\varepsilon A + L^T).$$

Using formula (4.133), it is easy to check that X_ε decreases as ε decreases to 0.

Rewrite now (4.145) as

$$A^T X_\varepsilon A - X_\varepsilon - F_\varepsilon^T (R + B^T X_\varepsilon B + \varepsilon I) F_\varepsilon + Q = 0. \tag{4.146}$$

Since (A, B) is controllable, we may choose $\widetilde{F}$ such that $\widetilde{A} := A + B\widetilde{F}$ is antistable. By Corollary 3.6.7, (4.146) is equivalent to

$$A^T X_\varepsilon A - X_\varepsilon - (F_\varepsilon - \widetilde{F})^T (R + B^T X_\varepsilon B + \varepsilon I)(F_\varepsilon - \widetilde{F}) + \widetilde{Q}_\varepsilon = 0, \tag{4.147}$$

where $\widetilde{Q}_\varepsilon := Q + L\widetilde{F} + \widetilde{F}^T L^T + \widetilde{F}^T (R + \varepsilon I)\widetilde{F}$. On the other hand, since $\widetilde{A}$ is antistable, Theorem 1.5.2 shows that the symmetric Stein equation

$$\widetilde{A}^T Z \widetilde{A} - Z + Q + L\widetilde{F} + \widetilde{F}^T L^T + \widetilde{F}^T R_1 \widetilde{F} = 0 \tag{4.148}$$

has a unique solution Z for every symmetric matrix R_1. Choose $R_1 \geq R + \varepsilon I$ and subtract (4.148) from (4.147). Then

$$\begin{aligned} & \widetilde{A}^T (X_\varepsilon - Z)\widetilde{A} - (X_\varepsilon - Z) \\ -(F_\varepsilon - \widetilde{F})^T (R + B^T X_\varepsilon B + \varepsilon I)(F_\varepsilon - \widetilde{F}) - \widetilde{F}^T (R_1 - R - \varepsilon I)\widetilde{F} &= 0. \end{aligned} \tag{4.149}$$

As $\widetilde{A}$ is antistable and

$$(F_\varepsilon - \widetilde{F})^T (R + B^T X_\varepsilon B + \varepsilon I)(F_\varepsilon - \widetilde{F}) + \widetilde{F}^T (R_1 - R - \varepsilon I)\widetilde{F} \geq 0,$$

(4.149) implies that $X_\varepsilon - Z \geq 0$. Hence

$$X_\varepsilon \geq Z, \tag{4.150}$$

for every $\varepsilon > 0$. Since X_ε decreases as ε decreases to 0, we conclude that there exists $X = X^T$ such that

$$\lim_{\varepsilon \downarrow 0} X_\varepsilon = X. \tag{4.151}$$

Since X_ε satisfies the DTARE(Σ_ε) and $R + B^T X_\varepsilon B + \varepsilon I > 0$, we deduce that

$$D_{\Sigma_\varepsilon}(X_\varepsilon) = \begin{bmatrix} A^T X_\varepsilon A - X_\epsilon + Q & L + A^T X_\varepsilon B \\ B^T X_\varepsilon A + L^T & R + B^T X_\varepsilon B + \varepsilon I \end{bmatrix} \geq 0.$$

By taking $\varepsilon \to 0$ we get

$$D_\Sigma(X) = \begin{bmatrix} A^T X A - X + Q & L + A^T X B \\ B^T X A + L^T & R + B^T X B \end{bmatrix} \geq 0.$$

Assume that $D_\Sigma(X)$ has $p \leq n + m$ strictly positive eigenvalues. By continuity, for ε sufficiently small, $D_{\Sigma_\varepsilon}(X_\varepsilon)$ has at least p strictly positive eigenvalues. Since $\operatorname{rank} D_{\Sigma_\varepsilon}(X_\varepsilon) = m$ for each ε as follows from (4.145), it is necessary that $p \leq m$, hence $\operatorname{rank} D_\Sigma(X) \leq m$. Accordingly, there exist $V \in \mathbb{R}^{m\times m}$ and $W \in \mathbb{R}^{m\times n}$ such that

$$D_\Sigma(X) = \begin{bmatrix} W^T \\ V^T \end{bmatrix} \begin{bmatrix} W & V \end{bmatrix},$$

which is equivalent to (4.144).

If V is nonsingular, then $R + B^T X B$ is nonsingular, and thus $\lim_{\varepsilon\downarrow 0} F_\varepsilon = -(R + B^T X B)^{-1}(B^T X + L) = -V^{-1}W$ and $A + BF_\varepsilon$ is stable for each $\varepsilon > 0$. Therefore $A - BV^{-1}W$ will have all the eigenvalues in the closed unit disk. ∎

4.8 Removing the stability assumption

Here follows the discrete–time version of Theorem 4.2.1.

Theorem 4.8.1 *Let $\Sigma = (A, B; Q, L, R)$ be a Popov triplet with (A, B) stabilizable. Then the following two assertions are equivalent:*

1. The DTARE(Σ) has a stabilizing solution.

2. There exists $F \in \mathbb{R}^{m \times n}$ such that $A+BF$ is stable and such that the F–equivalent of Σ, $\widetilde{\Sigma} = (\widetilde{A}, \widetilde{B}; \widetilde{Q}, \widetilde{L}, \widetilde{R})$, has an invertible associated Toeplitz operator $\mathcal{R}_{\widetilde{\Sigma}}$.

Let us state now the discrete–time counterpart of Corollary 4.2.2.

Corollary 4.8.2 *Let $\Sigma_s = (A, B; C^T C, 0, I)$ be the standard Popov triplet for control, and assume that the pairs (A, B) and (C, A) are stabilizable and detectable, respectively. Then the DTARE(Σ_s), known as the* standard discrete Riccati equation *for control,*

$$A^T X A - X - A^T X B (I + B^T X B)^{-1} B^T X A + C^T C = 0 \tag{4.152}$$

has a positive semidefinite stabilizing solution.

The proofs of these last two results are entirely analogous to their continuous–time counterparts, and are skipped.

4.9 Existence of the stabilizing solution: the inverse–time case

We begin with a result which stresses the essential difference between the DTARE associated with a Popov triplet with *direct–time* significance and the DTARE associated with a Popov triplet with *inverse–time* significance.

Theorem 4.9.1 *Consider the Popov triplet $\Sigma = (A, B; Q, L, R)$ with A stable and seen with inverse–time significance. Let $\Sigma_r = (A, B; 0, L_r = L + A^T X_r B, R_r = R + B^T X_r B)$ and $\Sigma_d = (A_d = A^T, B_d = L + A^T X_r B; 0, L_d = B, R_d = R + B^T X_r B)$ be the reduced equivalent and the dual of Σ, respectively, where X_r is the unique solution to the reducing symmetric Stein equation $A^T X A - X + Q = 0$. Then the following assertions are equivalent:*

1. The DTARE(Σ_d) has a stabilizing solution Z.

2. The Toeplitz operator $\widehat{\mathcal{R}}_{\Sigma_r}$ has a bounded inverse, where $\widehat{\mathcal{R}}_{\Sigma_r}$ is the operator (3.179) written for Σ_r (see also (3.182)).

If point 1 or 2 holds, then Z is given by the representation formula

$$Z = -\widehat{\Psi} \widehat{\mathcal{R}}_{\Sigma_r}^{-1} \widehat{\Psi}^*,$$

and $\widehat{\Psi}$ is the anticausal controllability operator (2.198) associated with the pair (A, B).

Proof. The proof follows by combining Lemma 3.8.6 with Theorem 4.7.1. ∎

4.10 Existence of the stabilizing solution: the antistable case

As in Section 4.3 for the continuous–time case, we consider now the case that A is antistable. In this case, A is of course invertible and the triplet Σ is reversible.

4.10.1 The main result

Theorem 4.10.1 *Let $\Sigma = (A, B; Q, L, R)$ be a Popov triplet with A antistable and seen with direct–time significance. Let $\Sigma_r = (A, B; 0, L_r = A^T X_r B + L, R_r = R + B^T X_r B)$ be the reduced equivalent of Σ, where X_r is the unique solution to the reducing Stein equation $A^T X A - X + Q = 0$. Then the following assertions are equivalent:*

1. The DTARE(Σ) has a stabilizing solution X such that $X - X_r$ is nonsingular.

2. The Toeplitz operator $\widehat{\mathcal{R}}_{\Sigma_r}$ is invertible, and $\widehat{\Psi}\widehat{\mathcal{R}}_{\Sigma_r}^{-1}\widehat{\Psi}^*$ is invertible, where $\widehat{\mathcal{R}}_{\Sigma_r}$ is the operator (3.179) written for Σ_r (see also (3.182)), and $\widehat{\Psi}$ is the anticausal controllability operator (2.198) associated with the pair (A, B).

Before proceeding with the proof we need some auxiliary steps. First of all, let us notice that assertion 1 is equivalent to the existence of a nonsingular stabilizing solution to the DTARE(Σ_r), where Σ_r is the reduced equivalent Popov triplet of Σ. Indeed, according to Corollary 3.6.7, X is a stabilizing solution to the DTARE(Σ) if and only if $X - X_r$ is a solution to the DTARE(Σ_r).

Since A is antistable, the Popov triplet $\Sigma_r = (A, B; 0, L_r = A^T X_r B + L, R_r = R + B^T X_r B)$ is reversible. Denote by $\widehat{\Sigma}_r$ the time–reversed Popov triplet of Σ_r with the entries given explicitly by (3.144) written for the actual data, and by $\widehat{\Sigma}_d = (A_d, B_d; 0, L_d, R_d)$ the dual of $\widehat{\Sigma}_r$, with the entries given explicitly by

$$A_d := A^{-T}, B_d := -A^{-T} L_r, L_d := -A^{-1} B, R_d := R_r - B^T A^{-T} L_r - L_r^T A^{-1} B. \tag{4.153}$$

Lemma 4.10.2 *Let $\Sigma_r = (A, B; 0, L_r, R)$ be the reduced equivalent of Σ with A antistable and $\widehat{\Sigma}_d = (A_d, B_d; 0, L_d, R_d)$ be the triplet with entries given in (4.153). Then the DTARE(Σ_r) has a nonsingular stabilizing solution X if and only if the DTARE($\widehat{\Sigma}_d$) has a nonsingular stabilizing solution Z and in this case*

$$X = -Z^{-1}.$$

Proof. Direct substitution shows that the following identity is true:

$$D_{\widehat{\Sigma}_d}(-X^{-1}) = \begin{bmatrix} A^{-1}X^{-1}A^{-T} & O \\ -B^T A^{-T} - L^T A^{-1} X^{-1} A^{-T} & I \end{bmatrix} D_{\Sigma_r}(X) \begin{bmatrix} X^{-1} & O \\ O & I \end{bmatrix}, \tag{4.154}$$

where $D_{\widehat{\Sigma}_d}$ and D_{Σ_r} are the dissipation matrices corresponding to the respective triplets.

Assume now that the DTARE(Σ_r) has a nonsingular stabilizing solution X. Then it also satisfies the DTARS(Σ_r), that is,

$$D_{\Sigma_r}(X) \begin{bmatrix} I \\ F_{Ricc} \end{bmatrix} = 0$$

where F_{Ricc} is the Riccati feedback associated with Σ_r. With (4.154), we deduce that

$$D_{\widehat{\Sigma}_d}(-X^{-1}) \begin{bmatrix} I \\ F_{Ricc} X^{-1} \end{bmatrix} = 0.$$

If we denote $F_d = F_{Ricc}X^{-1}$, we have

$$A_d + B_dF_d = A^{-T} - A^{-T}L_rF_{Ricc}X^{-1} = A^{-T}(X - L_rF_{Ricc})X^{-1}.$$

From the fact that X is a solution of the DTARE(Σ_r), we have

$$X - L_rF_{Ricc} = A^TX(A + BF_{Ricc}).$$

We conclude that

$$A_d + B_dF_d = X(A + BF_{Ricc})X^{-1},$$

and since $A+BF_{Ricc}$ is stable, it follows that $A_d+B_dF_d$ is also stable. In order to show that indeed $-X^{-1}$ is the stabilizing solution, we have to show that $R_d+B_d^T(-X^{-1})B_d$ is nonsingular. We have successively

$$\begin{aligned} R_d + B_d^T(-X^{-1})B_d &= R_r - B^TA^{-T}L_r - L_r^TA^{-1}B - L_r^TA^{-1}X^{-1}A^{-T}L_r \\ &= R_r + B^TXB - (B^TXA + L_r^T)(A^TXA)^{-1}(A^TXB + L_r). \end{aligned}$$

Using the DTARE(Σ_r), we have that

$$A^TXA = X - (A^TXB + L_r)F_{Ricc}.$$

Using this and the expression of F_{Ricc} we have

$$\begin{aligned} & R_d + B_d^T(-X^{-1})B_d \\ &= (R_r + B^TXB)\{I + F_{Ricc}[X - (A^TXB + L_r)F_{Ricc}]^{-1}(A^TXB + L_r)\} \\ &= (R_r + B^TXB)\{I + F_{Ricc}X^{-1}[I - (A^TXB + L_r)F_{Ricc}X^{-1}]^{-1}(A^TXB + L_r)\} \\ &= (R_r + B^TXB)\{I + [I - F_{Ricc}X^{-1}(A^TXB + L_r)]^{-1}F_{Ricc}X^{-1}(A^TXB + L_r)\} \\ &= (R_r + B^TXB)[I - F_{Ricc}X^{-1}(A^TXB + L_r)]^{-1}. \end{aligned}$$

As $R_r + B^TXB$ is invertible, it follows that $R_d + B_d^T(-X^{-1})B_d$ is also invertible, and $-X^{-1}$ is the stabilizing solution of the DTARE($\widehat{\Sigma}_d$).

The converse follows if we notice that the transformation from Σ_r to $\widehat{\Sigma}_d$ is involutive, that is, the roles of the two triplets are interchangeable. ■

Now we can proceed with the proof of Theorem 4.10.1.

Proof. 1. ⇒ 2. Assume that the DTARE(Σ) has a stabilizing solution X such that $X - X_r$ is nonsingular. This means that the DTARE(Σ_r) has a nonsingular solution. According to Lemma 4.10.2, this implies that the DTARE($\widehat{\Sigma}_d$) has a nonsingular stabilizing solution $Z = -X^{-1}$. By Theorem 4.7.1, the Toeplitz operator $\mathcal{R}_{\widehat{\Sigma}_d}$ is invertible. Now, by Lemma 3.8.8, we conclude that $\mathcal{R}_{\Sigma_r}$ is invertible. Furthermore, the stabilizing solution of the CTARE($\widehat{\Sigma}_d$) admits an operator representation like (4.133):

$$Z = \mathcal{P}_{o,\widehat{\Sigma}_d} - \mathcal{P}_{\widehat{\Sigma}_d}\mathcal{R}_{\widehat{\Sigma}_d}^{-1}\mathcal{P}_{\widehat{\Sigma}_d}^*.$$

But $\mathcal{P}_{o,\widehat{\Sigma}_d} = 0$ since $Q_d = 0$ and using again Lemma 3.8.8, we have

$$Z = -\widehat{\Psi}\mathcal{R}_{\Sigma_r}^{-1}\widehat{\Psi}^*. \tag{4.155}$$

Since Z is nonsingular, assertion 2 follows. In particular, with Lemma 4.10.2 we obtain the following operator representation:

$$X = X_r + (\widehat{\Psi}\mathcal{R}_{\Sigma_r}^{-1}\widehat{\Psi}^*)^{-1}. \tag{4.156}$$

2. $\Rightarrow$ 1. Assume now that $\mathcal{R}_{\Sigma_r}$ and $\widehat{\Psi}\mathcal{R}_{\Sigma_r}^{-1}\widehat{\Psi}^*$ are invertible. From Lemma 4.10.2, it follows that $\mathcal{R}_{\widehat{\Sigma}_d}$ is invertible. By Theorem 4.7.1, the DTARE($\widehat{\Sigma}_d$) has a stabilizing solution, that can be represented as in (4.155). Since we have assumed that $\widehat{\Psi}\mathcal{R}_{\Sigma_r}^{-1}\widehat{\Psi}^*$ is invertible, we deduce that the stabilizing solution of the DTARE($\widehat{\Sigma}_d$) is nonsingular. According to Lemma 4.10.2, DTARE(Σ_r) has a nonsingular stabilizing solution, thus the DTARE(Σ) has a stabilizing solution X with $X - X_r$ nonsingular. ■

4.10.2 The positiveness case

We explore now the positiveness case and show that the stabilizing solution that we have studied in Section 4.10.1 is closely related to a type of inverse–time linear quadratic optimal problem.

Corollary 4.10.3 *Let $\Sigma = (A, B; Q, L, R)$ be a Popov triplet with $A \in \mathbb{R}^{n\times n}$, $B \in \mathbb{R}^{n\times m}$, A antistable, and seen with direct–time significance. Assume that:*

a. *$\mathcal{R}_{e,\Sigma}$ is coercive.*

b. *The pair (A, B) is controllable.*

Then:

1. The DTARE(Σ) has a stabilizing solution X such that $X - X_r$ is positive definite.

2. We have

$$\min_{\substack{\widehat{\Psi}u = \xi \\ u \in \ell_+^{2,m}}} J_\Sigma(u) = \xi^T X \xi, \tag{4.157}$$

where J_Σ is the restricted quadratic index (3.106) associated with Σ.

Proof. 1. We prove this point by checking that point 2 of Theorem 4.10.1 holds.

By Proposition 3.8.1, $\mathcal{R}_{e,\Sigma} = \mathcal{R}_{e,\Sigma_r}$, where Σ_r is the reduced equivalent of Σ. From (3.199), we deduce that $\mathcal{R}_{\Sigma_r}$ is coercive, and thus invertible. Of course, its inverse is also coercive. On the other hand, by the controllability assumption, the anticausal controllability Gramian $\widehat{P} = \widehat{\Psi}\widehat{\Psi}^*$ is positive definite. Take a $\nu > 0$ such that $\xi^T\widehat{P}\xi \geq \nu\|\xi\|^2$ and a $\delta > 0$ such that $\langle u, \mathcal{R}_{\Sigma_r}^{-1}u\rangle \geq \delta\|u\|^2$. Then we have

$$\xi^T\widehat{\Psi}\mathcal{R}_{\Sigma_r}^{-1}\widehat{\Psi}^*\xi = \left\langle \widehat{\Psi}^*\xi, \mathcal{R}_{\Sigma_r}^{-1}\widehat{\Psi}^*\xi \right\rangle \geq \delta\|\widehat{\Psi}^*\xi\|^2$$

$$= \delta\left\langle \xi, \widehat{\Psi}\widehat{\Psi}^*\xi \right\rangle = \delta\xi^T\widehat{P}\xi \geq \delta\nu\|\xi\|^2, \ \forall \xi \in \mathbb{R}^n. \tag{4.158}$$

We deduce that $\widehat{\Psi}{\mathcal{R}_{\Sigma_r}}^{-1}\widehat{\Psi}^*$ is positive definite, and therefore invertible. The existence of the stabilizing solution follows from Theorem 4.10.1 and the sign of $X - X_r$ follows from the operator representation (4.156) and assertion 1 is proved.

2. For each $u \in \ell_+^{2,m}$ such that $\widehat{x}_0^u = \widehat{\Psi}u = \xi$, we have

$$J_\Sigma(u) = J_{\Sigma_r}(u) + \xi^T X_r \xi$$

and by (3.197),

$$J_{\Sigma_r}(u) = \langle u, \mathcal{R}_{\Sigma_r} u \rangle , \tag{4.159}$$

where $\mathcal{R}_{\Sigma_r} >> 0$. Thus the problem reduces to

$$\min_{\widehat{\Psi}u=\xi} J_{\Sigma_r}(u)$$

which is solved identically as in the proof of Corollary 4.3.4 and we get (4.157). ■

4.11 The signature condition

We present in this section the discrete–time counterparts of the results in Section 4.4. Here we briefly recall the setting of that section again, even though it hardly differs from the continuous case.

Let $\Sigma = (A, B; Q, L, R)$ be a Popov triplet with $A \in \mathbb{R}^{n\times n}$, $B \in \mathbb{R}^{n\times m}$, and A stable. Let the matrix B be partitioned as

$$B = [B_1 \;\; B_2],\; B_i \in \mathbb{R}^{n\times m_i},\; i = 1, 2,\; m_1 + m_2 = m, \tag{4.160}$$

and the matrices L and R be partitioned in accordance with B, i.e.,

$$L = [L_1 \;\; L_2],\;\; R = \begin{bmatrix} R_{11} & R_{12} \\ R_{12}^T & R_{22} \end{bmatrix}. \tag{4.161}$$

The solution to the initial value problem

$$\sigma x = Ax + B_1 u_1 + B_2 u_2 \;, \quad x(0) = \xi, \tag{4.162}$$

will be written as

$$x^{\xi,u_1,u_2} = \Phi\xi + \mathcal{L}_1 u_1 + \mathcal{L}_2 u_2,$$

where $\mathcal{L}_i$, $i = 1, 2$, are the operators (4.121) corresponding to the pairs (A, B_i).

Following (4.120), the operator $\mathcal{R}_\Sigma$ can be partitioned conformably as

$$\mathcal{R}_\Sigma = \begin{bmatrix} \mathcal{R}_{11} & \mathcal{R}_{12} \\ \mathcal{R}_{12}^* & \mathcal{R}_{22} \end{bmatrix}. \tag{4.163}$$

It is easy to see that $\mathcal{R}_{22}$ coincides with the Toeplitz operator $\mathcal{R}_{\Sigma_2}$ associated with the Popov triplet

$$\Sigma_2 = (A, B_2; Q, L_2, R_{22}). \tag{4.164}$$

The discrete–time counterparts of Theorem 4.4.1 and Theorem 4.4.2 look identical to their continuous–time versions. As their proofs involve only operator manipulations that are identical for the two cases, we omit them and give only their statements.

Theorem 4.11.1 *Let $\Sigma = (A, B; Q, L, R)$ be a Popov triplet with its entries partitioned as in (4.160)–(4.161) and A stable. If $\mathcal{R}_{22} >> 0$ and*

$$\mathcal{R}_{11}^{\times} = \mathcal{R}_{11} - \mathcal{R}_{12}\mathcal{R}_{22}^{-1}\mathcal{R}_{12}^{*} << 0, \tag{4.165}$$

(see (4.163)) then we have:

1. The DTARE(Σ_2) has a stabilizing solution X_2, where $\Sigma_2 = (A, B_2; Q, L_2, R_{22})$.

2. The DTARE(Σ) has a stabilizing solution X such that

$$X \geq X_2. \tag{4.166}$$

Theorem 4.11.2 *With the same notation and under the same hypotheses as in Theorem 4.11.1, let*

$$J_\Sigma(\xi, u_1, u_2) = \left\langle \begin{bmatrix} x^{\xi,u_1,u_2} \\ u_1 \\ u_2 \end{bmatrix}, \begin{bmatrix} Q & L_1 & L_2 \\ L_1^T & R_{11} & R_{12} \\ L_2^T & R_{12}^T & R_{22} \end{bmatrix} \begin{bmatrix} x^{\xi,u_1,u_2} \\ u_1 \\ u_2 \end{bmatrix} \right\rangle$$

be the direct–time quadratic index (3.102) for all $(\xi, u_1, u_2) \in \mathbb{R}^n \times \ell_+^{2,m_1} \times \ell_+^{2,m_2}$. Then:

1. For each $\xi \in \mathbb{R}^n$ and each $u_1 \in \ell_+^{2,m_1}$, there exists a unique u_2 in ℓ_+^{2,m_2}, that we denote by u_2^{ξ,u_1}, that minimizes $J_\Sigma(\xi, u_1, u_2)$, i.e.,

$$J_\Sigma(\xi, u_1, u_2) \geq J_\Sigma(\xi, u_1, u_2^{\xi,u_1}) \tag{4.167}$$

for all $u_2 \in \ell_+^{2,m_2}$.

2. For each ξ in $\mathbb{R}^n$, there exists a unique u_1, that we denote u_1^ξ, that maximizes $J_\Sigma(\xi, u_1, u_2^{\xi,u_1})$.

3. If we denote $u_2^\xi = u_2^{\xi,u_1^\xi}$, then

$$J_\Sigma(\xi, u_1^\xi, u_2^\xi) = \xi^T X \xi \tag{4.168}$$

for every $\xi \in \mathbb{R}^n$, where X is the stabilizing solution to the DTARE(Σ).

Moreover, if x^ξ is the solution of (4.162) corresponding to $u_1 = u_1^\xi$ and $u_2 = u_2^\xi$, then

$$u^\xi = \begin{bmatrix} u_1^\xi \\ u_2^\xi \end{bmatrix} = F_{Ricc}\, x^\xi, \tag{4.169}$$

where F_{Ricc} is the stabilizing Riccati feedback.

4.12 Frequency domain conditions

Let us now express our conditions for the existence of a solution to the DTARE in the frequency domain. The main difference lies in the fact that frequency domain conditions are in this case expressed with reference to the unit circle (disk) rather than with respect to the imaginary axis (left–half plane).

Definition 4.12.1 *Let $\Sigma = (A, B; Q, L, R)$ be a Popov triplet and let $\mathbf{\Pi}_\Sigma$ be the associated Popov function, that is*

$$\mathbf{\Pi}_\Sigma = \begin{bmatrix} B^T(z^{-1}I - A^T)^{-1} & I \end{bmatrix} \begin{bmatrix} Q & L \\ L^T & R \end{bmatrix} \begin{bmatrix} (zI - A)^{-1}B \\ I \end{bmatrix}. \tag{4.170}$$

If A is stable and there exist $\mathbf{\Xi}$ and $\mathbf{\Omega}$, two units in $RH^{\infty}_{+,m\times m}$ such that

$$\mathbf{\Pi}_{\Sigma}(e^{j\theta}) = [\mathbf{\Xi}(e^{j\theta})]^{*}\mathbf{\Omega}(e^{j\theta}), \tag{4.171}$$

then we shall say that $\mathbf{\Pi}_{\Sigma}$ is antianalytic factorizable.

If the pair (A, B) is stabilizable, and there exists F such that $A+BF$ is stable, and the Popov function $\mathbf{\Pi}_{\widetilde{\Sigma}}$ is antianalytic factorizable, where $\widetilde{\Sigma}$ is the F–equivalent of Σ, then we say that $\mathbf{\Pi}_{\Sigma}$ is antianalytic prefactorizable.

An easy consequence of Proposition 3.6.17 is the following.

Proposition 4.12.2 *Let $\Sigma = (A, B; Q, L, R)$ be a Popov triplet with the pair (A, B) stabilizable. If there exists an F such that $A + BF$ is stable and for which $\mathbf{\Pi}_{\widetilde{\Sigma}}$ is antianalytic factorizable, where $\widetilde{\Sigma}$ is the F–equivalent of Σ, then $\mathbf{\Pi}_{\widetilde{\Sigma}}$ is antianalytic factorizable for* every *F–equivalent triplet $\widetilde{\Sigma}$ for which $A + BF$ is stable.*

Theorem 4.12.3 *Let $\Sigma = (A, B; P)$ be a Popov triplet with A stable. Then the following two assertions are equivalent:*

1. *The DTARE(Σ) has a stabilizing solution.*
2. *The Popov function $\mathbf{\Pi}_{\Sigma}$ is antianalytic factorizable.*

Theorem 4.12.4 *Let $\Sigma = (A, B; Q, L, R)$ be a Popov triplet with (A, B) stabilizable. Then the following two assertions are equivalent:*

1. *The DTARE(Σ) has a stabilizing solution.*
2. *The Popov function $\mathbf{\Pi}_{\Sigma}$ is antianalytic prefactorizable.*

For the positiveness case we need the discrete–time counterpart of Corollary 4.5.5.

Corollary 4.12.5 *Let $\Sigma = (A, B; P)$ be a Popov triplet with A stable. Then:*
1. *$\mathbf{\Pi}_{\Sigma}(e^{j\theta}) > 0$ for each $\theta \in [0,\ 2\pi)$, if and only if $\mathcal{R}_{e,\Sigma}$ is coercive.*
2. *$\mathbf{\Pi}_{\Sigma}(e^{j\theta}) \geq 0$ for each $\theta \in [0,\ 2\pi)$, if and only if $\mathcal{R}_{e,\Sigma} \geq 0$.*

Proof. For $u \in \ell^{2,m}$, we denote by $\widehat{u}$ the function defined on $\partial\mathbb{D}$ by its Fourier series

$$\widehat{u}(e^{j\theta}) := \sum_{k=-\infty}^{\infty} u_k e^{jk\theta}\ , \quad \theta \in [0, 2\pi).$$

Then $\widehat{u}$ is square–integrable on $\partial\mathbb{D}$ and $\|\widehat{u}\|_2^2 = \frac{1}{2\pi}\|u\|_2^2$. By Parseval's identity we have

$$\langle u, \mathcal{R}_{e,\Sigma}u\rangle_{\ell^{2,m}} = \frac{1}{2\pi}\int_0^{2\pi} \langle \widehat{u}(e^{j\theta}), \mathbf{\Pi}_{\Sigma}(e^{j\theta})\widehat{u}(e^{j\theta})\rangle \, d\theta. \tag{4.172}$$

The rest of the proof is similarly to the proof of Corollary 4.5.5 where we simply replace the imaginary axis with the unit circle. ■

Corollary 4.12.6 *Corollaries 4.7.7 and 4.7.8 hold if conditions $\mathcal{R}_{e,\Sigma} >> 0$ and $\mathcal{R}_{e,\Sigma} \geq 0$ are replaced by their frequency domain counterparts $\mathbf{\Pi}_{\Sigma} > 0$ on $\partial\mathbb{D}$ and $\mathbf{\Pi}_{\Sigma} \geq 0$ on $\partial\mathbb{D}$, respectively.*

4.12.1 The positiveness case: A arbitrary

We remove now the stability condition and get the following discrete–time counterpart of Corollary 4.5.7.

Corollary 4.12.7 *Let $\Sigma = (A, B; Q, L, R)$ be a Popov triplet. Assume that:*
a. *The pair (A, B) is stabilizable.*
b. *The descriptor realization of $\mathbf{\Pi}_\Sigma$,*

$$\mathbf{\Pi}_\Sigma = \left[\begin{array}{cc|c} zI - A & O & -B \\ -Q & I - zA^T & -L \\ \hline L^T & zB^T & R \end{array}\right], \tag{4.173}$$

is $\partial\mathbb{D}$–controllable and $\partial\mathbb{D}$–observable.
c. $\mathbf{\Pi}_\Sigma(e^{j\theta}) > 0, \quad \forall\theta \in [0,\ 2\pi)$.
Then:
1. The DTARE(Σ) has a stabilizing solution X with $R + B^T X B > 0$.
2. We have

$$\min_{u \in \mathcal{U}^\xi} J_\Sigma(\xi, u) = \xi^T X \xi$$

where $\mathcal{U}_\xi := \{u : u \in \ell_+^{2,m}$ such that $x^{\xi,u} \in \ell_+^{2,n}\}$.
3. If, in addition,

$$\left[\begin{array}{cc} Q & L \\ L^T & R \end{array}\right] \geq 0, \tag{4.174}$$

then $X \geq 0$.

4.12.2 The signature condition in the frequency domain

We give now a frequency domain version of Theorem 4.11.1. Consider again a Popov triplet $\Sigma = (A, B; Q, L, R)$ with B, L and R partitioned as

$$B = \left[\begin{array}{cc} B_1 & B_2 \end{array}\right], L = \left[\begin{array}{cc} L_1 & L_2 \end{array}\right], R = \left[\begin{array}{cc} R_{11} & R_{12} \\ R_{12}^T & R_{22} \end{array}\right], \tag{4.175}$$

($B_i \in \mathbb{R}^{n \times m_i}, i = 1, 2, m = m_1 + m_2$), and let the Popov function $\mathbf{\Pi}_\Sigma$ be partitioned in accordance with R in (4.175), i.e.,

$$\mathbf{\Pi}_\Sigma = \left[\begin{array}{cc} \mathbf{\Pi}_{11} & \mathbf{\Pi}_{12} \\ \mathbf{\Pi}_{12}^* & \mathbf{\Pi}_{22} \end{array}\right]. \tag{4.176}$$

Introduce also the sign matrix

$$J = \left[\begin{array}{cc} -I_{m_1} & \\ & I_{m_2} \end{array}\right]. \tag{4.177}$$

Remember also the Popov triplet $\Sigma_2 = (A, B_2; Q, L_2, R_{22})$. It is easy to see that $\mathbf{\Pi}_{22} = \mathbf{\Pi}_{\Sigma_2}$.

Theorem 4.12.8 *Let $\Sigma = (A, B; Q, L, R)$ be a Popov triplet with A stable and the entries partitioned as in (4.175). Assume that $\mathbf{\Pi}_{22} > 0$ on $\partial\mathbb{D}$. If there exists an $\mathbf{S} \in RH^{\infty}_{+,m_2\times m_1}$ such that*

$$\begin{bmatrix} I & \mathbf{S}^* \end{bmatrix} \mathbf{\Pi}_{\Sigma} \begin{bmatrix} I \\ \mathbf{S} \end{bmatrix} < 0 \text{ on } \partial\mathbb{D} \tag{4.178}$$

then the KSPYS(Σ, J) has a stabilizing solution

$$(X, V, W) = (X, \begin{bmatrix} V_{11} & O \\ V_{21} & V_{22} \end{bmatrix}, \begin{bmatrix} W_1 \\ W_2 \end{bmatrix})$$

where the partitions of V and W are in accordance with J given in (4.177).

Proof. The existence of the stabilizing solution X to the DTARE(Σ) follows by similar arguments as for the proof of Theorem 4.5.8. Since $\mathbf{\Pi}_{22} > 0$ on $\partial\mathbb{D}$ it follows from Corollary 4.12.6 that the DTARE(Σ_2) has a stabilizing solution X_2 with $R_{22} + B_2^T X_2 B_2 > 0$. As $X \geq X_2$ it follows that

$$R_{22} + B_2^T X B_2 > 0. \tag{4.179}$$

Combining the positiveness of $\mathbf{\Pi}_{22}$ with (4.178) we deduce easily that $\operatorname{sgn} \mathbf{\Pi}_{\Sigma}(e^{j\theta}) = J, \forall\theta \in [0,\ 2\pi)$. Invoking (3.142) one concludes that $\operatorname{sgn}(R + B^T X B) = J$, and with (4.179) we get

$$R_{11} + B_1^T X B_1 - (R_{12} + B_1^T X B_2)(R_{22} + B_2^T X B_2)^{-1}(R_{12}^T + B_2^T X B_1) < 0. \tag{4.180}$$

From (4.179) and (4.180) it follows, analogously to the proof of Theorem 4.5.8, that $R + B^T X B$ can be factorized as $R + B^T X B = V^T J V$, with V of lower–left block triangular form. The remainder of the proof follows now from Remark 3.6.9. ■

We can also state the discrete–time counterpart of Theorem 4.5.9.

Theorem 4.12.9 *Let $\Sigma = (A, B; Q, L, R)$ be a Popov triplet with its entries partitioned as in (4.175). Assume that:*

a. *A is stable.*

b. $\mathbf{\Pi}_{22} > 0$ *on* $\partial\mathbb{D}$.

c.

$$\begin{bmatrix} Q & L_2 \\ L_2^T & R_{22} \end{bmatrix} \geq 0.$$

Then:

I. *The following assertions are equivalent:*

1. There exists $\mathbf{S} \in RH^{\infty}_{+,m_2\times m_1}$ which satisfies the signature inequality

$$\begin{bmatrix} I & \mathbf{S}^* \end{bmatrix} \mathbf{\Pi}_{\Sigma} \begin{bmatrix} I \\ \mathbf{S} \end{bmatrix} < 0 \text{ on } \partial\mathbb{D}. \tag{4.181}$$

2. The KSPYS(Σ, J) has a stabilizing solution

$$(X, V, W) = (X, \begin{bmatrix} V_{11} & O \\ V_{21} & V_{22} \end{bmatrix}, \begin{bmatrix} W_1 \\ W_2 \end{bmatrix})$$

with $X \geq 0$ and where the partitions of V and W are in accordance with J given in (4.177).

II. *If assertion 1 or 2 holds, then there exists a rational function*

$$\mathbf{G} = \begin{bmatrix} \mathbf{G}_1 & \mathbf{G}_2 \\ \mathbf{G}_3 & \mathbf{G}_4 \end{bmatrix}$$

partitioned in accordance with $\mathbf{\Pi}_\Sigma$ in (4.176) such that $\mathbf{G}$ and $\mathbf{G}_4$ are units in RH_+^∞ and

$$\mathbf{\Pi}_\Sigma = \mathbf{G}^* J \mathbf{G}, \tag{4.182}$$

(i.e., $\mathbf{G}$ is a J–spectral factor of $\mathbf{\Pi}_\Sigma$). The class of all $\mathbf{S}$ for which (4.181) holds is given by

$$\mathbf{S} = \mathbf{S}_2 \mathbf{S}_1^{-1}, \quad \begin{bmatrix} \mathbf{S}_1 \\ \mathbf{S}_2 \end{bmatrix} = \mathbf{G}^{-1} \begin{bmatrix} I \\ \boldsymbol{\theta} \end{bmatrix}, \tag{4.183}$$

where $\boldsymbol{\theta}$ is a free parameter in $RH_{+,m_2 \times m_1}^\infty$ with $\|\boldsymbol{\theta}\|_\infty < 1$.

We conclude the section with the discrete–time counterpart of Theorem 4.5.10.

Theorem 4.12.10 *Let $\Sigma = (A, B; Q, L, R)$ be a Popov triplet with its entries partitioned as in (4.175) (with A not necessarily stable). Assume that the KSPYS(Σ, J) has a stabilizing solution*

$$(X, V, W) = (X, \begin{bmatrix} V_{11} & O \\ V_{21} & V_{22} \end{bmatrix}, \begin{bmatrix} W_1 \\ W_2 \end{bmatrix})$$

with $X \geq 0$, $R_{22} + B_2^T X B_2 > 0$, and $R_{11} + B_1^T X B_1 < 0$. If the stabilizing Riccati feedback is partitioned in accordance with R, i.e.,

$$F_{Ricc} = \begin{bmatrix} F_1 \\ F_2 \end{bmatrix}$$

then we have:

1. *$u_2 = F_2 x$ stabilizes the system $\sigma x = Ax + B_1 u_1 + B_2 u_2$.*
2. *There exists $\rho > 0$ such that for $u_2 = F_2 x$, the following inequality holds*

$$J_{e,\Sigma}(u) \leq -\rho^2 \|u_1\|_2^2, \quad \forall u_1 \in \ell^{2,m_1}, \tag{4.184}$$

where $J_{e,\Sigma}$ is the extended quadratic index (3.104), and $u = \begin{bmatrix} u_1 \\ u_2 \end{bmatrix}$.

We omit the proofs of the above two theorems as they are entirely similar to their continuous–time counterparts.

4.13 Algebraic Riccati inequalities

Here follow the discrete–time counterparts of the results in Section 4.6. To this end we introduce first the following notation: for a fixed Popov triplet $\Sigma = (A, B; Q, L, R)$ and for $X = X^T$, denote

$$\text{DRICOP}(\Sigma, X) := A^T X A - X - (A^T X B + L)(R + B^T X B)^{-1}(B^T X A + L^T) + Q.$$

Theorem 4.13.1 *Let $\Sigma = (A, B; Q, L, R)$ be a Popov triplet with $Q \geq 0$. Then the following two assertions are equivalent:*

1. *A is stable and $\mathbf{\Pi}_\Sigma < 0$ on $\partial\mathbb{D}$.*
2. *There exists $X > 0$ such that $R + B^T XB < 0$ and* $\mathrm{DRICOP}(\Sigma, X) < 0$.

Proof. The proof runs by similar arguments as the proof of Theorem 4.6.1 and is therefore omitted. ∎

Assertion 2 in Theorem 4.13.1 is equivalent to

$$D_\Sigma(X) = \begin{bmatrix} A^T XA - X + Q & A^T XB + L \\ L^T + B^T XA & R + B^T XB \end{bmatrix} < 0$$

for $X > 0$.

Notes and references

The operator theoretic viewpoint in Riccati equation theory is due to Jonckheere and Silverman [JS78], and was extensively investigated for more than a decade by Jonckheere, Juang, Silverman and Verma [JS81, JV86, JJ88, JJS89]. However, in all these works the attention is restricted to the case in which the coefficient of the quadratic term is positive definite. Extensions under the weakest possible assumptions on the coefficients of the Riccati equation, including the quadratic term of indefinite sign, as well as sufficient solvability conditions in terms of the signature condition are due to Halanay, Ionescu and Weiss [IW93b, HI93, HI94d]. The necessary and sufficient result on the signature condition as appears in this chapter is due to Oară [Oar95, Oar96a]. The existence of the stabilizing solution in the inverse–time and antistable cases has been investigated by Ionescu and Halanay [HI94c, HI94a, Ion96]. Connections between time–domain and frequency domain existence conditions for the stabilizing solution have been established via the celebrated Gohberg and Feldman factorization theory [GF74]. A purely algebraic proof of Theorem 4.5.3 in a more general framework can be found in Oară [Oar98]. For more details on the classical results in the linear quadratic problem, initiated by Kalman [Kal60, Kal64], we refer to the works of Popov [Pop66], Willems [Wil71], Kučera [Kuc72, Kuc73], Molinari [Mol73], Anderson and Moore [AM71], Kwakernaak and Sivan [KS72]. The results on linear matrix inequalities are due to Ionescu [Ion97a] and they extend the results of Zhou and Kargonekaar [ZK88], and Ben–Artzi, Gohberg, and Kaashoek [BAGK95]. Extension of the Riccati equation theory with indefinite sign to the class of infinite dimensional systems was obtained by Weiss [Wei94a, Wei97].

5

Riccati Equations and Matrix Pencils: the Regular Case

In this chapter we develop a matrix pencil approach to the algebraic Riccati equations theory. As in previous chapters, we consider Riccati equations under very general assumptions on the matrix coefficients allowing for the inclusion of game–theoretic (H^∞–type) cases in which the quadratic coefficient is sign indefinite. However, in this chapter we restrict the developments to the *regular case* in which the underlying Popov function takes values of maximal rank, i.e., it is invertible as a rational matrix, and defer to the next chapter the much more technical *general case.*

In Section 5.1 we obtain a characterization of the particular eigenstructure of a regular extended Hamiltonian pencil. The main result in continuous–time setting is presented in Section 5.2 and consists of necessary and sufficient existence conditions and computable formulas for various solutions to the continuous–time algebraic Riccati equation expressed in terms of deflating subspaces of the extended Hamiltonian pencil. In Section 5.3 we derive some interesting features of the continuous–time Bernoulli equation that is a particular case of the Riccati equation. The corresponding results for the extended symplectic pencil, the discrete–time algebraic Riccati equations, and the discrete Bernoulli equation, are presented in Sections 5.4–5.7.

A. The continuous–time case

5.1 The eigenstructure of a regular EHP

In this section we outline several properties of the eigenstructure and of the deflating subspaces of a regular extended Hamiltonian pencil.

Let $\Sigma = (A, B; Q, L, R)$ be a Popov triplet with $A \in \mathbb{R}^{n\times n}$, $B \in \mathbb{R}^{n\times m}$, and associate with it the EHP(Σ) given by

$$sM_\Sigma - N_\Sigma := s\begin{bmatrix} I & O & O \\ O & I & O \\ O & O & O \end{bmatrix} - \begin{bmatrix} A & O & B \\ -Q & -A^T & -L \\ L^T & B^T & R \end{bmatrix} \tag{5.1}$$

and the Popov function given by

$$\mathbf{\Pi}_\Sigma(s) = \left[\begin{array}{cc|c} A & O & B \\ -Q & -A^T & -L \\ \hline L^T & B^T & R \end{array}\right]. \tag{5.2}$$

Recall that the EHP(Σ) in (5.1) is the transmission pencil associated with the realization (5.2) of the Popov function $\mathbf{\Pi}_\Sigma$, and for a regular EHP we have $\det(sM_\Sigma - N_\Sigma) \not\equiv 0$.

Denote by n_f^-, n_f^0, and n_f^+ the number of finite generalized eigenvalues in $\mathbb{C}_-$, $\mathbb{C}_0$, and in $\mathbb{C}_+$, respectively, and by n_∞ the number of infinite generalized eigenvalues of $sM_\Sigma - N_\Sigma$. Let π_0 be the number of zeros of $\mathbf{\Pi}_\Sigma(s^{-1})$ located at zero or, alternatively, the number of zeros of $\mathbf{\Pi}_\Sigma(s)$ at ∞. We have the following result.

Theorem 5.1.1 *For a regular EHP the following hold:*

1. $n_f^- = n_f^+$.
2. $\operatorname{rank}_{\mathbb{R}(s)}(sM_\Sigma - N_\Sigma) = 2n + m$, $\operatorname{rank}_{\mathbb{R}(s)}\mathbf{\Pi}_\Sigma(s) = m$.
3. $n_\infty = m + \pi_0$.

Proof. 1. The following identity can be checked directly

$$P(sM_\Sigma - N_\Sigma)^T P = -sM_\Sigma - N_\Sigma,$$

where

$$P := \begin{bmatrix} O & -I_n & O \\ I_n & O & O \\ O & O & I_m \end{bmatrix}.$$

Hence

$$\det(sM_\Sigma - N_\Sigma) = \det(-sM_\Sigma - N_\Sigma).$$

The EHP being regular, it follows that s is a generalized eigenvalue of $sM_\Sigma - N_\Sigma$ if and only if $-s$ is a generalized eigenvalue of $sM_\Sigma - N_\Sigma$ with the same multiplicity.

2. Since $sM_\Sigma - N_\Sigma$ is the transmission pencil of the Popov function $\mathbf{\Pi}_\Sigma(s)$ we have from Theorem 1.14.1 that

$$\operatorname{rank}_{\mathbb{R}(s)}(sM_\Sigma - N_\Sigma) = 2n + \operatorname{rank}_{\mathbb{R}(s)}\mathbf{\Pi}_\Sigma(s). \tag{5.3}$$

From the regularity of the pencil it follows that $\operatorname{rank}_{\mathbb{R}(s)}(sM_\Sigma - N_\Sigma) = 2n + m$ which combined with (5.3) leads to $\operatorname{rank}_{\mathbb{R}(s)}\mathbf{\Pi}_\Sigma(s) = m$.

3. For any s for which $I - sA$ and $I + sA^T$ are invertible we get the identity

$$M_\Sigma - sN_\Sigma = \begin{bmatrix} I_n & O & O \\ O & I_n & O \\ \mathbf{X}(s) & \mathbf{Y}(s) & -I_m \end{bmatrix} \begin{bmatrix} I_n - sA & O & sB \\ sQ & I_n + sA^T & -sL \\ O & O & s\mathbf{\Pi}_\Sigma(s) \end{bmatrix},$$

where

$$\begin{aligned} \mathbf{X}(s) &:= -s[L^T + sB^T(I_n + sA^T)^{-1}Q](I_n - sA)^{-1}, \\ \mathbf{Y}(s) &:= -sB^T(I_n + sA^T)^{-1}, \end{aligned}$$

have both no poles at $s = 0$. Since the pencil $sM_\Sigma - N_\Sigma$ is regular,

$$\det(M_\Sigma - sN_\Sigma) = (-1)^m s^m \det(I_n - sA)\det(I_n + sA^T)\det(\mathbf{\Pi}_\Sigma(s^{-1})).$$

By using the Smith–McMillan form of $\mathbf{\Pi}_\Sigma$, we get that $s = 0$ is a generalized eigenvalue of $M_\Sigma - sN_\Sigma$ with multiplicity $m + \pi_0$ which means that $s = \infty$ is a generalized eigenvalue of $sM_\Sigma - N_\Sigma$ with the same multiplicity $n_\infty = m + \pi_0$. ∎

Definition 5.1.2 *A regular EHP is called* dichotomic *if* $n_f^0 = 0$ *and* $n_\infty = m$ *(or, equivalently, if* $n_f^0 = 0$ *and* $\pi_0 = 0$*).*

Note that since $sM_\Sigma - N_\Sigma$ is the transmission pencil of the Popov function $\mathbf{\Pi}_\Sigma(s)$, Theorem 1.14.1 shows that the zeros of $\mathbf{\Pi}_\Sigma(s)$ are among the generalized eigenvalues of $sM_\Sigma - N_\Sigma$. Therefore, the dichotomy implies that $\mathbf{\Pi}_\Sigma(s)$ has no zeros on the imaginary axis including infinity and that $R = \mathbf{\Pi}_\Sigma(\infty)$ is nonsingular. Hence, all the partial multiplicities of the generalized eigenvalue at ∞ of the EHP (5.1) are equal to one.

The following result gives a characterization of the dichotomy in terms of deflating subspaces.

Proposition 5.1.3 *A regular EHP is dichotomic if and only if it has a* $\mathbb{C}_-$ *(or a* $\mathbb{C}_+$*)–deflating subspace of dimension* n.

Proof. For any regular EHP we have

$$n_f^- + n_f^+ + n_f^0 + n_\infty = 2n + m. \tag{5.4}$$

In particular, for a dichotomic EHP we get

$$2n = n_f^- + n_f^+ \tag{5.5}$$

and using point 1 of Theorem 5.1.1 we get $n_f^- = n_f^+ = n$. Therefore Corollary 1.8.3 shows that there exists a $\mathbb{C}_-$ (and a $\mathbb{C}_+$)–deflating subspace of dimension n.

Conversely, if the EHP has a $\mathbb{C}_-$ (or a $\mathbb{C}_+$)–deflating subspace of dimension n we get from Theorem 5.1.1 and Theorem 1.8.4 that

$$n_f^- = n_f^+ \geq n. \tag{5.6}$$

Using again Theorem 5.1.1 and (5.4) we get

$$2n = 2n_f^- + n_f^0 + \pi_0. \tag{5.7}$$

Combining (5.6) with (5.7) shows that $n_f^- = n_f^+ = n$ and $n_f^0 = \pi_0 = 0$ that implies the dichotomy of the EHP. ∎

In particular, the previous result implies that the EHP is dichotomic if and only if

$$M_\Sigma V S = N_\Sigma V \tag{5.8}$$

holds, where

$$V = \begin{bmatrix} V_1 \\ V_2 \\ V_3 \end{bmatrix} \begin{matrix} \}n \\ \}n \\ \}m \end{matrix} \tag{5.9}$$

is a basis matrix for *the* n–dimensional $\mathbb{C}_-$ (or $\mathbb{C}_+$)–deflating subspace, and S is an $n \times n$ stable (or antistable) matrix.

Definition 5.1.4 *A deflating subspace* $\mathcal{V}$ *of a regular EHP is called* disconjugate *if it admits a basis matrix (5.9) with* V_1 *having full column rank.*

Notice that disconjugacy of a deflating subspace is independent of the choice of the basis matrix for $\mathcal{V}$.

Definition 5.1.5 *A regular EHP is called* stable (antistable) disconjugate *if it is dichotomic and has a disconjugate $\mathbb{C}_-$ ($\mathbb{C}_+$)–deflating subspace of dimension n.*

Notice that a dichotomic EHP has a unique (maximal) n–dimensional $\mathbb{C}_-$ (and $\mathbb{C}_+$)–deflating subspace and thus the disconjugacy of the EHP is well defined. A dichotomic EHP has always an n–dimensional $\mathbb{C}_-$ and an n–dimensional $\mathbb{C}_+$–deflating subspace, each of which might be disconjugate or not.

We give now a property of deflating subspaces of the EHP that will be instrumental in the sequel.

Proposition 5.1.6 *If V is a basis matrix as in (5.9) of a $\mathbb{C}_-$ (or $\mathbb{C}_+$)–deflating subspace of a regular EHP then*

$$V_1^T V_2 = V_2^T V_1. \tag{5.10}$$

Proof. Writing (5.8) explicitly with (5.9) we obtain

$$\begin{array}{rcrcrcr} V_1 S & = & AV_1 & & & + & BV_3, \\ V_2 S & = & -QV_1 & - & A^T V_2 & - & LV_3, \\ 0 & = & L^T V_1 & + & B^T V_2 & + & RV_3, \end{array} \tag{5.11}$$

where, by assumption, S is stable (or antistable). Adding the first equation premultiplied by V_2^T to the second one transposed and postmultiplied by V_1 and using the third equation in (5.11) we obtain

$$V_2^T V_1 S + S^T V_2^T V_1 + V_1^T Q V_1 - V_2^T B V_3 - V_3^T B^T V_2 - V_3^T R V_3 = 0. \tag{5.12}$$

This equation is a Lyapunov equation in $V_2^T V_1$, having a symmetric free term. Since S is stable (or antistable) $V_2^T V_1$ is its unique symmetric solution, as Theorem 1.5.2 shows, and the conclusion follows. ∎

5.2 The CTARE and the EHP

In this section we give necessary and sufficient conditions for the existence of the stabilizing and antistabilizing solutions to the continuous–time algebraic Riccati equation

$$A^T X + XA - (XB + L)R^{-1}(L^T + B^T X) + Q = 0 \tag{5.13}$$

associated with a given Popov triplet $\Sigma = (A, B; Q, L, R)$, where $A \in \mathbb{R}^{n \times n}$, $B \in \mathbb{R}^{n \times m}$. The conditions are expressed in terms of deflating subspaces of the extended Hamiltonian pencil

$$sM_\Sigma - N_\Sigma = s \begin{bmatrix} I & O & O \\ O & I & O \\ O & O & O \end{bmatrix} - \begin{bmatrix} A & O & B \\ -Q & -A^T & -L \\ L^T & B^T & R \end{bmatrix}, \tag{5.14}$$

associated with Σ, and suggest a numerically reliable algorithm for checking the existence of and, if the case, for computing these solutions.

Theorem 5.2.1 *Let $\Sigma = (A, B; Q, L, R)$ be a Popov triplet. The following assertions are equivalent:*

1. The matrix R is invertible and the CTARE(Σ) in (5.13) has a stabilizing (antistabilizing) solution.

2. The EHP(Σ) in (5.14) is regular and stable (antistable) disconjugate.

Moreover, the stabilizing (antistabilizing) solution can be computed from

$$X = V_2 V_1^{-1} \tag{5.15}$$

and the stabilizing (antistabilizing) Riccati feedback can be computed from

$$F = V_3 V_1^{-1} \tag{5.16}$$

where V is a basis matrix partitioned as in (5.9) of the n-dimensional $\mathbb{C}_-$-deflating ($\mathbb{C}_+$-deflating) subspace of the EHP.

Proof. We prove only for the stabilizing solution since the antistabilizing solution can be treated analogously.

1. $\Rightarrow$ 2. Since R is invertible, we can derive the identity

$$Q(sM_\Sigma - N_\Sigma)Z = s\begin{bmatrix} I_{2n} & O \\ O & O \end{bmatrix} - \begin{bmatrix} H_\Sigma & O \\ O & I_m \end{bmatrix}, \tag{5.17}$$

where

$$H_\Sigma := \begin{bmatrix} A - BR^{-1}L^T & -BR^{-1}B^T \\ -Q + LR^{-1}L^T & -A^T + LR^{-1}B^T \end{bmatrix} \tag{5.18}$$

and

$$Q = \begin{bmatrix} I_n & O & -BR^{-1} \\ O & I_n & LR^{-1} \\ O & O & I_m \end{bmatrix}, \qquad Z = \begin{bmatrix} I_n & O & O \\ O & I_n & O \\ -R^{-1}L^T & -R^{-1}B^T & R^{-1} \end{bmatrix}.$$

The relation (5.17) shows that $sM_\Sigma - N_\Sigma$ is regular. Let X be the stabilizing solution to the CTARE(Σ) in (5.13) and let $F := -R^{-1}(B^TX + L^T)$ be the stabilizing Riccati feedback. Define

$$V = \begin{bmatrix} V_1 \\ V_2 \\ V_3 \end{bmatrix} := \begin{bmatrix} I_n \\ X \\ F \end{bmatrix}. \tag{5.19}$$

Then using (5.13) it can be easily checked that $M_\Sigma VS = N_\Sigma V$ holds for $S := A + BF$ which shows that $\mathcal{V} = \operatorname{Im} V$ is an n-dimensional (maximal) $\mathbb{C}_-$-deflating subspace for $sM_\Sigma - N_\Sigma$. From Proposition 5.1.3 we get further that $sM_\Sigma - N_\Sigma$ is dichotomic. Moreover, since $V_1 = I_n$ it results that $\mathcal{V}$ is disconjugate and finally that $sM_\Sigma - N_\Sigma$ is stable disconjugate.

2. $\Rightarrow$ 1. Notice first that the regularity and dichotomy of $sM_\Sigma - N_\Sigma$ imply that $\pi_0 = 0$ and thus $\mathbf{\Pi}_\Sigma(\infty) = R$ is invertible. Now, since $sM_\Sigma - N_\Sigma$ is stable disconjugate, it has a disconjugate n-dimensional $\mathbb{C}_-$-deflating subspace $\mathcal{V}$. Thus

$$M_\Sigma VS = N_\Sigma V \tag{5.20}$$

holds for a basis matrix V of $\mathcal{V}$ and a stable $n \times n$ matrix S. Partition V as in (5.9). Since $\mathcal{V}$ is disconjugate it follows that V_1 is invertible. Denote

$$X := V_2 V_1^{-1}, \qquad F := V_3 V_1^{-1}, \qquad \widehat{S} := V_1 S V_1^{-1},$$

where X is symmetric due to Proposition 5.1.6. With this notation, (5.20) can be rewritten as

$$\begin{array}{rclcrcl} \widehat{S} & = & A & & & + & BF, \\ X\widehat{S} & = & -Q & - & A^T X & - & LF, \\ 0 & = & L^T & + & B^T X & + & RF. \end{array} \tag{5.21}$$

Substituting $\widehat{S}$ from the first equation in the second one and using the third equation (5.21) we get that X and F satisfy the CTARS(Σ)

$$\begin{array}{rcrcl} A^T X + XA + Q & + & (XB + L)F & = & 0, \\ L^T + B^T X & + & RF & = & 0. \end{array} \tag{5.22}$$

As R is invertible, we can eliminate F in (5.22) and get (5.13). Finally, $\widehat{S}$ is stable which shows that X is the stabilizing solution and F is the stabilizing Riccati feedback. This concludes the proof. ■

The matrix H_Σ in (5.18) is a $2n \times 2n$ *Hamiltonian matrix* that by definition satisfies

$$H_\Sigma^T \widehat{J} + \widehat{J} H_\Sigma = 0, \quad \text{for} \quad \widehat{J} := \begin{bmatrix} O & -I_n \\ I_n & O \end{bmatrix}.$$

Corollary 5.2.2 *Let $\Sigma = (A, B; Q, L, R)$ be a Popov triplet with R invertible. The CTARE (5.13) has a stabilizing (antistabilizing) solution if and only if the Hamiltonian matrix H_Σ given in (5.18) has an n-dimensional $\mathbb{C}_-$ ($\mathbb{C}_+$)-invariant subspace that admits a basis matrix*

$$V = \underbrace{\begin{bmatrix} V_1 \\ V_2 \end{bmatrix}}_{n} \begin{matrix} \}n \\ \}n \end{matrix} \tag{5.23}$$

with V_1 invertible. In this case, the stabilizing (antistabilizing) solution can be computed from

$$X = V_2 V_1^{-1}$$

and the stabilizing (antistabilizing) Riccati feedback can be computed from

$$F = -R^{-1}(B^T X + L^T).$$

This corollary shows that the existence of the stabilizing (antistabilizing) solution to the CTARE can be checked by using alternatively the Hamiltonian matrix or the EHP. Moreover, the solutions can be computed by solving the eigenvalue problem for the Hamiltonian matrix which has dimension $2n$ instead of solving the more complicated generalized eigenvalue problem for the EHP which has dimension $2n+m$. Nevertheless, the use of the Hamiltonian matrix can provide very inaccurate solutions if the matrix R is ill conditioned with respect to numerical inversion. It is desirable in this last case to use the EHP instead of the Hamiltonian matrix. For computational details we refer the reader to Chapter 6.

5.3 The continuous–time Bernoulli equation

Let $\Sigma = (A, B; Q, L, R)$ be a Popov triplet. In the particular case in which $Q = 0, L = 0$, and R is nonsingular, the algebraic Riccati equation is known as the *Bernoulli equation* and takes the form

$$A^T X + XA - XBR^{-1}B^T X = 0. \tag{5.24}$$

Due to its particular form, the Bernoulli equation has several remarkable properties in addition to the Riccati equation. Here follow some of them.

Proposition 5.3.1 *Let $\Sigma = (A, B; O, O, R)$ be a Popov triplet with R invertible. If the Bernoulli equation (5.24) associated with Σ has a stabilizing (antistabilizing) solution then A is dichotomic.*

Proof. The EHP(Σ) is

$$sM_\Sigma - N_\Sigma = \begin{bmatrix} sI - A & O & -B \\ O & sI + A^T & O \\ O & -B^T & -R \end{bmatrix}. \tag{5.25}$$

Since R is invertible and (5.24) has a stabilizing (antistabilizing) solution it follows from Theorem 5.2.1 that $sM_\Sigma - N_\Sigma$ is regular and dichotomic. The regularity implies in particular that the finite generalized eigenvalues of (5.25) are given by the set $\Lambda(A) \cup \Lambda(-A)$. The dichotomy of $sM_\Sigma - N_\Sigma$ implies that $sM_\Sigma - N_\Sigma$ has all the finite generalized eigenvalues in $\mathbb{C} \setminus \mathbb{C}_0$. Therefore, A is dichotomic. ■

Proposition 5.3.2 *Let $\Sigma = (A, B; O, O, R)$ be a Popov triplet with $R < 0$. If the Bernoulli equation (5.24) associated with Σ has a positive semidefinite stabilizing solution X then A is stable.*

Proof. As $R < 0$, there exists a square and invertible matrix V such that $V^T V = -R$ and let $S := V^{-T} B^T X$. Then (5.24) becomes

$$A^T X + XA + S^T S = 0. \tag{5.26}$$

For the stabilizing solution X we have that $A - BR^{-1}B^T X = A - BV^{-1}S$ is stable and it follows that the pair (S, A) is detectable (or, equivalently, c–observable). Since $X \geq 0$, it follows from Theorem 1.5.5 that A is stable and the proof ends. ■

Theorem 5.3.3 *Let $\Sigma = (A, B; O, O, R)$ be a Popov triplet with A dichotomic, the pair (A, B) stabilizable, and $R > 0$. Then the Bernoulli equation (5.24) associated with Σ has a positive semidefinite stabilizing solution.*

Proof. Since R is invertible and A is dichotomic it follows that the EHP(Σ) in (5.25) is regular and dichotomic. We show that $sM_\Sigma - N_\Sigma$ is stable disconjugate. Denote by n the dimension of A. Since the EHP is dichotomic it has a stable deflating subspace $\mathcal{V}$ of dimension n which is equivalent to $M_\Sigma VS = N_\Sigma V$ where S is an $n \times n$ stable matrix and V is a basis of $\mathcal{V}$. Partitioning V conformably with $sM_\Sigma - N_\Sigma$ in (5.25) we obtain

$$\begin{bmatrix} I & O & O \\ O & I & O \\ O & O & O \end{bmatrix} \begin{bmatrix} V_1 \\ V_2 \\ V_3 \end{bmatrix} S = \begin{bmatrix} A & O & B \\ O & -A^T & O \\ O & B^T & R \end{bmatrix} \begin{bmatrix} V_1 \\ V_2 \\ V_3 \end{bmatrix}. \tag{5.27}$$

As R is invertible, we eliminate V_3 in (5.27) and by simple algebraic manipulations we get further

$$S^T V_1^T V_2 + V_1^T V_2 S + V_2^T B R^{-1} B^T V_2 = 0. \tag{5.28}$$

From the stability of S and $V_2^T B R^{-1} B^T V_2 \geq 0$ we conclude that

$$V_1^T V_2 = V_2^T V_1 \geq 0. \tag{5.29}$$

Denote by K a basis matrix for $\operatorname{Ker} V_1$. We pre– and post–multiply both sides of (5.28) by K^T and K, respectively, and we use (5.29) to get $K^T V_2^T B R^{-1} B^T V_2 K = 0$. As $R > 0$ it follows that

$$B^T V_2 K = 0 \tag{5.30}$$

and further from the third equation (5.27) that $V_3 K = 0$. We use this last equation in the first equation (5.27) postmultiplied by K to get $V_1 S K = 0$. Hence $\operatorname{Ker} V_1 = \operatorname{Im} K$ is an invariant subspace of S. Thus, there exists a stable matrix J such that

$$SK = KJ. \tag{5.31}$$

This together with the second equation (5.27) leads to

$$-A^T V_2 K = V_2 K J. \tag{5.32}$$

Since the pair (A, B) is stabilizable, there exists F such that $A + BF$ is stable. Then by subtracting (5.30) premultiplied by F^T from (5.32) we get

$$(A + BF)^T V_2 K + V_2 K J = 0. \tag{5.33}$$

As both $A + BF$ and J are stable, (5.33) yields $V_2 K = 0$ and the third equation in (5.27) shows that $V_3 K = 0$. Thus

$$VK = \begin{bmatrix} V_1 \\ V_2 \\ V_3 \end{bmatrix} K = 0.$$

Since V has full column rank it follows that $K = 0$ which shows that V_1 is invertible and thus $sM_\Sigma - N_\Sigma$ is stable disconjugate. By Theorem 5.2.1 we conclude that the Bernoulli equation has a stabilizing solution given by $X = V_2 V_1^{-1}$. Combining this formula with (5.29) we get $X \geq 0$ and the theorem is completely proved. ■

B. The discrete–time case

5.4 The eigenstructure of a regular ESP

In this section we provide an analysis of the eigenstructure and of the deflating subspaces of a regular extended symplectic pencil. The exposition follows closely Section 5.1, although some important differences will arise.

Let $\Sigma = (A, B; Q, L, R)$ be a Popov triplet with $A \in \mathbb{R}^{n\times n}$, $B \in \mathbb{R}^{n\times m}$, and associate with it the ESP(Σ) given by

$$zM_\Sigma - N_\Sigma := z\begin{bmatrix} I & O & O \\ O & -A^T & O \\ O & -B^T & O \end{bmatrix} - \begin{bmatrix} A & O & B \\ Q & -I & L \\ L^T & O & R \end{bmatrix} \tag{5.34}$$

and the discrete–time Popov function given by

$$\mathbf{\Pi}_\Sigma(z) = \left[\begin{array}{cc|c} zI_n - A & O & -B \\ -Q & I_n - zA^T & -L \\ \hline L^T & zB^T & R \end{array}\right]. \tag{5.35}$$

Recall that the ESP(Σ) (5.34) is the transmission pencil associated with the descriptor realization (5.35) of the Popov function $\mathbf{\Pi}_\Sigma$.

Denote by n_f^-, n_f^0, and n_f^+, the number of finite generalized eigenvalues of the ESP in $\mathbb{D}$, on the unit circle $\partial\mathbb{D}$, and in $\mathbb{C}\setminus\overline{\mathbb{D}}$, respectively, and by n_∞ the number of infinite generalized eigenvalues of $zM_\Sigma - N_\Sigma$. Let π_0 be the algebraic multiplicity of the generalized eigenvalue at zero and $\tilde{n}_f^- = n_f^- - \pi_0$, the number of nonzero generalized eigenvalues of $zM_\Sigma - N_\Sigma$ located in $\mathbb{D}$.

Theorem 5.4.1 *For a regular ESP the following hold.*

1. $\tilde{n}_f^- = n_f^+$.
2. $\operatorname{rank}_{\mathbb{R}(z)}(zM_\Sigma - N_\Sigma) = 2n + m, \quad \operatorname{rank}_{\mathbb{R}(z)}\mathbf{\Pi}_\Sigma(z) = m.$
3. $n_\infty = m + \pi_0$.

Proof. 1. A direct check shows that for any $z \neq 0$ we have

$$\mathbf{P}(z)(zM_\Sigma - N_\Sigma)^T\mathbf{P}(z) = \frac{1}{z}M_\Sigma - N_\Sigma,$$

where

$$\mathbf{P}(z) = \begin{bmatrix} O & \frac{1}{z}I_n & O \\ I_n & O & O \\ O & O & I_m \end{bmatrix}.$$

Thus

$$\det(zM_\Sigma - N_\Sigma) = z^{2n}\det(\frac{1}{z}M_\Sigma - N_\Sigma).$$

Since the ESP is regular, it follows that $z \neq 0$ is a generalized eigenvalue of $zM_\Sigma - N_\Sigma$ if and only if $\frac{1}{z}$ is a generalized eigenvalue of $zM_\Sigma - N_\Sigma$ with the same multiplicity.

2. Since $zM_\Sigma - N_\Sigma$ is the transmission pencil of the discrete–time Popov function $\mathbf{\Pi}_\Sigma(z)$, Theorem 1.14.2 shows

$$\operatorname{rank}_{\mathbb{R}(z)}(zM_\Sigma - N_\Sigma) = 2n + \operatorname{rank}_{\mathbb{R}(z)}\mathbf{\Pi}_\Sigma(z). \tag{5.36}$$

From the regularity of the pencil it follows that $\operatorname{rank}_{\mathbb{R}(z)}(zM_\Sigma - N_\Sigma) = 2n+m$ which combined with (5.36) leads to $\operatorname{rank}_{\mathbb{R}(z)} \mathbf{\Pi}_\Sigma(z) = m$.

3. For any z for which $I - zA$ is invertible we get the identity

$$(M_\Sigma - zN_\Sigma)\mathbf{X}(z) = \mathbf{Y}(z)z(zM_\Sigma - N_\Sigma)^T\mathbf{Z}(z) \tag{5.37}$$

where

$$\mathbf{X}(z) := \begin{bmatrix} zI_n & O & O \\ O & zI_n & O \\ O & O & I_m \end{bmatrix},$$

$$\mathbf{Y}(z) := \begin{bmatrix} O & I_n & O \\ I_n & (z-1)Q(I-zA)^{-1} & O \\ O & (1-z)L^T(I-zA)^{-1} & I_m \end{bmatrix},$$

$$\mathbf{Z}(z) := \begin{bmatrix} O & I_n & O \\ I_n & O & (z-1)(I-zA)^{-1}B \\ O & O & I_m \end{bmatrix}.$$

From (5.37) we deduce that

$$\det(M_\Sigma - zN_\Sigma) = z^m \det(zM_\Sigma - N_\Sigma).$$

This shows that the number of infinite generalized eigenvalues of $zM_\Sigma - N_\Sigma$, which by definition is the multiplicity of the zero generalized eigenvalue of $M_\Sigma - zN_\Sigma$, equals m plus the multiplicity of the generalized eigenvalue of $zM_\Sigma - N_\Sigma$ at zero, i.e., $n_\infty = m + \pi_0$. ■

Definition 5.4.2 *A regular ESP is called* dichotomic *if* $n_f^0 = 0$ *and* strongly dichotomic *if in addition* $\pi_0 = 0$.

Since Theorem 1.14.2 shows that the zeros of $\mathbf{\Pi}_\Sigma(z)$ are among the generalized eigenvalues of $zM_\Sigma - N_\Sigma$, it follows that the dichotomy of the ESP implies that the Popov function has no zeros on the unit circle whereas *strong dichotomy* implies that the Popov function has neither zeros at zero.

We give now a useful characterization of dichotomy in terms of deflating subspaces.

Proposition 5.4.3 *1. A regular ESP is dichotomic if and only if it has a* $\mathbb{D}$*–deflating subspace of dimension* n.

2. A regular ESP is strongly dichotomic if and only if it has a $\mathbb{C} \setminus \overline{\mathbb{D}}$*–deflating subspace of dimension* n.

Proof. 1. For any regular ESP we have

$$n_f^- + n_f^+ + n_f^0 + n_\infty = 2n + m. \tag{5.38}$$

In particular, for a dichotomic ESP we get

$$n_f^- + n_f^+ + n_\infty = 2n + m \tag{5.39}$$

and using Theorem 5.4.1 we get $n_f^- = \widetilde{n}_f^- + \pi_0 = n$. Therefore, Corollary 1.8.3 shows that there exists a $\mathbb{D}$–deflating subspace of dimension n. Conversely, if the ESP has a

$\mathbb{D}$–deflating subspace of dimension n we get from Theorem 1.8.4 and Theorem 5.4.1 that

$$n_f^- \geq n. \tag{5.40}$$

Using Theorem 5.4.1 and (5.38) we get

$$2n = 2(\tilde{n}_f^- + \pi_0) + n_f^0 = 2n_f^- + n_f^0. \tag{5.41}$$

Combining (5.40) with (5.41) shows that $n_f^0 = 0$ and $n_f^- = \tilde{n}_f^- + \pi_0 = n$ that implies the dichotomy of the ESP.

2. This part follows immediately if we use the fact that $n_f^- = n_f^+ + \pi_0$. ■

In particular, the previous result implies that the ESP is (strongly) dichotomic if and only if

$$M_\Sigma V S = N_\Sigma V \tag{5.42}$$

holds, where

$$V = \begin{bmatrix} V_1 \\ V_2 \\ V_3 \end{bmatrix} \begin{matrix} \}n \\ \}n \\ \}m \end{matrix} \tag{5.43}$$

is a basis matrix for the n–dimensional $\mathbb{D}$ ($\mathbb{C}\setminus\overline{\mathbb{D}}$)–deflating subspace, and S is an $n\times n$ stable (antistable) matrix. Unlike the continuous–time case, the ESP could have an n–dimensional stable deflating subspace while having a maximal antistable deflating subspace of lower dimension. However, the symmetry with respect to the boundary of the stability domain is recaptured if we define the maximal antistable deflating subspace as the deflating subspace corresponding to the set of all antistable finite eigenvalues, together with the infinite eigenvalues of the ESP due to the infinite zeros of the Popov function $\mathbf{\Pi}_\Sigma(z)$.

Definition 5.4.4 *A deflating subspace $\mathcal{V}$ of a regular ESP is called* disconjugate *if it admits a basis matrix (5.43) with V_1 having full column rank.*

Disconjugacy is independent of the choice of the basis matrix for $\mathcal{V}$.

Definition 5.4.5 *A regular ESP is called stable (antistable)* disconjugate *if it is (strongly) dichotomic and has a disconjugate $\mathbb{D}$ ($\mathbb{C}\setminus\overline{\mathbb{D}}$)–deflating subspace of dimension n.*

Notice that a (strongly) dichotomic ESP has a unique maximal n–dimensional $\mathbb{D}$ ($\mathbb{C}\setminus\overline{\mathbb{D}}$)–deflating subspace and thus the disconjugacy of the ESP is well defined.

Proposition 5.4.6 *If V is a basis matrix as in (5.43) of a $\mathbb{D}$ (or $\mathbb{C}\setminus\overline{\mathbb{D}}$)–deflating subspace of a regular ESP then*

$$V_1^T V_2 = V_2^T V_1. \tag{5.44}$$

Proof. Writing (5.42) explicitly with (5.43) we obtain

$$\begin{array}{rcllr} V_1 S & = & AV_1 & & +BV_3, \\ -A^T V_2 S & = & QV_1 & -V_2 & +LV_3, \\ -B^T V_2 S & = & L^T V_1 & & +RV_3, \end{array} \tag{5.45}$$

where, by assumption, S is stable (or antistable). Adding the second equation premultiplied by V_1^T to the third one premultiplied by V_3^T, and using the first equation in (5.45) we obtain

$$S^T V_1^T V_2 S - V_1^T V_2 + V_1^T Q V_1 + V_1^T L V_3 + V_3^T L^T V_1 + V_3^T R V_3 = 0. \tag{5.46}$$

This equation is a symmetric Stein equation in $V_1^T V_2$. Since S is stable (antistable) $V_1^T V_2$ is its unique symmetric solution and the conclusion follows. ■

5.5 The DTARE and the ESP

In this section we give necessary and sufficient conditions for the existence of the stabilizing and antistabilizing solutions to the discrete–time algebraic Riccati equation

$$A^T X A - X - (A^T X B + L)(R + B^T X B)^{-1}(L^T + B^T X A) + Q = 0 \tag{5.47}$$

associated with a given Popov triplet $\Sigma = (A, B; Q, L, R)$, where $A \in \mathbb{R}^{n \times n}, B \in \mathbb{R}^{n \times m}$. The conditions are expressed in terms of deflating subspaces of the extended symplectic pencil

$$zM_\Sigma - N_\Sigma = z \begin{bmatrix} I & O & O \\ O & -A^T & O \\ O & -B^T & O \end{bmatrix} - \begin{bmatrix} A & O & B \\ Q & -I & L \\ L^T & O & R \end{bmatrix} \tag{5.48}$$

associated with Σ.

Theorem 5.5.1 *Let $\Sigma = (A, B; Q, L, R)$ be a Popov triplet.*
I. *The following assertions are equivalent:*
1. The DTARE(Σ) in (5.47) has a stabilizing solution.
2. The ESP(Σ) in (5.48) is regular and stable disconjugate.
Moreover, the stabilizing solution can be computed from

$$X = V_2 V_1^{-1} \tag{5.49}$$

and the stabilizing Riccati feedback can be computed from

$$F = V_3 V_1^{-1} \tag{5.50}$$

where V is a basis matrix partitioned as in (5.43) of the n–dimensional $\mathbb{D}$–deflating subspace of the ESP.
II. *The following assertions are equivalent:*
1. The DTARE(Σ) in (5.47) has an antistabilizing solution.
2. The ESP(Σ) in (5.48) is regular and antistable disconjugate.
Moreover, the antistabilizing solution and the antistabilizing Riccati feedback can be computed by the same formulas as their stabilizing counterparts, where now V is a basis matrix of the n–dimensional maximal $\mathbb{C} \setminus \overline{\mathbb{D}}$–deflating subspace of the ESP.

Proof. We prove only for the stabilizing solution since the antistabilizing solution can be treated in a much similar way.

1. $\Rightarrow$ 2. Let X be the stabilizing solution to the DTARE(Σ) in (5.47) and let $F = -(R + B^T XB)^{-1}(L^T + B^T XA)$ be the stabilizing Riccati feedback. Then the spectral factorization identity

$$\mathbf{\Pi}_\Sigma(z) = \mathbf{S}^*(z)(R + B^T XB)\mathbf{S}(z) \tag{5.51}$$

holds where $\mathbf{S}(z) := I - F(zI - A)^{-1}B$. Since $R + B^T XB$ is invertible, the identity (5.51) shows that $\operatorname{rank}_{\mathbb{R}(z)} \mathbf{\Pi}_\Sigma(z) = m$ and therefore the ESP(Σ) is regular. Define

$$V = \begin{bmatrix} V_1 \\ V_2 \\ V_3 \end{bmatrix} := \begin{bmatrix} I_n \\ X \\ F \end{bmatrix}. \tag{5.52}$$

By using the DTARE (5.47) it can be easily checked that $M_\Sigma VS = N_\Sigma V$ holds for $S := A + BF$ which shows that $\mathcal{V} = \operatorname{Im} V$ is a maximal n–dimensional $\mathbb{D}$–deflating subspace for the ESP. Thus the ESP is dichotomic. Finally, since $V_1 = I_n$ it emerges that $\mathcal{V}$ is disconjugate which shows that the ESP is stable disconjugate.

2. $\Rightarrow$ 1. Let V be a basis matrix for the disconjugate $\mathbb{D}$–deflating subspace $\mathcal{V}$ of the ESP. Then

$$M_\Sigma VS = N_\Sigma V \tag{5.53}$$

holds for a stable $n \times n$ matrix S. Partition V as in (5.43). Since $\mathcal{V}$ is disconjugate V_1 is invertible. Denote

$$X := V_2 V_1^{-1}, \qquad F := V_3 V_1^{-1}, \qquad \widehat{S} = V_1 S V_1^{-1},$$

where X is symmetric due to Proposition 5.4.6. With this notation (5.53) can be rewritten as

$$\begin{array}{rclcl} \widehat{S} & = & A & & + BF, \\ -A^T X\widehat{S} & = & Q & - X & + LF, \\ -B^T X\widehat{S} & = & L^T & & + RF. \end{array} \tag{5.54}$$

Substituting $\widehat{S}$ from the first equation in the second and third ones in (5.54) we get that X and F satisfy the DTARS(Σ)

$$\begin{array}{rcrcl} A^T XA - X + Q & + & (A^T XB + L)F & = & 0, \\ L^T + B^T XA & + & (R + B^T XB)F & = & 0. \end{array} \tag{5.55}$$

Therefore, we can again derive the spectral factorization identity (5.51) (see Chapter 3). As the ESP is regular, Theorem 5.4.1 shows that

$$\operatorname{rank}_{\mathbb{R}(z)} \mathbf{\Pi}_\Sigma(z) = m,$$

from which $R + B^T XB$ in (5.51) is invertible. Thus F in (5.55) could be eliminated and we arrive at the DTARE (5.47). Finally, since $\widehat{S}$ is stable it emerges that X is the stabilizing solution and F is the stabilizing Riccati feedback. ■

Remark 5.5.2 Notice that for the existence of the antistabilizing solution to the DTARE it is necessary that $\pi_0 = 0$, which is equivalent, for a regular ESP, to the invertibility of the matrix N_Σ in (5.48). Further, N_Σ is invertible if and only if the matrix

$$\begin{bmatrix} A & B \\ L^T & R \end{bmatrix} \tag{5.56}$$

is invertible.

The next corollary shows that if some additional conditions are met then the stabilizing and antistabilizing solutions to the DTARE can be computed by solving lower dimension eigenvalue problems. But first a definition. A $2n \times 2n$ matrix Z is called *symplectic* if it satisfies

$$Z^T \widehat{J} Z - \widehat{J} = 0, \quad \text{for} \quad \widehat{J} := \begin{bmatrix} O & -I_n \\ I_n & O \end{bmatrix}.$$

Corollary 5.5.3 *Let $\Sigma = (A, B; Q, L, R)$ be a Popov triplet.*

I. *Assume that R is invertible and denote*

$$\widehat{A} := A - BR^{-1}L^T, \quad \widehat{Q} := Q - LR^{-1}L^T. \tag{5.57}$$

The following assertions are equivalent:

1. *The DTARE(Σ) in (5.47) has a stabilizing (antistabilizing) solution.*
2. *The regular* $2n \times 2n$ symplectic pencil

$$z\widehat{M_\Sigma} - \widehat{N}_\Sigma = z\begin{bmatrix} I_n & BR^{-1}B^T \\ O & -\widehat{A}^T \end{bmatrix} - \begin{bmatrix} \widehat{A} & O \\ \widehat{Q} & -I_n \end{bmatrix}$$

associated with Σ has a $\mathbb{D}$ ($\mathbb{C} \setminus \overline{\mathbb{D}}$)-deflating subspace of dimension n that admits a basis matrix

$$V = \underbrace{\begin{bmatrix} V_1 \\ V_2 \end{bmatrix}}_{n} \begin{matrix} \}n \\ \}n \end{matrix} \tag{5.58}$$

with V_1 invertible.

The stabilizing (antistabilizing) solution can be computed from

$$X = V_2 V_1^{-1}$$

and the stabilizing (antistabilizing) Riccati feedback can be computed from

$$F = -(R + B^T X B)^{-1}(L^T + B^T X A).$$

II. *Assume R is invertible and $\widehat{A}$ in (5.57) is invertible. Define $\widehat{Q}$ as in (5.57). The following assertions are equivalent:*

1. *The DTARE(Σ) in (5.47) has a stabilizing (antistabilizing) solution.*
2. *The $2n \times 2n$ symplectic matrix*

$$Z_\Sigma = \begin{bmatrix} \widehat{A} - BR^{-1}B^T\widehat{A}^{-T}\widehat{Q} & BR^{-1}B^T\widehat{A}^{-T} \\ -\widehat{A}^{-T}\widehat{Q} & \widehat{A}^{-T} \end{bmatrix}$$

associated with Σ has a $\mathbb{D}$ $(\mathbb{C} \setminus \overline{\mathbb{D}})$–invariant subspace of dimension n that admits a basis matrix (5.58) with V_1 invertible.

The stabilizing (antistabilizing) solution can be computed by the same formulas as in part I.

Proof. This follows easily from Theorem 5.5.1 by using strict equivalence operations on matrix pencils. ■

This corollary shows that the stabilizing (antistabilizing) solution to the DTARE can be characterized by using alternatively the reduced order symplectic pencil or the symplectic matrix. However, in contrast to the continuous–time case these $2n$ order eigenvalue problems can be used instead of the $2n+m$ generalized eigenvalue problem only if some additional conditions are fulfilled. Whenever R or $A - BR^{-1}L^T$ are singular, or badly conditioned with respect to numerical inversion, it is necessary to stick to the ESP.

5.6 The discrete–time Bernoulli equation

Let $\Sigma = (A, B; Q, L, R)$ be a Popov triplet. In the particular case in which $Q = 0, L = 0$, the discrete–time Riccati equation is known as the (discrete–time) *Bernoulli equation* and takes the form

$$A^T XA - X - A^T XB(R + B^T XB)^{-1}B^T XA = 0. \tag{5.59}$$

We give below several important properties of the discrete–time Bernoulli equation.

Proposition 5.6.1 *Let $\Sigma = (A, B; O, O, R)$ be a Popov triplet.*

I. *If the discrete–time Bernoulli equation in (5.59) associated with Σ has a stabilizing solution X then A is dichotomic, R is nonsingular, and*

$$\operatorname{sgn}(R + B^T XB) = \operatorname{sgn}(R). \tag{5.60}$$

II. *If the discrete–time Bernoulli equation (5.59) associated with Σ has an antistabilizing solution X then A and R are nonsingular, A is dichotomic, and (5.60) holds.*

Proof. I. The ESP(Σ) is given by

$$zM_\Sigma - N_\Sigma = \begin{bmatrix} zI - A & O & -B \\ O & I - zA^T & O \\ O & -zB^T & -R \end{bmatrix}. \tag{5.61}$$

Since (5.59) has a stabilizing solution it follows from Theorem 5.5.1 that $zM_\Sigma - N_\Sigma$ is regular and dichotomic. The regularity implies that the finite generalized eigenvalues of (5.61) are given by the set $\Lambda(A) \cup \Lambda_f(A^T, I)$, and that R is invertible. Here $\Lambda_f(A^T, I)$ stands for the finite spectrum of the pencil $zA^T - I$. The dichotomy of $zM_\Sigma - N_\Sigma$ implies that (5.61) has no generalized eigenvalues on the unit circle. Thus A is dichotomic. It only remains to prove (5.60). A direct check shows that the discrete–time Popov function (5.35) associated with Σ is $\mathbf{\Pi}_\Sigma = R$. From the spectral factorization identity (3.142) we get

$$R = \mathbf{S}_F^*(R + B^T XB)\mathbf{S}_F \tag{5.62}$$

where

$$\mathbf{S}_F := \left[\begin{array}{c|c} A & B \\ \hline -F & I \end{array}\right]$$

and F is the stabilizing Riccati feedback. It is easy to check that the zeros of $\mathbf{S}_F$ are given by the eigenvalues of $A + BF$ from where it follows that $\mathbf{S}_F$ has neither poles nor zeros on the unit circle. Hence (5.62) leads to (5.60).

II. If the Bernoulli equation has an antistabilizing solution then it follows from Theorem 5.5.1 that the ESP associated with Σ is regular and strongly dichotomic. Strong dichotomy implies in particular that the ESP has no generalized eigenvalues at 0. From (5.61) we conclude that A is invertible. The rest of the proof is entirely similar as for part I. ■

Proposition 5.6.2 *Let* $\Sigma = (A, B; O, O, R)$ *be a Popov triplet with* $R < 0$. *If the discrete–time Bernoulli equation associated with* Σ *has a positive semidefinite stabilizing solution* X *then* A *is stable.*

Proof. The proof follows analogously as for Proposition 5.3.2, by replacing continuous–time Lyapunov equations and corresponding properties with their discrete–time counterparts. ■

Theorem 5.6.3 *Let* $\Sigma = (A, B; O, O, R)$ *be a Popov triplet with* A *dichotomic, the pair* (A, B) *stabilizable, and* $R > 0$. *Then the discrete–time Bernoulli equation associated with* Σ *has a positive semidefinite stabilizing solution.*

Proof. Since R is invertible and A is dichotomic it follows that the ESP(Σ) in (5.61) is regular and dichotomic. We show that $zM_\Sigma - N_\Sigma$ is stable disconjugate. Denote by n the dimension of A. Since the ESP is dichotomic it has a stable deflating subspace $\mathcal{V}$ of dimension n which is equivalent to $M_\Sigma V S = N_\Sigma V$ where S is an $n \times n$ stable matrix and V is a basis of $\mathcal{V}$. Partitioning V conformably with $zM_\Sigma - N_\Sigma$ in (5.61) we obtain

$$\begin{bmatrix} I & O & O \\ O & -A^T & O \\ O & -B^T & O \end{bmatrix} \begin{bmatrix} V_1 \\ V_2 \\ V_3 \end{bmatrix} S = \begin{bmatrix} A & O & B \\ O & -I & O \\ O & O & R \end{bmatrix} \begin{bmatrix} V_1 \\ V_2 \\ V_3 \end{bmatrix}. \tag{5.63}$$

As R is invertible, we eliminate V_3 in (5.63) and by simple algebraic manipulations we get further that

$$S^T V_1^T V_2 S - V_1^T V_2 + S^T V_2^T B R^{-1} B^T V_2 S = 0. \tag{5.64}$$

From the stability of S and $S^T V_2^T B R^{-1} B^T V_2 S \geq 0$ we conclude that

$$V_1^T V_2 = V_2^T V_1 \geq 0. \tag{5.65}$$

Denote by K a basis matrix for Ker V_1. We pre– and post–multiply both sides of (5.64) by K^T and K, respectively, to get

$$K^T S^T V_2^T B R^{-1} B^T V_2 S K = 0.$$

As $R > 0$ it follows that

$$B^T V_2 S K = 0 \tag{5.66}$$

and further from the third equation in (5.63) that $V_3 K = 0$. We use this last equation in the first equation (5.63) postmultiplied by K to get $V_1 S K = 0$. Hence $\operatorname{Ker} V_1 = \operatorname{Im} K$ is an invariant subspace of S. Thus, there exists a stable matrix J such that

$$SK = KJ. \tag{5.67}$$

This together with the second equation (5.63) leads to

$$A^T V_2 K J \quad = \quad V_2 K. \tag{5.68}$$

Since the pair (A, B) is stabilizable, there exists F such that $A + BF$ is stable. Then by adding (5.66) premultiplied by F^T to (5.68) and using (5.67) we get

$$(A + BF)^T V_2 K J - V_2 K = 0. \tag{5.69}$$

As both $A + BF$ and J are stable, (5.69) yields $V_2 K = 0$ and the third equation in (5.63) shows that $V_3 K = 0$. Thus

$$VK = \begin{bmatrix} V_1 \\ V_2 \\ V_3 \end{bmatrix} K = 0.$$

Since V has full column rank it follows that $K = 0$. Thus V_1 is invertible and $zM_\Sigma - N_\Sigma$ is stable disconjugate. By Theorem 5.5.1 we conclude that the Bernoulli equation has a stabilizing solution given by $X = V_2 V_1^{-1}$. Combining this formula with (5.65) we get $X \geq 0$ and the theorem is completely proved. ■

5.7 The time–reversed DTARE

Recall from Section 3.7 that we have associated with a reversible Popov triplet $\Sigma = (A, B; Q, L, R)$ (i.e., a Popov triplet with A invertible) a time–reversed Popov triplet $\widehat{\Sigma} = (\widehat{A}, \widehat{B}; \widehat{Q}, \widehat{L}, \widehat{R})$ given explicitly by

$$\begin{aligned} \widehat{A} &= A^{-1}, \\ \widehat{B} &= -A^{-1}B, \\ \widehat{Q} &= A^{-T}QA^{-1}, \\ \widehat{L} &= A^{-T}L - A^{-T}QA^{-1}B, \\ \widehat{R} &= R - L^T A^{-1} B - B^T A^{-T} L + B^T A^{-T} Q A^{-1} B. \end{aligned} \tag{5.70}$$

In this section we study solutions to the time–reversed DTARE associated with Σ which in fact is the DTARE associated with $\widehat{\Sigma}$. A straightforward approach is to apply the theory of the DTARE and the ESP with respect to the time–reversed Popov triplet $\widehat{\Sigma}$. However, both for theoretical and computational reasons it is desirable to obtain necessary and sufficient existence conditions and computable formulas for the solutions to the time–reversed DTARE expressed in terms of the ESP associated with the original Popov triplet Σ. It turns out that the stabilizing solution to the time–reversed DTARE is connected to deflating subspaces for the transposed ESP, also called *deflating subspaces to the left* for the ESP.

The time–reversed DTARE associated with a reversible Popov triplet Σ is given explicitly by

$$X + A^{-T}(Q-X)A^{-1} - A^{-T}[L-(Q-X)A^{-1}B][R - L^TA^{-1}B - B^TA^{-T}L + \\ + B^TA^{-T}(Q-X)A^{-1}B]^{-1}[L^T - B^TA^{-T}(Q-X)]A^{-1} = 0, \tag{5.71}$$

where $X \in \mathbb{R}^{n\times n}$. A matrix $X = X^T$ is a stabilizing solution to the time–reversed DTARE if the symmetric matrix

$$G(X) := R - L^TA^{-1}B - B^TA^{-T}L + B^TA^{-T}(Q-X)A^{-1}B \tag{5.72}$$

is invertible, X satisfies (5.71), and $A^{-1}(I - BF)$ is stable, where

$$\begin{aligned} F \;:=\; & -[R - L^TA^{-1}B - B^TA^{-T}L + B^TA^{-T}(Q-X)A^{-1}B]^{-1} \\ & \times[L^T - B^TA^{-T}(Q-X)]A^{-1}. \end{aligned} \tag{5.73}$$

Notice that if the stabilizing solution to the time–reversed DTARE exists, then it is unique. This follows from the uniqueness of the stabilizing solution to the DTARE.

We give now the time–reversed version of the spectral factorization identity. We introduce first the following notation. For any $X = X^T \in \mathbb{R}^{n\times n}$ define

$$H(X) := A^{-T}[L - (Q-X)A^{-1}B], \tag{5.74}$$

$$K(X) := X + A^{-T}(Q-X)A^{-1} = K^T(X). \tag{5.75}$$

With (5.72), (5.74), and (5.75), equations (5.71) and (5.73) can be rewritten as

$$\begin{aligned} K(X) - F^TG(X)F &= 0, \\ G(X)F + H^T(X) &= 0. \end{aligned} \tag{5.76}$$

Notice that the system (5.76) in the unknowns X and F is more general than equation (5.71), and reduces to the time–reversed DTARE under the assumption that $G(X)$ is invertible. The system (5.76) may be seen as a time–reversed version of the discrete–time algebraic Riccati system (3.108).

Proposition 5.7.1 *If $X = X^T$ and F satisfy (5.76) then the Popov function associated with Σ has the following expression:*

$$\mathbf{\Pi}_\Sigma(z) = \mathbf{S}_F(z)G(X)\mathbf{S}_F^*(z) \tag{5.77}$$

where

$$\mathbf{S}_F(z) := I - B^T(I - zA^T)^{-1}F^T. \tag{5.78}$$

Proof. Evaluating the right–hand side of (5.77), we get successively

$$\mathbf{S}_F(z)G(X)\mathbf{S}_F^*(z)$$

$$\overset{(5.78)}{=} G(X) - B^T(I - zA^T)^{-1}F^TG(X) - G(X)F(I - \frac{1}{z}A)^{-1}B$$

$$
\begin{aligned}
& +B^T(I - zA^T)^{-1}F^TG(X)F(I - \frac{1}{z}A)^{-1}B \\
\overset{(5.76)}{=} \quad & G(X) + B^T(I - zA^T)^{-1}H(X) + H^T(X)(I - \frac{1}{z}A)^{-1}B \\
& +B^T(I - zA^T)^{-1}K(X)(I - \frac{1}{z}A)^{-1}B \\
& \text{by using (5.72) – (5.75)} \\
= \quad & R + B^T[(I - zA^T)^{-1} - I]A^{-T}L + L^TA^{-1}[(I - \frac{1}{z}A)^{-1} - I]B \\
& -B^T(I - zA^T)^{-1}A^{-T}(Q - X)A^{-1}B \\
& -B^TA^{-T}(Q - X)A^{-1}(I - \frac{1}{z}A)^{-1}B \\
& +B^T(I - zA^T)^{-1}[X + A^{-T}(Q - X)A^{-1}](I - \frac{1}{z}A)^{-1}B \\
& +B^TA^{-T}(Q - X)A^{-1}B \\
= \quad & R + B^T(\frac{1}{z}I - A^T)^{-1}L + L^T(zI - A)^{-1}B \\
& +B^T(\frac{1}{z}I - A^T)^{-1}\{X + [I - (I - zA^T)]A^{-T}(Q - X)A^{-1} \\
& \times[I - (I - \frac{1}{z}A)]\}(zI - A)^{-1}B \\
= \quad & \mathbf{\Pi}_\Sigma(z).
\end{aligned}
$$

Thus (5.77) holds. ■

We outline now some properties of the ESP relevant to the study of the time–reversed DTARE.

Proposition 5.7.2 *If*

$$
V = \begin{bmatrix} V_1 \\ V_2 \\ V_3 \end{bmatrix} \begin{matrix} \}n \\ \}n \\ \}m \end{matrix} \tag{5.79}
$$

is a column basis matrix of a left $\mathbb{D}$*–deflating subspace for the ESP (5.48), then*

$$
V_1^T(AV_2 + BV_3) = (AV_2 + BV_3)^TV_1.
$$

Proof. Since $\mathcal{V} = \operatorname{Im} V$ is a $\mathbb{D}$–deflating subspace for $zM_\Sigma^T - N_\Sigma^T$, it follows that

$$
M_\Sigma^TVS = N_\Sigma^TV \tag{5.80}
$$

holds for a stable matrix S or, in explicit form,

$$
\begin{aligned}
V_1S &= A^TV_1 + QV_2 + LV_3, \\
(AV_2 + BV_3)S &= V_2, \\
0 &= B^TV_1 + L^TV_2 + RV_3.
\end{aligned} \tag{5.81}
$$

Adding the first equation premultiplied by V_2^T and the second one transposed and postmultiplied by V_1S to the third one premultiplied by V_3^T in (5.81), we get

$$
\begin{aligned}
& S^T(AV_2 + BV_3)^TV_1S - (AV_2 + BV_3)^TV_1 \\
& -V_2^TQV_2 - V_2^TLV_3 - V_3^TL^TV_2 - V_3^TRV_3 = 0.
\end{aligned} \tag{5.82}
$$

This equation is a Stein equation in $(AV_2 + BV_3)^T V_1$ having a symmetric free term. Since S is stable, $(AV_2 + BV_3)^T V_1$ is the unique symmetric solution. ∎

Recall that a regular and dichotomic ESP always has a $\mathbb{D}$–deflating subspace of dimension n. Then it is easy to see by Corollary 1.8.3 that the ESP also has a left $\mathbb{D}$–deflating subspace $\mathcal{V}$ of dimension n.

Definition 5.7.3 *A deflating subspace to the left $\mathcal{V}$ of the ESP is called* time–reversed disconjugate *if it admits a column basis matrix V partitioned as in (5.79) with $AV_2 + BV_3$ having full column rank.*

Notice that the time–reversed disconjugacy of $\mathcal{V}$ is independent of the particular choice of the basis matrix V.

Definition 5.7.4 *A regular ESP is called* time–reversed disconjugate *if it is dichotomic and has a time–reversed disconjugate $\mathbb{D}$–deflating subspace to the left of dimension n.*

Similar to the notion of disconjugacy of the ESP, we note that *time–reversed* disconjugacy is well defined.

The next theorem contains the main result of the section.

Theorem 5.7.5 *Let $\Sigma = (A, B; Q, L, R)$ be a reversible Popov triplet. The following two assertions are equivalent.*

1. *The time–reversed DTARE(Σ) given in (5.71) has a stabilizing solution.*
2. *The ESP(Σ) in (5.48) is regular and time–reversed disconjugate.*

Moreover, the stabilizing solution can be computed from

$$X := V_1(AV_2 + BV_3)^{-1} \tag{5.83}$$

and the stabilizing Riccati feedback can be computed from

$$F := V_3(AV_2 + BV_3)^{-1} \tag{5.84}$$

where V is a column basis matrix partitioned as in (5.79) for the n-dimensional $\mathbb{D}$-deflating subspace to the left of the ESP(Σ).

Proof. 1. ⇒ 2. Let X be the stabilizing solution to the time–reversed DTARE, let F be defined by (5.73) and

$$S := A^{-1}(I - BF). \tag{5.85}$$

Since $G(X)$ in (5.72) is nonsingular, it follows from the time–reversed version of the spectral factorization identity (5.77) that $\operatorname{rank}_{\mathbb{R}(z)} \mathbf{\Pi}_\Sigma(z) = m$ which shows that the ESP(Σ) is regular. Denote

$$V := \begin{bmatrix} X \\ S \\ F \end{bmatrix}. \tag{5.86}$$

Notice first that V is a basis matrix, i.e., it has full column rank. This follows immediately from (5.85). Further, we show that V is a column basis matrix of the

n–dimensional left $\mathbb{D}$–deflating subspace of $zM_\Sigma - N_\Sigma$ or, equivalently, of the $\mathbb{D}$–deflating subspace of $zM_\Sigma^T - N_\Sigma^T$. To this end we prove that (5.80) or, equivalently, (5.81) holds for a stable matrix S. First rewrite (5.85) as

$$S = (AS + BF)S \tag{5.87}$$

and the time–reversed DTARE as

$$X + A^{-T}(Q - X)A^{-1} + A^{-T}[L - (Q - X)A^{-1}B]F = 0. \tag{5.88}$$

Premultiplying (5.88) by A^T and using (5.85) results in

$$A^T X + QS + LF = XS. \tag{5.89}$$

Premultiplying (5.88) with B^T and adding the second equation (5.76), we obtain with (5.72) and (5.74)

$$B^T X + L^T A^{-1} + (R - L^T A^{-1} B)F = 0$$

or, equivalently, using (5.85) we get

$$B^T X + L^T S + RF = 0. \tag{5.90}$$

From (5.89), (5.87) and (5.90) we conclude that (5.80) holds for V given in (5.86). As S defined through (5.85) is stable we obtain that the ESP(Σ) is dichotomic. Moreover, $AS + BF = I$ shows that the ESP is time–reversed disconjugate.

2. $\Rightarrow$ 1. Assume that the ESP(Σ) is time–reversed disconjugate. Then we have that (5.80) holds for a stable matrix S and a basis matrix V, partitioned as in (5.79), with $AV_2 + BV_3$ invertible. Denote

$$\begin{array}{rcl} \widetilde{S} & := & V_2(AV_2 + BV_3)^{-1}, \\ X & := & V_1(AV_2 + BV_3)^{-1}, \\ F & := & V_3(AV_2 + BV_3)^{-1}. \end{array} \tag{5.91}$$

Notice that Proposition 5.7.2 shows that $X = X^T$. Now from (5.80) written explicitly as in (5.81), we get with (5.91)

$$\begin{array}{rcl} A^T X + Q\widetilde{S} + LF & = & X\widetilde{S}, \\ B^T X + L^T\widetilde{S} + RF & = & 0, \\ A^{-1}(I - BF) & = & \widetilde{S}, \end{array} \tag{5.92}$$

where for the first equation (5.92) we have used the second equation (5.81) written as

$$(AV_2 + BV_3)S(AV_2 + BV_3)^{-1} = \widetilde{S}, \tag{5.93}$$

while the third equation (5.92) follows by using the notation (5.91) in the identity $(AV_2 + BV_3)(AV_2 + BV_3)^{-1} = I$. Since S is stable, (5.93) shows that $\widetilde{S}$ is stable as well.

Substituting $\widetilde{S}$ from the third into the first and the second equation in (5.92), we get

$$\begin{array}{rcl} A^T X + LF + (Q - X)A^{-1}(I - BF) & = & 0, \\ L^T A^{-1}(I - BF) + B^T X + RF & = & 0. \end{array} \tag{5.94}$$

Premultiplying the first equation (5.94) by A^{-T} and using (5.74), we obtain

$$X + A^{-T}(Q - X)A^{-1} + H(X)F = 0. \tag{5.95}$$

Subtracting equation (5.95) premultiplied by B^T from the second equation (5.94) and using (5.72) and (5.74) we obtain

$$H^T(X) + G(X)F = 0. \tag{5.96}$$

Premultiplying (5.96) by F^T and using (5.75) we get from (5.95) that

$$K(X) - F^T G(X)F = 0. \tag{5.97}$$

The equations (5.96) and (5.97) prove that the system (5.76) is fulfilled for X and F defined through (5.91). Finally, since the ESP is regular, $G(X)$ in (5.77) is invertible, and therefore F can be substituted from (5.96) into (5.97). We conclude that X satisfies the time–reversed DTARE (5.71). Moreover, (5.96) and the third equation (5.92) show that F is the time–reversed stabilizing Riccati feedback. ∎

We conclude this section with a result giving the relation between the stabilizing solution to the time–reversed DTARE and the antistabilizing solution to the DTARE.

Theorem 5.7.6 *Let $\Sigma = (A, B; Q, L, R)$ be a reversible Popov triplet.*

I. *If the DTARE(Σ) has an antistabilizing solution X, then the time–reversed DTARE associated with Σ has a stabilizing solution which equals X.*

II. *If the time–reversed DTARE(Σ) has a stabilizing solution X then the DTARE(Σ) has an antistabilizing solution if and only if*

$$E := R - L^T A^{-1} B \tag{5.98}$$

is invertible. In this case X is the antistabilizing solution to the DTARE(Σ).

Proof. Notice first that E invertible is necessary for the existence of the antistabilizing solution to the DTARE(Σ). Indeed, following Remark 5.5.2, the matrix (5.56) is invertible which is equivalent, since A is invertible, to the invertibility of its Schur complement which is E.

I. Suppose that the DTARE(Σ) has an antistabilizing solution X and let F be the corresponding antistabilizing Riccati feedback. Let $zM_\Sigma - N_\Sigma$ be the ESP(Σ). Then $M_\Sigma V S = N_\Sigma V$ holds for

$$V := \begin{bmatrix} I \\ X \\ F \end{bmatrix},$$

and $S := A + BF$, $\Lambda(S) \subset \mathbb{C} \setminus \overline{\mathbb{D}}$. Define

$$\widetilde{V} := \begin{bmatrix} \widetilde{V_1} \\ \widetilde{V_2} \\ \widetilde{V_3} \end{bmatrix} = \begin{bmatrix} XS \\ I \\ F \end{bmatrix}.$$

Since S is antistable, it is invertible and a simple computation shows that

$$M_\Sigma^T \widetilde{V} \widetilde{S} = N_\Sigma^T \widetilde{V}$$

holds for $\widetilde{S} = S^{-1}$. Consequently, $\widetilde{V}$ is a basis matrix for an n–dimensional $\mathbb{D}$–deflating subspace for $zM_\Sigma^T - N_\Sigma^T$ and $A\widetilde{V_2} + B\widetilde{V_3} = S$ is invertible. It emerges that $zM_\Sigma - N_\Sigma$ is time–reversed disconjugate. Moreover, from Theorem 5.7.5 we have that the stabilizing solution to the time–reversed DTARE is given by $\widetilde{X} = \widetilde{V_1}(A\widetilde{V_2} + B\widetilde{V_3})^{-1} = X$.

II. Suppose that the time–reversed DTARE has a stabilizing solution X and let F be the time–reversed stabilizing Riccati feedback. If E is invertible, it follows that the ESP $zM_\Sigma - N_\Sigma$ has N_Σ invertible and therefore it has no generalized eigenvalues at zero. Then (5.80) holds for V and S defined through (5.86) and (5.85), respectively, with S stable and invertible. Define

$$\widetilde{V} := \begin{bmatrix} \widetilde{V_1} \\ \widetilde{V_2} \\ \widetilde{V_3} \end{bmatrix} = \begin{bmatrix} S \\ XS \\ F \end{bmatrix}$$

which has full column rank. A direct check shows that

$$M_\Sigma \widetilde{V} \widetilde{S} = N_\Sigma \widetilde{V}$$

holds for $\widetilde{S} := S^{-1}$ and $\Lambda(\widetilde{S}) \subset \mathbb{C} \setminus \overline{\mathbb{D}}$. Moreover, $\widetilde{V_1} = S$ is invertible and the ESP is antistable disconjugate. As $\widetilde{V_2}\widetilde{V_1}^{-1} = X$, we conclude that X is also the antistabilizing solution to the DTARE. ■

The following example outlines the differences between the antistabilizing solution to the DTARE and the stabilizing solution to the time–reversed DTARE.

Example 5.7.7 Consider the reversible Popov triplet $\Sigma = (1, 1; 1, 1, 1)$. Then the DTARE(Σ) reduces to $X = 0$ and the Riccati feedback is $F = -1$ while the time–reversed DTARE reduces to $X + 1 = 0$ and the Riccati feedback is $F = 1$. Hence there is no antistabilizing solution to the DTARE(Σ) while $X = -1$ is the stabilizing solution to the time–reversed DTARE.

In conclusion, we note that the significant differences between the continuous–time and discrete–time cases originate in the differences that occur in the intimate eigenstructure of the EHP and the corresponding ESP. Contrary to the EHP, the ESP has, in general, an asymmetry of the finite eigenstructure with respect to the boundary of the stability domain (i.e., the stable and antistable eigenstructure).

Notes and references

The connection between Riccati equations and eigenvalue (or invariant subspace) problems has been known for about a hundred years. The idea to use invariant subspaces of the Hamiltonian matrix to compute solutions to the continuous–time Riccati equation was applied first by Laub [Lau79]. The discrete–time counterpart is far more tricky. In fact, historically, there have been three discrete–time counterparts, in increasing degree of generality. The first attempt made by Laub [Lau79] was to assume that R and $\widehat{A} := A - BR^{-1}L^T$ are nonsingular and to replace the Hamiltonian matrix by the symplectic matrix. The second step was to remove the assumption that $\widehat{A}$ is invertible. Inherently, one has to consider instead a generalized eigenvalue problem as in Pappas, Laub, and Sandell [PLS80] where the symplectic matrix is replaced with a $2n \times 2n$ symplectic pencil. However, the proofs stick on some unnecessary local positiveness assumptions. The third step was to remove the assumption on the invertibility of R – a restrictive condition in discrete–time since the Riccati equation requires only that $R+B^TXB$ be invertible. It was noticed by Van Dooren [Van81] that it is natural to associate with the DTARE the extended symplectic pencil of dimension $(2n+m) \times (2n+m)$ and it was conjectured by him that this pencil can be used for computing the stabilizing solution in the general case. This is indeed true and was proved by Ionescu and Weiss [IW92b]. The connection between the stabilizing solution to the reverse DTARE and the antistabilizing solution to the DTARE and left deflating subspaces of the ESP is due to Oară [Oar95, Oar96b]. Most of the proofs in this chapter are new and the presentation of Sections 5.1–5.6 is mainly based on Ionescu, Oară, and Weiss [IOW97], while Section 5.7 is based on Oară [Oar96b]. Extension to the time–varying ESP are due to Ionescu and Oară [IO96c, Ion98b]. Applications of the EHP and ESP techniques in robust control of time–delay systems can be found in Niculescu and Ionescu [NI96, NI97].

6

Riccati Systems and Matrix Pencils: the General Case

The matrix pencil approach to the algebraic Riccati equation theory presented in Chapter 5 is extended here to the general case in which the underlying Popov function does not necessary take values of maximal rank, i.e., it is not necessary invertible as a rational matrix. In this so–called singular case the extended Hamiltonian and symplectic pencils are not regular, and the algebraic Riccati equations have to be replaced by more general algebraic Riccati systems.

The chapter is arranged in the following way. We investigate the particular Kronecker eigenstructure of the extended Hamiltonian pencil in Section 6.1. In Section 6.2 we introduce the generalized continuous–time algebraic Riccati system which incorporates as particular cases the algebraic Riccati system and the algebraic Riccati equation, and we give necessary and sufficient solvability conditions and computable formulas for its stabilizing and antistabilizing solutions in terms of proper deflating subspaces of the extended Hamiltonian pencil.

The discrete–time counterpart of the results presented in Sections 6.1–6.2 are discussed in Sections 6.3–6.4, respectively. In Section 6.5 we sketch a general–purpose numerically sound algorithm for computing with high accuracy various solutions to algebraic Riccati equations and systems.

A. The continuous–time case

6.1 The eigenstructure of a general EHP

In this section we investigate the Kronecker structure of a general (possibly singular) extended Hamiltonian pencil. As in the case in which the pencil is regular, the more complicated eigenstructure of a general extended Hamiltonian pencil plays a foremost role in deriving the existence of the solutions to the algebraic Riccati systems.

Consider a Popov triplet $\Sigma = (A, B; Q, L, R)$, where $A \in \mathbb{R}^{n \times n}$ and $B \in \mathbb{R}^{n \times m}$. We associate with Σ the EHP(Σ) given by

$$sM_\Sigma - N_\Sigma := s \begin{bmatrix} I & O & O \\ O & I & O \\ O & O & O \end{bmatrix} - \begin{bmatrix} A & O & B \\ -Q & -A^T & -L \\ L^T & B^T & R \end{bmatrix} \tag{6.1}$$

and the Popov function

$$\mathbf{\Pi}_\Sigma(s) = \left[\begin{array}{cc|c} A & O & B \\ -Q & -A^T & -L \\ \hline L^T & B^T & R \end{array}\right]. \tag{6.2}$$

Recall that the EHP(Σ) given in (6.1) is the transmission pencil associated with the realization (6.2) of the Popov function $\mathbf{\Pi}_\Sigma$. In this chapter we allow both the Popov function $\mathbf{\Pi}_\Sigma$ and the EHP(Σ) to be singular. A singular Popov function has not full rank and for a singular EHP we have $\det(sM_\Sigma - N_\Sigma) \equiv 0$. The more complicated eigenstructure of a singular pencil is revealed by the Kronecker canonical form (see Section 1.10), and we expect the proper deflating subspaces to play an important role in the derivations (see Section 1.11).

Denote by n_f^-, n_f^0, n_f^+, and n_∞ the number of generalized eigenvalues of the EHP located in $\mathbb{C}_-$, $\mathbb{C}_0$, $\mathbb{C}_+$, and at ∞, respectively. Let n_r and n_ℓ be, respectively, the sum of the right and left Kronecker indices of the EHP. Let $\rho = \operatorname{rank}_{\mathbb{R}(s)} \mathbf{\Pi}_\Sigma(s)$ and let π_0 be the number of zeros of $\mathbf{\Pi}_\Sigma(s^{-1})$ located in the origin (or, alternatively, the number of zeros at ∞ of $\mathbf{\Pi}_\Sigma(s)$). The next result is our main tool for investigating the eigenstructure of a singular EHP.

Lemma 6.1.1 *The EHP has a proper $\mathbb{C}_-$-deflating subspace to the right if and only if it has a proper $\mathbb{C}_+$-deflating subspace to the left of the same dimension. Similarly, it has a proper $\mathbb{C}_+$-deflating subspace to the right if and only if it has a proper $\mathbb{C}_-$-deflating subspace to the left of the same dimension.*

Proof. We prove only the first part, since the second follows by duality. Using (6.1), it can be checked that

$$V = \begin{bmatrix} V_1 \\ V_2 \\ V_3 \end{bmatrix} \begin{matrix} \}n \\ \}n \\ \}m \end{matrix} \tag{6.3}$$

is a basis matrix for a stable proper deflating subspace to the right if and only if

$$W = \begin{bmatrix} V_2^T & -V_1^T & V_3^T \end{bmatrix}$$

is a row basis matrix for an antistable proper deflating subspace to the left. Indeed, $M_\Sigma VS = N_\Sigma V$ is equivalent to $(-S^T)WM_\Sigma = WN_\Sigma$ and $M_\Sigma V$ has full column rank if and only if WM_Σ has full row rank, and the two subspaces have the same dimension. ■

The following result extends the characterization of the eigenstructure of a regular EHP given in Theorem 5.1.1 to the singular case.

Theorem 6.1.2 *For an arbitrary EHP the following hold:*

1. $n_r = n_\ell$.
2. $n_f^- = n_f^+$.
3. $\operatorname{rank}_{\mathbb{R}(s)}(sM_\Sigma - N_\Sigma) = 2n + \rho$.
4. $n_\infty = \rho + \pi_0$.

Proof. 1. and 2. According to Theorem 1.11.3, there exists a proper $\mathbb{C}_-$–deflating subspace to the right with dimension $n_r + n_f^-$. By Lemma 6.1.1, there exists a proper $\mathbb{C}_+$–deflating subspace to the left with the same dimension. Applying again Theorem 1.11.3, the dimension of a proper $\mathbb{C}_+$–deflating subspace to the left cannot exceed $n_\ell + n_f^+$. Therefore

$$\begin{aligned} n_r + n_f^- &\leq n_\ell + n_f^+, \\ n_r + n_f^+ &\leq n_\ell + n_f^-, \\ n_\ell + n_f^- &\leq n_r + n_f^+, \\ n_\ell + n_f^+ &\leq n_r + n_f^-, \end{aligned}$$

where the last three inequalities are obtained by repeating the arguments. Adding the first two inequalities, we get $n_r \leq n_\ell$ and adding the last two we get $n_\ell \leq n_r$ and point 1 is justified. For point 2, add the first and the third inequalities, and the second and the fourth, and the conclusion follows.

3. and 4. For any value s for which $I - sA$ and $I + sA^T$ are invertible we have

$$M_\Sigma - sN_\Sigma = \mathbf{F}(s)\mathbf{G}(s)\mathbf{H}(s) \tag{6.4}$$

where

$$\begin{aligned} \mathbf{F}(s) &:= \begin{bmatrix} I_n & O & O \\ sQ(I - sA)^{-1} & I_n & O \\ -sL^T(I - sA)^{-1} & -sB^T(I + sA^T)^{-1} & I_m \end{bmatrix}, \\ \mathbf{G}(s) &:= \begin{bmatrix} I - sA & O & O \\ O & I + sA^T & O \\ O & O & -s\mathbf{\Pi}_\Sigma(s^{-1}) \end{bmatrix}, \\ \mathbf{H}(s) &:= \begin{bmatrix} I_n & O & -s(I - sA)^{-1}B \\ O & I_n & (I + sA^T)^{-1}[sL + s^2Q(I - sA)^{-1}B] \\ O & O & I_m \end{bmatrix}. \end{aligned}$$

From (6.4) we deduce that

$$\operatorname{rank}_{\mathbb{R}(s)}(sM_\Sigma - N_\Sigma) = \operatorname{rank}_{\mathbb{R}(s)}(M_\Sigma - sN_\Sigma) = 2n + \rho.$$

Moreover, the multiplicity of $s = 0$ as a generalized eigenvalue of $M_\Sigma - sN_\Sigma$ (which equals the multiplicity of ∞ as a generalized eigenvalue of the reciprocal pencil $sM_\Sigma - N_\Sigma$) coincides with the multiplicity of the null zero of $s\mathbf{\Pi}_\Sigma(s^{-1})$. Finally, using the Smith–McMillan form for $\mathbf{\Pi}_\Sigma(s^{-1})$ we find that $s\mathbf{\Pi}_\Sigma(s^{-1})$ has exactly $\operatorname{rank}_{\mathbb{R}(s)}\mathbf{\Pi}_\Sigma(s) + \pi_0$ null zeros, which concludes the proof. ∎

Our next result says that a proper $\mathbb{C}_-$–deflating subspace of dimension n of the EHP is maximal and exists if and only if a certain type of *dichotomy* of the EHP holds.

Theorem 6.1.3 *The dimension of a proper $\mathbb{C}_-$ ($\mathbb{C}_+$)–deflating subspace of the EHP cannot exceed n. This upper bound is attained if and only if $n_f^0 = 0$ and $n_\infty = \rho$.*

Proof. Due to Lemma 6.1.1 it is enough to prove the statement for a proper $\mathbb{C}_-$–deflating subspace. According to Theorem 1.11.3, we have to show that $n_r + n_f^- \leq n$ and that equality takes place if and only if the EHP has no generalized eigenvalues on the imaginary axis and it has precisely ρ eigenvalues at ∞. From Theorem 6.1.2 we have $\text{rank}_{\mathbb{R}(s)}(sM_\Sigma - N_\Sigma) = 2n + \rho$. On the other hand, for any matrix pencil we have from (1.49) that $\text{rank}_{\mathbb{R}(s)}(sM_\Sigma - N_\Sigma) = n_r + n_\ell + n_f^- + n_f^+ + n_f^0 + n_\infty$. Using now, points 1, 2 and 4 of Theorem 6.1.2, we find

$$2n + \rho = 2(n_r + n_f^-) + n_f^0 + \rho + \pi_0.$$

Hence

$$n_r + n_f^- \leq n$$

and the equality holds if and only if $n_f^0 = 0$ and $n_\infty = \rho$ (or, equivalently, $n_f^0 = \pi_0 = 0$). As explained before, this completes the proof. ∎

Definition 6.1.4 *An EHP is called* dichotomic *if* $n_f^0 = 0$ *and* $n_\infty = \rho$ *(or, equivalently,* $n_f^0 = \pi_0 = 0$*).*

In particular, Theorem 6.1.3 implies that the EHP is dichotomic if and only if

$$M_\Sigma V S = N_\Sigma V \tag{6.5}$$

holds with $M_\Sigma V$ of full column rank, where

$$V = \begin{bmatrix} V_1 \\ V_2 \\ V_3 \end{bmatrix} \begin{matrix} \}n \\ \}n \\ \}m \end{matrix} \tag{6.6}$$

is a basis matrix for the n–dimensional $\mathbb{C}_-$ (or $\mathbb{C}_+$)–deflating subspace, and $S \in \mathbb{R}^{n\times n}$ is a stable (antistable) matrix.

We introduce now the notion of disconjugacy in the case of singular pencils in a way very similar to the regular case.

Definition 6.1.5 *A deflating subspace $\mathcal{V}$ of an EHP is called* disconjugate *if it admits a basis matrix V partitioned as in (6.6) where V_1 has full column rank.*

Notice that a disconjugate deflating subspace is automatically proper. Our next result says that if one maximal proper $\mathbb{C}_-$ (or $\mathbb{C}_+$)–deflating subspace is disconjugate then all such subspaces are disconjugate.

Theorem 6.1.6 *If the EHP has a maximal proper $\mathbb{C}_-$ ($\mathbb{C}_+$)–deflating subspace which is disconjugate, then* each *maximal proper $\mathbb{C}_-$ ($\mathbb{C}_+$)–deflating subspace is disconjugate.*

Proof. We consider only the case of $\mathbb{C}_-$–deflating subspaces. Let $\mathcal{V}$ and $\overline{\mathcal{V}}$ be two maximal $\mathbb{C}_-$–deflating subspaces and let

$$V = \begin{bmatrix} V_1 \\ V_2 \\ V_3 \end{bmatrix}, \quad \overline{V} = \begin{bmatrix} \overline{V}_1 \\ \overline{V}_2 \\ \overline{V}_3 \end{bmatrix}$$

be basis matrices for $\mathcal{V}$ and $\overline{\mathcal{V}}$, respectively. Then

$$M_\Sigma V = \begin{bmatrix} V_1 \\ V_2 \\ O \end{bmatrix}, \quad M_\Sigma \overline{V} = \begin{bmatrix} \overline{V}_1 \\ \overline{V}_2 \\ O \end{bmatrix},$$

have full column rank and, by Theorem 1.11.4, their images coincide. Hence there exists an invertible constant matrix G such that

$$\begin{bmatrix} V_1 \\ V_2 \\ O \end{bmatrix} = \begin{bmatrix} \overline{V}_1 \\ \overline{V}_2 \\ O \end{bmatrix} G.$$

Thus if V_1 has full column rank, then $\overline{V}_1$ has too, and conversely. ■

The previous result justifies the following definition.

Definition 6.1.7 *An EHP is called stable (antistable)* disconjugate *if it is dichotomic and has a disconjugate $\mathbb{C}_-$ ($\mathbb{C}_+$)-deflating subspace of dimension n.*

Thus the disconjugacy of an EHP can be established by finding one arbitrary maximal proper deflating subspace with an appropriate spectrum and checking its disconjugacy.

We give now a property of deflating subspaces of the EHP that will be important in proving the symmetry of X in a solution to the algebraic Riccati system.

Proposition 6.1.8 *If V is a basis matrix as in (5.9) of a proper $\mathbb{C}_-$ (or $\mathbb{C}_+$)-deflating subspace of an arbitrary EHP then*

$$V_1^T V_2 = V_2^T V_1. \tag{6.7}$$

Proof. The proof is similar to the regular case (see Proposition 5.1.6) and is therefore omitted. ■

6.2 The generalized continuous–time Riccati system and the EHP

In this section we introduce the generalized continuous–time algebraic Riccati system (GCTARS), which incorporates as special cases the algebraic Riccati equation and the Riccati system, and we characterize its stabilizing and antistabilizing solutions in terms of proper deflating subspaces of the EHP.

Definition 6.2.1 *Let $\Sigma = (A, B; Q, L, R)$ be a Popov triplet, where $A \in \mathbb{R}^{n\times n}, B \in \mathbb{R}^{n\times m}$. The* generalized continuous–time algebraic Riccati system (GCTARS) *associated with Σ is*

$$\begin{bmatrix} A^TX + XA + Q & XB + L \\ L^T + B^TX & R \end{bmatrix} \begin{bmatrix} I \\ F \end{bmatrix} V = 0, \tag{6.8}$$

where $X \in \mathbb{R}^{n\times n}, V \in \mathbb{R}^{n\times r}$ has full column rank with $r \le n$, r unfixed, and F such that $(A+BF)V = VS$ for an appropriate matrix $S \in \mathbb{R}^{r\times r}$.

A quadruple (X, F, r, V) for which (6.8) holds and $V^TXV = V^TX^TV$, is called a V–symmetric solution. *A V-symmetric solution is called* stabilizing (antistabilizing) *if in addition S is stable (antistable).*

Notice that if (X, F, r, V) is a stabilizing solution to the GCTARS then $\mathcal{V} = \operatorname{Im} V$ is an (A, B) invariant subspace, F is such that $(A+BF)\mathcal{V} \subset \mathcal{V}$ and $\Lambda(A+BF)|_{\mathcal{V}} \subset \mathbb{C}_-$.

If V is invertible we may assume that $V = I$ and (6.8) reduces to

$$\begin{array}{rcccl} A^T X + XA + Q & + & (XB + L)F & = & 0, \\ L^T + B^T X & + & RF & = & 0, \end{array} \tag{6.9}$$

which is exactly the CTARS(Σ) in (3.11). In this case a stabilizing (antistabilizing) solution is a pair (X, F) satisfying (6.9) with $X = X^T$ and $A+BF$ stable (antistable). Further, for $X = X^T$ we get

$$\begin{array}{rcccl} A^T X + XA + Q & - & F^T RF & = & 0, \\ L^T + B^T X & + & RF & = & 0, \end{array}$$

or, equivalently,

$$\begin{array}{rcl} A^T X + XA - (XB + L)R^\dagger(B^T X + L^T) + Q & = & 0, \\ \operatorname{Ker} R \subset \operatorname{Ker}(XB + L), & & \end{array} \tag{6.10}$$

which is known as the constrained continuous–time algebraic Riccati equation and reduces, when R is invertible, to

$$A^T X + XA - (XB + L)R^{-1}(L^T + B^T X) + Q = 0 \tag{6.11}$$

which is the CTARE(Σ).

The following uniqueness result holds for the stabilizing (antistabilizing) solution to the GCTARS.

Proposition 6.2.2 *If (X, F, r, V) and $(\widetilde{X}, \widetilde{F}, r, V)$ are two V-symmetric stabilizing (antistabilizing) solutions to (6.8) then*

$$V^T X V = V^T \widetilde{X} V.$$

Proof. We prove only for stabilizing solutions since antistabilizing solutions can be treated analogously. We have

$$(A + BF)V = VS, \quad (A + B\widetilde{F})V = V\widetilde{S}$$

where S and $\widetilde{S}$ are both stable and thus

$$A^T XV + X(A + BF)V + QV + LFV = A^T XV + XVS + QV + LFV.$$

With this we rewrite (6.8) as

$$V^T A^T XV + V^T XVS + V^T QV + V^T LFV = 0, \tag{6.12}$$

$$V^T \widetilde{F}^T (L^T V + B^T XV + RFV) = 0. \tag{6.13}$$

Adding (6.12) and (6.13) yields

$$\widetilde{S}^T V^T XV + V^T XVS + V^T(LF + \widetilde{F}^T L^T)V + V^T \widetilde{F}^T RFV + V^T QV = 0. \tag{6.14}$$

If we interchange in the above reasoning the role of X and F with the role of $\widetilde{X}$ and $\widetilde{F}$, respectively, we get

$$S^T V^T \widetilde{X} V + V^T \widetilde{X} V \widetilde{S} + V^T (L\widetilde{F} + F^T L^T) V + V^T F^T R \widetilde{F} V + V^T Q V = 0. \tag{6.15}$$

Since $V^T \widetilde{X} V = V^T \widetilde{X}^T V$, we transpose (6.15) and get

$$V^T \widetilde{X} V S + \widetilde{S}^T V^T \widetilde{X} V + V^T (\widetilde{F}^T L^T + LF) V + V^T \widetilde{F}^T R F V + V^T Q V = 0. \tag{6.16}$$

Subtracting (6.16) from (6.14) results in

$$\widetilde{S}^T (V^T X V - V^T \widetilde{X} V) + (V^T X V - V^T \widetilde{X} V) S = 0 \tag{6.17}$$

which is a Sylvester equation in the unknown $V^T X V - V^T \widetilde{X} V$. As S and $\widetilde{S}$ are both stable it follows that $V^T X V - V^T \widetilde{X} V = 0$ is the unique solution. ■

The next theorem contains the main result of the section.

Theorem 6.2.3 *Let $\Sigma = (A, B; Q, L, R)$ be a Popov triplet. The following assertions are equivalent:*

1. The GCTARS(Σ) in (6.8) has a stabilizing (antistabilizing) solution (X, F, r, V).

2. The EHP(Σ) in (6.1) has a disconjugate proper $\mathbb{C}_-$ ($\mathbb{C}_+$)-deflating subspace $\mathcal{V}$ of dimension r.

If assertion 2 holds, let V be a basis matrix partitioned as in (6.6) of $\mathcal{V}$. Then a stabilizing (antistabilizing) solution is (X, F, r, V_1) where

$$X = V_2 V_1^\dagger, \quad F = V_3 V_1^\dagger. \tag{6.18}$$

Proof. We prove for the stabilizing solution only since the antistable case follows identically.

1. $\Rightarrow$ 2. Let (X, F, r, V_1) be a stabilizing solution to (6.8). Denoting

$$V = \begin{bmatrix} V_1 \\ V_2 \\ V_3 \end{bmatrix} := \begin{bmatrix} V_1 \\ XV_1 \\ FV_1 \end{bmatrix}$$

it can be easily checked that the following equations

$$\begin{array}{rrrcl} AV_1 & & +BV_3 & = & V_1 S, \\ -QV_1 & -A^T V_2 & -LV_3 & = & V_2 S, \\ L^T V_1 & +B^T V_2 & +RV_3 & = & 0, \end{array} \tag{6.19}$$

hold for a stable matrix S or, more compactly, the EHP(Σ) in (6.1) satisfies $M_\Sigma V S = N_\Sigma V$. Now $M_\Sigma V$ has full column rank since V_1 has full column rank. Thus $\mathcal{V} = \operatorname{Im} V$ is a disconjugate proper $\mathbb{C}_-$-deflating subspace (of dimension r) for the EHP.

2. $\Rightarrow$ 1. Let $\mathcal{V}$ be a disconjugate proper $\mathbb{C}_-$-deflating subspace of dimension r, with a basis matrix V partitioned as in (6.6), for the EHP(Σ) in (6.1). Then $M_\Sigma V S = N_\Sigma V$ holds for a stable matrix S which can be rewritten as in (6.19). Define X and F through (6.18). Then (6.19) provides

$$\begin{array}{rcl} (A + BF)V_1 & = & V_1 S, \\ (-Q - A^T X - LF)V_1 & = & XV_1 S, \\ (L^T + B^T X + RF)V_1 & = & 0. \end{array} \tag{6.20}$$

We eliminate $V_1 S$ in the first two equations (6.20), and get (6.8) written with $V = V_1$. Clearly, (X, F, r, V_1) is a solution to the GCTARS(Σ). Finally, we have with Proposition 6.1.8 that

$$V_1^T X V_1 = V_1^T V_2 V_1^\dagger V_1 = V_1^T V_2 = V_2^T V_1 = V_1^T (V_1^\dagger)^T V_2^T V_1 = V_1^T X^T V_1$$

from where we conclude that (X, F, r, V_1) is a V_1 symmetric solution which is stabilizing, as S is stable. This concludes the proof. ■

From Theorem 6.2.3 we obtain immediately a characterization of the stabilizing (antistabilizing) solution to the continuous–time algebraic Riccati system.

Theorem 6.2.4 *Let $\Sigma = (A, B; Q, L, R)$ be a Popov triplet. The following assertions are equivalent:*

1. The CTARS(Σ) in (6.9) has a stabilizing (antistabilizing) solution (X, F).

2. The EHP(Σ) in (6.1) is dichotomic and stable (antistable) disconjugate.

If assertion 2 holds, let V be a basis matrix partitioned as in (6.6) of an n–dimensional maximal proper $\mathbb{C}_-$ ($\mathbb{C}_+$)–deflating subspace of the EHP(Σ). Then a stabilizing (antistabilizing) solution is (X, F) where

$$X = V_2 V_1^{-1}, \quad F = V_3 V_1^{-1}. \tag{6.21}$$

It is important to notice that the nonuniqueness of the maximal proper $\mathbb{C}_-$ ($\mathbb{C}_+$)–deflating subspace of the EHP does not affect the computational procedures for the solutions of the CTARS. Specifically, the solution and the corresponding Riccati feedback can be computed by formulas (6.21) where V is a basis matrix of *any* maximal proper deflating subspace with an appropriate spectrum. This follows from Theorem 1.11.4, and thus the uniqueness of X stated in Proposition 3.2.3 is recovered.

B. The discrete–time case

6.3 The eigenstructure of a singular ESP

In this section we investigate the Kronecker structure of an arbitrary extended symplectic pencil. Unlike the extended Hamiltonian pencil, the extended symplectic pencil presents in general an asymmetry of its eigenstructure that is reflected also in the existence conditions for the solution to the discrete–time algebraic Riccati systems.

Consider a Popov triplet $\Sigma = (A, B; Q, L, R)$, where $A \in \mathbb{R}^{n \times n}$ and $B \in \mathbb{R}^{n \times m}$. We associate with Σ the ESP(Σ) given by

$$zM_\Sigma - N_\Sigma := z \begin{bmatrix} I & O & O \\ O & -A^T & O \\ O & -B^T & O \end{bmatrix} - \begin{bmatrix} A & O & B \\ -Q & -I & L \\ L^T & O & R \end{bmatrix} \tag{6.22}$$

and the discrete–time Popov function

$$\mathbf{\Pi}_\Sigma(z) = \left[\begin{array}{cc|c} zI_n - A & O & -B \\ -Q & I_n - zA^T & -L \\ \hline L^T & zB^T & R \end{array} \right]. \tag{6.23}$$

Recall that the ESP(Σ) is the transmission pencil associated with the descriptor realization (6.23) of the Popov function $\mathbf{\Pi}_\Sigma$. We allow further both the Popov function $\mathbf{\Pi}_\Sigma$ and the ESP(Σ) to be singular.

Denote by n_f^-, n_f^0, n_f^+, and n_∞ the numbers of generalized eigenvalues of the ESP located in $\mathbb{D}$, $\partial\mathbb{D}$, $\mathbb{C} \setminus \overline{\mathbb{D}}$, and at ∞, respectively. Let n_r and n_ℓ be, respectively, the sum of the right and left Kronecker indices of the ESP. Also denote by π_0 the algebraic multiplicity of the zero generalized eigenvalue of $zM_\Sigma - N_\Sigma$ and let $\widetilde{n}_f^- = n_f^- - \pi_0$ be the number of nonzero generalized eigenvalues in $\mathbb{D}$. Let $\rho = \operatorname{rank}_{\mathbb{R}(z)} \mathbf{\Pi}_\Sigma(z)$.

Unlike the continuous–time case, the existence of stable and antistable subspaces of the ESP has to be addressed separately. We start with a general result on deflating subspaces of the ESP.

Lemma 6.3.1 *1. If the ESP has a proper $\mathbb{D} \setminus \{0\}$–deflating subspace to the right then it has a proper $\mathbb{C} \setminus \overline{\mathbb{D}}$–deflating subspace to the left of the same dimension.*

2. If the ESP has a proper $\mathbb{C} \setminus \overline{\mathbb{D}}$–deflating subspace to the right then it has a proper $\mathbb{D} \setminus \{0\}$–deflating subspace to the left of the same dimension.

Proof. We prove only the first assertion since the second follows similarly. Let $\mathcal{V}$ be a proper $\mathbb{D} \setminus \{0\}$–deflating subspace to the right of the ESP (6.22) with a basis matrix

$$V = \begin{bmatrix} V_1 \\ V_2 \\ V_3 \end{bmatrix} \begin{matrix} \}n \\ \}n \\ \}m \end{matrix} . \tag{6.24}$$

Then there exists a stable and invertible matrix S such that $M_\Sigma V S = N_\Sigma V$. Define

$$W := \begin{bmatrix} S^T V_2^T & V_1^T & V_3^T \end{bmatrix}.$$

From simple computations we get

$$S^{-T} W M_\Sigma = W N_\Sigma.$$

Moreover, since

$$W M_\Sigma = S^T \begin{bmatrix} V_2^T & -V_1^T & O \end{bmatrix}$$

as can be easily checked, it follows that $W M_\Sigma$ has full row rank because

$$\begin{bmatrix} V_1 \\ V_2 \end{bmatrix}$$

has full column rank. Thus $\mathcal{W} = \operatorname{Im} W$ is a proper $\mathbb{C} \setminus \overline{\mathbb{D}}$–deflating subspace to the left, which proves the first part of the theorem. ∎

Remark 6.3.2 In contrast to the continuous–time case (see Lemma 6.1.1), Lemma 6.3.1 is not an "if and only if" result, i.e., it does not hold if we interchange in its statement *deflating subspaces to the right* with *deflating subspaces to the left.* To see where this lack of symmetry originates from, assume that

$$W = \begin{bmatrix} W_1 & W_2 & W_3 \end{bmatrix}$$

is a row basis for a proper $\mathbb{C}\setminus\overline{\mathbb{D}}$–deflating subspace to the left $\mathcal{W}$ for the ESP(Σ). This means that $TWM_\Sigma = N_\Sigma V$ holds, where T is antistable (and therefore invertible), and WM_Σ has full row rank. It can be easily checked that

$$V := \begin{bmatrix} W_2^T \\ W_1^T T^T \\ W_3^T \end{bmatrix}$$

is a column basis matrix, $M_\Sigma V S = N_\Sigma V$ holds for $S = T^{-T}$, and S is stable and invertible. Moreover,

$$WM_\Sigma = \begin{bmatrix} W_1 & -T^{-1}W_2 & O \end{bmatrix}$$

has full row rank and thus

$$\begin{bmatrix} W_2^T T^{-T} \\ W_1^T \end{bmatrix}$$

has full column rank. Further we have

$$M_\Sigma V = \begin{bmatrix} I & O \\ O & -A^T \\ O & -B^T \end{bmatrix} \begin{bmatrix} W_2^T T^{-T} \\ W_1^T \end{bmatrix} T^T. \tag{6.25}$$

Since the two rightmost matrices in (6.25) have both full column rank their product has full column rank as well. It follows that the whole product in the right–hand side of (6.25) has full column rank provided that

$$\begin{bmatrix} I & O \\ O & -A^T \\ O & -B^T \end{bmatrix}$$

has full column rank as well. But this happens if the pair (A, B) is $\{0\}$–controllable. Thus we conclude that the converse of Lemma 6.3.1 is true as well if we assume additionally that the pair (A, B) is controllable at zero. Notice also that if the ESP is regular then the converse holds without any additional assumption.

We are ready now to give a characterization of the eigenstructure of a singular ESP.

Theorem 6.3.3 *For an arbitrary ESP the following hold:*

1. $n_r \le n_\ell$.
2. $\widetilde{n}_f^- = n_f^+$.
3. $\text{rank}_{\mathbb{R}(z)}(zM_\Sigma - N_\Sigma) = 2n + \rho$.
4. $n_\infty \ge \rho + \pi_0$.

Proof. 1. According to Theorem 1.11.3, there exists a proper $\mathbb{D}\setminus\{0\}$–deflating subspace to the right of dimension $n_r + \widetilde{n}_f^-$. By Lemma 6.3.1, there exists a proper $\mathbb{C}\setminus\overline{\mathbb{D}}$–deflating subspace to the left of the same dimension. Applying again Theorem 1.11.3, the dimension of a proper $\mathbb{C}\setminus\overline{\mathbb{D}}$–deflating subspace to the left cannot exceed $n_\ell + n_f^+$. Therefore $n_r + \widetilde{n}_f^- \le n_\ell + n_f^+$. In a similar manner, we obtain that $n_r + n_f^+ \le n_\ell + \widetilde{n}_f^-$. Adding these inequalities, we obtain $n_r \le n_\ell$.

2. For any $z \neq 0$ we have

$$\frac{1}{z}M_\Sigma - N_\Sigma = \begin{bmatrix} O & \frac{1}{z}I_n & O \\ I_n & O & O \\ O & O & I_m \end{bmatrix} (zM_\Sigma - N_\Sigma)^T \begin{bmatrix} O & \frac{1}{z}I_n & O \\ I_n & O & O \\ O & O & I_m \end{bmatrix} \tag{6.26}$$

from where it follows that $z \neq 0$ is in the spectrum of the ESP if and only if $\frac{1}{z}$ is in the spectrum of the ESP with the same multiplicity.

3. follows from the identity below which holds for any z for which $zI - A$ and $I - zA^T$ are both invertible,

$$M_\Sigma - zN_\Sigma = \mathbf{F}(z)\mathbf{G}(z)\mathbf{H}(z)$$

where

$$\begin{aligned} \mathbf{F}(z) &:= \begin{bmatrix} I_n & O & O \\ -Q(zI - A)^{-1} & I_n & O \\ -L^T(zI - A)^{-1} & -zB^T(I - zA^T)^{-1} & I_m \end{bmatrix}, \\ \mathbf{G}(z) &:= \begin{bmatrix} zI - A & O & O \\ O & I - zA^T & O \\ O & O & \mathbf{\Pi}_\Sigma(z) \end{bmatrix}, \\ \mathbf{H}(z) &:= \begin{bmatrix} I_n & O & -(zI - A)^{-1}B \\ O & I_n & -(I - zA^T)^{-1}(L + Q(zI - A)^{-1}B) \\ O & O & I_m \end{bmatrix}. \end{aligned}$$

4. Denote by $\text{no}\,(\mathbf{K}(z))$ the multiplicity of $z = 0$ as a zero of the rational matrix $\mathbf{K}(z)$. From the identity

$$(M_\Sigma - zN_\Sigma)\mathbf{X}(z) = \mathbf{Y}(z)z(zM_\Sigma - N_\Sigma)^T\mathbf{Z}(z),$$

where

$$\begin{aligned} \mathbf{X}(z) &:= \begin{bmatrix} zI_n & O & O \\ O & zI_n & O \\ O & O & I_m \end{bmatrix}, \\ \mathbf{Y}(z) &:= \begin{bmatrix} O & I_n & O \\ I_n & (1 - z)Q(I - zA)^{-1} & O \\ O & (z - 1)L^T(I - zA)^{-1} & I_m \end{bmatrix}, \\ \mathbf{Z}(z) &:= \begin{bmatrix} O & I_n & O \\ I_n & O & (z - 1)(I - zA)^{-1}B \\ O & O & I_m \end{bmatrix}, \end{aligned}$$

we deduce that

$$\text{no}\,(z(zM_\Sigma - N_\Sigma)^T) \leq \text{no}\,(M_\Sigma - zN_\Sigma) + \text{no}\,(\mathbf{X}(z)).$$

Combining this with

$$\text{no}\,(\mathbf{X}(z)) = 2n, \quad n_\infty = \text{no}\,(M_\Sigma - zN_\Sigma)$$

and the equality

$$\text{no}\,(z(zM_\Sigma - N_\Sigma)^T) = \pi_0 + \text{rank}_{\mathbb{R}(z)}(zM_\Sigma - N_\Sigma) = \pi_0 + 2n + \rho,$$

that is a consequence of the Smith–McMillan form and point 3, we deduce that

$$\pi_0 + 2n + \rho \leq n_\infty + 2n$$

form which the conclusion follows. ■

Remark 6.3.4 The inequalities in Theorem 6.3.3 can be strict as we illustrate with the following simple example. Consider the Popov triplet $\Sigma = (0, 0; 0, 1, 0)$. Then the associated ESP is

$$zM_\Sigma - N_\Sigma = z \begin{bmatrix} 1 & 0 & 0 \\ 0 & 0 & 0 \\ 0 & 0 & 0 \end{bmatrix} - \begin{bmatrix} 0 & 0 & 0 \\ 0 & -1 & 1 \\ 1 & 0 & 0 \end{bmatrix}$$

which is strictly equivalent to

$$z \begin{bmatrix} 1 & 0 & 0 \\ 0 & 0 & 0 \\ 0 & 0 & 0 \end{bmatrix} - \begin{bmatrix} 0 & 0 & 0 \\ 1 & 0 & 0 \\ 0 & 0 & 1 \end{bmatrix}.$$

This shows that $n_r = 0, n_\ell = 1, n_\infty = 1, \pi_0 = 0$ and $\rho = 0$, because $\mathbf{\Pi}_\Sigma(z) = 0$. Hence $n_\ell > n_r$ and $n_\infty > \rho + \pi_0$.

Remark 6.3.5 If we assume additionally that the pair (A, B) is $\{0\}$–controllable (see also Remark 6.3.2) then the inequalities at points 1 and 4 of Theorem 6.3.3 become equalities. We show first $n_r = n_\ell$. Indeed, with Lemma 6.3.1 and Remark 6.3.2 we get, similarly as in the proof of Theorem 6.3.3, that

$$\begin{aligned} n_\ell + \widetilde{n}_f^- &\leq n_r + n_f^+, \\ n_\ell + n_f^+ &\leq n_r + \widetilde{n}_f^-. \end{aligned}$$

Adding these inequalities results in $n_\ell \leq n_r$. This together with point 1 in Theorem 6.3.3 shows that $n_r = n_\ell$. We show further that $n_\infty = \rho + \pi_0$. Since the pair (A, B) is $\{0\}$–controllable there exists a feedback matrix $\widetilde{F}$ such that $\widetilde{A} = A + BF$ is invertible. Let $\widetilde{\Sigma} = (\widetilde{A}, \widetilde{B}; \widetilde{Q}, \widetilde{L}, \widetilde{R})$ denote the Popov triplet which is $\widetilde{F}$–equivalent to Σ, and let $zM_{\widetilde{\Sigma}} - N_{\widetilde{\Sigma}}$ and $\mathbf{\Pi}_{\widetilde{\Sigma}}(z)$ be the ESP and the Popov function, respectively, associated with $\widetilde{\Sigma}$. For any z for which $zI - \widetilde{A}^T$ and $I - z\widetilde{A}$ are both invertible we can prove by a direct check the following identity:

$$M_{\widetilde{\Sigma}} - zN_{\widetilde{\Sigma}} = \widetilde{\mathbf{X}}(z)\widetilde{\mathbf{Y}}(z)\widetilde{\mathbf{Z}}(z) \tag{6.27}$$

where

$$\widetilde{\mathbf{X}}(z) := \begin{bmatrix} O & I & O \\ I & (1-z)\widetilde{Q}(I-z\widetilde{A})^{-1} & O \\ \widetilde{B}^T(zI-\widetilde{A}^T)^{-1} & -[\widetilde{B}^T(zI-\widetilde{A}^T)^{-1}\widetilde{Q}+z\widetilde{L}^T](I-z\widetilde{A})^{-1} & I \end{bmatrix},$$
$$\widetilde{\mathbf{Y}}(z) := \begin{bmatrix} zI-\widetilde{A} & O & O \\ O & I-z\widetilde{A}^T & O \\ O & O & z\mathbf{\Pi}_{\widetilde{\Sigma}}(z) \end{bmatrix},$$
$$\widetilde{\mathbf{Z}}(z) := \begin{bmatrix} -(zI-\widetilde{A}^T)^{-1}\widetilde{Q} & I & -(zI-\widetilde{A}^T)^{-1}[(z-1)\widetilde{Q}(I-z\widetilde{A})^{-1}\widetilde{B}+\widetilde{L}] \\ I & O & -(I-z\widetilde{A})^{-1}\widetilde{B} \\ O & O & I \end{bmatrix}. \tag{6.28}$$

Since the ESPs associated with Σ and $\widetilde{\Sigma}$ are strictly equivalent, and $\widetilde{A}$ is invertible, we get from (6.27) that

$$n_\infty = \mathrm{no}\,(M_\Sigma - zN_\Sigma) = \mathrm{no}\,(M_{\widetilde{\Sigma}} - zN_{\widetilde{\Sigma}}) = \rho + \mathrm{no}\,(\mathbf{\Pi}_{\widetilde{\Sigma}}(z)). \tag{6.29}$$

On the other side, since $zM_{\widetilde{\Sigma}} - N_{\widetilde{\Sigma}}$ is the transmission pencil of the Popov function $\mathbf{\Pi}_{\widetilde{\Sigma}}$, we have also

$$\mathrm{no}\,(\mathbf{\Pi}_{\widetilde{\Sigma}}(z)) = \mathrm{no}\,(zM_{\widetilde{\Sigma}} - N_{\widetilde{\Sigma}}) = \mathrm{no}\,(zM_\Sigma - N_\Sigma) = \pi_0$$

which together with (6.29) proves that $n_\infty = \rho + \pi_0$.

Our next result says that a proper $\mathbb{D}$ ($\mathbb{C}\setminus\overline{\mathbb{D}}$)–deflating subspace of dimension n of the ESP is maximal and it exists if and only if certain types of dichotomy of the ESP holds.

Theorem 6.3.6 *1. The dimension of a proper $\mathbb{D}$–deflating subspace of the ESP cannot exceed n. This upper bound is attained if and only if $n_f^0 = 0$, $n_\infty = \rho + \pi_0$, and $n_r = n_\ell$.*

2. The dimension of a proper $\mathbb{C}\setminus\overline{\mathbb{D}}$–deflating subspace of the ESP cannot exceed n. This upper bound is attained if and only if $n_f^0 = 0$, $n_\infty = \rho + \pi_0$, $n_r = n_\ell$, and $\pi_0 = 0$.

Proof. For any matrix pencil we have

$$\mathrm{rank}_{\,\mathbb{R}(z)}(zM_\Sigma - N_\Sigma) = n_r + n_\ell + n_f^- + n_f^+ + n_f^0 + n_\infty$$

and from point 3 of Theorem 6.3.3, we deduce further

$$n_r + n_\ell + n_f^- + n_f^+ + n_f^0 + n_\infty = 2n + \rho.$$

Using points 1, 2 and 4 of Theorem 6.3.3, and the fact that $n_f^- = \widetilde{n}_f^- + \pi_0$, we obtain that $2n + \rho \geq 2(n_r + n_f^-) + \rho + n_f^0$ and hence $n \geq n_r + n_f^-$. Equality is possible only when all the inequalities that we used become equalities, i.e., $n_r = n_\ell$, $n_\infty = \rho + \pi_0$, and $n_f^0 = 0$. Invoking Theorem 1.11.3, the first part of the theorem is proved.

The second part follows immediately, if we use the fact that $n_f^- = n_f^+ + \pi_0$. It is also clear that equality, in this case, requires additionally that $\pi_0 = 0$. ■

Compare Theorem 6.3.6 with its continuous–time counterpart Theorem 6.1.3, and notice the presence of the extra condition on the Kronecker indices and the distinction made between the stable and antistable cases. These are all characteristic features of the discrete–time case.

Definition 6.3.7 *An arbitrary ESP is called* dichotomic *if* $n_r = n_\ell$, $n_\infty = \rho + \pi_0$ *and* $n_f^0 = 0$, *and* strongly dichotomic *if in addition* $\pi_0 = 0$.

In particular, Theorem 6.3.6 shows that the ESP is (strongly) dichotomic if and only if

$$M_\Sigma V S = N_\Sigma V \tag{6.30}$$

holds with $M_\Sigma V$ of full column rank, where

$$V = \begin{bmatrix} V_1 \\ V_2 \\ V_3 \end{bmatrix} \begin{matrix} \}n \\ \}n \\ \}m \end{matrix} \tag{6.31}$$

is a basis matrix for the n–dimensional proper $\mathbb{D}$ ($\mathbb{C} \setminus \overline{\mathbb{D}}$)–deflating subspace, and $S \in \mathbb{R}^{n\times n}$ is a stable (antistable) matrix.

We introduce now the notion of disconjugacy of a deflating subspace of the ESP.

Definition 6.3.8 *A deflating subspace* $\mathcal{V}$ *of an arbitrary ESP is called* disconjugate *if it admits a basis matrix* V *partitioned as in (6.31), with* V_1 *of full column rank.*

Notice that a disconjugate deflating subspace is automatically proper. Here follows the counterpart of Theorem 6.1.6.

Theorem 6.3.9 *If the ESP has a maximal proper* $\mathbb{D}$ *(*$\mathbb{C} \setminus \overline{\mathbb{D}}$*)–deflating subspace* $\mathcal{V}$ *which is disconjugate, then* each *maximal proper* $\mathbb{D}$ *(*$\mathbb{C} \setminus \overline{\mathbb{D}}$*)–deflating subspace is disconjugate.*

Proof. We consider only the case of a $\mathbb{D}$–deflating subspace. Let $\mathcal{V}$ and $\overline{\mathcal{V}}$ be two maximal $\mathbb{D}$–deflating subspaces and let

$$V = \begin{bmatrix} V_1 \\ V_2 \\ V_3 \end{bmatrix}, \quad \overline{V} = \begin{bmatrix} \overline{V}_1 \\ \overline{V}_2 \\ \overline{V}_3 \end{bmatrix}$$

be some basis matrices for $\mathcal{V}$ and $\overline{\mathcal{V}}$, respectively. Then

$$M_\Sigma V = \begin{bmatrix} V_1 \\ -A^T V_2 \\ -B^T V_2 \end{bmatrix}, \quad M_\Sigma \overline{V} = \begin{bmatrix} \overline{V}_1 \\ -A^T \overline{V}_2 \\ -B^T \overline{V}_2 \end{bmatrix}$$

have full column rank and, by Theorem 1.11.4, their images coincide. Hence there exists an invertible constant matrix G such that

$$\begin{bmatrix} V_1 \\ -A^T V_2 \\ -B^T V_2 \end{bmatrix} = \begin{bmatrix} \overline{V}_1 \\ -A^T \overline{V}_2 \\ -B^T \overline{V}_2 \end{bmatrix} G.$$

Thus if V_1 has full column rank, then $\overline{V}_1$ has too. ■

The previous result justifies the following definition.

Definition 6.3.10 *An arbitrary ESP is called stable (antistable)* disconjugate *if it is (strongly) dichotomic and has a disconjugate $\mathbb{D}$ ($\mathbb{C}\backslash\overline{\mathbb{D}}$)-deflating subspace of dimension n.*

We give now a property of deflating subspaces of the ESP that will be important in proving the symmetry of X in a solution to the discrete–time algebraic Riccati system.

Proposition 6.3.11 *If V is a basis matrix as in (5.9) of a proper $\mathbb{D}$ (or $\mathbb{C}\setminus\mathbb{D}$)-deflating subspace of an arbitrary ESP then*

$$V_1^T V_2 = V_2^T V_1. \tag{6.32}$$

Proof. The proof is entirely similar to the regular case (see Proposition 5.4.6) and is therefore omitted. ■

6.4 The generalized discrete–time Riccati system and the ESP

In this section we introduce the generalized discrete–time algebraic Riccati system (GDTARS), which incorporates as special cases the discrete–time algebraic Riccati equation and system, and we characterize its stabilizing and antistabilizing solutions in terms of proper deflating subspaces of the associated ESP.

Definition 6.4.1 *Let $\Sigma = (A, B; Q, L, R)$ be a Popov triplet where $A \in \mathbb{R}^{n\times n}, B \in \mathbb{R}^{n\times m}$. The* generalized discrete–time algebraic Riccati system (GDTARS) *associated with Σ is*

$$\begin{bmatrix} A^TXA - X + Q & A^TXB + L \\ L^T + B^TXA & R + B^TXB \end{bmatrix} \begin{bmatrix} I \\ F \end{bmatrix} V = 0, \tag{6.33}$$

where $X \in \mathbb{R}^{n\times n}, V \in \mathbb{R}^{n\times r}$ has full column rank with $r \leq n$, r is unfixed, and F is such that $(A + BF)V = VS$ for an appropriate matrix $S \in \mathbb{R}^{r\times r}$.

A quadruple (X, F, r, V) for which (6.33) holds and $V^TXV = V^TX^TV$, is called a V–symmetric solution. *A V-symmetric solution is called* stabilizing (antistabilizing) *if in addition S is stable (antistable).*

If (X, F, r, V) is a stabilizing solution to the GDTARS then $\mathcal{V} = \operatorname{Im} V$ is an (A, B)–invariant subspace, F is such that $(A + BF)\mathcal{V} \subset \mathcal{V}$ and $\Lambda(A + BF)|\mathcal{V} \subset \mathbb{D}$.

If V is invertible we may assume that $V = I$ and (6.33) reduces to

$$\begin{aligned} A^TXA - X + Q \quad &+ \quad (A^TXB + L)F &= 0, \\ L^T + B^TXA \quad &+ \quad (R + B^TXB)F &= 0, \end{aligned} \tag{6.34}$$

which is exactly the DTARS(Σ) in (3.108). In this case a stabilizing (antistabilizing) solution is a pair (X, F) satisfying (6.34) with $X = X^T$ and $A+BF$ stable (antistable). Further, for $X = X^T$ we get

$$\begin{aligned} A^TXA - X \quad &- \quad F^T(R + B^TXB)F + Q &= 0, \\ L^T + B^TXA \quad &+ \quad (R + B^TXB)F &= 0, \end{aligned} \tag{6.35}$$

or, equivalently,

$$\begin{aligned} A^TXA - X - (A^TXB + L)(R + B^TXB)^{\dagger}(B^TXA + L^T) + Q &= 0, \\ \operatorname{Ker}(R + B^TXB) \subset \operatorname{Ker}(A^TXB + L), & \end{aligned}$$

which is known as the constrained discrete–time algebraic Riccati equation. If in addition $R + B^T XB$ is assumed invertible we obtain

$$A^T XA - X - (A^T XB + L)(R + B^T XB)^{-1}(B^T XA + L^T) + Q = 0 \tag{6.36}$$

which is the DTARE(Σ).

The following uniqueness result holds for the stabilizing (antistabilizing) solution to the GDTARS.

Proposition 6.4.2 *If (X, F, r, V) and $(\widetilde{X}, \widetilde{F}, r, V)$ are two V–symmetric stabilizing (antistabilizing) solutions to (6.33) then $V^T XV = V^T \widetilde{X} V$.*

Proof. We prove for the stabilizing solution only since antistabilizing solutions can be treated analogously. We have $(A + BF)V = VS$ and $(A + B\widetilde{F})V = V\widetilde{S}$, where S and $\widetilde{S}$ are both stable. Using this we rewrite the first equation (6.33) as

$$V^T A^T XVS - V^T XV + V^T QV + V^T LFV \quad = \quad 0. \tag{6.37}$$

Using the second equation (6.33) we get

$$\begin{aligned} 0 &= V^T \widetilde{F}^T (L^T + B^T XA)V + V^T \widetilde{F}^T (R + B^T XB)FV \\ &= V^T \widetilde{F}^T B^T X(A + BF)V + V^T \widetilde{F}^T L^T V + V^T F^T RFV \\ &= V^T \widetilde{F}^T B^T XVS + V^T \widetilde{F}^T L^T V + V^T \widetilde{F}^T RFV. \end{aligned} \tag{6.38}$$

Adding (6.37) and the last equality in (6.38) yields

$$\widetilde{S}^T V^T XVS - V^T XV + V^T (LF + \widetilde{F}^T L^T)V + V^T \widetilde{F}^T RFV + V^T QV = 0. \tag{6.39}$$

If we interchange in the above reasoning the role of X and F with the role of $\widetilde{X}$ and $\widetilde{F}$, respectively, we get similarly to (6.39)

$$S^T V^T \widetilde{X} V\widetilde{S} - V^T \widetilde{X} V + V^T (L\widetilde{F} + F^T L^T)V + V^T F^T R\widetilde{F} V + V^T QV = 0. \tag{6.40}$$

As $V^T \widetilde{X} V = V^T \widetilde{X}^T V$, we get with (6.40) transposed

$$\widetilde{S}^T V^T \widetilde{X} VS - V^T \widetilde{X} V + V^T (\widetilde{F}^T L^T + LF)V + V^T \widetilde{F}^T RFV + V^T QV = 0 \tag{6.41}$$

and subtracting (6.41) from (6.39) this results in

$$\widetilde{S}^T (V^T XV - V^T \widetilde{X} V)S - (V^T XV - V^T \widetilde{X} V) = 0 \tag{6.42}$$

which is a Stein equation in the unknown $V^T XV - V^T \widetilde{X} V$. Since S and $\widetilde{S}$ are both stable this implies $V^T XV - V^T \widetilde{X} V = 0$ is the unique solution. ∎

We state now the main result of the section.

Theorem 6.4.3 *Let $\Sigma = (A, B; Q, L, R)$ be a Popov triplet. The following assertions are equivalent:*

1. The GDTARS(Σ) in (6.33) has a stabilizing (antistabilizing) solution (X, F, r, V).

2. The ESP(Σ) in (6.22) has a disconjugate proper $\mathbb{D}$ ($\mathbb{C} \setminus \overline{\mathbb{D}}$)–deflating subspace $\mathcal{V}$ of dimension r.

If assertion 2 holds, let V be a basis matrix of $\mathcal{V}$ partitioned as in (6.31). Then a stabilizing (antistabilizing) solution is (X, F, r, V_1) where

$$X = V_2 V_1^\dagger, \quad F = V_3 V_1^\dagger, \tag{6.43}$$

and $V_1^\dagger$ is an inverse to the left of V_1.

Proof. We prove for the stabilizing solution only since the antistable case follows by duality.

1. $\Rightarrow$ 2. Let (X, F, r, V_1) be a stabilizing solution to (6.33). Denoting

$$V = \begin{bmatrix} V_1 \\ V_2 \\ V_3 \end{bmatrix} := \begin{bmatrix} V_1 \\ XV_1 \\ FV_1 \end{bmatrix}$$

it can be easily checked that the equations

$$\begin{array}{rcl} (A+BF)V_1 & = & V_1 S, \\ (Q-X+LF)V_1 & = & -A^T X V_1 S, \\ (L^T+RF)V_1 & = & -B^T X V_1 S, \end{array} \tag{6.44}$$

hold for a stable matrix S or, more compactly, the ESP(Σ) in (6.22) satisfies $M_\Sigma V S = N_\Sigma V$. Now $M_\Sigma V$ has full column rank since V_1 has full column rank. Thus $\mathcal{V} = \operatorname{Im} V$ is a disconjugate proper $\mathbb{D}$–deflating subspace of dimension r for the ESP.

2. $\Rightarrow$ 1. Let $\mathcal{V}$ be a disconjugate proper $\mathbb{D}$–deflating subspace of dimension r, with a basis matrix V partitioned as in (6.31), for the ESP(Σ) in (6.22). Then $M_\Sigma V S = N_\Sigma V$ holds for a stable matrix S or, in explicit form,

$$\begin{array}{rrrcl} AV_1 & & +BV_3 & = & V_1 S, \\ QV_1 & -V_2 & +LV_3 & = & -A^T V_2 S, \\ L^T V_1 & +B^T V_2 & +RV_3 & = & -B^T V_2 S. \end{array} \tag{6.45}$$

Define X and F through (6.43). Then from (6.45) we arrive at (6.44). Eliminating $V_1 S$ in the first two equations (6.44), we get (6.33) written with $V = V_1$. Finally, from Proposition 6.4.2 we have

$$V_1^T X V_1 = V_1^T V_2 V_1^\dagger V_1 = V_1^T V_2 = V_2^T V_1 = V_1^T (V_1^T)^\dagger V_2^T V_1 = V_1^T X^T V_1$$

form where we conclude that (X, F, r, V_1) is a V_1 symmetric solution which is stabilizing, as S is stable. This concludes the proof. ∎

Theorem 6.4.3 has the following important consequence.

Theorem 6.4.4 *Let $\Sigma = (A, B; Q, L, R)$ be a Popov triplet. The following assertions are equivalent:*

1. *The DTARS(Σ) in (6.35) has a stabilizing (antistabilizing) solution (X, F).*
2. *The ESP(Σ) in (6.22) is dichotomic and stable (antistable) disconjugate.*

If assertion 2 holds, let V be a basis matrix partitioned as in (6.31) of an n–dimensional maximal proper $\mathbb{D}$ $(\mathbb{C} \setminus \overline{\mathbb{D}})$–deflating subspace of the ESP(Σ). Then a stabilizing (antistabilizing) solution is (X, F) where

$$X = V_2 V_1^{-1}, \quad F = V_3 V_1^{-1}. \tag{6.46}$$

As in the continuous–time case, the nonuniqueness of the maximal deflating subspace does not affect the computational procedures for the solutions of the discrete–time algebraic Riccati system.

6.5 Numerical algorithms

In this section we present a general–purpose numerically sound algorithm for computing (proper) deflating subspaces with a specified spectrum for an arbitrary matrix pencil. The key tools are the computation of an ordered generalized Schur form of a regular pencil and the computation of the Kronecker–like form of an arbitrary pencil. The algorithm for computing deflating subspaces can be applied further to the computation of solutions to various Riccati equations and systems, either in the continuous or discrete–time setting.

6.5.1 Basic algorithms

The Schur form of a matrix

An alternative form to the Jordan canonical form which displays the eigenvalues of the matrix but not the detailed information about the elementary Jordan indices is the Schur form. It may be obtained by a restricted class of similarity matrices U that satisfy $U^*U = I$, i.e., unitary matrices. As for the Jordan canonical form, we obtain real or complex Schur forms by taking U with real or complex elements. Precisely, a matrix $A \in \mathbb{C}^{n\times n}$ can always be reduced by a unitary similarity to a triangular form

$$U^*AU = A_S := \begin{bmatrix} \lambda_1 & \cdots & \star \\ & \ddots & \vdots \\ 0 & & \lambda_n \end{bmatrix} \tag{6.47}$$

where the diagonal elements $\lambda_1, \ldots, \lambda_n$ are the eigenvalues of A. The matrix A_S is called a *complex Schur form* of A. Alternatively, if $A \in \mathbb{R}^{n\times n}$ and U is restricted to have real elements, i.e., U is now orthogonal $U^TU = I$, we obtain the *real Schur form*

$$U^TAU = A_S := \begin{bmatrix} A_{11} & \cdots & A_{1k} \\ & \ddots & \vdots \\ O & & A_{kk} \end{bmatrix} \tag{6.48}$$

where the diagonal blocks $A_{11}, \ldots, A_{kk}$ are of dimension 1×1 or 2×2. Each 1×1 block contains a real eigenvalue of A, and each 2×2 block contains a pair of complex conjugate eigenvalues of A. Moreover, the eigenvalues on the diagonal in (6.47) or (6.48) can be arranged in any order. Schur forms have the advantage over Jordan canonical forms that they can be obtained by using numerically reliable algorithms. This follows from the use of unitary similarity matrices U for which $||U|| = 1$. The Schur forms are obtained by using a recursive scheme known as the Schur algorithm which has variants for the real and complex cases. An ordered Schur form may be obtained by first bringing the matrix to the Schur form and then interchanging eigenvalues on the main diagonal by using updating orthogonal (unitary) similarity transformations.

The generalized Schur form of a regular pencil

As for the matrix case, there exist numerically appealing alternative forms to the Weierstrass canonical forms that display the generalized eigenvalues of a regular pencil and which can be obtained by unitary (orthogonal) transformations only. Precisely, a regular matrix pencil $\lambda M - N$ with $M, N \in \mathbb{C}^{n\times n}$, can always be reduced by unitary strict equivalence transformations $Q, Z \in \mathbb{C}^{n\times n}$, to an upper triangular form

$$Q(\lambda M - N)Z = \lambda M_S - N_S := \lambda \begin{bmatrix} m_1 & \cdots & \star \\ & \ddots & \vdots \\ 0 & & m_n \end{bmatrix} - \begin{bmatrix} n_1 & \cdots & \star \\ & \ddots & \vdots \\ 0 & & n_n \end{bmatrix}, \qquad (6.49)$$

where the diagonal pairs (m_i, n_i), $(i = 1, \ldots, n)$, determine the generalized eigenvalues of $\lambda M - N$ as follows: $\lambda_i = n_i/m_i$ if $m_i \neq 0$ and $\lambda_i = \infty$ if $m_i = 0$. The pencil $\lambda M_S - N_S$ is called a *complex generalized Schur form* of $\lambda M - N$.

Alternatively, the pencil $\lambda M - N$, with $M, N \in \mathbb{R}^{n\times n}$, can always be reduced by orthogonal strict equivalence transformations $Q, Z \in \mathbb{R}^{n\times n}$ to a block upper triangular form

$$Q(\lambda M - N)Z = \lambda M_S - N_S := \lambda \begin{bmatrix} M_{11} & \cdots & M_{1k} \\ & \ddots & \vdots \\ O & & M_{kk} \end{bmatrix} - \begin{bmatrix} N_{11} & \cdots & N_{1k} \\ & \ddots & \vdots \\ O & & N_{kk} \end{bmatrix} \qquad (6.50)$$

where the diagonal pencils $\lambda M_{ii} - N_{ii}$, $(i = 1, \ldots, k)$, are of dimension 1×1 or 2×2, each 1×1 pencil contains either a finite real or an infinite generalized eigenvalue of $\lambda M - N$ and each 2×2 pencil contains a pair of finite complex conjugate generalized eigenvalues of $\lambda M - N$. Moreover, each M_{ii} is upper triangular. The pencil $\lambda M_S - N_S$ is called a *generalized real Schur form* of $\lambda M - N$. The generalized eigenvalues on the diagonal in (6.49) or (6.50) can be arranged in any order. The generalized Schur forms are obtained by using a recursive scheme known as the *QZ algorithm* which has variants for the real and complex cases. An ordered generalized Schur form may be obtained by first bringing the matrix to the generalized Schur form and then interchanging the generalized eigenvalues on the main diagonal by using updating orthogonal (unitary) transformation matrices.

The Kronecker–like form of a general pencil

As for the regular pencil case, instead of computing the Kronecker canonical form which is a poor numerical tool, it is recommended to use the so–called Kronecker–like form of an arbitrary (possibly singular) pencil. The Kronecker–like form is computed by using orthogonal (unitary) matrices only and displays essentially the same information as the Kronecker canonical form. Precisely, any matrix pencil $\lambda M - N$ with $M, N \in \mathbb{C}^{n\times n}$, can always be reduced by unitary equivalence transformations $Q, Z \in \mathbb{C}^{n\times n}$ to the block upper triangular form, called *Kronecker–like form*,

$$\begin{aligned} Q(\lambda M - N)Z & \\ = \ & \lambda M_K - N_K \\ := \ & \begin{bmatrix} \lambda M_\epsilon - N_\epsilon & \star & \star & \star \\ O & \lambda M_\infty - N_\infty & \star & \star \\ O & O & \lambda M_f - N_f & \star \\ O & O & O & \lambda M_\eta - N_\eta \end{bmatrix}, \end{aligned} \qquad (6.51)$$

where:

1. The regular part of the pencil is determined by $\lambda M_f - N_f$ and $\lambda M_\infty - N_\infty$ which are square and regular, and contain the finite and infinite generalized eigenvalues, respectively, M_f and N_∞ are invertible, and M_∞ is nilpotent.
2. The singular part of the pencil is determined by $\lambda M_\epsilon - N_\epsilon$ which contains the right Kronecker indices and has full row rank for all $\lambda \in \mathbb{C}$, and M_ϵ has full row rank, and by $\lambda M_\eta - N_\eta$ which contains the left Kronecker indices and has full column rank for all $\lambda \in \mathbb{C}$, and M_η has full column rank.

The main step in computing the Kronecker–like form of a singular pencil is the so–called staircase algorithm. The staircase algorithm uses a sequence of alternate rank decompositions on M and N to obtain orthogonal matrices Q and Z such that

$$Q(\lambda M - N)Z = \begin{bmatrix} \lambda M_{\epsilon\infty} - N_{\epsilon\infty} & \star \\ O & \lambda M_{f\eta} - N_{f\eta} \end{bmatrix}$$

$$= \underbrace{\begin{bmatrix} -N_{11} & \lambda M_{12} - N_{12} & \cdots & \lambda M_{1,\ell} - N_{1,\ell} & \lambda M_{1,\ell+1} - N_{1,\ell+1} \\ O & -N_{22} & \cdots & \lambda M_{2,\ell} - N_{2,\ell} & \lambda M_{2,\ell+1} - N_{2,\ell+1} \\ \vdots & \vdots & \ddots & \vdots & \vdots \\ O & O & \cdots & -N_{\ell,\ell} & \lambda M_{\ell,\ell+1} - N_{\ell,\ell+1} \\ O & O & \cdots & O & \lambda M_{\ell+1} - N_{\ell+1} \end{bmatrix}}_{\mu_1 \quad \mu_2 \quad \cdots \quad \mu_\ell \quad n_{\ell+1}} \begin{matrix} \}\nu_1 \\ \}\nu_2 \\ \\ \}\nu_\ell \\ \}m_{\ell+1} \end{matrix} \tag{6.52}$$

with $\lambda M_{\ell+1} - N_{\ell+1} := \lambda M_{f\eta} - N_{f\eta}$ and where:

1. $\lambda M_{\epsilon\infty} - N_{\epsilon\infty}$ contains only right Kronecker indices and infinite generalized eigenvalues and $\lambda M_{f\eta} - N_{f\eta}$ contains only left Kronecker indices and finite generalized eigenvalues.
2. $M_{\ell+1}$ has full column rank.
3. The blocks N_{ii} have full row rank $\nu_i, (i = 1, \cdots, \ell)$.
4. The blocks $M_{i-1,i}$ have full column rank $\mu_i, (i = 2, \cdots, \ell)$.

With $\mu_{\ell+1} = 0$ and defining

$$\begin{aligned} e_i &:= \mu_i - \nu_i, (i = 1, \cdots, \ell), \\ d_i &:= \nu_i - \mu_{i+1}, (i = 1, \cdots, \ell), \end{aligned}$$

we have for the decomposition (6.52)

$$0 \leq \text{rank}\,(M_{\ell+1}) = n_{\ell+1} \leq m_{\ell+1}, \ell \geq 0,$$

$$\mu_1 \geq \nu_1 \geq \mu_2 \geq \cdots \geq \mu_\ell \geq \nu_\ell \geq \nu_{\ell+1} = 0.$$

The indices e_i and d_i completely determine the right Kronecker structure and the eigenstructure of the infinite eigenvalue as follows:

1. There are d_i nilpotent Jordan blocks of dimension i, $(i = 1, \cdots, \ell)$ corresponding to the infinite eigenvalue (see (1.30)).
2. There are e_i elementary right Kronecker blocks of size $(i-1) \times i, (i = 1, \cdots, \ell)$.

We show now how the staircase algorithm can be used to obtain the generalized Schur form of an arbitrary pencil. To this end we introduce the *pertranspose* A^P of an $m \times n$ matrix A which is by definition

$$A^P := P_n A^T P_m,$$

where for $k \in \mathbb{N}$, P_k denotes the $k \times k$ permutation matrix

$$P_k := \begin{bmatrix} 0 & 0 & \cdots & 0 & 1 \\ 0 & 0 & \cdots & 1 & 0 \\ & & \vdots & & \\ 0 & 1 & \cdots & 0 & 0 \\ 1 & 0 & \cdots & 0 & 0 \end{bmatrix}.$$

For the pertranspose we can prove the following simple but useful properties.

1. $(A^P)^P = A$.
2. $(AB)^P = B^P A^P$.
3. $\begin{bmatrix} A_{11} & A_{12} \\ O & A_{22} \end{bmatrix}^P = \begin{bmatrix} A_{22}^P & A_{12}^P \\ O & A_{11}^P \end{bmatrix}$.

We explain now the computation of the Kronecker–like form based on the staircase algorithm. Given a pencil $\lambda M - N$ apply first the staircase algorithm to bring it to the form (6.52). Further, apply twice the staircase algorithm to the petransposed diagonal subpencils in (6.52) $\lambda M_{\epsilon\infty}^P - N_{\epsilon\infty}^P$ and $\lambda M_{f\eta}^P - N_{f\eta}^P$, respectively, and pertranspose back the result to get (6.51). A better approach is to use two special customized staircase algorithms that fully exploit the particular structure of the diagonal subpencils in (6.52) (see the references at the end of this chapter).

Assignment of generalized eigenvalues by feedback

Consider the regular pencil $\lambda M - N$, where $M, N \in \mathbb{R}^{n\times n}$ and let $B \in \mathbb{R}^{n\times m}$. Suppose that for the triplet (M, N, B) all finite eigenvalues are controllable which means

$$\text{rank} \begin{bmatrix} \lambda M - N & B \end{bmatrix} = n, \quad \forall \lambda \in \mathbb{C}.$$

It follows that there exists a feedback matrix $F \in \mathbb{R}^{m\times n}$ such that the finite generalized eigenvalues of $\lambda M - N - BF$ can be assigned to arbitrary locations while keeping the infinite spectrum unchanged with respect to $\lambda M - N$.

We describe here a numerically reliable algorithm that computes the real feedback matrix F in a recursive fashion, such that at each step one eigenvalue (or a pair of complex conjugated eigenvalues) in a "bad" location is moved to a desired position while leaving the rest in place. Let $\mathbb{C} = \mathbb{C}_b \cup \mathbb{C}_g$ be a disjoint partition of the complex plane in two symmetric (with respect to the imaginary axis) sets, the good and the bad one.

We handle simultaneously the case of moving a single real eigenvalue and the case of moving a pair of complex conjugated eigenvalues by taking in the algorithm below k=1 and k=2, respectively.

Algorithm 1

1. Use an ordered QZ algorithm to determine Q and Z such that

$$\begin{aligned}
&\lambda\overline{M} - \overline{N} \\
&:= Q(\lambda M - N)Z \\
&= \lambda \begin{bmatrix} M_1 & \star \\ O & M_2 \end{bmatrix} - \begin{bmatrix} N_1 & \star \\ O & N_2 \end{bmatrix} \}n_\infty \\
&= \lambda \begin{bmatrix} M_1 & \star & \star \\ O & M_{21} & \star \\ O & O & M_{22} \end{bmatrix} - \begin{bmatrix} N_1 & \star & \star \\ O & N_{21} & \star \\ O & O & N_{22} \end{bmatrix} \begin{matrix} \}n_\infty \\ \}n_1 \\ \}n_2 \end{matrix}
\end{aligned} \tag{6.53}$$

where $\lambda M_1 - N_1$ contains only infinite generalized eigenvalues, and the pencils $\lambda M_{21} - N_{21}$ and $\lambda M_{22} - N_{22}$ contain only finite generalized eigenvalues located in $\mathbb{C}_g$ and $\mathbb{C}_b$, respectively. The number of generalized eigenvalues to be assigned is therefore n_2. Update

$$B \leftarrow QB = \begin{bmatrix} B_1 \\ B_2 \end{bmatrix} = \begin{bmatrix} B_1 \\ B_{21} \\ B_{22} \end{bmatrix} \begin{matrix} \}n_\infty \\ \}n_1 \\ \}n_2 \end{matrix} . \tag{6.54}$$

From a numerical viewpoint, a better approach to obtain (6.53) is to separate first the finite and infinite eigenvalues with the staircase algorithm and apply then an ordered QZ algorithm to the subpencil $\lambda M_2 - N_2$ only. This last approach avoids the ill conditioned "computation" of the infinite eigenvalues performed by the QZ algorithm in the case of multiple infinite eigenvalues. Set $t = n_\infty + n_1, F = 0$.
2. If $t = n$ then **STOP**.
3. Consider now in (6.53) and (6.54) that $n_2 = k$. Recall that either $k = 1$ or $k = 2$, depending if the pencil $\lambda M_{22} - N_{22}$ contains either a real or a pair of complex conjugated generalized eigenvalues. Choose an appropriate $k \times k$ real matrix A having the desired k eigenvalues to be assigned located in the desired position in $\mathbb{C}_g$. Compute

$$f = \begin{bmatrix} O & B_2^\dagger(M_2 A - N_2) \end{bmatrix}$$

where $B_2^\dagger$ is a right inverse of B_2 which is well defined due to the controllability assumption. Update the pencil and the feedback

$$\overline{N} \leftarrow \overline{N} + Bf, \quad F \leftarrow F + fZ^T.$$

4. Using an algorithm for swapping eigenvalues in the generalized Schur form, determine two orthogonal matrices Q_1 and Z_1 to move the last k generalized eigenvalues of $\lambda\overline{M} - \overline{N}$ to the position $t+1$. Set

$$\lambda\overline{M} - \overline{N} \leftarrow Q_1(\lambda\overline{M} - \overline{N})Z_1, B \leftarrow Q_1 B, Z \leftarrow ZZ_1, Q \leftarrow Q_1 Q, t \leftarrow t + k$$

and go to Step 2.
5. **END**

We have obtained

$$\overline{M} = \begin{bmatrix} M_1 & \star \\ O & \overline{M_2} \end{bmatrix}, \quad \overline{N} = \begin{bmatrix} N_1 & \star \\ O & \overline{N_2} \end{bmatrix} \begin{matrix} \}n_\infty \\ \}n_1 + n_2 \end{matrix}$$

where now $\lambda M_1 - N_1$ has the same infinite generalized eigenvalues as the original pencil $\lambda M - N$ and $\lambda \overline{M_2} - \overline{N_2}$ is in a generalized Schur form and has only finite generalized eigenvalues located in $\mathbb{C}_g$.

Notice that Algorithm 1 moves recursively the generalized eigenvalues located in $\mathbb{C}_b$ without affecting those located in $\mathbb{C}_g$. In the case in which M is invertible an alternative approach is to take the controllable pair $(\widetilde{N}, \widetilde{B}) := (M^{-1}N, M^{-1}B)$ and to determine F to place the eigenvalues of the matrix $\widetilde{N} + \widetilde{B}F$ in $\mathbb{C}_g$. This can be achieved by any algorithm for eigenvalue assignment. However, even when M is invertible, Algorithm 1 for assignment of generalized eigenvalues has several potential advantages:

1. If $\lambda M - N$ is already brought into the generalized Schur form it needs fewer computations.
2. It provides accurate solutions even when M is ill conditioned with respect to numerical inversion.

6.5.2 *Computation of proper deflating subspaces*

Let

$$\mathbb{C} = \mathbb{C}_g \cup \mathbb{C}_b \tag{6.55}$$

be a partition of the complex plane in two symmetric (with respect to the imaginary axis) disjoint sets. We describe further an algorithm for computing a proper $\mathbb{C}_g$–deflating subspace of maximal dimension for an arbitrary matrix pencil $\lambda M - N$.

Algorithm 2

1. Apply the staircase algorithm to

$$\lambda\overline{M} - \overline{N} := \lambda M^P - N^P$$

to get the orthogonal matrices Q_1 and Z_1 such that

$$Q_1(\lambda\overline{M} - \overline{N})Z_1 = \begin{bmatrix} \lambda M_{\epsilon\infty} - N_{\epsilon\infty} & \star \\ O & \lambda M_{\eta f} - N_{\eta f} \end{bmatrix}. \tag{6.56}$$

Set

$$\begin{aligned} \lambda\overline{M} - \overline{N} \quad &\leftarrow \quad \begin{bmatrix} \lambda M_{\epsilon f} - N_{\epsilon f} & \star \\ O & \lambda M_{\eta\infty} - N_{\eta\infty} \end{bmatrix} \\ &:= \quad \begin{bmatrix} \lambda M_{\epsilon\infty} - N_{\epsilon\infty} & \star \\ O & \lambda M_{\eta f} - N_{\eta f} \end{bmatrix}^P . \end{aligned} \tag{6.57}$$

The pencil $\lambda M_{\epsilon f} - N_{\epsilon f}$ contains the right Kronecker indices and all the finite generalized eigenvalues while $\lambda M_{\eta\infty} - N_{\eta\infty}$ contains the left Kronecker indices and all the infinite generalized eigenvalues. The structure of (6.57) follows from the properties of the pertranspose. Notice that pertransposing a pencil preserves the generalized eigenvalues (finite and infinite) and interchanges the right with the left singular Kronecker structure. Set

$$Q \leftarrow Z_1^P, Z \leftarrow Q_1^P.$$

2. Apply the staircase algorithm to the subpencil $\lambda M_{\epsilon f} - N_{\epsilon f}$ to determine Q_2 and Z_2 such that

$$Q_2(\lambda M_{\epsilon f} - N_{\epsilon f})Z_2 = \begin{bmatrix} \lambda M_\epsilon - N_\epsilon & \star \\ O & \lambda M_f - N_f \end{bmatrix} \begin{matrix} \}n_r \\ \}n_f \end{matrix}. \tag{6.58}$$

Compute

$$Q \leftarrow \begin{bmatrix} Q_2 & O \\ O & I \end{bmatrix} Q, \quad Z \leftarrow Z \begin{bmatrix} Z_2 & O \\ O & I \end{bmatrix}$$

and set

$$\lambda \overline{M} - \overline{N} \leftarrow \begin{bmatrix} Q_2(\lambda M_{\epsilon f} - N_{\epsilon f})Z_2 & \star \\ O & \lambda M_{\eta\infty} - N_{\eta\infty} \end{bmatrix}. \tag{6.59}$$

The pencil $\lambda M_{\eta\infty} - N_{\eta\infty}$ is in a dual staircase form (6.52), outlining the left Kronecker indices and the infinite eigenvalues, $\lambda M_\epsilon - N_\epsilon$ is in the staircase form with square blocks on the second upper diagonal, and $\lambda M_f - N_f$ is regular with only finite generalized eigenvalues with M_f upper triangular.

3. Denote by n_g the number of generalized eigenvalues of $\lambda M_f - N_f$ located in $\mathbb{C}_g$. According to Theorem 1.11.3, the maximum dimension of a proper $\mathbb{C}_g$–deflating subspace to the right of the pencil $\lambda M - N$ is $n_r + n_g$. Using an algorithm for swapping generalized eigenvalues, determine orthogonal Q_3 and Z_3 such that

$$Q_3(\lambda M_f - N_f)Z_3 = \begin{bmatrix} \lambda M_1 - N_1 & \star \\ O & \lambda M_2 - N_2 \end{bmatrix} \begin{matrix} \}n_1 \\ \}n_2 \end{matrix} \tag{6.60}$$

and $\Lambda(M_1, N_1) \subset \mathbb{C}_g, \Lambda(M_2, N_2) \subset \mathbb{C}_b$. Set

$$\lambda \overline{M} - \overline{N} \leftarrow \begin{bmatrix} \lambda M_\epsilon - N_\epsilon & \star & \star & \star \\ O & \lambda M_1 - N_1 & \star & \star \\ O & O & \lambda M_2 - N_2 & \star \\ O & O & O & \lambda M_{\eta\infty} - N_{\eta\infty} \end{bmatrix}$$

and

$$Q \leftarrow \begin{bmatrix} I & O & O \\ O & Q_3 & O \\ O & O & I \end{bmatrix} Q, \quad Z \leftarrow Z \begin{bmatrix} I & O & O \\ O & Z_3 & O \\ O & O & I \end{bmatrix}.$$

4. The pencil $\lambda M_\epsilon - N_\epsilon$ has full row rank for all $\lambda \in \mathbb{C}$. Actually, both M_ϵ and N_ϵ have full row rank and they have the particular form

$$\lambda M_\epsilon - N_\epsilon = \lambda \left[\begin{array}{c|cccc} O & M_{12} & \cdots & M_{1\ell} & M_{1\ell+1} \\ O & O & \cdots & M_{2\ell} & M_{2\ell+1} \\ \vdots & \vdots & \ddots & \vdots & \vdots \\ O & O & \cdots & O & M_{\ell\ell+1} \end{array}\right] - \left[\begin{array}{c|cccc} N_{11} & N_{12} & \cdots & N_{1\ell} & N_{1\ell+1} \\ O & N_{22} & \cdots & N_{2\ell} & N_{2\ell+1} \\ \vdots & \vdots & \ddots & \vdots & \vdots \\ O & O & \cdots & N_{\ell\ell} & N_{\ell\ell+1} \end{array}\right] \quad (6.61)$$

where $\nu_\ell = 0$ and $M_{i,i+1}, i = 1, \cdots, \ell$ are square and nonsingular. The pencil in (6.61) outlines the following partition $\lambda M_\epsilon - N_\epsilon = \lambda [\ O \mid M_{\epsilon 2}\ ] - [\ N_{\epsilon 1} \mid N_{\epsilon 2}\ ]$, where $M_{\epsilon 2}$ is square, block upper triangular and nonsingular. Therefore $\lambda M_{\epsilon 2} - N_{\epsilon 2}$ is regular and has only finite generalized eigenvalues. Moreover, the pair $(M_{\epsilon 2}^{-1} N_{\epsilon 2}, M_{\epsilon 2}^{-1} N_{\epsilon 1})$ is controllable. Therefore, there exists a feedback F such that

$$\lambda M_{\epsilon 2} - N_{\epsilon 2} - N_{\epsilon 1} F \quad (6.62)$$

is regular and has generalized eigenvalues located exclusively in $\mathbb{C}_g$. The feedback F can be computed using a simplified version of Algorithm 1 ($n_\infty = 0$). Notice that the effect of the feedback on the whole pencil $\lambda \overline{M} - \overline{N}$ is that of a strict equivalence. Set

$$Z_4 = \begin{bmatrix} I & F \\ O & I \end{bmatrix}, \quad Z \leftarrow Z \begin{bmatrix} Z_4 & O \\ O & I \end{bmatrix},$$

$$\lambda \overline{M} - \overline{N} \leftarrow \begin{bmatrix} -N_{\epsilon 1} & \lambda M_{\epsilon 2} - N_{\epsilon 1} F - N_{\epsilon 2} & \star & \star \\ O & O & \lambda M_1 - N_1 & \star \\ O & O & O & \star \\ O & O & O & \star \end{bmatrix}.$$

After adequate column permutation, we get finally

$$\lambda \overline{M} - \overline{N} = \begin{bmatrix} \lambda M_{\epsilon 2} - N_{\epsilon 1} F - N_{\epsilon 2} & \lambda X_1 - X_2 & -N_{\epsilon 1} & \star \\ O & \lambda M_1 - N_1 & O & \star \\ O & O & O & \star \\ O & O & O & \star \end{bmatrix}.$$

Set

$$Z_5 = \begin{bmatrix} O & I \\ I & O \end{bmatrix}, \quad Z \leftarrow Z \begin{bmatrix} Z_5 & O \\ O & I \end{bmatrix}.$$

Denoting the first $n_s := n_r + n_g$ columns of Z by V we conclude that $MVS = NV$ holds with MV of full column rank,

$$S := \begin{bmatrix} M_{\epsilon 2}^{-1}(N_{\epsilon 1} F + N_{\epsilon 2}) & M_{\epsilon 2}^{-1}(X_2 - X_1 M_1^{-1} N_1) \\ O & M_1^{-1} N_1 \end{bmatrix},$$

$\Lambda(S) \subset \mathbb{C}_g$. Thus $\mathcal{V} = \operatorname{Im} V$ is a maximal proper $\mathbb{C}_g$–deflating subspace.

5. **END**

The transformations Q and Z produced by the above algorithm are, in general, not orthogonal only because of Step 4 where we solve a pole assignment problem. However, notice that the final location of the poles is not fixed, and a suitable choice within the stability domain could avoid a high gain feedback F and the possible ill conditioned transformation of the pencil. When the pencil is regular, Step 2 and Step 4 are not necessary, and therefore a basis matrix for a deflating subspace can be computed by using orthogonal transformations only.

In the regular case Step 1 and Step 3 can be condensed into a single one by using directly the QZ algorithm in conjunction with an algorithm for swapping generalized eigenvalues which also handles infinite eigenvalues. A better approach is to preserve Step 1 as above, i.e., to separate first the finite and infinite eigenvalues with the staircase algorithm and then to apply the QZ algorithm with reordering only to $\lambda M_f - N_f$. This approach avoids ill conditioned eigenvalue computation in the case of multiple infinite eigenvalues. Moreover, in contrast to the continuous–time case, no *a priori* information about the regularity of the ESP is available, without imposing any additional conditions on the input matrices. Such information becomes available after the first step of the above algorithm. Preserving Step 1 of the algorithm is supported also by the fact that a direct application of the QZ algorithm to a singular pencil can compromise the results irrevocably.

A dual version of the above algorithm can be applied to compute a maximal proper $\mathbb{C}_g$–deflating subspace to the left. As we have seen, such subspaces are related to solutions of the reverse discrete–time Riccati equation.

6.5.3 Computation of solutions to Riccati equations and systems

In this section we indicate how the general–purpose algorithm for computing proper deflating subspaces could be applied for solving different types of algebraic Riccati equations.

The continuous–time Riccati equation and system

Consider the Popov triplet $\Sigma = (A, B; Q, L, R)$ and associate with it the EHP(Σ) denoted $sM_\Sigma - N_\Sigma$. The regular and singular cases may be distinguished through the invertibility of the matrix R. However, notice that even for a singular R matrix, the EHP(Σ) can still be regular, yet the associated CTARE(Σ) has no stabilizing or antistabilizing solution since the EHP(Σ) is not dichotomic. Below we give a prototype algorithm for computing stabilizing solutions to the CTARE(Σ), CTARS(Σ), and GCTARS(Σ).

Algorithm 3
Define $\mathbb{C}_g := \mathbb{C}_-$ and $\mathbb{C}_b = \mathbb{C} \setminus \mathbb{C}_g$.

1. Apply Steps 1–3 of Algorithm 2 to $sM_\Sigma - N_\Sigma$ in order to deflate the "left" and "right" structure of the pencil. At this step we decide if the EHP is regular by checking if $\nu_r = 0$.
2. If $n_r + n_1 < n$ then there is no stabilizing (antistabilizing) solution to the associated CTARS (or to the CTARE, in the case $\nu_r = 0$).

3. If $n_r = 0$ we have already computed at Steps 1–3 a basis matrix V for the maximal proper $\mathbb{C}_g$–deflating subspace, and go to the next step. If $n_r \neq 0$, compute a basis V for the maximal proper $\mathbb{C}_g$–deflating subspace by using Step 4 of Algorithm 2.
4. Partition V as in (6.31). If V_1 has not full column rank (is not invertible), there exists no solution of maximal dimension to the GCTARS (no solution to the CTARS or CTARE), and **STOP**.
5. If $r := n_r + n_g = n$ the stabilizing solution to the CTARS (or CTARE) is $X = V_2 V_1^{-1}$ and the stabilizing feedback is $F = V_3 V_1^{-1}$. If $r < n$ a maximal stabilizing solution to the GCTARS is (X, F, r, V_1), where now $X = V_2 V_1^{\dagger}$ and $F = V_3 V_1^{\dagger}$.
6. **END**

For the antistabilizing solution we simply put $\mathbb{C}_g := \mathbb{C}_+, \mathbb{C}_b := \mathbb{C} \setminus \mathbb{C}_g$.

The discrete–time algebraic Riccati equation and system

Consider the Popov triplet $\Sigma = (A, B; Q, L, R)$ and associate with it the ESP(Σ) denoted $zM_\Sigma - N_\Sigma$. Set $\mathbb{C}_g = \mathbb{D}$ and $\mathbb{C}_b = \mathbb{C} \setminus \mathbb{C}_g$. The same algorithm as in the continuous–time case can be applied now to $zM_\Sigma - N_\Sigma$ in order to compute the stabilizing solution to the DTARE(Σ), DTARS(Σ) and GDTARS(Σ). In contrast to the continuous–time case, no *a priori* information about the regularity of the ESP is in general available. The regularity of the pencil can be decided at Step 1 of Algorithm 3. For the antistabilizing solution set $\mathbb{C}_g = \mathbb{C} \setminus \overline{\mathbb{D}}$ and $\mathbb{C}_b = \mathbb{C} \setminus \mathbb{C}_g$.

The time–reversed discrete–time algebraic Riccati equation

For computing the stabilizing solution to the time–reversed DTARE(Σ) a similar algorithm can be applied to $zM_\Sigma^T - N_\Sigma^T$ or, alternatively, one can use the left transformation Q to obtain a column basis matrix V as in (5.79) for the left $\mathbb{D}$–deflating subspace. Then the stabilizing solution and the stabilizing Riccati feedback can be computed by formulas (5.83) and (5.84), respectively.

Numerical refinements

The QZ and Schur algorithms have a "natural" tendency to deflate out eigenvalues from smallest to largest in magnitude. Simple "tricks" can avoid the reordering of the (generalized) eigenvalues at Step 3 of Algorithm 2 (as needed at Step 1 of Algorithm 3). For computing the antistabilizing solution in discrete–time no reordering of eigenvalues is required since the Schur and QZ algorithms provide in general the appropriate ordering. For computing the stabilizing solution in discrete–time use the reciprocal pencil $\lambda N - M$ instead of $\lambda M - N$. This has no effect on the basis of proper deflating subspaces and simply "inverts" the eigenvalues. In continuous–time, to compute the stabilizing solution use the pencil $\lambda(N+M) - (N-M)$ instead of $\lambda M - N$. Again, the deflating subspace remains unchanged and the generalized eigenvalues are mapped by the bilinear transformation $\gamma = \frac{\lambda - 1}{\lambda + 1}$. A similar procedure works for the antistabilizing solution.

In the regular case, the feedback matrix F is usually computed using the formulas $F = -R^{-1}(B^TX + L^T)$ and $F = -(R + B^TXB)^{-1}(B^TXA + L^T)$ for the continuous and discrete–time cases, respectively. Due to the possibility that the matrices R or $R + B^TXB$ are ill-conditioned with respect to numerical inversion, a better approach is to compute F directly using the formula $F = V_3V_1^{-1}$, which also avoids the computation of the matrix X, when this is unnecessary.

Finally, notice that the prototype Riccati solver based on the algorithm proposed above approaches all cases in a unitary manner, with no initial testing of controllability, positiveness, invertibility or without any inelegant preliminary transformation, both in the continuous and discrete–time cases. The main steps of such an algorithm for solving algebraic Riccati equations and systems can be summarized as follows:

1. Construct the extended pencil (EHP or ESP).
2. Solve the generalized eigenstructure problem with reordering for this pencil.
3. Decide whether the solution corresponding to a certain spectrum (stable, antistable, etc.) exists or not.
4. Compute that solution, if it exists.

The general algorithm for computing proper deflating subspaces with a specified spectrum has many other applications, as for example the computation of the structural invariants of a system.

Notes and references

The extension of the Riccati theory to the singular case as presented in this chapter is due to Ionescu and Oară [Oar95, IO96d, IO96e]. An excellent reference on general numerical algorithms is Golub and Van Loan [GV89]. Numerically sound algorithms for computing the Kronecker–like forms have been proposed by Van Dooren [Van79], Beelen and Van Dooren [BV88], and were recently improved by Oară and Van Dooren [OV97]. Algorithms for generalized eigenvalue assignment were proposed by Varga [Var95]. Numerical algorithms for computing proper deflating subspaces and solutions to Riccati equations and systems were developed by Oară [Oar94, Oar95, Oar96b]. An overview of both theoretical and numerical aspects of the singular algebraic Riccati theory is given in Ionescu, Oară, and Weiss [IOW97]. Descriptor Riccati equations have been studied by Mehrmann [Meh91], Katayama [Kat94], and Ionescu and Oară [IO97].

Part III

Applications to Systems Theory and Robust Control

7

Applications to Systems Theory

In this chapter we apply the generalized Riccati theory presented in Part II to obtain new straightforward proofs or extensions of several classical results in systems theory. In Section 7.1 we present the bounded real lemma in a general setting. We pay a special attention to the CTARE associated with the positiveness Popov triplet and study the convergence of a family of such CTAREs in Section 7.2. We derive explicit formulas for normalized coprime factorizations in Section 7.3. In Section 7.4 we give a state–space proof for the Small Gain Theorem. The topics of Sections 7.2–7.4 will play a major part in the solution to the robust stabilization problems presented in Chapter 11. We discuss in Sections 7.5 and 7.6 two applications of the Riccati systems to the optimal control with constrained dynamics and the spectral and inner–outer factorizations of rational matrices of arbitrary rank. The discrete–time counterpart results are given in Sections 7.7–7.12.

A. The continuous–time case

7.1 The bounded real lemma

In this section we give an extension of a classical result, known as bounded real lemma, which gives necessary and sufficient conditions for a system to be bounded in H^∞–norm by a $\gamma > 0$.

Lemma 7.1.1 (Generalized bounded real lemma) *Let* $\gamma > 0$ *and*

$$\mathbf{G} = \left[\begin{array}{c|c} A & B \\ \hline C & D \end{array}\right] \tag{7.1}$$

be given. Let $\Sigma_c = (A, B; C^T C, C^T D, -\gamma^2 I + D^T D)$ *be the* γ*–contractiveness Popov triplet associated with* $\mathbf{G}$.

I. *The following statements are equivalent:*

1. A is stable and $\|\mathbf{G}\|_\infty < \gamma$.

2. The KPYS$(\Sigma_c, -I)$ *given by*

$$\begin{aligned} -\gamma^2 I + D^T D &= -V^T V, \\ C^T D + XB &= -W^T V, \\ C^T C + A^T X + XA &= -W^T W, \end{aligned} \tag{7.2}$$

has a stabilizing solution (X, V, W) *with* $X \geq 0$.

II. *The following statements are equivalent:*
1. A is dichotomic, the pair (A, B) is stabilizable, and $\|\mathbf{G}\|_\infty < \gamma$.
2. The KPYS($\Sigma_c, -I$) (7.2) has a stabilizing solution (X, V, W).
Moreover, if point 2 holds then we have the following relations between the inertia of X and A: $\delta_c(X) + \pi_c(X) = \nu_c(A)$; $\nu_c(X) = \pi_c(A)$.

Proof. I. 1. $\Rightarrow$ 2. Consider the Popov triplet $\Sigma_c = (A, B; C^TC, C^TD, -\gamma^2 I + D^TD)$. The Popov function $\mathbf{\Pi}_{\Sigma_c}(j\omega) = -\gamma^2 I + \mathbf{G}(j\omega)^*\mathbf{G}(j\omega)$ is negative definite for each $\omega \in \overline{\mathbb{R}}$ by hypothesis. In accordance with Corollary 4.5.5, we can apply Corollary 4.1.7 to Σ_c to obtain the existence of the stabilizing solution to the CTARE(Σ_c). The KPYS($\Sigma_c, -I$) given in (7.2) is simply another way of writing the Riccati equation in the case that $R < 0$ as we explained in Remark 3.2.9. Further, since A is stable, the last equation in (7.2) implies that $X \geq 0$.

2. $\Rightarrow$ 1. Notice first that $A - BV^{-1}W$ is stable. Hence the pair (W, A) is detectable and so is $(\begin{bmatrix} W \\ C \end{bmatrix}, A)$. By combining this conclusion with the fact that the last equation in (7.2) is fulfilled for $X \geq 0$ the stability of A follows by Theorem 1.5.5. Further, from the spectral factorization identity (3.39) we can write

$$\mathbf{\Pi}_{\Sigma_c} = -\left[\begin{array}{c|c} A & B \\ \hline V & W \end{array}\right]^* \left[\begin{array}{c|c} A & B \\ \hline V & W \end{array}\right],$$

i.e., $\mathbf{\Pi}_{\Sigma_c}(j\omega) < 0$ for $\omega \in \overline{\mathbb{R}}$ and the conclusion follows.

II. 1. $\Rightarrow$ 2. Let F be such that $\widetilde{A} = A + BF$ is stable and let $\widetilde{\Sigma}_c$ be the F–equivalent of Σ_c. Using (3.37) one gets

$$\mathbf{\Pi}_{\Sigma_c} = \mathbf{S}_F^* \mathbf{\Pi}_{\widetilde{\Sigma}_c} \mathbf{S}_F \tag{7.3}$$

where

$$\mathbf{S}_F = \left[\begin{array}{c|c} A & B \\ \hline -F & I \end{array}\right]$$

is a unit in $RL^\infty_{+,m\times m}$ since A is dichotomic and $A + BF$ is stable.

As $\mathbf{\Pi}_{\Sigma_c}(j\omega) < 0$ for $\omega \in \overline{\mathbb{R}}$ it follows from (7.3) that $\mathbf{\Pi}_{\widetilde{\Sigma}_c}(j\omega) < 0$ for $\omega \in \overline{\mathbb{R}}$. Further, we use the same scheme of proof as for implication 1. $\Rightarrow$ 2. in part I, in which we replace the Popov triplet Σ_c by its equivalent $\widetilde{\Sigma}_c$, to prove that the KYPS($\widetilde{\Sigma}_c, -I$) has a stabilizing solution. From Theorem 3.2.13 we conclude that the KPYS($\Sigma_c, -I$) has a stabilizing solution, say (X, V, W).

2. $\Rightarrow$ 1. Since the KPYS($\Sigma_c, -I$) has a stabilizing solution (X, V, W) we have that $A - BV^{-1}W$ is stable, from where it follows that the pair (A, B) is stabilizable and the pair $(\begin{bmatrix} W \\ C \end{bmatrix}, A)$ is detectable. With this and Theorem 1.5.5, we infer from the last equation (7.2) that A is dichotomic. The rest of the proof is similar as for part I.

Finally, if the KPYS(Σ_c) has a stabilizing solution, we get that $A - BV^{-1}W$ is stable, and it follows that the pair $(\begin{bmatrix} W \\ C \end{bmatrix}, A)$ is detectable. From the last equation in (7.2) seen as a Lyapunov equation we infer with Theorem 1.5.5 that $\delta_c(X) + \pi_c(X) = \nu_c(A)$, and $\nu_c(X) = \pi_c(A)$, and the proof ends. ■

Corollary 7.1.2 *Let $\gamma > 0$ and*

$$\mathbf{G} = \left[\begin{array}{c|c} A & B \\ \hline C & D \end{array}\right]$$

be given. Then the following two assertions are equivalent:

1. *A is stable and $\|\mathbf{G}\|_\infty < \gamma$.*
2. *$-\gamma^2 I + D^T D < 0$ and there exists $X > 0$ such that*

$$\begin{bmatrix} A^T X + XA + C^T C & XB + C^T D \\ D^T C + B^T X & -\gamma^2 I + D^T D \end{bmatrix} < 0.$$

Proof. Assertion 1 is equivalent to the following assertion:
$1'$. A is stable and $-\gamma^2 I + \mathbf{G}^* \mathbf{G} < 0$ on $\overline{\mathbb{C}}_0$.
The equivalence between assertions $1'$ and 2 follows by applying Theorem 4.6.1 and Remark 4.6.2 to the triplet $\Sigma_c = (A, B; C^T C, D^T C, -\gamma^2 I + D^T D)$ since $\mathbf{\Pi}_{\Sigma_c} = -\gamma^2 I + \mathbf{G}^* \mathbf{G}$. ■

Corollary 7.1.2 is usually known as the *strict bounded real lemma.*

7.2 The CTARE associated with the positiveness Popov triplet

In this section we study some convergence properties of a family of CTAREs associated with positiveness Popov triplets. The main result will be a basic ingredient for solving the robust stabilization problem with multiplicative uncertainties in Chapter 11.

Proposition 7.2.1 *Let*

$$\mathbf{G} = \left[\begin{array}{c|c} A & B \\ \hline C & D \end{array}\right]$$

be given. Let $\Sigma_p = (A, B; C^T C, C^T D, D^T D)$ be the positiveness Popov triplet associated with $\mathbf{G}$. Assume that:

a. *The pair (A, B) is stabilizable.*

b. *D is of full column rank.*

c. $\begin{bmatrix} j\omega I - A & -B \\ C & D \end{bmatrix}$ *is of full column rank for all $\omega \in \mathbb{R}$.*

Then the CTARE(Σ_p) has a positive semidefinite stabilizing solution.

Proof. The existence of the stabilizing solution follows by checking that Corollary 4.5.7 is applicable to Σ_p. It is also easy to check that condition (4.72) is true, hence the solution is positive semidefinite. ■

Remark 7.2.2 By Proposition 7.2.1, the proof of Corollary 4.2.2 follows straightforwardly. Indeed, for

$$\mathbf{G}_0 := \left[\begin{array}{c|c} A & B \\ \hline C & O \\ O & I \end{array}\right]$$

the associated positiveness Popov triplet is $\Sigma_p = \Sigma_s = (A, B; C^TC, 0, I)$, i.e., coincides with the standard Popov triplet for control (see (3.83)). For this triplet the hypotheses of Proposition 7.2.1 are fulfilled. As the CTARE(Σ_p) is exactly (4.33), the conclusion follows automatically.

We investigate now the dependence on the small parameter ε of the stabilizing solution to the CTARE(Σ_ε), where

$$\Sigma_\varepsilon = (A, B, Q, L, R_\varepsilon) = (A, B; C^TC, C^TD, \varepsilon^2 I + D^TD). \tag{7.4}$$

Notice that Σ_ε is exactly the positiveness Popov triplet associated with the system (see Section 3.4)

$$\mathbf{G}_{p,\varepsilon} = \left[\begin{array}{c|c} A & B \\ \hline C & D \\ O & \varepsilon I \end{array}\right].$$

Theorem 7.2.3 *Assume that for the stabilizable and detectable system*

$$\mathbf{G} = \left[\begin{array}{c|c} A & B \\ \hline C & D \end{array}\right], \tag{7.5}$$

where $A \in \mathbb{R}^{n\times n}$, $B \in \mathbb{R}^{n\times m}$, $C \in \mathbb{R}^{p\times n}$, and $D \in \mathbb{R}^{p\times m}$, the following hypothesis holds:

(H) *There exists a matrix $F \in \mathbb{R}^{m\times n}$ such that $A + BF$ is stable and*

$$\mathbb{T}_{\widetilde{\mathcal{G}}}\mathbb{T}^*_{\widetilde{\mathcal{G}}} > 0 \tag{7.6}$$

where $\mathbb{T}_{\widetilde{\mathcal{G}}}$ denotes the Toeplitz operator associated with the linear bounded input–output operator $\widetilde{\mathcal{G}} : L^{2,m} \to L^{2,p}$ of the stable system

$$\widetilde{\mathbf{G}} = \left[\begin{array}{c|c} \widetilde{A} & B \\ \hline \widetilde{C} & D \end{array}\right] = \left[\begin{array}{c|c} A+BF & B \\ \hline C+DF & D \end{array}\right], \tag{7.7}$$

where $\widetilde{A} := A + BF$ and $\widetilde{C} = C + DF$. Define the Popov triplet (7.4). Then the following two assertions hold:

1. *The CTARE(Σ_ε) has a positive semidefinite stabilizing solution X_ε, $\forall \varepsilon > 0$.*
2. *$\lim_{\varepsilon\downarrow 0} X_\varepsilon = 0$.*

Before proceeding with the proof of the theorem we make two remarks.

Remark 7.2.4 If (7.6) holds for a certain $F = \widehat{F}$ that makes $A + B\widehat{F}$ stable, then it holds for each stabilizing feedback matrix F. Indeed, let an arbitrary $\widehat{F}$ be such that $A + B\widehat{F}$ is stable and let $\widehat{\mathcal{G}}$ be the input–output operator of the system

$$\widehat{\mathbf{G}} = \left[\begin{array}{c|c} A+B\widehat{F} & B \\ \hline C+D\widehat{F} & D \end{array}\right]. \tag{7.8}$$

Then it is easy to check the identity

$$\widehat{\mathcal{G}} = \widetilde{\mathcal{G}}\mathcal{N}, \tag{7.9}$$

where $\mathcal{N}$ is the input–output operator of the system

$$\left[\begin{array}{c|c} A+B\widehat{F} & B \\ \hline \widehat{F}-F & I \end{array}\right]. \tag{7.10}$$

Since $\widetilde{\mathcal{G}}$ and $\mathcal{N}$ are operators associated with causal systems, we can apply Corollary 2.4.13 to conclude that

$$\mathbb{T}_{\widehat{\mathcal{G}}} = \mathbb{T}_{\widetilde{\mathcal{G}}}\mathbb{T}_{\mathcal{N}}. \tag{7.11}$$

On the other hand $\mathcal{N}$ is invertible and its inverse, which is the input–output operator of the system

$$\left[\begin{array}{c|c} A+BF & B \\ \hline F-\widehat{F} & I \end{array}\right], \tag{7.12}$$

is causal. According to Corollary 2.4.14, $\mathbb{T}_{\mathcal{N}}$ is boundedly invertible. We deduce that

$$\mathbb{T}_{\widehat{\mathcal{G}}}\mathbb{T}^*_{\widehat{\mathcal{G}}} = \mathbb{T}_{\widetilde{\mathcal{G}}}\mathbb{T}_{\mathcal{N}}\mathbb{T}^*_{\mathcal{N}}\mathbb{T}^*_{\widetilde{\mathcal{G}}} > 0.$$

Remark 7.2.5 Notice that the positiveness condition (7.6) is weaker than the coerciveness condition $\mathbb{T}_{\widetilde{\mathcal{G}}}\mathbb{T}^*_{\widetilde{\mathcal{G}}} \geq c_o I, c_o > 0$. In fact condition (7.6) is equivalent to

$$\operatorname{Ker} \mathbb{T}^*_{\widetilde{\mathcal{G}}} = \{0\}$$

or to

$$\overline{\operatorname{Im} \mathbb{T}_{\widetilde{\mathcal{G}}}} = L^{2,p}_{+}.$$

Proof of Theorem 7.2.3.

1. Since (C, A) is detectable, the matrix

$$\begin{bmatrix} j\omega I - A & -B \\ C & D \\ O & \varepsilon I \end{bmatrix}$$

has full column rank for all $\omega \in \mathbb{R}$ and $\varepsilon > 0$. Further, the matrix

$$\begin{bmatrix} D \\ \varepsilon I \end{bmatrix}$$

has full column rank. By combining these facts with the stabilizability of the pair (A, B) it follows from Proposition 7.2.1 that the CTARE(Σ_ε)

$$A^T X + XA - (XB + C^T D)(\varepsilon^2 I + D^T D)^{-1}(B^T X + D^T C) + C^T C = 0 \tag{7.13}$$

has a positive semidefinite stabilizing solution X_ε, for all $\varepsilon > 0$.

2. Assume that for some F hypothesis **(H)** holds. Let $\widetilde{\Sigma}_\varepsilon = (\widetilde{A}, B; \widetilde{Q}, \widetilde{L}, R_\varepsilon)$ be the F–equivalent of Σ_ε, where

$$\begin{aligned}
\widetilde{A} &:= A + BF, \\
\widetilde{Q} &:= Q + LF + F^T L^T + F^T R_\varepsilon F \\
&= C^T C + C^T DF + F^T D^T C + F^T D^T DF + \varepsilon^2 F^T F \\
&= (C + DF)^T (C + DF) + \varepsilon^2 F^T F = \widetilde{C}^T \widetilde{C} + \varepsilon^2 F^T F \\
&= \begin{bmatrix} \widetilde{C}^T & \varepsilon F^T \end{bmatrix} \begin{bmatrix} \widetilde{C} \\ \varepsilon F \end{bmatrix}, \\
\widetilde{L} &:= L + F^T R_\varepsilon = C^T D + F^T(\varepsilon^2 I + D^T D) = \widetilde{C}^T D + \varepsilon^2 F^T.
\end{aligned} \tag{7.14}$$

According to Corollary 3.2.7, the stabilizing solution X_ε of the CTARE(Σ_ε) is also the stabilizing solution of the CTARE($\widetilde{\Sigma}_\varepsilon$) and $\widetilde{A}$ is stable. Therefore we may use the representation formula (4.15) of X_ε,

$$X_\varepsilon = \mathcal{P}_{o,\widetilde{\Sigma}_\varepsilon} - \mathcal{P}_{\widetilde{\Sigma}_\varepsilon} \mathcal{R}_{\widetilde{\Sigma}_\varepsilon}^{-1} \widetilde{\mathcal{P}}^*_{\widetilde{\Sigma}_\varepsilon}, \tag{7.15}$$

where

$$\begin{aligned} \mathcal{P}_{o,\widetilde{\Sigma}_\varepsilon} &:= \widetilde{\Phi}^* \widetilde{Q} \widetilde{\Phi} = \widetilde{\Phi}^* \begin{bmatrix} \widetilde{C}^T & \varepsilon F^T \end{bmatrix} \begin{bmatrix} \widetilde{C} \\ \varepsilon F \end{bmatrix} \widetilde{\Phi}, \\ \mathcal{R}_{\widetilde{\Sigma}_\varepsilon} &:= R + \widetilde{L}^T \widetilde{\mathcal{L}} + \widetilde{\mathcal{L}}^* \widetilde{L} + \widetilde{\mathcal{L}}^* \widetilde{Q} \widetilde{\mathcal{L}}, \\ \mathcal{P}_{\widetilde{\Sigma}_\varepsilon} &:= \Phi^* (\widetilde{Q} \widetilde{\mathcal{L}} + \widetilde{L}), \end{aligned} \tag{7.16}$$

with

$$(\widetilde{\Phi} x)(t) = e^{\widetilde{A} t} x, \quad t \geq 0,$$

and

$$(\widetilde{\mathcal{L}} u)(t) = \int_0^t e^{\widetilde{A}(t-\tau)} B u(\tau) \mathrm{d}\tau, \quad t \geq 0, \quad \forall u \in L_+^{2,m}.$$

We compute

$$\begin{aligned} \mathcal{P}_{\widetilde{\Sigma}_\varepsilon} &= \widetilde{\Phi}^* (\widetilde{Q} \widetilde{\mathcal{L}} + \widetilde{L}) \\ &= \widetilde{\Phi}^* [(\widetilde{C}^T \widetilde{C} + \varepsilon^2 F^T F) \widetilde{\mathcal{L}} + \widetilde{C}^T D + \varepsilon^2 F^T] \\ &= \widetilde{\Phi}^* [\widetilde{C}^T (\widetilde{C} \widetilde{\mathcal{L}} + D) + \varepsilon^2 F^T (F \widetilde{\mathcal{L}} + I)] \\ &= \widetilde{\Phi}^* \begin{bmatrix} \widetilde{C}^T & \varepsilon F^T \end{bmatrix} \begin{bmatrix} \widetilde{C} \widetilde{\mathcal{L}} + D \\ \varepsilon (F \widetilde{\mathcal{L}} + I) \end{bmatrix} \end{aligned} \tag{7.17}$$

and

$$\begin{aligned} \mathcal{R}_{\widetilde{\Sigma}_\varepsilon} &= R + \widetilde{L}^T \widetilde{\mathcal{L}} + \widetilde{\mathcal{L}}^* \widetilde{L} + \widetilde{\mathcal{L}}^* \widetilde{Q} \widetilde{\mathcal{L}} \\ &= \varepsilon^2 I + D^T D + (D^T \widetilde{C} + \varepsilon^2 F) \widetilde{\mathcal{L}} + \widetilde{\mathcal{L}}^* (\widetilde{C}^T D + \varepsilon^2 F) + \widetilde{\mathcal{L}}^* (\widetilde{C}^T \widetilde{C} + \varepsilon^2 F^T F) \widetilde{\mathcal{L}} \\ &= \varepsilon^2 (F \widetilde{\mathcal{L}} + I)^* (F \widetilde{\mathcal{L}} + I) + (\widetilde{C} \widetilde{\mathcal{L}} + D)^* (\widetilde{C} \widetilde{\mathcal{L}} + D) \\ &= \begin{bmatrix} \widetilde{C} \widetilde{\mathcal{L}} + D \\ \varepsilon (F \widetilde{\mathcal{L}} + I) \end{bmatrix}^* \begin{bmatrix} \widetilde{C} \widetilde{\mathcal{L}} + D \\ \varepsilon (F \widetilde{\mathcal{L}} + I) \end{bmatrix}. \end{aligned} \tag{7.18}$$

Notice that $\widetilde{C} \widetilde{\mathcal{L}} + D$ coincides with the Toeplitz operator $\mathbb{T}_{\widetilde{\mathcal{G}}}$ associated with the input–output operator $\widetilde{\mathcal{G}}$ of the system (7.7).

If we denote by $\mathcal{S}$ the input–output operator of the system

$$\left[\begin{array}{c|c} \widetilde{A} & B \\ \hline F & I \end{array} \right],$$

then $F \widetilde{\mathcal{L}} + I$ coincides with the Toeplitz operator $\mathbb{T}_{\mathcal{S}}$. Introducing the notation

$$\mathcal{V}_\varepsilon := \begin{bmatrix} \mathbb{T}_{\widetilde{\mathcal{G}}} \\ \varepsilon \mathbb{T}_{\mathcal{S}} \end{bmatrix}, \tag{7.19}$$

we see from (7.17) that

$$\mathcal{P}_{\widetilde{\Sigma}_\varepsilon} = \widetilde{\Phi}^* \begin{bmatrix} \widetilde{C}^T & \varepsilon F^T \end{bmatrix} \mathcal{V}_\varepsilon$$

and from (7.18) that

$$\mathcal{R}_{\widetilde{\Sigma}_\varepsilon} = \mathcal{V}_\varepsilon^* \mathcal{V}_\varepsilon.$$

Introducing these expressions in (7.15), we have

$$X_\varepsilon = \widetilde{\Phi}^* \begin{bmatrix} \widetilde{C}^T & \varepsilon F^T \end{bmatrix} [I - \mathcal{V}_\varepsilon (\mathcal{V}_\varepsilon^* \mathcal{V}_\varepsilon)^{-1} \mathcal{V}_\varepsilon^*] \begin{bmatrix} \widetilde{C} \\ \varepsilon F \end{bmatrix} \widetilde{\Phi}. \tag{7.20}$$

Consider the operator

$$\mathcal{W}_\varepsilon := I - \mathcal{V}_\varepsilon (\mathcal{V}_\varepsilon^* \mathcal{V}_\varepsilon)^{-1} \mathcal{V}_\varepsilon^*. \tag{7.21}$$

It is easy to check that $\mathcal{W}_\varepsilon = \mathcal{W}_\varepsilon^*$ and $\mathcal{W}_\varepsilon^2 = \mathcal{W}_\varepsilon$, that is $\mathcal{W}_\varepsilon$ is an orthogonal projection operator. In particular, $\mathcal{W}_\varepsilon \geq 0$ and $\|\mathcal{W}_\varepsilon\| = 1$.

We write $\mathcal{W}_\varepsilon$ in a partitioned form in accordance with $\mathcal{V}_\varepsilon$ in (7.19) as

$$\mathcal{W}_\varepsilon = \begin{bmatrix} \mathcal{W}_\varepsilon^{11} & \mathcal{W}_\varepsilon^{12} \\ (\mathcal{W}_\varepsilon^{12})^* & \mathcal{W}_\varepsilon^{22} \end{bmatrix} \tag{7.22}$$

where

$$\mathcal{W}_\varepsilon^{11} = I - \mathbb{T}_{\widetilde{\mathcal{G}}} (\mathcal{V}_\varepsilon^* \mathcal{V}_\varepsilon)^{-1} \mathbb{T}_{\widetilde{\mathcal{G}}}^*. \tag{7.23}$$

The operator $\mathcal{W}_\varepsilon^{11}$ is positive semidefinite since $\mathcal{W}_\varepsilon \geq 0$. Notice also that $\mathcal{V}_\varepsilon^* \mathcal{V}_\varepsilon = \mathbb{T}_{\widetilde{\mathcal{G}}}^* \mathbb{T}_{\widetilde{\mathcal{G}}} + \varepsilon^2 \mathbb{T}_{\mathcal{S}}^* \mathbb{T}_{\mathcal{S}}$ decreases monotonically as $\varepsilon \downarrow 0$ and therefore, as (7.21) reveals, $\mathcal{W}_\varepsilon^{11}$ decreases also monotonically as $\varepsilon \downarrow 0$. We conclude that $\mathcal{W}_\varepsilon^{11}$ has a limit as $\varepsilon \downarrow 0$ and let us denote

$$\lim_{\varepsilon \to 0} \mathcal{W}_\varepsilon^{11} = \mathcal{W}^{11} \geq 0. \tag{7.24}$$

Now, from (7.21), $\mathcal{V}_\varepsilon^* \mathcal{W}_\varepsilon = 0$, or

$$\begin{bmatrix} \mathbb{T}_{\widetilde{\mathcal{G}}}^* & \varepsilon \mathbb{T}_{\mathcal{S}}^* \end{bmatrix} \begin{bmatrix} \mathcal{W}_\varepsilon^{11} & \mathcal{W}_\varepsilon^{12} \\ (\mathcal{W}_\varepsilon^{12})^* & \mathcal{W}_\varepsilon^{22} \end{bmatrix} = 0,$$

which leads to

$$\mathbb{T}_{\widetilde{\mathcal{G}}}^* \mathcal{W}_\varepsilon^{11} + \varepsilon \mathbb{T}_{\mathcal{S}}^* (\mathcal{W}_\varepsilon^{12})^* = 0. \tag{7.25}$$

Since $\|\mathcal{W}_\varepsilon^{12}\| \leq 1$, by taking $\varepsilon \downarrow 0$ in (7.25), we get

$$\mathbb{T}_{\widetilde{\mathcal{G}}}^* \mathcal{W}^{11} = 0. \tag{7.26}$$

As $\mathbb{T}_{\widetilde{\mathcal{G}}}^*$ is one to one (see Remark 7.2.5) we deduce that $\mathcal{W}^{11} = 0$.

With these considerations, we rewrite X_ε in (7.20) as

$$X_\varepsilon = \widetilde{\Phi}^* \begin{bmatrix} \widetilde{C}^T & F^T \end{bmatrix} \begin{bmatrix} \mathcal{W}_\varepsilon^{11} & \varepsilon \mathcal{W}_\varepsilon^{12} \\ \varepsilon (\mathcal{W}_\varepsilon^{12})^* & \varepsilon^2 \mathcal{W}_\varepsilon^{22} \end{bmatrix} \begin{bmatrix} \widetilde{C} \\ F \end{bmatrix} \widetilde{\Phi}.$$

As $\|\mathcal{W}_\varepsilon^{12}\| \leq 1$ and $\|\mathcal{W}_\varepsilon^{22}\| \leq 1$, the limit as $\varepsilon \downarrow 0$ in the previous relation exists and equals 0 which concludes the proof. ■

7.3 Normalized coprime factorizations

In this section we give explicit formulas in terms of realizations for normalized coprime factorizations of a system.

Theorem 7.3.1 *Consider the system*

$$\mathbf{G} = \left[\begin{array}{c|c} A & B \\ \hline C & O \end{array}\right],$$

where $A \in \mathbb{R}^{n\times n}$, $B \in \mathbb{R}^{n\times m}$ and $C \in \mathbb{R}^{p\times n}$, and the pair (A, B) is stabilizable and the pair (C, A) is detectable. Let

$$\Sigma_X = (A, B; C^TC, 0, I), \quad \Sigma_Y = (A^T, C^T; BB^T, 0, I)$$

be the standard Popov triplets for control and estimation, respectively, and let X and Y be the stabilizing solutions to the CTARE(Σ_X) and CTARE(Σ_Y). Define $F := -B^TX$, $K := -YC^T$, $A_F := A + BF$, $A_K := A + KC$, and

$$\begin{array}{ll}
\mathbf{M} = \left[\begin{array}{c|c} A_F & B \\ \hline F & I \end{array}\right], & \mathbf{N} = \left[\begin{array}{c|c} A_F & B \\ \hline C & O \end{array}\right], \\
\mathbf{U} = \left[\begin{array}{c|c} A_K & B \\ \hline -F & I \end{array}\right], & \mathbf{W} = \left[\begin{array}{c|c} A_K & K \\ \hline F & O \end{array}\right], \\
\widetilde{\mathbf{M}} = \left[\begin{array}{c|c} A_K & K \\ \hline C & I \end{array}\right], & \widetilde{\mathbf{N}} = \left[\begin{array}{c|c} A_K & B \\ \hline C & O \end{array}\right], \\
\widetilde{\mathbf{U}} = \left[\begin{array}{c|c} A_F & -K \\ \hline C & I \end{array}\right], & \widetilde{\mathbf{W}} = \left[\begin{array}{c|c} A_F & K \\ \hline F & O \end{array}\right],
\end{array} \tag{7.27}$$

which are all in RH_+^∞. Then we have:

1. The doubly coprime factorization identity

$$\begin{bmatrix} -\mathbf{W} & \mathbf{U} \\ \widetilde{\mathbf{M}} & \widetilde{\mathbf{N}} \end{bmatrix} \begin{bmatrix} -\mathbf{N} & \widetilde{\mathbf{U}} \\ \mathbf{M} & \widetilde{\mathbf{W}} \end{bmatrix} = \begin{bmatrix} I & O \\ O & I \end{bmatrix} \tag{7.28}$$

holds.

2. The pair $(\mathbf{N}, \mathbf{M})$ defines a normalized right coprime factorization *of* $\mathbf{G}$ *over RH_+^∞, i.e.,*

$$\mathbf{G} = \mathbf{N}\mathbf{M}^{-1}, \qquad \mathbf{N}^*\mathbf{N} + \mathbf{M}^*\mathbf{M} = I. \tag{7.29}$$

3. The pair $(\widetilde{\mathbf{N}}, \widetilde{\mathbf{M}})$ defines a normalized left coprime factorization *of* $\mathbf{G}$ *over RH_+^∞, i.e.,*

$$\mathbf{G} = \widetilde{\mathbf{M}}^{-1}\widetilde{\mathbf{N}}, \qquad \widetilde{\mathbf{N}}\widetilde{\mathbf{N}}^* + \widetilde{\mathbf{M}}\widetilde{\mathbf{M}}^* = I. \tag{7.30}$$

Proof. Notice first that the pairs (A, B) and (C, A) being stabilizable and detectable, respectively, it follows with Corollary 4.2.2 that the CTARE(Σ_X) and the CTARE(Σ_Y) have stabilizing solutions $X \geq 0$ and $Y \geq 0$, respectively, and therefore the matrices A_F and A_K are stable.

1. The proof of (7.28) is a matter of direct substitutions and straightforward computations.

2. The first identity in (7.29) follows by direct substitution. With the spectral factorization identity (3.39) we get for the Popov function associated with Σ_X,

$$\mathbf{\Pi}_{\Sigma_X} = I + \mathbf{G}^*\mathbf{G} = \mathbf{S}_F^*\mathbf{S}_F \tag{7.31}$$

where

$$\mathbf{S}_F = \left[\begin{array}{c|c} A & B \\ \hline -F & I \end{array}\right]. \tag{7.32}$$

It is easy to check that

$$\mathbf{M} = \mathbf{S}_F^{-1} = \left[\begin{array}{c|c} A_F & B \\ \hline F & I \end{array}\right]. \tag{7.33}$$

With (7.31) we get

$$\mathbf{S}_F^{-*}\mathbf{S}_F^{-1} + \mathbf{S}_F^{-*}\mathbf{G}^*\mathbf{G}\mathbf{S}_F^{-1} = I \tag{7.34}$$

from which we see that the second identity in (7.29) is nothing but a mere rewriting of (7.34) with (7.33) and the first identity in (7.29).

Assertion 3 of the theorem follows now by duality. ■

7.4 The Small Gain Theorem

In this section we give a state–space proof a classical result, known as the Small Gain Theorem, which is a very useful tool to assert the stability of a feedback closed–loop system.

Theorem 7.4.1 (Small Gain Theorem) *Let*

$$\mathbf{G}_i = \left[\begin{array}{c|c} A_i & B_i \\ \hline C_i & D_i \end{array}\right], \quad y_i = \mathbf{G}_i u_i, \quad i = 1, 2,$$

be two systems with A_i stable, $i = 1, 2$, and the TFMs $\mathbf{G}_1$ and $\mathbf{G}_2$ of dimensions $p \times m$ and $m \times p$, respectively. Assume that $S_c := I_m - D_2 D_1$ (or, equivalently $\widehat{S}_c := I_p - D_1 D_2$) is nonsingular. If $\|\mathbf{G}_1\|_\infty < \frac{1}{\gamma}$ and $\|\mathbf{G}_2\|_\infty \leq \gamma$ for some $\gamma > 0$, then the closed–loop feedback system of $\mathbf{G}_1$ and $\mathbf{G}_2$ is (internally) stable.

Proof. To show the internal stability of the feedback closed–loop system (see Section 2.1.5), it suffices to prove that its state matrix

$$A_R = \begin{bmatrix} A_1 + B_1 S_c^{-1} D_2 C_1 & B_1 S_c^{-1} C_2 \\ B_2 \widehat{S}_c^{-1} C_1 & A_2 + B_2 \widehat{S}_c^{-1} D_1 C_2 \end{bmatrix} \tag{7.35}$$

is stable, i.e., $\Lambda(A_R) \subset \mathbb{C}_-$.

From the hypotheses, there exists $\widehat{\gamma} > \gamma$ such that $\|\mathbf{G}_2\| < \widehat{\gamma}$ and $\|\mathbf{G}_1\| < \frac{1}{\widehat{\gamma}}$. Without restricting the generality we may set $\widehat{\gamma} = 1$ by performing a mere scaling on $\mathbf{G}_1$ and $\mathbf{G}_2$. Further, since both $\mathbf{G}_i$, $i = 1, 2$, are stable and $\|\mathbf{G}_i\|_\infty < 1$, we may apply Lemma 7.1.1 and obtain that the following KPYSs

$$\begin{aligned} -I + D_1^T D_1 &= -V_1^T V_1, \\ C_1^T D_1 + X_1 B_1 &= -W_1^T V_1, \\ C_1^T C_1 + A_1^T X_1 + X_1 A_1 &= -W_1^T W_1, \end{aligned} \tag{7.36}$$

and

$$\begin{aligned} -I + D_2^T D_2 &= -V_2^T V_2, \\ C_2^T D_2 + X_2 B_2 &= -W_2^T V_2, \\ C_2^T C_2 + A_2^T X_2 + X_2 A_2 &= -W_2^T W_2, \end{aligned} \tag{7.37}$$

have stabilizing solutions (X_1, V_1, W_1) with $X_1 \geq 0$, and (X_2, V_2, W_2) with $X_2 \geq 0$, respectively. Define

$$X_R := \begin{bmatrix} X_1 & O \\ O & X_2 \end{bmatrix} \geq 0,$$

$$C_R := \begin{bmatrix} W_1 + V_1 S_c^{-1} D_2 C_1 & V_1 S_c^{-1} C_2 \\ V_2 \widehat{S}_c^{-1} C_1 & W_2 + V_2 \widehat{S}_c^{-1} D_1 C_2 \end{bmatrix},$$

$$K_R := \begin{bmatrix} -B_1 V_1^{-1} & O \\ O & -B_2 V_2^{-1} \end{bmatrix}.$$

It is easy to check that

$$A_R^T X_R + X_R A_R + C_R^T C_R = 0 \tag{7.38}$$

and

$$A_R + K_R C_R = \begin{bmatrix} A_1 - B_1 V_1^{-1} W_1 & O \\ O & A_2 - B_2 V_2^{-1} W_2 \end{bmatrix} \tag{7.39}$$

which is stable because (X_1, V_1, W_1) and (X_2, V_2, W_2) are stabilizing solutions to the respective KPYSs. Thus the pair (C_R, A_R) is detectable. Using this conclusion in the Lyapunov equation (7.38) which is fulfilled for $X_R \geq 0$, it follows from Theorem 1.5.5 that $\Lambda(A_R) \subset \mathbb{C}_-$, and the proof is completed. ∎

7.5 A quadratic index with constrained dynamics

In this section we discuss a relevant application of the GCTARS: the quadratic control problem with constrained dynamics.

Let $\Sigma = (A, B; Q, L, R)$ be a Popov triplet and associate with it:

1. The linear system

$$\dot{x} = Ax + Bu, \quad x(0) = x_0, \tag{7.40}$$

and the integral quadratic index

$$J_\Sigma(x_0, u) = \int_0^\infty \begin{bmatrix} x \\ u \end{bmatrix}^T \begin{bmatrix} Q & L \\ L^T & R \end{bmatrix} \begin{bmatrix} x \\ u \end{bmatrix} \mathrm{d}t. \tag{7.41}$$

2. The GCTARS(Σ) given in (6.8) and written in explicit form for $V = V_1$ as

$$\begin{aligned} (A^T X + XA + Q)V_1 &+ (XB + L)FV_1 &= 0, \\ (L^T + B^T X)V_1 &+ RFV_1 &= 0. \end{aligned} \tag{7.42}$$

Assume that the GCTARS(Σ) has a stabilizing solution (X, F, r, V_1) for which (7.42) is fulfilled. Then

$$X_1 := V_1^T X V_1 = V_1^T X^T V_1 = X_1^T \tag{7.43}$$

and

$$(A + BF)V_1 = V_1 S \tag{7.44}$$

holds with S stable. In particular, $\mathcal{V}_1 = \operatorname{Im} V_1$ is a stable (A, B)–invariant subspace of dimension r. Hence, for any initial condition $x_0 = V_1\xi_0 \in \mathcal{V}_1$, $\xi_0 \in \mathbb{R}^r$, we can force the system (7.40) to have its evolution exclusively located in $\mathcal{V}_1$, i.e.,

$$V_1\dot{\xi} = AV_1\xi + Bu, \ \xi(0) = \xi_0 \in \mathbb{R}^r. \tag{7.45}$$

In this section we shall evaluate the quadratic index in (7.41) over the trajectories located in $\mathcal{V}_1$ described by (7.45).

Define

$$Q_1 := V_1^T Q V_1, \ L_1 := V_1^T L, \ F_1 := FV_1. \tag{7.46}$$

With (7.44) and (7.46), (7.42) becomes

$$A^T X V_1 + X V_1 S + Q V_1 + L F_1 = 0. \tag{7.47}$$

Premultiplying (7.47) with V_1^T and using (7.44) transposed and (7.46) we get

$$S^T X_1 - F_1^T B^T X V_1 + X_1 S + Q_1 + L_1 F_1 = 0. \tag{7.48}$$

From (7.42) we have also

$$F_1^T B^T X V_1 + F_1^T L_1^T + F_1^T R F_1 = 0. \tag{7.49}$$

Using (7.49) in (7.48) yields

$$S^T X_1 + X_1 S + Q_1 + L_1 F_1 + F_1^T L_1^T + F_1^T R F_1 = 0. \tag{7.50}$$

Notice that if V_1 is nonsingular then we may take $V_1 = I_n$ in (7.42). Then (7.50) becomes

$$(A + BF)^T X + X(A + BF) + Q + LF + F^T L^T + F^T RF = 0 \tag{7.51}$$

with $A + BF$ stable, which is the "closed–loop" form of the algebraic Riccati equation.

Now we are ready for the main result of this section.

Proposition 7.5.1 *Suppose that the GCTARS(Σ) in (7.42) has a stabilizing solution. Then the quadratic index (7.41) takes the form*

$$J_\Sigma(V_1\xi_0, u) = \xi_0^T X_1 \xi_0 + \int_0^\infty (u - F_1\xi)^T R (u - F_1\xi) \mathrm{d}t \tag{7.52}$$

for any pair (ξ, u) that satisfies (7.45), where $X_1 := V_1^T X V_1$ and $F_1 := FV_1$. If in addition $R \geq 0$, then the minimum of $J_\Sigma(V_1\xi_0, u)$ over the class $\mathcal{U}_{\xi_0}$ of all control inputs which steer ξ_0 into the origin is attained (nonuniquely) for $u = F_1\xi$.

Proof. Since the GCTARS(Σ) has a stabilizing solution (X, F, r, V_1), we have from (7.44) and (7.45)

$$V_1\dot{\xi} = V_1 S\xi - BF_1\xi + Bu.$$

Premultiplying both sides by $V_1^T X^T$ we get with (7.43)

$$X_1\dot{\xi} = X_1 S\xi - V_1^T X^T B F_1 \xi + V_1^T X^T B u. \tag{7.53}$$

Equation (7.53) can be seen as the evolution of the system (7.40) constrained to $\mathcal{V}_1$. If the evolution is located in $\mathcal{V}_1$, the quadratic index (7.41) becomes

$$J_\Sigma(V_1\xi_0, u) = \int_0^\infty \begin{bmatrix} \xi \\ u \end{bmatrix}^T \begin{bmatrix} Q_1 & L_1 \\ L_1^T & R \end{bmatrix} \begin{bmatrix} \xi \\ u \end{bmatrix} \mathrm{d}t, \tag{7.54}$$

where we have used $x = V_1\xi$ and (7.46). With (7.50) and (7.53), we have successively

$$\begin{aligned}
&\begin{bmatrix} \xi \\ u \end{bmatrix}^T \begin{bmatrix} Q_1 & L_1 \\ L_1^T & R \end{bmatrix} \begin{bmatrix} \xi \\ u \end{bmatrix} \\
= \;& \xi^T Q_1 \xi + 2u^T L_1^T \xi + u^T R u \\
= \;& -\xi^T (S^T X_1 + X_1 S + L_1 F_1 + F_1^T L_1^T + F_1^T R F_1)\xi + 2u^T L_1^T \xi + u^T R u \\
= \;& -\xi^T (X_1\dot{\xi} + V_1^T X^T B F_1 \xi - V_1^T X^T B u) - (\dot{\xi}^T X_1 + \xi^T F_1^T B^T X V_1 \\
& -u^T B^T X V_1)\xi - 2\xi^T F_1^T L_1^T \xi - \xi^T F_1^T R F_1 \xi + 2u^T L_1^T \xi + u^T R u \\
= \;& -\frac{\mathrm{d}}{\mathrm{d}t}(\xi^T X_1 \xi) - 2\xi^T (F_1^T B^T X V_1 + F_1^T L_1^T + F_1^T R F_1)\xi \\
& +2u^T (B^T X V_1 + L_1^T)\xi + u^T R u + \xi^T F_1^T R F_1 \xi \\
= \;& -\frac{\mathrm{d}}{\mathrm{d}t}(\xi^T X_1 \xi) + (u - F_1\xi)^T R (u - F_1\xi)
\end{aligned}$$

where in the last equality we have used (7.49) and (7.42). Since $\mathcal{V}_1$ is a stable (A, B)–invariant subspace it follows that the class $\mathcal{U}_{\xi_0}$ of all control inputs which steer ξ_0 in the origin is nonempty. Hence we get

$$J_\Sigma(V_1\xi_0, u) = \xi_0^T X_1 \xi_0 + \int_0^\infty (u - F_1\xi)^T R (u - F_1\xi)\mathrm{d}t \tag{7.55}$$

for all $\xi_0 \in \mathbb{R}^r$ and all $u \in \mathcal{U}_{\xi_0}$. If $R \geq 0$, the minimum of $J_\Sigma(V_1\xi_0, u)$ with respect to $u \in \mathcal{U}_{\xi_0}$ equals $\xi_0^T X_1 \xi_0$ and it is attained (nonuniquely) by $u = F_1\xi$. Finally, notice that the Riccati feedback F_1 also provides stabilization with respect to the stable (A, B)–invariant subspace $\mathcal{V}_1$. ■

7.6 Spectral and inner–outer factorizations: the arbitrary rank case

In this section we solve the spectral and inner–outer factorization problems for a system with a proper TFM of constant rank (not necessarily maximal) on the extended imaginary axis. The factorization problems are reduced to the computation of the stabilizing solution to the CTARS. We start with the precise definitions.

Definition 7.6.1 *Let* $\mathbf{G}(s) \in RH^\infty_{+,p\times m}$ *without zeros on the extended imaginary axis, i.e.,*

$$\text{rank } \mathbf{G}(j\omega) = k \; (= \text{rank}_{\mathbb{R}(s)} \mathbf{G}(s)), \; \forall \omega \in \overline{\mathbb{R}}. \tag{7.56}$$

1. A matrix function $\mathbf{G}_o(s) \in RH^\infty_{+,k\times m}$ *with a right inverse in* $RH^\infty_{+,m\times k}$ *such that*

$$\mathbf{G}^*(s)\mathbf{G}(s) = \mathbf{G}_o^*(s)\mathbf{G}_o(s) \tag{7.57}$$

is called a spectral factor *of* $\mathbf{G}$.

2. Let $\mathbf{G}_i := \mathbf{G}\mathbf{G}_o^\dagger \in RH^\infty_{+,p\times k}$ *where* $\mathbf{G}_o^\dagger$ *is a right inverse of* $\mathbf{G}_o$ *in* RH^∞_+*. Then* $\mathbf{G}_i$ *is* inner, *i.e., is in* RH^∞_+ *and satisfies*

$$\mathbf{G}_i^*(s)\mathbf{G}_i(s) = I \tag{7.58}$$

and

$$\mathbf{G}(s) = \mathbf{G}_i(s)\mathbf{G}_o(s). \tag{7.59}$$

The relation (7.59) defines an inner–outer factorization *of* $\mathbf{G}$.

The following theorem gives the theoretical support for performing spectral and inner–outer factorizations in the general case through realizations.

Theorem 7.6.2 *Let*

$$\mathbf{G} = \left[\begin{array}{c|c} A & B \\ \hline C & D \end{array}\right] \tag{7.60}$$

be given, where $A \in \mathbb{R}^{n\times n}, B \in \mathbb{R}^{n\times m}, C \in \mathbb{R}^{p\times n}, D \in \mathbb{R}^{p\times m}$. *Assume* $\mathbf{G}$ *has no zeros on the extended imaginary axis (i.e., satisfies (7.56)) and the realization (7.60) is stabilizable and* $\mathbb{C}_0$*–observable. Define the positiveness Popov triplet* $\Sigma = (A, B; C^TC, C^TD, D^TD)$. *Then we have:*

1. The CTARS(Σ*) has a stabilizing solution* (X, F).

2. Let H *be a full row rank matrix such that* $H^TH = D^TD$. *Define*

$$\begin{aligned} \mathbf{G}_o &:= \left[\begin{array}{c|c} A & B \\ \hline -HF & H \end{array}\right], \\ \mathbf{G}_i &:= \left[\begin{array}{c|c} A+BF & BH^\dagger \\ \hline C+DF & DH^\dagger \end{array}\right], \end{aligned} \tag{7.61}$$

where $H^\dagger$ *is an inverse to the right of* H. *Then* $\mathbf{G}_o$ *satisfies (7.57),* $\mathbf{G}_i$ *is inner and together they satisfy (7.59). If, in addition,* A *is stable (i.e.,* $\mathbf{G} \in RH^\infty_{+,p\times m}$*), then* $\mathbf{G}_o$ *is the spectral and* $\mathbf{G}_i$ *is the inner factor of* $\mathbf{G}$.

Proof. 1. We make intensive use of the notation given in Section 6.1. Using Theorem 6.2.4 it is enough to prove that the EHP(Σ) written explicitly as

$$sM_\Sigma - N_\Sigma := s\begin{bmatrix} I & O & O \\ O & I & O \\ O & O & O \end{bmatrix} - \begin{bmatrix} A & O & B \\ -C^TC & -A^T & -C^TD \\ D^TC & B^T & D^TD \end{bmatrix} \tag{7.62}$$

is dichotomic and, moreover, stable disconjugate. Notice that this EHP could be singular. We begin by proving that $sM_\Sigma - N_\Sigma$ is dichotomic which means $n_f^0 = 0$ and $\pi_0 = 0$.

Proof of $n_f^0 = 0$. Since Σ is the positiveness Popov triplet associated with $\mathbf{G}$ (see Section 3.4), the Popov function associated with Σ takes the form $\mathbf{\Pi}_\Sigma(s) = \mathbf{G}^*(s)\mathbf{G}(s)$ and has a realization given by

$$\mathbf{\Pi}_\Sigma = \left[\begin{array}{c|c} A_R & B_R \\ \hline C_R & D_R \end{array}\right] := \left[\begin{array}{cc|c} A & O & B \\ -C^TC & -A^T & -C^TD \\ \hline D^TC & B^T & D^TD \end{array}\right].$$

Since the pair (A, B) is assumed stabilizable and the pair (C, A) is assumed $\mathbb{C}_0$–observable, we use Definition 1.4.1 to show that the matrix pairs (A_R, B_R) and (C_R, A_R) are $\mathbb{C}_0$–controllable and $\mathbb{C}_0$–observable, respectively. This, together with the assumption that $\mathbf{\Pi}_\Sigma(s)$ has no zeros on the extended imaginary axis, shows that $sM_\Sigma - N_\Sigma$, which is the transmission pencil of $\mathbf{\Pi}_\Sigma(s)$, has no generalized eigenvalues on the imaginary axis. Therefore $n_f^0 = 0$.

Proof of $\pi_0 = 0$. Evaluating $\mathbf{\Pi}_\Sigma(s^{-1})$ in zero or, equivalently, $\mathbf{\Pi}_\Sigma(s)$ at infinity, we get $\mathbf{\Pi}_\Sigma(\infty) = R = D^TD$. By the assumption that $\mathbf{G}$ has no zeros at infinity, neither has $\mathbf{\Pi}_\Sigma(s)$ and therefore $\pi_0 = 0$.

Thus the EHP(Σ) is dichotomic and by Theorem 6.1.3 there exists a proper $\mathbb{C}_-$–deflating subspace $\mathcal{V}$ of dimension n. Let V be a basis matrix for $\mathcal{V}$ partitioned conformably with the EHP(Σ) as

$$V = \begin{bmatrix} V_1 \\ V_2 \\ V_3 \end{bmatrix}.$$

We prove now that $\mathcal{V} = \operatorname{Im} V$ is disconjugate which means that V_1 is invertible. We show this by proving that $\operatorname{Ker} V_1 = 0$.

Since $\mathcal{V}$ is a proper $\mathbb{C}_-$–deflating subspace we have $M_\Sigma VS = N_\Sigma V$ for a stable S or, explicitly,

$$\begin{array}{ccccc} AV_1 & & +BV_3 & = & V_1S, \\ -C^TCV_1 & -A^TV_2 & -C^TDV_3 & = & V_2S, \\ D^TCV_1 & +B^TV_2 & +D^TDV_3 & = & 0. \end{array} \tag{7.63}$$

We prove first that

$$\operatorname{Ker} V_1 \subset \operatorname{Ker} V_1^TV_2 \subset \operatorname{Ker} V_2. \tag{7.64}$$

Adding the second and the third equations premultiplied by $-V_1^T$ and V_3^T, respectively, and using the first one in (7.63) we obtain

$$V_1^TV_2S + S^TV_1^TV_2 + (CV_1 + DV_3)^T(CV_1 + DV_3) = 0. \tag{7.65}$$

Proposition 6.1.8 shows that $V_1^TV_2 = V_2^TV_1$ and the first inclusion in (7.64) follows immediately. Let $x \in \operatorname{Ker} V_1^TV_2$. Pre– and post–multiplying (7.65) by x^T and x, respectively, we get $(CV_1 + DV_3)x = 0$. Hence

$$\operatorname{Ker} V_1^TV_2 \subset \operatorname{Ker}(CV_1 + DV_3). \tag{7.66}$$

Further, postmultiplying (7.65) by x gives $V_1^TV_2Sx = 0$ which shows that $\operatorname{Ker} V_1^TV_2$ is an invariant subspace of S. Let K be a basis matrix for $\operatorname{Ker} V_1^TV_2$. Then there exists a square matrix J such that

$$SK = KJ, \quad \Lambda(J) \subset \Lambda(S). \tag{7.67}$$

From (7.66) we get $(CV_1+DV_3)K=0$. Postmultiplying the second and third equations (7.63) by K and using (7.67) we obtain

$$\begin{aligned} -A^TV_2K &= V_2KJ, \\ B^TV_2K &= 0. \end{aligned} \tag{7.68}$$

Let $\widetilde{F}$ be a stabilizing feedback for the stabilizable pair (A,B). From (7.68) we obtain

$$(A+B\widetilde{F})^TV_2K+V_2KJ=0 \tag{7.69}$$

where $A+B\widetilde{F}$ and J are both stable. Thus the unique solution to the Sylvester equation (7.69) is $V_2K=0$. So $\operatorname{Ker}V_1^TV_2\subset\operatorname{Ker}V_2$ and we conclude that (7.64) holds.

Now let $y\in\operatorname{Ker}V_1$. Then $V_1y=0$ and $V_2y=0$. Since $\mathcal{V}$ is a proper deflating subspace,

$$MV=\begin{bmatrix} V_1 \\ V_2 \end{bmatrix}$$

has full column rank and we conclude that $y=0$. Thus $\operatorname{Ker}V_1=\{0\}$ which shows that $\mathcal{V}$ is disconjugate.

In conclusion, we have shown that the EHP(Σ) is dichotomic and stable disconjugate which implies by Theorem 6.2.4 that the CTARS(Σ) has a stabilizing solution (X,F).

2. It is easy to check that a stable right inverse of $\mathbf{G}_o$ is

$$\mathbf{G}_o^\dagger=\left[\begin{array}{c|c} A+BF & BH^\dagger \\ \hline F & H^\dagger \end{array}\right]. \tag{7.70}$$

Then using (7.70) and (7.61) we can check by using the spectral factorization identity (3.39) that (7.57), (7.58), and (7.59) all hold.

Finally, if $\mathbf{G}$ is stable, then A and $A+BF$ are both stable and therefore $\mathbf{G}_o$ is the spectral factor and $\mathbf{G}_i$ is the inner factor of $\mathbf{G}$. ∎

B. The discrete–time case

7.7 The bounded real lemma

In this section we give the bounded real lemma for discrete–time systems.

Lemma 7.7.1 (Generalized bounded real lemma) *Let $\gamma>0$ and*

$$\mathbf{G}=\left[\begin{array}{c|c} A & B \\ \hline C & D \end{array}\right] \tag{7.71}$$

be given. Let $\Sigma_c=(A,B;C^TC,C^TD,-\gamma^2I+D^TD)$ be the γ–contractiveness Popov triplet associated with $\mathbf{G}$.

I. *The following statements are equivalent:*

1. A is stable and $\|\mathbf{G}\|_\infty<\gamma$.

2. The KSPYS($\Sigma_c,-I$) given by

$$\begin{aligned} -\gamma^2I+D^TD+B^TXB &= -V^TV, \\ C^TD+A^TXB &= -W^TV, \\ C^TC+A^TXA-X &= -W^TW, \end{aligned} \tag{7.72}$$

has a stabilizing solution (X, V, W) with $X \geq 0$.

II. *The following statements are equivalent:*

1. A is dichotomic, the pair (A, B) is stabilizable, and $\|\mathbf{G}\|_\infty < \gamma$.

2. The KSPYS($\Sigma_c, -I$) (7.72) has a stabilizing solution (X, V, W).

Moreover, if point 2 holds then we have the following relations between the inertia of X and A: $\delta_c(X) + \pi_c(X) = \nu_d(A)$; $\nu_c(X) = \pi_d(A)$.

The proof is identical to the continuous–time counterpart and is therefore omitted.

Corollary 7.7.2 (Strict bounded real lemma) *Let $\gamma > 0$ and*

$$\mathbf{G} = \left[\begin{array}{c|c} A & B \\ \hline C & D \end{array}\right]$$

be given. Then the following two assertions are equivalent:

1. A is stable and $\|\mathbf{G}\|_\infty < \gamma$.

2. There exists $X > 0$ such that $-\gamma^2 I + D^T D + B^T X B < 0$ and

$$\begin{bmatrix} A^T X A - X + C^T C & A^T X B + C^T D \\ D^T C + B^T X A & -\gamma^2 I + D^T D + B^T X B \end{bmatrix} < 0.$$

7.8 The DTARE associated with the positiveness Popov triplet

In this section we study some convergence properties of a family of DTAREs associated with positiveness Popov triplets. This result will serve as a basic ingredient for solving the robust stabilization problem with multiplicative uncertainties in Chapter 11.

Proposition 7.8.1 *Let*

$$\mathbf{G} = \left[\begin{array}{c|c} A & B \\ \hline C & D \end{array}\right]$$

be given. Let $\Sigma_p = (A, B; C^T C, C^T D, D^T D)$ be the positiveness Popov triplet associated with $\mathbf{G}$. Assume that:

a. *The pair (A, B) is stabilizable.*

b. $\begin{bmatrix} e^{j\theta} I - A & -B \\ C & D \end{bmatrix}$ *is of full column rank for all $\theta \in [0,\ 2\pi)$.*

Then the DTARE(Σ_p) has a positive semidefinite stabilizing solution.

Notice the difference with respect to the continuous–time counterpart which is Proposition 7.2.1 and where D was additionally assumed to have full column rank.

Remark 7.8.2 By Proposition 7.8.1, the proof of Corollary 4.8.2 follows straightforwardly. Indeed, for

$$\mathbf{G}_0 = \left[\begin{array}{c|c} A & B \\ \hline C & O \\ O & I \end{array}\right]$$

the associated positiveness Popov triplet is $\Sigma_p = \Sigma_s = (A, B; C^T C, 0, I)$ (see (3.83)), i.e., coincides with the standard Popov triplet for control, and the DTARE(Σ_p) is exactly (4.152).

Finally, we notice that the discrete counterpart of Theorem 7.2.3 is identical, with obvious updating, and therefore we omit the details.

7.9 Normalized coprime factorizations

In this section we give explicit formulas in terms of realizations for normalized coprime factorizations of a discrete–time system.

Theorem 7.9.1 *Consider the discrete–time system*

$$\mathbf{G} = \left[\begin{array}{c|c} A & B \\ \hline C & O \end{array}\right],$$

where $A \in \mathbb{R}^{n\times n}$, $B \in \mathbb{R}^{n\times m}$ and $C \in \mathbb{R}^{p\times n}$, and the pair (A,B) is stabilizable and the pair (C,A) is detectable. Let

$$\Sigma_X = (A, B; C^TC, 0, I), \quad \Sigma_Y = (A^T, C^T; BB^T, 0, I)$$

be the standard Popov triplets for control and estimation, respectively, and let X and Y be the stabilizing solutions to the DTARE(Σ_X) and DTARE(Σ_Y). Let $F := -(I + B^TXB)^{-1}B^TXA$, $K := -AYC^T(I + CYC^T)^{-1}$, $V^TV := I + B^TXB$, $\widehat{V}\widehat{V}^T := I + CYC^T$, $A_F := A + BF$, $A_K := A + KC$, and

$$\begin{array}{ll} \mathbf{M} = \left[\begin{array}{c|c} A_F & BV^{-1} \\ \hline F & V^{-1} \end{array}\right], & \mathbf{N} = \left[\begin{array}{c|c} A_F & BV^{-1} \\ \hline C & O \end{array}\right], \\ \mathbf{U} = \left[\begin{array}{c|c} A_K & B \\ \hline -VF & V \end{array}\right], & \mathbf{W} = \left[\begin{array}{c|c} A_K & K \\ \hline VF & O \end{array}\right], \\ \widetilde{\mathbf{M}} = \left[\begin{array}{c|c} A_K & K \\ \hline \widehat{V}^{-1}C & \widehat{V}^{-1} \end{array}\right], & \widetilde{\mathbf{N}} = \left[\begin{array}{c|c} A_K & B \\ \hline \widehat{V}^{-1}C & O \end{array}\right], \\ \widetilde{\mathbf{U}} = \left[\begin{array}{c|c} A_F & -K\widehat{V} \\ \hline C & \widehat{V} \end{array}\right], & \widetilde{\mathbf{W}} = \left[\begin{array}{c|c} A_F & K\widehat{V} \\ \hline F & O \end{array}\right], \end{array} \tag{7.73}$$

which are all in RH^∞_+. Then we have:

1. The doubly coprime factorization identity

$$\begin{bmatrix} -\mathbf{W} & \mathbf{U} \\ \widetilde{\mathbf{M}} & \widetilde{\mathbf{N}} \end{bmatrix} \begin{bmatrix} -\mathbf{N} & \widetilde{\mathbf{U}} \\ \mathbf{M} & \widetilde{\mathbf{W}} \end{bmatrix} = \begin{bmatrix} I & O \\ O & I \end{bmatrix}. \tag{7.74}$$

holds.

2. The pair $(\mathbf{N},\mathbf{M})$ defines a normalized right coprime factorization *of $\mathbf{G}$ over RH^∞_+, i.e.,*

$$\mathbf{G} = \mathbf{N}\mathbf{M}^{-1}, \qquad \mathbf{N}^*\mathbf{N} + \mathbf{M}^*\mathbf{M} = I. \tag{7.75}$$

3. The pair $(\widetilde{\mathbf{N}},\widetilde{\mathbf{M}})$ defines a normalized left coprime factorization *of $\mathbf{G}$ over RH^∞_+, i.e.,*

$$\mathbf{G} = \widetilde{\mathbf{M}}^{-1}\widetilde{\mathbf{N}}, \qquad \widetilde{\mathbf{N}}\widetilde{\mathbf{N}}^* + \widetilde{\mathbf{M}}\widetilde{\mathbf{M}}^* = I. \tag{7.76}$$

Proof. The proof runs similarly to the proof of Theorem 7.3.1 and is therefore omitted. ∎

7.10 The Small Gain Theorem

We give in this section a state–space proof of the Small Gain Theorem for discrete–time systems.

Theorem 7.10.1 (Small Gain Theorem) *Let*

$$\mathbf{G}_i = \left[\begin{array}{c|c} A_i & B_i \\ \hline C_i & D_i \end{array}\right], \quad y_i = \mathbf{G}_i u_i \quad i = 1,2,$$

be two discrete–time stable systems with A_i, $i = 1,2$ stable, and the TFMs $\mathbf{G}_1$ and $\mathbf{G}_2$ of dimensions $p \times m$ and $m \times p$, respectively. Assume that $S_c := I_m - D_2 D_1$ (or, equivalently, $\widehat{S}_c := I_p - D_1 D_2$) is nonsingular. If $\|\mathbf{G}_1\|_\infty < \frac{1}{\gamma}$ and $\|\mathbf{G}_2\|_\infty \leq \gamma$ for some $\gamma > 0$, then the closed loop feedback configuration of $\mathbf{G}_2$ and $\mathbf{G}_1$ is (internally) stable.

Proof. The state matrix of the closed–loop system is (see Section 2.1.5)

$$A_R = \left[\begin{array}{cc} A_1 + B_1 S_c^{-1} D_2 C_1 & B_1 S_c^{-1} C_2 \\ B_2 \widehat{S}_c^{-1} C_1 & A_2 + B_2 \widehat{S}_c^{-1} D_1 C_2 \end{array}\right] \tag{7.77}$$

and we have to show that $\Lambda(A_R) \subset \mathbb{D}$. Similar as we did in the proof of Theorem 7.4.1, we can assume, without restricting the generality, that $\gamma = 1$ and $\|\mathbf{G}_i\|_\infty < 1$, $i = 1,2$. It follows from Lemma 7.7.1 that there exist stabilizing solutions (X_1, V_1, W_1) and (X_2, V_2, W_2), with $X_1 \geq 0$ and $X_2 \geq 0$, to the following KSPYSs

$$\begin{array}{rcl} -I + D_1^T D_1 + B_1^T X_1 B_1 & = & -V_1^T V_1, \\ C_1^T D_1 + A_1^T X_1 B_1 & = & -W_1^T V_1, \\ C_1^T C_1 + A_1^T X_1 A_1 - X_1 & = & -W_1^T W_1, \end{array} \tag{7.78}$$

and

$$\begin{array}{rcl} -I + D_2^T D_2 + B^T X_2 B_2 & = & -V_2^T V_2, \\ C_2^T D_2 + A_2^T X_2 B_2 & = & -W_2^T V_2, \\ C_2^T C_2 + A_2^T X_2 A_2 - X_2 & = & -W_2^T W_2, \end{array} \tag{7.79}$$

respectively. Let

$$X_R := \left[\begin{array}{cc} X_1 & O \\ O & X_2 \end{array}\right] \geq 0,$$

$$C_R := \left[\begin{array}{cc} W_1 + V_1 S_c^{-1} D_2 C_1 & V_1 S_c^{-1} C_2 \\ V_2 \widehat{S}_c^{-1} C_1 & W_2 + V_2 \widehat{S}_c^{-1} D_1 C_2 \end{array}\right],$$

$$K_R := \left[\begin{array}{cc} -B_1 V_1^{-1} & O \\ O & -B_2 V_2^{-1} \end{array}\right].$$

It is easy to check that

$$A_R^T X_R A_R - X_R + C_R^T C_R = 0 \tag{7.80}$$

and

$$A_R + K_R C_R = \left[\begin{array}{cc} A_1 - B_1 V_1^{-1} W_1 & O \\ O & A_2 - B_2 V_2^{-1} W_2 \end{array}\right]$$

which is stable because (X_1, V_1, W_1) and (X_2, V_2, W_2) are stabilizing solutions to the respective KSPYSs. Thus the pair (C_R, A_R) is detectable. Using this conclusion in the symmetric Stein equation (7.80) which is fulfilled for $X_R \geq 0$, it follows from Theorem 1.5.5 that $\Lambda(A_R) \subset \mathbb{D}$, and the proof is completed. ■

7.11 A quadratic index with constrained dynamics

In this section we discuss the problem of quadratic control with constrained dynamics in discrete–time.

Let $\Sigma = (A, B; Q, L, R)$ be a Popov triplet and associate with it:

1. The discrete–time linear system

$$\sigma x = Ax + Bu, \quad x(0) = x_0, \tag{7.81}$$

and the quadratic index

$$J_\Sigma(x_0, u) = \sum_{k=0}^{\infty} \begin{bmatrix} x_k \\ u_k \end{bmatrix}^T \begin{bmatrix} Q & L \\ L^T & R \end{bmatrix} \begin{bmatrix} x_k \\ u_k \end{bmatrix}. \tag{7.82}$$

2. The GDTARS(Σ) given in (6.33) and written in explicit form for $V = V_1$ as

$$\begin{array}{rcrcl} (A^TXA - X + Q)V_1 & + & (A^TXB + L)FV_1 & = & 0, \\ (L^T + B^TXA)V_1 & + & (R + B^TXB)FV_1 & = & 0. \end{array} \tag{7.83}$$

Assume that the GDTARS(Σ) has a stabilizing solution (r, X, V_1, F) for which (7.83) is fulfilled. Then

$$X_1 := V_1^T X V_1 = V_1^T X^T V_1 = X_1^T, \tag{7.84}$$

and

$$(A + BF)V_1 = V_1 S \tag{7.85}$$

holds with S stable. In particular, $\mathcal{V}_1 = \operatorname{Im} V_1$ is a stable (A, B)–invariant subspace of dimension r. Hence, for any initial condition $x_0 = V_1\xi_0 \in \mathcal{V}_1$, $\xi_0 \in \mathbb{R}^r$, and an appropriate control input $u \in \ell_+^{2,m}$ we can force the system (7.81) to have its evolution exclusively located in $\mathcal{V}_1$. Thus we can write

$$V_1 \sigma \xi = AV_1 \xi + Bu, \ \xi(0) = \xi_0 \in \mathbb{R}^r. \tag{7.86}$$

In this section we evaluate the quadratic index (7.82) along the trajectories located in $\mathcal{V}_1$ described by (7.86), with $u \in \ell_+^{2,m}$.

Define

$$Q_1 := V_1^T Q V_1, L_1 := V_1^T L, F_1 := FV_1. \tag{7.87}$$

With (7.87) and (7.85), (7.83) becomes

$$A^TXV_1S - XV_1 + QV_1 + LF_1 = 0, \tag{7.88}$$

$$L_1^T + B^TXV_1S + RF_1 = 0. \tag{7.89}$$

Premultiplying (7.88) by V_1^T and using (7.85) transposed we get

$$S^T X_1 S - F_1^T B^T X V_1 S - X_1 + Q_1 + L_1 F_1 = 0. \tag{7.90}$$

Premultiplying (7.89) with F_1^T we obtain

$$F_1^T L_1^T + F_1^T B^T X V_1 S + F_1^T R F_1 = 0. \tag{7.91}$$

Substituting $F_1^T B^T X V_1 S$ from (7.91) in (7.90) we have

$$S^T X_1 S - X_1 + Q_1 + L_1 F_1 + F_1^T L_1^T + F_1^T R F_1 = 0. \tag{7.92}$$

Notice that if V_1 is nonsingular then we may take $V_1 = I_n$ in (7.83) and (7.85). Then (7.92) becomes

$$(A + BF)^T X (A + BF) - X + Q + LF + F^T L^T + F^T RF = 0, \tag{7.93}$$

with $A + BF$ stable, which is the "closed–loop" form of the discrete–time algebraic Riccati equation.

Now we are ready for the main result of this section.

Proposition 7.11.1 *Suppose that the GDTARS(Σ) in (7.83) has a stabilizing solution (X, F, r, V_1). Then the quadratic index (7.82) takes the form*

$$J_\Sigma(V_1\xi_0, u) = \xi_0^T X_1 \xi_0 + \sum\nolimits_{k=0}^{\infty} (u_k - F_1\xi_k)^T [R + \frac{1}{2} B^T (X + X^T) B](u_k - F_1 \xi_k) \tag{7.94}$$

for any pair (ξ, u) linked by (7.86), where $X_1 := V_1^T X V_1$ and $F_1 := F V_1$. If $R + \frac{1}{2}B^T(X + X^T)B \geq 0$ holds, then the minimum of $J_\Sigma(V_1\xi_0, u)$ over $u \in \ell_+^{2,m}$ is attained (nonuniquely) for $u = F_1\xi$.

Proof. Since the GDTARS(Σ) has a stabilizing solution (X, F, r, V_1) we have from (7.85), (7.86), and (7.87)

$$V_1 \sigma \xi = V_1 S \xi - B F_1 \xi + B u. \tag{7.95}$$

Premultiplying both sides by $V_1^T X^T$ we get with (7.84)

$$X_1 \sigma \xi = X_1 S \xi - V_1^T X^T B F_1 \xi + V_1^T X^T B u. \tag{7.96}$$

Equation (7.96) can be seen as the evolution of the system (7.81) constrained to $\mathcal{V}_1$. If the evolution is located in $\mathcal{V}_1$, the quadratic index (7.82) becomes

$$\begin{aligned} J_\Sigma(V_1\xi_0, u) &= \sum\nolimits_{k=0}^{\infty} \begin{bmatrix} \xi_k \\ u_k \end{bmatrix}^T \begin{bmatrix} Q_1 & L_1 \\ L_1^T & R \end{bmatrix} \begin{bmatrix} \xi_k \\ u_k \end{bmatrix} \\ &= \left\langle \begin{bmatrix} \xi \\ u \end{bmatrix}, \begin{bmatrix} Q_1 & L_1 \\ L_1^T & R \end{bmatrix} \begin{bmatrix} \xi \\ u \end{bmatrix} \right\rangle_{\ell_+^{2,r} \times \ell_+^{2,m}} \end{aligned}$$

where we have used $x = V_1\xi$ and (7.87), and $(\xi, u) \in \ell_+^{2,r} \times \ell_+^{2,m}$ are linked by (7.86). With (7.92), (7.89), (7.95) and (7.96), we obtain after some lengthy but routine computations

$$\begin{aligned}
&\begin{bmatrix} \xi \\ u \end{bmatrix}^T \begin{bmatrix} Q_1 & L_1 \\ L_1^T & R \end{bmatrix} \begin{bmatrix} \xi \\ u \end{bmatrix} \\
&= -(\sigma\xi)^T X_1 \sigma\xi + \xi^T X_1 \xi + (u - F_1\xi)^T [R + \frac{1}{2}B^T(X + X^T)B](u - F_1\xi).
\end{aligned}$$

As $\xi \in \ell_+^{2,r}$ then $\langle \xi, X_1\xi\rangle - \langle \sigma\xi, X_1\sigma\xi\rangle = \xi_0^T X_1 \xi_0$ and it follows from the identity above that

$$J_\Sigma(V_1\xi_0, u) = \xi_0^T X_1 \xi_0 + \left\langle u - F_1\xi, [R + \frac{1}{2}B^T(X + X^T)B](u - F_1\xi)\right\rangle$$

from which the conclusion follows easily. ■

7.12 Spectral and inner–outer factorization with respect to the unit circle

In this section we solve the spectral and inner–outer factorization problems for a discrete–time system with a proper TFM of constant rank (not necessarily maximal) on the unit circle. The factorization problems are reduced to the computation of the stabilizing solution to the DTARS. We start with the precise definitions.

Definition 7.12.1 *Let* $\mathbf{G}(z) \in RH^\infty_{+,p\times m}$ *without zeros on the unit circle, i.e.,*

$$\operatorname{rank} \mathbf{G}(e^{j\theta}) = k \ (= \operatorname{rank}_{\mathbb{R}(z)} \mathbf{G}(z)),\ \forall\theta \in [0, 2\pi). \qquad (7.97)$$

1. A matrix function $\mathbf{G}_o(z) \in RH^\infty_{+,k\times m}$ *with a right inverse in* $RH^\infty_{+,m\times k}$ *such that*

$$\mathbf{G}^*(z)\mathbf{G}(z) = \mathbf{G}_o^*(z)\mathbf{G}_o(z) \qquad (7.98)$$

is called a spectral factor *of* $\mathbf{G}$.

2. Let $\mathbf{G}_i := \mathbf{G}\mathbf{G}_o^\dagger \in RH^\infty_{+,p\times k}$ *where* $\mathbf{G}_o^\dagger$ *is a stable right inverse of* $\mathbf{G}_o$*. Then* $\mathbf{G}_i$ *is* inner, *i.e., is in* RH^∞_+ *and satisfies*

$$\mathbf{G}_i^*(z)\mathbf{G}_i(z) = I, \qquad (7.99)$$

and

$$\mathbf{G}(z) = \mathbf{G}_i(z)\mathbf{G}_o(z). \qquad (7.100)$$

Relation (7.100) defines an inner–outer factorization *of* $\mathbf{G}$.

Notice that if $\mathbf{G}$ is given by a minimal realization with state matrix A, then A is invertible if and only if $\mathbf{G}^*(z)$ is proper.

The following theorem gives the theoretical support for performing spectral and inner–outer factorization in the general case through realizations.

Theorem 7.12.2 *Let*

$$\mathbf{G} = \left[\begin{array}{c|c} A & B \\ \hline C & D \end{array}\right] \tag{7.101}$$

be given, where $A \in \mathbb{R}^{n\times n}, B \in \mathbb{R}^{n\times m}, C \in \mathbb{R}^{p\times n}, D \in \mathbb{R}^{p\times m}$. *Assume* $\mathbf{G}$ *has no zeros on the unit circle (i.e., satisfies (7.97)) and suppose that the realization (7.101) is stabilizable,* $\{0\}$*–controllable, and* $\partial\mathbb{D}$*–observable. Define the positiveness Popov triplet* $\Sigma = (A, B; C^TC, C^TD, D^TD)$. *Then we have:*

1. *The DTARS(*Σ*) has a stabilizing solution* (X, F).
2. *Let* H *be a full row rank matrix such that* $H^TH = D^TD + B^TXB$ *and let*

$$\mathbf{G}_o := \left[\begin{array}{c|c} A & B \\ \hline -HF & H \end{array}\right],$$

$$\mathbf{G}_i := \left[\begin{array}{c|c} A+BF & BH^\dagger \\ \hline C+DF & DH^\dagger \end{array}\right], \tag{7.102}$$

where $H^\dagger$ *is an inverse to the right of* H. *Then* $\mathbf{G}_o$ *satisfies (7.98),* $\mathbf{G}_i$ *is inner and together they satisfy (7.100). If in addition* A *is stable (i.e.,* $\mathbf{G} \in RH^\infty_{+,p\times m}$*), then* $\mathbf{G}_o$ *is the spectral and* $\mathbf{G}_i$ *is the inner factor of* $\mathbf{G}$.

Proof. 1. We make intensive use of the notation in Section 6.3. Using Theorem 6.4.4 it is enough to prove that the ESP(Σ) written explicitly as

$$zM_\Sigma - N_\Sigma = z\begin{bmatrix} I & O & O \\ O & -A^T & O \\ O & -B^T & O \end{bmatrix} - \begin{bmatrix} A & O & B \\ C^TC & -I & C^TD \\ D^TC & O & D^TD \end{bmatrix} \tag{7.103}$$

is dichotomic and, moreover, stable disconjugate. We begin by proving that $zM_\Sigma - N_\Sigma$ is dichotomic which means $n_r = n_\ell$, $n_\infty = \rho + \pi_0$, and $n_f^0 = 0$.

Proof of $n_r = n_\ell$, $n_\infty = \rho + \pi_0$. Since the pair (A, B) is $\{0\}$–controllable, Remark 6.3.5 shows that the nonstrict inequalities in Theorem 6.3.3 can be strengthened to equalities, i.e., $n_r = n_\ell$ and $n_\infty = \rho + \pi_0$.

Proof of $n_f^0 = 0$. Since Σ is the positiveness Popov triplet associated with $\mathbf{G}$, the Popov function associated with Σ is $\mathbf{\Pi}_\Sigma(z) = \mathbf{G}^*(z)\mathbf{G}(z)$. By assumption $\mathbf{G}(z)$ has no zeros on the unit circle which implies that neither has $\mathbf{\Pi}_\Sigma(z)$. Since the pair (A, B) is assumed stabilizable and the pair (C, A) is assumed $\partial\mathbb{D}$–observable, it follows that the ESP(Σ) which is the transmission pencil of the Popov function $\mathbf{\Pi}_\Sigma(z)$ has no eigenvalues on the unit circle. Therefore $n_f^0 = 0$.

Thus the ESP(Σ) is dichotomic and by Theorem 6.3.6 there exists a proper $\mathbb{D}$–deflating subspace of dimension n. Let V be a basis matrix for it partitioned conformably with the ESP(Σ) as

$$V = \begin{bmatrix} V_1 \\ V_2 \\ V_3 \end{bmatrix}.$$

We prove now that $\mathcal{V} = \operatorname{Im} V$ is disconjugate which means that V_1 is invertible. We show this by proving that $\operatorname{Ker} V_1 = 0$.

Since $\mathcal{V}$ is a proper $\mathbb{D}$–deflating subspace we have that $M_\Sigma VS = N_\Sigma V$ holds for a stable S or, explicitly,

$$\begin{array}{lclcl} AV_1 & & +BV_3 & = & V_1S, \\ C^TCV_1 & -V_2 & +C^TDV_3 & = & -A^TV_2S, \\ D^TCV_1 & & +D^TDV_3 & = & -B^TV_2S. \end{array} \tag{7.104}$$

We prove first that

$$\operatorname{Ker} V_1 \subset \operatorname{Ker} V_1^TV_2 \subset \operatorname{Ker} V_2. \tag{7.105}$$

Adding the second and the third equations (7.104) premultiplied by V_1^T and V_3^T, respectively, and using the first one we obtain

$$S^TV_1^TV_2S - V_1^TV_2 + (CV_1 + DV_3)^T(CV_1 + DV_3) = 0. \tag{7.106}$$

Proposition 6.3.11 shows that $V_1^TV_2 = V_2^TV_1$ and the first inclusion in (7.105) follows immediately. Let $x \in \operatorname{Ker} V_1^TV_2$. Pre– and postmultiplying (7.106) by x^T and x, respectively, we get that $\operatorname{Ker} V_1^TV_2$ is an invariant subspace of S and

$$\operatorname{Ker} V_1^TV_2 \subset \operatorname{Ker}(CV_1 + DV_3). \tag{7.107}$$

Let K be a basis matrix for $\operatorname{Ker} V_1^TV_2$. Then there exists a square matrix J such that

$$SK = KJ, \qquad \Lambda(J) \subset \Lambda(S). \tag{7.108}$$

From (7.107) we have $(CV_1 + DV_3)K = 0$. Postmultiplying the second and third equations (7.104) by K we obtain

$$\begin{array}{rcl} A^TV_2KJ & = & V_2K, \\ B^TV_2KJ & = & 0. \end{array} \tag{7.109}$$

Let $\widetilde{F}$ be a stabilizing feedback for the pair (A, B). From (7.109) we have

$$(A + B\widetilde{F})^TV_2KJ - V_2K = 0 \tag{7.110}$$

where $A+B\widetilde{F}$ and J are both stable matrices. Thus the unique solution to the discrete–time Sylvester equation (7.110) is $V_2K = 0$. So $\operatorname{Ker} V_1^TV_2 \subset \operatorname{Ker} V_2$ and we conclude that (7.105) holds.

Let now $y \in \operatorname{Ker} V_1$. Then $V_1y = 0$ and $V_2y = 0$. Since $\mathcal{V}$ is a proper deflating subspace

$$MV = \begin{bmatrix} V_1 \\ -A^TV_2 \\ -B^TV_2 \end{bmatrix}$$

has full column rank from which we conclude that

$$\begin{bmatrix} V_1 \\ V_2 \end{bmatrix}$$

has full column rank as well. Therefore $y = 0$ and we get $\operatorname{Ker} V_1 = \{0\}$ which shows that $\mathcal{V}$ is disconjugate.

In conclusion we have proved that the ESP(Σ) is dichotomic and stable disconjugate. which implies by Theorem 6.4.4 that the DTARS(Σ) has a stabilizing solution (X, F).

2. It is easy to check that a stable right inverse of $\mathbf{G}_o$ is

$$\mathbf{G}_o^{\dagger} = \left[\begin{array}{c|c} A + BF & BH^{\dagger} \\ \hline F & H^{\dagger} \end{array}\right]. \tag{7.111}$$

Then using (7.111) and (7.102) we can check with the spectral factorization identity (3.142) that (7.98), (7.99), and (7.100) all hold.

Finally, if $\mathbf{G}$ is stable, then A and $A + BF$ are both stable and therefore $\mathbf{G}_o$ is the spectral factor and $\mathbf{G}_i$ is the inner factor of $\mathbf{G}$. ■

Notice that the assumption on the controllability in zero of the pair (A, B) is essential for the proposed method of factorization. However, all assumptions in Theorem 7.12.2 on the state–space realization of $\mathbf{G}$ are automatically fulfilled if we start with a minimal realization (7.101).

Notes and references

The bounded real lemma is a classical result in systems theory and can be found for example in Zhou, Doyle and Glover [ZDG95]. However, the result given in this chapter is more general. The convergence of the family of algebraic Riccati equations associated with positiveness Popov triplets was obtained by Ionescu [Ion97b]. Coprime and normalized coprime factorizations in systems theory were introduced by Vidyasagar [Vid85]. The solutions to the spectral and inner–outer factorizations follow Weiss [Wei94b], Oară [Oar95], and Ionescu and Oară [IO96f]. Solutions to the most general spectral, inner–outer, coprime, and minimal factorizations have been recently obtained by Oară [Oar98], and Oară and Varga [OV99a, OV99b]. Quadratic indices with constrained dynamics were studied by Ionescu and Oară [Oar95, IO96d, IO96e].

8

The Four Block Nehari Problem

In this chapter we solve the Nehari problem which consists of approximating in L^∞–norm an antistable system by a stable one. We consider here the general case in which the approximant adds only to a "corner" of the transfer matrix function of the system to be approximated and which is also known as a four block problem. Our approach is to reduce the four block Nehari problem to the signature condition which can be solved by the general results of Chapter 4.

In Section 8.1 we formulate the four block Nehari problem and describe in some detail the idea behind our approach. We solve in Section 8.2 a very particular four block problem in which the system to be approximated is a constant matrix and which will serve as a basic ingredient in our approach. Necessary and sufficient solvability conditions and explicit formulas for the class of solutions are provided in Section 8.3 in terms of two coupled KPYSs. We show that the two coupled KPYSs are equivalent to two uncoupled KPYSs plus a signature condition on their solutions in Section 8.4. The discrete–time case is presented in Section 8.5.

A. The continuous–time case

8.1 The Nehari problem and the signature condition

Consider an antistable continuous–time system

$$\begin{aligned} \dot{x} &= Ax + B_1u_1 + B_2u_2, \\ y_1 &= C_1x + D_{11}u_1 + D_{12}u_2, \\ y_2 &= C_2x + D_{21}u_1 + D_{22}u_2, \end{aligned} \tag{8.1}$$

having the TFM given by

$$\mathbf{T} = \left[\begin{array}{c|c} A & B \\ \hline C & D \end{array}\right] = \left[\begin{array}{c|cc} A & B_1 & B_2 \\ \hline C_1 & D_{11} & D_{12} \\ C_2 & D_{21} & D_{22} \end{array}\right] = \left[\begin{array}{cc} \mathbf{T}_{11} & \mathbf{T}_{12} \\ \mathbf{T}_{21} & \mathbf{T}_{22} \end{array}\right] \in RH^\infty_{-,p\times m}, \tag{8.2}$$

where $A \in \mathbb{R}^{n\times n}$ is antistable, $B_i \in \mathbb{R}^{n\times m_i}$, $C_j \in \mathbb{R}^{p_j\times n}$, $D_{ij} \in \mathbb{R}^{p_i\times m_j}$ $(i,j = 1,2,$ $m = m_1+m_2$, $p = p_1+p_2)$. We shall refer to a *one block type problem* if $m_2 = p_2 = 0$, to a *two block type problem* if $m_2 = 0, p_2 > 0$, to a *dual two block type problem* if $m_2 > 0, p_2 = 0$, and to a *four block problem* if $m_2 > 0, p_2 > 0$. These names come from the number of nonvanishing blocks of $\mathbf{T}$ in (8.2).

We define an *approximant* for the system $\mathbf{T}$ as being a stable system

$$\begin{aligned} \dot{x}_S &= A_S x_S + B_S u_S, \\ y_S &= C_S x_S + D_S u_S, \end{aligned} \tag{8.3}$$

having the TFM given by

$$\mathbf{S} = \left[\begin{array}{c|c} A_S & B_S \\ \hline C_S & D_S \end{array} \right] \in RH^{\infty}_{+,p_1 \times m_1}, \tag{8.4}$$

where A_S is stable.

Given the antistable system $\mathbf{T}$ and a threshold $\gamma > 0$, the *suboptimal four block Nehari problem (4–NP)* consists in finding all (stable) approximants $\mathbf{S}$, when any exists, such that

$$\left\| \begin{bmatrix} \mathbf{T}_{11} + \mathbf{S} & \mathbf{T}_{12} \\ \mathbf{T}_{21} & \mathbf{T}_{22} \end{bmatrix} \right\|_{\infty} < \gamma. \tag{8.5}$$

The *optimal four block Nehari problem* consists in finding all (stable) approximants $\mathbf{S}$ such that

$$\min_{\widehat{\mathbf{S}} \in RH^{\infty}_{+,p_1 \times m_1}} \left\| \begin{bmatrix} \mathbf{T}_{11} + \widehat{\mathbf{S}} & \mathbf{T}_{12} \\ \mathbf{T}_{21} & \mathbf{T}_{22} \end{bmatrix} \right\|_{\infty} = \left\| \begin{bmatrix} \mathbf{T}_{11} + \mathbf{S} & \mathbf{T}_{12} \\ \mathbf{T}_{21} & \mathbf{T}_{22} \end{bmatrix} \right\|_{\infty} = \gamma_{\min} \tag{8.6}$$

which is equivalent to finding $\gamma_{\min}$ and all stable $\mathbf{S}$ that satisfy (8.5) with the strict inequality relaxed to a nonstrict one and $\gamma = \gamma_{\min}$. The strict inequality in (8.5) avoids the intricacies related to the optimal cases. Further on we shall treat the suboptimal case only.

Note that there is no loss of generality in assuming $D_{11} = 0$ and $\gamma = 1$. Indeed, setting D_{11} to zero comes to a shift of the solution $\mathbf{S}$ to $\mathbf{S} - D_{11}$, while putting $\gamma = 1$ implies an appropriate scaling of $\mathbf{T}$ only. We assume hereafter that these two conditions hold. No additional hypothesis on the state–space realization (8.2) is assumed.

Roughly speaking, our approach to the suboptimal 4–NP consists of reducing it to a signature condition written for an appropriate Popov function, which could be solved by invoking the general result of Theorem 4.5.9. In this respect, notice that (8.5) is equivalent to a two block problem

$$\left\| \begin{bmatrix} \mathbf{T}_1 + \overline{\mathbf{S}} \\ \mathbf{T}_2 \end{bmatrix} \right\|_{\infty} < 1 \tag{8.7}$$

with the structure constraint $\overline{\mathbf{S}} = \begin{bmatrix} \mathbf{S} & O \end{bmatrix}$, where $\mathbf{T}_1 := \begin{bmatrix} \mathbf{T}_{11} & \mathbf{T}_{12} \end{bmatrix}$, $\mathbf{T}_2 := \begin{bmatrix} \mathbf{T}_{21} & \mathbf{T}_{22} \end{bmatrix}$. Further, (8.7) is equivalent to

$$\begin{aligned} (\mathbf{T}_1 + \overline{\mathbf{S}})^*(\mathbf{T}_1 + \overline{\mathbf{S}}) + \mathbf{T}_2^*\mathbf{T}_2 - I &= \begin{bmatrix} \mathbf{T}_1 \\ \mathbf{T}_2 \end{bmatrix}^* \begin{bmatrix} \mathbf{T}_1 \\ \mathbf{T}_2 \end{bmatrix} + \overline{\mathbf{S}}^*\mathbf{T}_1 + \mathbf{T}_1^*\overline{\mathbf{S}} - I \\ &= \mathbf{T}^*\mathbf{T} + \overline{\mathbf{S}}^*\mathbf{T}_1 + \mathbf{T}_1^*\overline{\mathbf{S}} - I < 0, \text{ on } \overline{\mathbb{C}}_0. \end{aligned} \tag{8.8}$$

Rewriting (8.8) compactly results in the signature condition

$$\begin{bmatrix} I & \overline{\mathbf{S}}^* \end{bmatrix} \mathbf{\Pi}_{\overline{\Sigma}} \begin{bmatrix} I \\ \overline{\mathbf{S}} \end{bmatrix} < 0, \text{ on } \overline{\mathbb{C}}_0 \tag{8.9}$$

for the Popov function

$$\begin{aligned} \mathbf{\Pi}_{\overline{\Sigma}} &= \begin{bmatrix} \overline{\mathbf{\Pi}}_{11} & \overline{\mathbf{\Pi}}_{12} \\ \overline{\mathbf{\Pi}}_{12}^* & I \end{bmatrix} := \begin{bmatrix} \mathbf{T}^*\mathbf{T} - I_m & \mathbf{T}_1^* \\ \mathbf{T}_1 & I_{p_1} \end{bmatrix} \\ &= \begin{bmatrix} \mathbf{T}_1 & I_{p_1} \\ \mathbf{T}_2 & O \end{bmatrix}^* \begin{bmatrix} \mathbf{T}_1 & I_{p_1} \\ \mathbf{T}_2 & O \end{bmatrix} - \begin{bmatrix} I & O \\ O & O \end{bmatrix} \end{aligned} \tag{8.10}$$

associated with the Popov triplet

$$\begin{aligned} \overline{\Sigma} &= (\overline{A}, \overline{B}; \overline{Q}, \overline{L}, \overline{R}) \\ &:= (A, \begin{bmatrix} B & O_{p_1} \end{bmatrix}; C^T C, \begin{bmatrix} C^T D & C_1^T \end{bmatrix}, \begin{bmatrix} D^T D - I & D_1^T \\ D_1 & I_{p_1} \end{bmatrix}), \end{aligned} \tag{8.11}$$

where $D_1 := \begin{bmatrix} O & D_{12} \end{bmatrix}$. A direct check shows that $\mathbf{\Pi}_{\overline{\Sigma}}$ defined by (8.10) is indeed the Popov function associated with $\overline{\Sigma}$. Thus, the 4–NP has a solution $\mathbf{S}$ if and only if the signature condition (8.9) holds for $\overline{\mathbf{S}} = \begin{bmatrix} \mathbf{S} & O \end{bmatrix}$.

However, we can not apply directly the result given in Theorem 4.5.9 due to two reasons. Firstly, $\overline{A}$ in $\overline{\Sigma}$ is antistable while Theorem 4.5.9 requires it to be stable. Secondly, the solution $\overline{\mathbf{S}}$ to the inequality (8.9) should feature a particular zero pattern which is not automatically provided by this theorem.

To circumvent the first difficulty we shall first transform the *antistable* triplet $\overline{\Sigma}$ into a *stable* one by employing the general technique described in Chapter 4. More precisely, we define first a particular equivalent Popov triplet of $\overline{\Sigma}$, denoted $\widehat{\Sigma} = (\widehat{A}, \widehat{B}; \widehat{Q}, \widehat{L}, \widehat{R})$, for which $\widehat{A}$ is antistable and $\widehat{Q} = 0$. Further, we take its dual, say $\widetilde{\Sigma} = (\widetilde{A}, \widetilde{B}; \widetilde{Q}, \widetilde{L}, \widetilde{R})$, for which $\widetilde{A}$ is stable and $\widetilde{Q} = 0$. To define the suitable equivalent $\widehat{\Sigma}$ we use the solution to the Riccati equation that expresses the contraction property

$$\left\| \begin{bmatrix} \mathbf{T}_{12} \\ \mathbf{T}_{22} \end{bmatrix} \right\|_\infty < 1. \tag{8.12}$$

Note that (8.12) is a necessary solvability condition for the 4–NP. As we shall prove, transforming the Popov triplet Σ into $\widehat{\Sigma}$ and taking the dual $\widetilde{\Sigma}$ has the effect of transforming the original problem of solving for $\overline{\mathbf{S}}$ the signature condition (8.9) into an equivalent one of solving for $\overline{\mathbf{S}}$ a signature condition written for $\mathbf{\Pi}_{\widetilde{\Sigma}}$.

Finally, the solutions to the signature condition for $\mathbf{\Pi}_{\widetilde{\Sigma}}$ will be obtained from an updated version of Theorem 4.5.9 that provides the particular form of the solution $\overline{\mathbf{S}}$, thus circumventing the second mentioned difficulty. The key consists in a J spectral factorization of $\mathbf{\Pi}_{\widetilde{\Sigma}}$ with prescribed "zero" pattern. Here follows the precise statement of this result.

Theorem 8.1.1 *Let*

$$\mathbf{\Pi} := \begin{bmatrix} \mathbf{\Pi}_{11} & \mathbf{\Pi}_{12} \\ \mathbf{\Pi}_{12}^* & \mathbf{\Pi}_{22} \end{bmatrix} = \mathbf{\Pi}^*, \tag{8.13}$$

be a $(m + p_1) \times (m + p_1)$ *rational matrix with* $\mathbf{\Pi}_{22} > 0$ *on* $\overline{\mathbb{C}}_0$. *Assume that* $\mathbf{\Pi}$ *admits a* J *spectral factorization*

$$\mathbf{\Pi} = \mathbf{G}^* J \mathbf{G}, \tag{8.14}$$

where $J := \text{diag}(-I_m, I_{p_1}) = \text{diag}(-I_{m_1}, -I_{m_2}, I_{p_1})$, *the spectral factor has the particular structure*

$$\mathbf{G} = \left[\begin{array}{c|c} \mathbf{G}_1 & \mathbf{G}_2 \\ \hline \mathbf{G}_3 & \mathbf{G}_4 \end{array}\right] = \left[\begin{array}{cc|c} \mathbf{G}_{11} & O & \mathbf{G}_{13} \\ \mathbf{G}_{21} & \mathbf{G}_{22} & \mathbf{G}_{23} \\ \hline \mathbf{G}_{31} & O & \mathbf{G}_{33} \end{array}\right], \tag{8.15}$$

$\mathbf{G}$ *is a unit in* $\in RH^{\infty}_{+,(m+p_1)\times(m+p_1)}$ *and* $\mathbf{G}_4 (= \mathbf{G}_{33})$ *is a unit in* $RH^{\infty}_{+,p_1\times p_1}$.

Then the class of $\overline{\mathbf{S}} \in RH^{\infty}_{+,p_1\times m}$, $(\overline{\mathbf{S}} = [\ \mathbf{S} \quad O_{p_1\times m_2}\ ]$, $\mathbf{S} \in RH^{\infty}_{+,p_1\times m_1})$, *which satisfy the signature condition*

$$[\ I \quad \overline{\mathbf{S}}^*\ ]\,\mathbf{\Pi} \left[\begin{array}{c} I \\ \overline{\mathbf{S}} \end{array}\right] < 0, \quad \text{on } \overline{\mathbb{C}}_0, \tag{8.16}$$

is given by

$$\overline{\mathbf{S}} := \overline{\mathbf{S}}_2 \overline{\mathbf{S}}_1^{-1}, \quad \left[\begin{array}{c} \overline{\mathbf{S}}_1 \\ \overline{\mathbf{S}}_2 \end{array}\right] = \mathbf{G}^{-1} \left[\begin{array}{c} I_m \\ \overline{\boldsymbol{\theta}} \end{array}\right], \quad \overline{\boldsymbol{\theta}} = [\ \boldsymbol{\theta} \quad O\ ], \tag{8.17}$$

where $\boldsymbol{\theta}$ *is an arbitrary element of* $RH^{\infty}_{+,p_1\times m_1}$ *with* $\|\boldsymbol{\theta}\|_\infty < 1$.

Proof. The proof uses essentially the ideas of the proof of the second part of Theorem 4.5.9 combined with a careful exploitation of the particular form of the spectral factor which leads to the zero pattern of the solution $\overline{\mathbf{S}} = [\ \mathbf{S} \quad O\ ]$. We illustrate this last fact only. We show first that the solution $\overline{\mathbf{S}}$ given by (8.17) takes the form $\overline{\mathbf{S}} = [\ \mathbf{S} \quad O\ ]$. Indeed,

$$\left[\begin{array}{c} \overline{\mathbf{S}}_1 \\ \hline \overline{\mathbf{S}}_2 \end{array}\right] = \left[\begin{array}{cc} \overline{\mathbf{G}}_{11} + \overline{\mathbf{G}}_{13}\boldsymbol{\theta} & O \\ \overline{\mathbf{G}}_{21} + \overline{\mathbf{G}}_{23}\boldsymbol{\theta} & \overline{\mathbf{G}}_{22} \\ \hline \overline{\mathbf{G}}_{31} + \overline{\mathbf{G}}_{33}\boldsymbol{\theta} & O \end{array}\right]$$

where

$$\overline{\mathbf{G}} := \mathbf{G}^{-1} = \left[\begin{array}{ccc} \overline{\mathbf{G}}_{11} & O & \overline{\mathbf{G}}_{13} \\ \overline{\mathbf{G}}_{21} & \overline{\mathbf{G}}_{22} & \overline{\mathbf{G}}_{23} \\ \overline{\mathbf{G}}_{31} & O & \overline{\mathbf{G}}_{33} \end{array}\right] \tag{8.18}$$

has the same zero pattern as $\mathbf{G}$. Consequently

$$\overline{\mathbf{S}} = \overline{\mathbf{S}}_2 \overline{\mathbf{S}}_1^{-1} = [\ (\overline{\mathbf{G}}_{31} + \overline{\mathbf{G}}_{33}\boldsymbol{\theta})(\overline{\mathbf{G}}_{11} + \overline{\mathbf{G}}_{13}\boldsymbol{\theta})^{-1} \quad O\ ] =: [\ \mathbf{S} \quad O\ ].$$

Conversely, we show now that any $\overline{\mathbf{S}} = [\ \mathbf{S} \quad O\ ]$ that satisfies the signature condition (8.16) can be written, for an appropriate $\boldsymbol{\theta}$, in the form (8.17). Indeed, according to the second part of the proof of Theorem 4.5.9 we define

$$\left[\begin{array}{c} \overline{\boldsymbol{\theta}}_1 \\ \overline{\boldsymbol{\theta}}_2 \end{array}\right] = \mathbf{G} \left[\begin{array}{c} I \\ \overline{\mathbf{S}} \end{array}\right]$$

and by substituting (8.15) and $\overline{\mathbf{S}} = [\ \mathbf{S} \quad O\ ]$ we get that $\overline{\boldsymbol{\theta}} := \overline{\boldsymbol{\theta}}_2 \overline{\boldsymbol{\theta}}_1^{-1} = [\ (\mathbf{G}_{31} + \mathbf{G}_{33}\mathbf{S})(\mathbf{G}_{11} + \mathbf{G}_{13}\mathbf{S})^{-1} \quad O\ ] =: [\ \boldsymbol{\theta} \quad O\ ]$ has the desired zero pattern. ■

The next corollary gives equivalent parameterizations of the class of solutions to the signature condition (8.16).

Corollary 8.1.2 *Assume that all conditions stated in Theorem 8.1.1 hold. Then the class of* $\mathbf{S} \in RH_+^\infty$ *for which* $\overline{\mathbf{S}} = \begin{bmatrix} \mathbf{S} & O \end{bmatrix}$ *satisfies (8.16) is given by one of the following equivalent forms:*

1.

$$\mathbf{S} = (\overline{\mathbf{G}}_{31} + \overline{\mathbf{G}}_{33}\boldsymbol{\theta})(\overline{\mathbf{G}}_{11} + \overline{\mathbf{G}}_{13}\boldsymbol{\theta})^{-1},$$

where $\overline{\mathbf{G}}_{ij}$*,* $(i, j = 1, 2, 3)$ *are the entries of* $\overline{\mathbf{G}} := \mathbf{G}^{-1}$ *depicted in (8.18).*

2.

$$\mathbf{S} = -(\mathbf{G}_{33} + \boldsymbol{\theta}\mathbf{G}_{13})^{-1}(\mathbf{G}_{31} + \boldsymbol{\theta}\mathbf{G}_{11}).$$

3.

$$\mathbf{S} = \mathrm{LFT}(\mathbf{H}, \boldsymbol{\theta}) = \mathbf{H}_{11} + \mathbf{H}_{12}\boldsymbol{\theta}(I - \mathbf{H}_{22}\boldsymbol{\theta})^{-1}\mathbf{H}_{21},$$

where the coefficients $\mathbf{H}_{ij}$ *of the linear fractional transformation are given by* $\mathbf{H}_{11} := -\mathbf{G}_{33}^{-1}\mathbf{G}_{31}$, $\mathbf{H}_{12} := \mathbf{G}_{33}^{-1}$, $\mathbf{H}_{21} := \mathbf{G}_{11} - \mathbf{G}_{13}\mathbf{G}_{33}^{-1}\mathbf{G}_{31}$, $\mathbf{H}_{22} := \mathbf{G}_{13}\mathbf{G}_{33}^{-1}$.

Proof. Point 1 is exactly (8.17), while points 2 and 3 follow by simple computations. ■

8.2 Parrott's problem

In this section we solve the four block Nehari problem for the constant matrix case – also known as *Parrott's problem.* We need first the following auxiliary result.

Lemma 8.2.1 *Consider the symmetric matrix* $R \in \mathbb{R}^{n\times n}$*, partitioned as*

$$R = \begin{bmatrix} R_{11} & R_{12} & R_{13} \\ R_{12}^T & R_{22} & R_{23} \\ R_{13}^T & R_{23}^T & R_{33} \end{bmatrix} \tag{8.19}$$

where $R_{ij} \in \mathbb{R}^{n_i \times n_j}, (i, j = 1, 2, 3)$, $n_1 + n_2 + n_3 = n$. *Assume that*

(i) $R_{22} < 0$.

(ii) $R_{33} > 0$.

(iii) $\mathrm{sgn}\,(R) = J$, *where* $J = \mathrm{diag}\,(-I_{n_1}, -I_{n_2}, I_{n_3})$.

Then there exists a J*-factorization of* R *in the form*

$$R = V^T J V, \tag{8.20}$$

where the factor $V \in \mathbb{R}^{n\times n}$ *has the zero pattern*

$$V = \begin{bmatrix} V_{11} & O & O \\ V_{21} & V_{22} & V_{23} \\ V_{31} & O & V_{33} \end{bmatrix} \tag{8.21}$$

and $V_{ij} \in \mathbb{R}^{n_i \times n_j}, (i, j = 1, 2, 3)$.

Proof. The proof is constructive. Substituting (8.21) in (8.20) and identifying the entries in both sides leads to

$$(2,2) \quad : \quad V_{22}^T V_{22} = -R_{22} \overset{(i)}{>} 0, \tag{8.22}$$

$$(2,3) \quad : \quad V_{23} = -V_{22}^{-T} R_{23}, \tag{8.23}$$

$$(3,3) \quad : \quad V_{33}^T V_{33} = R_{33} + V_{23}^T V_{23} \overset{(ii)}{>} 0, \tag{8.24}$$

$$(2,1) \quad : \quad V_{21} = -V_{22}^{-T} R_{12}^T, \tag{8.25}$$

$$(3,1) \quad : \quad V_{31} = V_{33}^{-T}(V_{23}^T V_{21} + R_{13}), \tag{8.26}$$

$$(1,1) \quad : \quad V_{11}^T V_{11} = -V_{21}^T V_{21} + V_{31}^T V_{31} - R_{11}$$

$$= -(R_{11} - \begin{bmatrix} R_{12} & R_{13} \end{bmatrix} \begin{bmatrix} R_{22} & R_{23} \\ R_{23}^T & R_{33} \end{bmatrix}^{-1} \begin{bmatrix} R_{12}^T \\ R_{13}^T \end{bmatrix}) \overset{(iii)}{>} 0. \tag{8.27}$$

This last relation follows by substituting all the entries V_{ij} from (8.22)-(8.26) in (8.27). Note that the parenthesized term in (8.27) is exactly the Schur complement of

$$\begin{bmatrix} R_{22} & R_{23} \\ R_{23}^T & R_{33} \end{bmatrix}.$$

We conclude that V_{22}, V_{33} and V_{11} can be chosen as Cholesky factors from (8.22), (8.24) and (8.27), respectively, and V_{23}, V_{21} and V_{31} can be defined through (8.23), (8.25) and (8.26), respectively. ■

Theorem 8.2.2 (Parrott's problem) *Let $D \in \mathbb{R}^{p \times m}$ be partitioned as*

$$D = \begin{bmatrix} O & D_{12} \\ D_{21} & D_{22} \end{bmatrix}, \tag{8.28}$$

where $D_{ij} \in \mathbb{R}^{p_i \times m_j}$, $i, j = 1, 2$. Then there exists a matrix $S \in \mathbb{R}^{p_1 \times m_1}$ such that

$$\| \begin{bmatrix} S & D_{12} \\ D_{21} & D_{22} \end{bmatrix} \| < 1 \tag{8.29}$$

if and only if the following two conditions hold:

a. $\|D_2\| < 1$, *where* $D_2 := \begin{bmatrix} D_{21} & D_{22} \end{bmatrix}$.

b. $\|\overline{D}_2\| < 1$, *where* $\overline{D}_2 := \begin{bmatrix} D_{12} \\ D_{22} \end{bmatrix}$.

If conditions a and b hold, then

$$R := \left[\begin{array}{c|c} D^T D - I & D_1^T \\ \hline D_1 & I \end{array}\right] = \left[\begin{array}{cc|c} D_{21}^T D_{21} - I & D_{21}^T D_{22} & O \\ D_{22}^T D_{21} & \overline{D}_2^T \overline{D}_2 - I & D_{12}^T \\ \hline O & D_{12} & I \end{array}\right], \tag{8.30}$$

has a J factorization

$$R = V^T J V, \tag{8.31}$$

where $D_1 := \begin{bmatrix} O & D_{12} \end{bmatrix}$ and V can be chosen of the form

$$V = \begin{bmatrix} V_{11} & O & O \\ V_{21} & V_{22} & V_{23} \\ V_{31} & O & V_{33} \end{bmatrix}. \tag{8.32}$$

The class of all matrices S that satisfy (8.29) is given by

$$S = -V_{33}^{-1}(V_{31} + \theta V_{11}), \ \forall \theta \in \mathbb{R}^{p_1 \times m_1}, \text{ with } \|\theta\| < 1. \tag{8.33}$$

Proof. The necessity of conditions a and b is straightforward. Now, we prove the sufficiency. To this end, we invoke Lemma 8.2.1. Note that (ii) of Lemma 8.2.1 is fulfilled, while (i) and (iii) follow by conditions b and a just proved. Therefore, by using the construction in the proof of Lemma 8.2.1, we find a V of the form (8.32) that satisfies (8.31). The rest of the proof runs by similar arguments as the proof of Theorem 8.1.1. ■

8.3 Necessary and sufficient conditions

In order to derive the necessary and sufficient solvability conditions for the suboptimal 4–NP we need first the following lemma that expresses the contraction property of two subsystems of $\mathbf{T}$ in terms of the stabilizing solutions to two appropriate KPYSs.

Lemma 8.3.1 *Let* $\mathbf{T}$ *be as in (8.2). Define the Popov triplets*

$$\Sigma_1 = (-A, -B_2; C^T C, C^T \overline{D}_2, \overline{D}_2^T \overline{D}_2 - I), \quad \overline{D}_2 := \begin{bmatrix} D_{12} \\ D_{22} \end{bmatrix}, \tag{8.34}$$

and

$$\Sigma_2 = (-A^T, -C_2^T; BB^T, BD_2^T, D_2 D_2^T - I), \quad D_2 := \begin{bmatrix} D_{21} & D_{22} \end{bmatrix}. \tag{8.35}$$

I. *The following assertions are equivalent:*

1. $\left\| \begin{bmatrix} \mathbf{T}_{12} \\ \mathbf{T}_{22} \end{bmatrix} \right\|_\infty < 1.$

2. The KPYS($\Sigma_1, -I$) given by

$$\begin{aligned} \overline{D}_2^T \overline{D}_2 - I &= -V^T V, \\ C^T \overline{D}_2 - X B_2 &= -W^T V, \\ C^T C - A^T X - XA &= -W^T W, \end{aligned} \tag{8.36}$$

has a stabilizing solution (X_1, V_1, W_1)*, with* $X_1 \geq 0$.

II. *The following assertions are equivalent:*

1. $\| \begin{bmatrix} \mathbf{T}_{21} & \mathbf{T}_{22} \end{bmatrix} \|_\infty < 1.$

2. The KPYS($\Sigma_2, -I$)

$$\begin{aligned} D_2 D_2^T - I &= -V^T V, \\ BD_2^T - XC_2^T &= -W^T V, \\ BB^T - AX - XA^T &= -W^T W, \end{aligned} \tag{8.37}$$

has a stabilizing solution (X_2, V_2, W_2)*, with* $X_2 \geq 0$.

Proof. We show only part I since part II follows by duality. Let $\overline{\mathbf{T}}_2 := \begin{bmatrix} \mathbf{T}_{12} \\ \mathbf{T}_{22} \end{bmatrix}$. Notice from (3.82) that Σ_1 is a γ–contractiveness Popov triplet (with $\gamma = 1$) associated with the realization of $\overline{\mathbf{T}}_2(-s)$ given by

$$\overline{\mathbf{T}}_2(-s) = \left[\begin{array}{c|c} -A & -B_2 \\ \hline C_1 & D_{12} \\ C_2 & D_{22} \end{array} \right]$$

and $\|\overline{\mathbf{T}}_2(-s)\|_\infty = \|\overline{\mathbf{T}}_2(s)\|_\infty$. Since $-A$ is stable, the equivalence between assertions 1 and 2 is exactly the bounded real lemma 7.1.1 written for the stable $\overline{\mathbf{T}}_2(-s)$. ■

As explained earlier, we use now the stabilizing solution to the KPYS$(\Sigma_1, -I)$ to define the equivalent triplet $\widehat{\Sigma} = (\widehat{A}, \widehat{B}; \widehat{Q}, \widehat{L}, \widehat{R})$ of $\overline{\Sigma}$ given in (8.11). More precisely, we construct $\widehat{\Sigma}$ to be the $(-X_1, \widehat{F})$–equivalent Popov triplet of $\overline{\Sigma}$, where

$$\widehat{F} := \begin{bmatrix} O \\ F_1 \\ \hline O \end{bmatrix} \begin{matrix} \}m_1 \\ \\ \}p_1 \end{matrix} = \begin{bmatrix} \overline{F}_1 \\ \hline O \end{bmatrix}, \tag{8.38}$$

and

$$F_1 := -(\overline{D}_2^T \overline{D}_2 - I)^{-1}(-B_2^T X_1 + \overline{D}_2^T C) \tag{8.39}$$

is the stabilizing feedback of the KPYS(Σ_1). Substituting formulas (8.11) and (8.38) in the general expressions that are satisfied by two equivalent Popov triplets (3.3) we obtain the following explicit expressions:

$$\begin{aligned}
\widehat{A} &= A + \begin{bmatrix} B & O \end{bmatrix} \begin{bmatrix} \overline{F}_1 \\ O \end{bmatrix} = A + B_2 F_1, \\
\widehat{B} &= \begin{bmatrix} B_1 & B_2 & O \end{bmatrix}, \\
\widehat{L} &= \begin{bmatrix} C^T D & C_1^T \end{bmatrix} + \begin{bmatrix} \overline{F}_1^T & O \end{bmatrix} \begin{bmatrix} D^T D - I & D_1^T \\ D_1 & I \end{bmatrix} - X_1 \begin{bmatrix} B & O \end{bmatrix} \\
&\overset{(8.38)}{=} \begin{bmatrix} (C_2^T + F_1^T D_{22}^T) D_{21} - X_1 B_1 & O & C_1^T + F_1^T D_{12}^T \end{bmatrix}, \\
\widehat{R} &= \overline{R} = \begin{bmatrix} D^T D - I & D_1^T \\ D_1 & I \end{bmatrix} = \begin{bmatrix} D_{21}^T D_{21} - I & D_{21}^T D_{22} & O \\ D_{22}^T D_{21} & \overline{D}_2^T \overline{D}_2 - I & D_{12}^T \\ O & D_{12} & I \end{bmatrix}, \\
\widehat{Q} &= -\widehat{A}^T X_1 - X_1 \widehat{A} + \widehat{F}^T \overline{R} \widehat{F} + \overline{L} \widehat{F} + \widehat{F}^T \overline{L}^T + \overline{Q} \\
&= -(A + B_2 F_1)^T X_1 - X_1 (A + B_2 F_1) + F_1^T (\overline{D}_2^T \overline{D}_2 - I) F_1 + C^T \overline{D}_2 F_1 \\
&\quad + F_1^T \overline{D}_2^T C + C^T C = 0,
\end{aligned} \tag{8.40}$$

where the last equality is a simple rewriting of the CTARE(Σ_1) with X_1 its stabilizing solution and F_1 its stabilizing Riccati feedback.

Let $\widetilde{\Sigma} := (\widetilde{A}, \widetilde{B}; \widetilde{Q}, \widetilde{L}, \widetilde{R})$ be the dual triplet Popov of $\widehat{\Sigma}$. As $\widehat{Q} = 0$ and $\widehat{A}$ is antistable the reducing Lyapunov equation $\widehat{A}^T X + X \widehat{A} + \widehat{Q} = 0$ has a unique solution $X_r = 0$ and we get for the entries of the dual Popov triplet $\widetilde{\Sigma}$ the following expressions (see (3.69)):

$$\begin{aligned}
\widetilde{A} &:= -\widehat{A}^T = -(A + B_2 F_1)^T, \\
\widetilde{B} &:= -\widehat{L} = -\begin{bmatrix} (C_2^T + F_1 D_{22}^T) D_{21} - X_1 B_1 & O & C_1^T + F_1^T D_{12}^T \end{bmatrix} \\
&=: \begin{bmatrix} \widetilde{B}_1 & O & \widetilde{B}_3 \end{bmatrix}, \\
\widetilde{Q} &:= \widehat{Q} = 0, \\
\widetilde{L} &:= \widehat{B} = \begin{bmatrix} B_1 & B_2 & O \end{bmatrix}, \\
\widetilde{R} &:= \overline{R} = \begin{bmatrix} D^T D - I & D_1^T \\ D_1 & I \end{bmatrix}.
\end{aligned} \tag{8.41}$$

We indicate now how the signature inequalities written for the Popov functions associated with $\overline{\Sigma}$ and $\widetilde{\Sigma}$ relate to each other.

Lemma 8.3.2 *Let $\overline{\Sigma}$ and $\widetilde{\Sigma}$ be the Popov triplets defined by (8.11) and (8.41). The signature condition*

$$\begin{bmatrix} I & \overline{\mathbf{S}}^* \end{bmatrix} \mathbf{\Pi}_{\overline{\Sigma}} \begin{bmatrix} I \\ \overline{\mathbf{S}} \end{bmatrix} < 0, \text{ on } \overline{\mathbb{C}}_0 \tag{8.42}$$

holds for $\overline{\mathbf{S}} = \begin{bmatrix} \mathbf{S} & O \end{bmatrix}$ if and only if the signature condition

$$\begin{bmatrix} I & \overline{\mathbf{S}}^* \end{bmatrix} \mathbf{\Pi}_{\widetilde{\Sigma}} \begin{bmatrix} I \\ \overline{\mathbf{S}} \end{bmatrix} < 0, \text{ on } \overline{\mathbb{C}}_0 \tag{8.43}$$

holds for the same $\overline{\mathbf{S}}$.

Proof. Since $\widehat{\Sigma}$ and $\widetilde{\Sigma}$ are dual Popov triplets we have by Lemma 3.3.5

$$\mathbf{\Pi}_{\widehat{\Sigma}} = \mathbf{\Pi}_{\widetilde{\Sigma}}. \tag{8.44}$$

Further, since the Popov triplet $\overline{\Sigma}$ is $(-X_1, \widehat{F})$–equivalent to the Popov triplet $\widehat{\Sigma}$, we have from Proposition 3.2.16 that the following identity holds

$$\mathbf{\Pi}_{\overline{\Sigma}} = \mathbf{S}^*_{\widehat{F}} \mathbf{\Pi}_{\widehat{\Sigma}} \mathbf{S}_{\widehat{F}} \overset{(8.44)}{=} \mathbf{S}^*_{\widehat{F}} \mathbf{\Pi}_{\widetilde{\Sigma}} \mathbf{S}_{\widehat{F}}, \tag{8.45}$$

where

$$\mathbf{S}_{\widehat{F}} := \left[\begin{array}{c|ccc} A & B_1 & B_2 & O \\ \hline O & I & O & O \\ -F_1 & O & I & O \\ O & O & O & I \end{array}\right] = \begin{bmatrix} I & O & O \\ \mathbf{S}_{\widehat{F},21} & \mathbf{S}_{\widehat{F},22} & O \\ O & O & I \end{bmatrix}. \tag{8.46}$$

Using (8.45) in the left–hand side of (8.42) and taking into account the special forms of $\mathbf{S}_{\widehat{F}}$ and $\overline{\mathbf{S}}$, we get succesively

$$\begin{aligned}
&\begin{bmatrix} I & \overline{\mathbf{S}}^* \end{bmatrix} \mathbf{S}^*_{\widehat{F}} \mathbf{\Pi}_{\widetilde{\Sigma}} \mathbf{S}_{\widehat{F}} \begin{bmatrix} I \\ \overline{\mathbf{S}} \end{bmatrix} = \\
&= \begin{bmatrix} I & O \\ \mathbf{S}_{\widehat{F},21} & \mathbf{S}_{\widehat{F},22} \\ \mathbf{S} & O \end{bmatrix}^* \mathbf{\Pi}_{\widetilde{\Sigma}} \begin{bmatrix} I & O \\ \mathbf{S}_{\widehat{F},21} & \mathbf{S}_{\widehat{F},22} \\ \mathbf{S} & O \end{bmatrix} \\
&= \begin{bmatrix} I & O \\ \mathbf{S}_{\widehat{F},21} & \mathbf{S}_{\widehat{F},22} \end{bmatrix}^* \begin{bmatrix} I & O \\ O & I \\ \mathbf{S} & O \end{bmatrix}^* \mathbf{\Pi}_{\widetilde{\Sigma}} \begin{bmatrix} I & O \\ O & I \\ \mathbf{S} & O \end{bmatrix} \begin{bmatrix} I & O \\ \mathbf{S}_{\widehat{F},21} & \mathbf{S}_{\widehat{F},22} \end{bmatrix} \\
&= \mathbf{S}^*_{\widehat{F},2} \begin{bmatrix} I & \overline{\mathbf{S}}^* \end{bmatrix} \mathbf{\Pi}_{\widetilde{\Sigma}} \begin{bmatrix} I \\ \overline{\mathbf{S}} \end{bmatrix} \mathbf{S}_{\widehat{F},2}
\end{aligned} \tag{8.47}$$

where

$$\mathbf{S}_{\widehat{F},2} := \begin{bmatrix} I & O \\ \mathbf{S}_{\widehat{F},21} & \mathbf{S}_{\widehat{F},22} \end{bmatrix}.$$

Finally, since $\mathbf{S}_{\widehat{F}}$ is a unit in RH_{-}^{∞}, it follows from (8.46) that $\mathbf{S}_{\widehat{F},2}$ is a unit in RH_{-}^{∞} as well. Therefore, we conclude by (8.47) that (8.42) holds if and only if (8.43) holds and the proof ends. ■

Actually, from the above result we get that the 4–NP has a solution $\mathbf{S}$ if and only if the signature condition (8.43) written for the Popov function $\mathbf{\Pi}_{\widetilde{\Sigma}}$ holds for $\overline{\mathbf{S}} = \begin{bmatrix} \mathbf{S} & O \end{bmatrix}$. In view of Theorem 8.1.1, it only remains to show that $\mathbf{\Pi}_{\widetilde{\Sigma}}$ has a J spectral factorization with the spectral factor satisfying the hypotheses of Theorem 8.1.1. The final result is encapsulated in the following theorem.

Theorem 8.3.3 *Let* $\mathbf{T}$ *be given by (8.2) and let* Σ_1 *and* $\widetilde{\Sigma}$ *be the Popov triplets defined in (8.34) and (8.41), respectively. Then the suboptimal 4–NP (8.5) (with* $\gamma = 1$*) has a solution* $\mathbf{S}$ *if and only if the following two conditions hold:*

a. *The KPYS(*$\Sigma_1, -I$*) in (8.36) has a stabilizing solution* (X_1, V_1, W_1)*, and* $X_1 \geq 0$.

b. *The KPYS(*$\widetilde{\Sigma}, J$*)*

$$\begin{aligned} \widetilde{R} &= V^T J V, \\ \widetilde{L} + \widetilde{X} B &= W^T J V, \\ \widetilde{Q} + \widetilde{A}^T X + X \widetilde{A} &= W^T J W, \end{aligned} \tag{8.48}$$

with

$$J = \operatorname{diag}(-I_{m_1}, -I_{m_2}, I_{p_1}), \tag{8.49}$$

has a stabilizing solution $(\widetilde{X}, V, W)$*, and* $\widetilde{X} \geq 0$.

If conditions a and b hold, then the Popov function $\mathbf{\Pi}_{\widetilde{\Sigma}}$ *has a J spectral factorization*

$$\mathbf{\Pi}_{\widetilde{\Sigma}} = \widetilde{\mathbf{G}}^* J \widetilde{\mathbf{G}}, \tag{8.50}$$

where the J spectral factor has the form

$$\widetilde{\mathbf{G}} = \begin{bmatrix} \widetilde{\mathbf{G}}_{11} & O & \widetilde{\mathbf{G}}_{12} \\ \widetilde{\mathbf{G}}_{21} & \widetilde{\mathbf{G}}_{22} & \widetilde{\mathbf{G}}_{23} \\ \widetilde{\mathbf{G}}_{31} & O & \widetilde{\mathbf{G}}_{33} \end{bmatrix} \tag{8.51}$$

and satisfies all the conditions in Theorem 8.1.1 written for $\mathbf{\Pi}_{\widetilde{\Sigma}}$*. Moreover, the class of all solutions to the 4–NP is given by any of the equivalent expressions 1–3 in Corollary 8.1.2.*

Proof. “Only if ”. Since the 4–NP has a solution, say $\mathbf{S}$, (8.5) and (8.9) hold for $\overline{\Sigma}$ defined in (8.11). Then, the necessity of condition a follows directly from

$$\left\| \begin{bmatrix} \mathbf{T}_{12} \\ \mathbf{T}_{22} \end{bmatrix} \right\|_\infty \leq \left\| \begin{bmatrix} \mathbf{T}_{11} + \mathbf{S} & \mathbf{T}_{12} \\ \mathbf{T}_{21} & \mathbf{T}_{22} \end{bmatrix} \right\|_\infty < 1$$

and part I of Lemma 8.3.1. Further, from point a just proved, we conclude that $\widetilde{\Sigma}$ is well defined and with Lemma 8.3.2 we get that (8.43) holds as well.

We prove the necessity of condition b by invoking Theorem 4.5.9 for the triplet $\widetilde{\Sigma}$. First we check the hypotheses of this theorem. Indeed, the matrix $\widetilde{A}$ is stable as F_1 is the stabilizing feedback of the KPYS$(\Sigma_1, -I)$, $\mathbf{\Pi}_{\widetilde{\Sigma},22} = I > 0$, on $\overline{\mathbb{C}}_0$, and $\begin{bmatrix} O & O \\ O & I \end{bmatrix} \geq 0$. Since (8.43) holds, it follows from part I of Theorem 4.5.9 that the KPYS$(\widetilde{\Sigma}, J)$ has a stabilizing solution $(\widetilde{X}, V, W)$, with $\widetilde{X} \geq 0$ and point b follows.

"If ". It is enough to prove that points a and b imply the existence of a solution $\mathbf{S}$ to the signature condition (8.9) or, by Lemma 8.3.2, to the signature condition (8.43). In this respect, by invoking Theorem 8.1.1 it suffices to construct a J spectral factorization (8.50) of $\mathbf{\Pi}_{\widetilde{\Sigma}}$, where the spectral factor satisfies all the conditions in Theorem 8.1.1.

Since $(\widetilde{X}, V, W)$ is a stabilizing solution to the KPYS$(\widetilde{\Sigma}, J)$, the spectral factorization identity holds (see (3.39))

$$\mathbf{\Pi}_{\widetilde{\Sigma}} = \mathbf{S}^*_{\widetilde{F}} \widetilde{R} \mathbf{S}_{\widetilde{F}}, \tag{8.52}$$

where

$$\mathbf{S}_{\widetilde{F}} := \left[\begin{array}{c|c} \widetilde{A} & \widetilde{B} \\ \hline -\widetilde{F} & I \end{array}\right] \tag{8.53}$$

and $\widetilde{F} := -V^{-1}W$ is the stabilizing feedback of the KPYS$(\widetilde{\Sigma}, J)$. Moreover, $\widetilde{R}$ satisfies the assumptions of Lemma 8.2.1. Therefore there exists a solution V to the factorization problem

$$\widetilde{R} = R = V^T J V, \tag{8.54}$$

with V invertible and of the form

$$V = \begin{bmatrix} V_{11} & O & O \\ V_{21} & V_{22} & V_{23} \\ V_{31} & O & V_{33} \end{bmatrix}. \tag{8.55}$$

The required spectral factor is given by

$$\widetilde{\mathbf{G}} := V \mathbf{S}_{\widetilde{F}} = \left[\begin{array}{c|c} \widetilde{A} & \widetilde{B} \\ \hline W & V \end{array}\right]. \tag{8.56}$$

Now, using (8.56) and (8.54) in (8.52) we get that $\mathbf{\Pi}_{\widetilde{\Sigma}}$ has a J spectral factorization (8.50). It only remains to prove that the spectral factor $\widetilde{\mathbf{G}}$ satisfies all the conditions imposed in Theorem 8.1.1.

We prove first that $\widetilde{\mathbf{G}}$ has the zero pattern depicted in (8.51). We partition (8.56) conformably to (8.51) to get

$$\widetilde{\mathbf{G}} = \left[\begin{array}{c|cc:c} \widetilde{A} & \widetilde{B}_1 & O & \widetilde{B}_3 \\ \hline W_1 & V_{11} & O & O \\ W_2 & V_{21} & V_{22} & V_{23} \\ \hdashline W_3 & V_{31} & O & V_{33} \end{array}\right] = \left[\begin{array}{cc:c} \widetilde{\mathbf{G}}_{11} & O & \widetilde{\mathbf{G}}_{13} \\ \widetilde{\mathbf{G}}_{21} & \widetilde{\mathbf{G}}_{22} & \widetilde{\mathbf{G}}_{23} \\ \hdashline \widetilde{\mathbf{G}}_{31} & O & \widetilde{\mathbf{G}}_{33} \end{array}\right] = \left[\begin{array}{c:c} \widetilde{\mathbf{G}}_1 & \widetilde{\mathbf{G}}_2 \\ \hdashline \widetilde{\mathbf{G}}_3 & \widetilde{\mathbf{G}}_4 \end{array}\right], \tag{8.57}$$

where $W := \begin{bmatrix} W_1^T & W_2^T & W_3^T \end{bmatrix}^T$ has been partitioned conformably as well.

We prove now that $\widetilde{\mathbf{G}}$ is a unit in RH_+^∞. Clearly $\widetilde{\mathbf{G}} \in RH_+^\infty$ since $\widetilde{A} = -A - B_2 F_1$ is stable and F_1 is the stabilizing feedback to the KPYS$(\Sigma_1, -I)$. Now, $\mathbf{S}_{\widetilde{F}}^{-1} \in RH_+^\infty$

since $\widetilde{F}$ is the stabilizing feedback to KPYS$(\widetilde{\Sigma}, J)$ and $\widetilde{A} + \widetilde{B}\widetilde{F}$ is stable. Hence $\widetilde{\mathbf{G}}^{-1} = \mathbf{S}_{\widetilde{F}}^{-1} V^{-1} \in RH_+^\infty$.

Finally, we proceed to prove that $\widetilde{\mathbf{G}}_4$ is a unit in RH_+^∞. The proof follows along the lines of the proof given for the second part of Theorem 4.5.9, albeit more intricate reasoning is necessary. Clearly $\widetilde{\mathbf{G}}_4 \in RH_+^\infty$. Now we show $\widetilde{\mathbf{G}}_4^{-1} \in RH_+^\infty$ by proving that its state matrix $\widetilde{A} - \widetilde{B}_3 V_{33}^{-1} W_3$ is stable. Since $(\widetilde{X}, V, W)$ is a solution to the KPYS$(\widetilde{\Sigma})$ we have

$$\widetilde{A}^T \widetilde{X} + \widetilde{X}\widetilde{A} - W^T J W = 0, \tag{8.58}$$

where

$$W = -V\widetilde{F} = V\widetilde{R}^{-1}(\widetilde{B}^T \widetilde{X} + \widetilde{L}^T) \overset{(8.54)}{=} JV^{-T}(\widetilde{B}^T \widetilde{X} + \widetilde{L}^T). \tag{8.59}$$

The idea of the proof is to write the Lyapunov equation (8.58) in an appropriate form to use standard stability results. To this end we need first to perform several computations. Denote

$$V^{-1} =: \begin{bmatrix} \overline{V}_{11} & O & O \\ \overline{V}_{21} & \overline{V}_{22} & \overline{V}_{23} \\ \overline{V}_{31} & O & \overline{V}_{33} \end{bmatrix}, \tag{8.60}$$

where

$$\overline{V}_{22} = V_{22}^{-1},\ \overline{V}_{23} = -V_{22}^{-1} V_{23} V_{33}^{-1},\ \overline{V}_{33} = V_{33}^{-1}. \tag{8.61}$$

The rest of the elements of V^{-1} have no relevance for further computations. Substituting (8.60) in (8.59) we get

$$W = \begin{bmatrix} W_1 \\ W_2 \\ W_3 \end{bmatrix} = \begin{bmatrix} -\overline{V}_{11}^T & -\overline{V}_{21}^T & -\overline{V}_{31}^T \\ O & -\overline{V}_{22}^T & O \\ O & \overline{V}_{23}^T & \overline{V}_{33}^T \end{bmatrix} \begin{bmatrix} \widetilde{B}_1^T \widetilde{X} - B_1^T \\ -B_2^T \\ \widetilde{B}_3^T \widetilde{X} \end{bmatrix}. \tag{8.62}$$

Picking the 2nd and 3rd entries of (8.62) shows that

$$W_2 = \overline{V}_{22}^T B_2^T, \tag{8.63}$$

$$W_3 = -\overline{V}_{23}^T B_2^T + \overline{V}_{33}^T \widetilde{B}_3^T \widetilde{X}. \tag{8.64}$$

Note that by (8.61) and (8.64) we have

$$\begin{aligned} & W_3^T W_3 \\ = \quad & W_3^T(-\overline{V}_{23}^T B_2^T + V_{33}^{-T} \widetilde{B}_3^T \widetilde{X}) \\ \overset{(8.64)}{=} \quad & W_3^T V_{33}^{-T} \widetilde{B}_3^T \widetilde{X} + \widetilde{X}\widetilde{B}_3 V_{33}^{-1} W_3 - \widetilde{X}\widetilde{B}_3 V_{33}^{-1}(-\overline{V}_{23}^T B_2^T + V_{33}^{-T} \widetilde{B}_3^T \widetilde{X}) \\ & + (-\widetilde{X}\widetilde{B}_3 V_{33}^{-1} + B_2 \overline{V}_{23})\overline{V}_{23}^T B_2^T \\ = \quad & W_3^T V_{33}^{-T} \widetilde{B}_3^T \widetilde{X} + \widetilde{X}\widetilde{B}_3 V_{33}^{-1} W_3 - \widetilde{X}\widetilde{B}_3 V_{33}^{-1} V_{33}^{-T} \widetilde{B}_3^T \widetilde{X} \\ & + B_2 \overline{V}_{23} \overline{V}_{23}^T B_2^T. \end{aligned} \tag{8.65}$$

Now we compute

$$\begin{aligned} W_2^T W_2 - B_2 \overline{V}_{23} \overline{V}_{23}^T B_2^T & \\ \stackrel{(8.63)}{=} \quad & B_2 \overline{V}_{22} \overline{V}_{22}^T B_2^T - B_2 \overline{V}_{23} \overline{V}_{23}^T B_2^T \\ \stackrel{(8.61)}{=} \quad & B_2 V_{22}^{-1} (I - V_{23} V_{33}^{-1} V_{33}^{-T} V_{23}^T) V_{22}^{-T} B_2^T \\ \stackrel{(8.54),(8.55)}{=} \quad & B_2 V_{22}^{-1} (I - V_{23} (I + V_{23}^T V_{23})^{-1} V_{23}^T) V_{22}^{-T} B_2^T \\ = \quad & B_2 V_{22}^{-1} (I + V_{23}^T V_{23})^{-1} V_{22}^{-T} B_2^T . \end{aligned} \tag{8.66}$$

From (8.58) we get

$$\widetilde{A}^T \widetilde{X} + \widetilde{X} \widetilde{A} + W_1^T W_1 + W_2^T W_2 - W_3^T W_3 = 0,$$

and by (8.65) we obtain

$$(\widetilde{A} - \widetilde{B}_3 V_{33}^{-1} W_3)^T \widetilde{X} + \widetilde{X} (A - \widetilde{B}_3 V_{33}^{-1} W_3) + W_1^T W_1 + \widetilde{X} \widetilde{B}_3 V_{33}^{-1} V_{33}^{-T} \widetilde{B}_3^T \widetilde{X}$$
$$+ W_2^T W_2 - B_2 \overline{V}_{23} \overline{V}_{23}^T B_2^T = 0.$$

By (8.66) this implies

$$(\widetilde{A} - \widetilde{B}_3 V_{33}^{-1} W_3)^T \widetilde{X} + \widetilde{X} (A - \widetilde{B}_3 V_{33}^{-1} W_3) + W_1^T W_1 + \widetilde{X} \widetilde{B}_3 V_{33}^{-1} V_{33}^{-T} \widetilde{B}_3^T \widetilde{X}$$
$$+ B_2 V_{22}^{-1} (I + V_{23}^T V_{23})^{-1} V_{22}^{-T} B_2^T = 0. \tag{8.67}$$

Since $\widetilde{F}$ is the stabilizing feedback, we have that

$$\widetilde{A} + \widetilde{B}\widetilde{F} = \widetilde{A} - \widetilde{B} V^{-1} W = \widetilde{A} - \begin{bmatrix} \widetilde{B}_1 & O & \widetilde{B}_3 \end{bmatrix} \begin{bmatrix} \overline{V}_{11} & O & O \\ \overline{V}_{21} & \overline{V}_{22} & \overline{V}_{23} \\ \overline{V}_{31} & O & \overline{V}_{33} \end{bmatrix} \begin{bmatrix} W_1 \\ W_2 \\ W_3 \end{bmatrix}$$
$$= \widetilde{A} - \widetilde{B}_3 V_{33}^{-1} W_3 - (\widetilde{B}_1 \overline{V}_{11} + \widetilde{B}_3 \overline{V}_{31}) W_1 \text{ is stable,} \tag{8.68}$$

from where we infer that the pair $(W_1,\ A - \widetilde{B}_3 V_{33}^{-1} W_1)$ is detectable. We also have that $\widetilde{X} \geq 0$ is the solution to the Lyapunov equation (8.67), which has a positive semidefinite symmetric free term, and we conclude with Theorem 1.5.5 that $A - \widetilde{B}_3 V_{33}^{-1} W_3$ is stable. Therefore $\widetilde{\mathbf{G}}_4$ is a unit in RH_+^∞ and all of the conditions of Theorem 8.1.1 have been checked.

The rest of the proof follows directly from Theorem 8.1.1 and Corollary 8.1.2. ■

We conclude this section by giving the explicit expression of the class of solutions to the 4–NP.

Corollary 8.3.4 *The class of solutions to the 4–NP is given by any of the equivalent forms 1–3 in Corollary 8.1.2, where a state–space representation of the intervening coefficients is given by*

$$\begin{bmatrix} \mathbf{G}_{11} & \mathbf{G}_{13} \\ \mathbf{G}_{31} & \mathbf{G}_{33} \end{bmatrix} := \begin{bmatrix} \widetilde{\mathbf{G}}_{11} & \widetilde{\mathbf{G}}_{13} \\ \widetilde{\mathbf{G}}_{31} & \widetilde{\mathbf{G}}_{33} \end{bmatrix} = \left[\begin{array}{c|cc} \widetilde{A} & \widetilde{B}_1 & \widetilde{B}_3 \\ \hline W_1 & V_{11} & O \\ W_3 & V_{31} & V_{33} \end{array} \right],$$

$$\begin{bmatrix} \overline{\mathbf{G}}_{11} & \overline{\mathbf{G}}_{13} \\ \overline{\mathbf{G}}_{31} & \overline{\mathbf{G}}_{33} \end{bmatrix} := \begin{bmatrix} \widetilde{\mathbf{G}}_{11} & \widetilde{\mathbf{G}}_{13} \\ \widetilde{\mathbf{G}}_{31} & \widetilde{\mathbf{G}}_{33} \end{bmatrix}^{-1} = \left[\begin{array}{c|cc} \widetilde{A}+\widetilde{B}\widetilde{F} & \widetilde{B}_1\overline{V}_{11}+\widetilde{B}_3\overline{V}_{31} & \widetilde{B}_3\overline{V}_{33} \\ \hline \widetilde{F}_1 & \overline{V}_{11} & O \\ \widetilde{F}_3 & \overline{V}_{31} & \overline{V}_{33} \end{array}\right],$$

where $\widetilde{F} := -V^{-1}W = \begin{bmatrix} \widetilde{F}_1^T & \widetilde{F}_2^T & \widetilde{F}_3^T \end{bmatrix}^T$ *has been partitioned conformably.*

8.4 Uncoupled solvability conditions

In this section we refine further the characterization of all solutions, and show that the 4–NP is solvable if and only if the two KPYSs in Lemma 8.3.1 have stabilizing solutions, or, equivalently, that the corresponding Riccati equations have stabilizing solutions, and an additional coupling condition holds. We make heavy use of the notation introduced in the previous section.

Theorem 8.4.1 *Let* $\mathbf{T}$ *be given by (8.2) and let* Σ_1, Σ_2, *and* $\widetilde{\Sigma}$ *be the Popov triplets defined in (8.34), (8.35), and (8.41), respectively.*

I. *The suboptimal 4–NP (8.5) (with* $\gamma = 1$*) has a solution* $\mathbf{S}$ *if and only if the following three conditions hold:*

a. *The KPYS(*$\Sigma_1, -I$*) in (8.36) has a stabilizing solution* (X_1, V_1, W_1), *with* $X_1 \geq 0$.

b. *The KPYS(*$\Sigma_2, -I$*) in (8.37) has a stabilizing solution* (X_2, V_2, W_2), *with* $X_2 \geq 0$.

c. $\rho(X_1X_2) < 1$.

II. *If conditions a, b and c hold, then the KPYS(*$\widetilde{\Sigma}, J$*) in (8.48) has a stabilizing solution* $(\widetilde{X}, V, W)$, *with* $\widetilde{X} \geq 0$ *satisfying* $\widetilde{X} = X_2(I - X_1X_2)^{-1}$.

Therefore, the class of solutions to the 4–NP given in Corollary 8.3.4 can be expressed in terms of the solutions X_1 *and* X_2 *corresponding to the KPYS(*$\Sigma_1, -I$*) and the KPYS(*$\Sigma_2, -I$*), respectively, whose coefficients are given in terms of the state-space realization (8.2) of the original system* $\mathbf{T}$.

Proof. Notice first that assertion a is exactly assertion a of Theorem 8.3.3. Therefore, by invoking Theorem 8.3.3, it is enough to prove that if assertion a holds, then assertions b and c are equivalent to assertion b of Theorem 8.3.3. The proof runs by performing equivalence operations on extended Hamiltonian pencils.

"Condition b of Theorem 8.3.3 ⇒ conditions b and c". Recall first that for an arbitrary Popov triplet $\Sigma = (A, B; Q, L, R)$ and a sign matrix J the KPYS(Σ, J) is equivalent to the CTARE(Σ) plus the signature equality $\operatorname{sgn}(R) = J$ (see Section 3.2.4).

First, $\operatorname{sgn}(D_2^TD_2 - I) = -I$ follows from $\operatorname{sgn}(\widetilde{R}) = J$. It only remains to prove that the CTARE(Σ_2) has a stabilizing solution $X_2 \geq 0$ and that condition c is satisfied.

Since $\widetilde{X} \geq 0$ is the stabilizing solution to the CTARE($\widetilde{\Sigma}$), Theorem 5.2.1 shows that

$$N_{\widetilde{\Sigma}}V_{\widetilde{\Sigma}} = M_{\widetilde{\Sigma}}V_{\widetilde{\Sigma}}S_{\widetilde{\Sigma}}, \quad V_{\widetilde{\Sigma}} := \begin{bmatrix} I \\ \widetilde{X} \\ \widetilde{F} \end{bmatrix}, \quad S_{\widetilde{\Sigma}} \text{ is stable} \tag{8.69}$$

where

$$sM_{\widetilde{\Sigma}} - N_{\widetilde{\Sigma}} := s\begin{bmatrix} I & O & O \\ O & I & O \\ O & O & O \end{bmatrix} - \begin{bmatrix} \widetilde{A} & O & \widetilde{B} \\ O & -\widetilde{A}^T & -\widetilde{L} \\ \widetilde{L}^T & \widetilde{B}^T & \widetilde{R} \end{bmatrix} \tag{8.70}$$

is the EHP($\widetilde{\Sigma}$) and $\widetilde{F}$ is the corresponding stabilizing feedback. We write explicitly the partition of the entries in (8.70) and extend the pencil to get

$$N_{\widetilde{\Sigma},e} V_{\widetilde{\Sigma},e} = M_{\widetilde{\Sigma},e} V_{\widetilde{\Sigma},e} S_{\widetilde{\Sigma}}, \tag{8.71}$$

where

$$sM_{\widetilde{\Sigma},e} - N_{\widetilde{\Sigma},e} := s \begin{bmatrix} I & O & O & O & O \\ O & I & O & O & O \\ O & O & O & O & O \\ O & O & O & O & O \\ O & O & O & O & O \end{bmatrix} - \begin{bmatrix} \widetilde{A} & O & \widetilde{B}_{12} & \widetilde{B}_3 & O \\ O & -\widetilde{A}^T & B & O & O \\ -B^T & \widetilde{B}_{12}^T & D^T D - I & D_1^T & O \\ O & \widetilde{B}_3^T & D_1 & I & O \\ O & O & O & O & I_{p_2} \end{bmatrix}, \tag{8.72}$$

and

$$V_{\widetilde{\Sigma},e} = \begin{bmatrix} I \\ \widetilde{X} \\ \widetilde{F}_{12} \\ \widetilde{F}_3 \\ O \end{bmatrix}, \quad \widetilde{F} = \begin{bmatrix} \widetilde{F}_{12} \\ \widetilde{F}_3 \end{bmatrix} = \begin{bmatrix} \widetilde{F}_1 \\ \widetilde{F}_2 \\ \widetilde{F}_3 \end{bmatrix}, \quad \begin{aligned} \widetilde{B} &= \begin{bmatrix} \widetilde{B}_{12} & \widetilde{B}_3 \end{bmatrix} \\ &= \begin{bmatrix} \widetilde{B}_1 & O & \widetilde{B}_3 \end{bmatrix}. \end{aligned}$$

Let $\widehat{D}_2 := D_2^T D_2 - I$ and $\widetilde{D}_2 := D_2 D_2^T - I$ and define

$$U := \begin{bmatrix} I & X_1 & -\overline{F}_1^T - C_2^T D_2 \widehat{D}_2^{-1} & -C_1^T + C_2^T D_2 \widehat{D}_2^{-1} D_1^T & -C_2^T \widetilde{D}_2^{-1} \\ O & I & -B\widehat{D}_2^{-1} & B\widehat{D}_2^{-1} D_1^T & -BD_2^T \widetilde{D}_2^{-1} \\ O & O & O & O & I \\ O & O & I & -D_1^T & O \\ O & O & O & I & O \end{bmatrix},$$

$$Z := \begin{bmatrix} I & -X_1 & O & O & O \\ O & I & O & O & O \\ \widehat{D}_2^{-1} B^T & -\overline{F}_1 - \widehat{D}_2^{-1} D_2^T C_2 & O & I & O \\ -D_1 \widehat{D}_2^{-1} B^T & -C_1 + D_1 \widehat{D}_2^{-1} D_2^T C_2 & O & -D_1 & I \\ D_2 B^T & -C_2 & \widetilde{D}_2 & O & O \end{bmatrix},$$

which are both nonsingular and where $\overline{F}_1$ is defined in (8.38). Premultiplying both sides of (8.71) with U results in

$$N_{\Sigma_2,e} V_{\Sigma_2,e} = M_{\Sigma_2,e} V_{\Sigma_2,e} S_{\widetilde{\Sigma}}, \tag{8.73}$$

where

$$M_{\Sigma_2,e} := U M_{\widetilde{\Sigma},e} Z = \begin{bmatrix} I & O & O & O & O \\ O & I & O & O & O \\ O & O & O & O & O \\ O & O & O & O & O \\ O & O & O & O & O \end{bmatrix}, \tag{8.74}$$

$$N_{\Sigma_2,e} := UN_{\widetilde{\Sigma},e}Z = \begin{bmatrix} -A^T & O & -C_2^T & O & O \\ -BB^T & A & -BD_2^T & O & O \\ D_2B^T & -C_2 & \widetilde{D_2} & O & O \\ O & O & O & \widehat{D_2} & O \\ O & O & O & O & I \end{bmatrix}, \tag{8.75}$$

$$V_{\Sigma_2,e} := Z^{-1}V_{\widetilde{\Sigma},e} = \begin{bmatrix} I + X_1\widetilde{X} \\ \widetilde{X} \\ \star \\ \star \\ \star \end{bmatrix}. \tag{8.76}$$

Picking up the first 3×3 entries in (8.73) we get

$$N_{\Sigma_2}V_{\Sigma_2} = M_{\Sigma_2}V_{\Sigma_2}S_{\widetilde{\Sigma}}, \tag{8.77}$$

where $sM_{\Sigma_2} - N_{\Sigma_2}$ is exactly the EHP(Σ_2) and

$$V_{\Sigma_2} = \begin{bmatrix} I + X_1\widetilde{X} \\ \widetilde{X} \\ \star \end{bmatrix}.$$

Since $\widetilde{X} \geq 0$ and $X_1 \geq 0$, clearly $I + X_1\widetilde{X} > 0$ and (8.77) gives

$$N_{\Sigma_2}\begin{bmatrix} I \\ \widetilde{X}(I + X_1\widetilde{X})^{-1} \\ \star \end{bmatrix} = M_{\Sigma_2}\begin{bmatrix} I \\ \widetilde{X}(I + X_1\widetilde{X})^{-1} \\ \star \end{bmatrix} S_2, \tag{8.78}$$

where $S_2 := (I - X_1\widetilde{X})S_{\widetilde{\Sigma}}(I - X_1\widetilde{X})^{-1}$ is stable. Invoking Theorem 5.2.1 and the uniqueness of the stabilizing solution to the Riccati equation shows that

$$X_2 := \widetilde{X}(I + X_1\widetilde{X})^{-1} \tag{8.79}$$

is the stabilizing solution to the CTARE(Σ_2). Finally, by (8.79) we get successively $I - X_1X_2 = I - X_1\widetilde{X}(I + X_1\widetilde{X})^{-1} = (I + X_1\widetilde{X})^{-1} > 0$, which implies $\rho(X_1X_2) < 1$ and the proof of this part ends.

"Conditions b and c $\Rightarrow$ condition b of Theorem 8.3.3". We apply the same scheme of proof as for the first part. From assertion b we have that (8.37) holds and the last equation of the KPYS$(\Sigma_2, -I)$ shows immediately that sgn $(\widetilde{R}) = J$. It only remains to prove that point b and c imply the existence of a stabilizing solution $\widetilde{X} \geq 0$ to the CTARE$(\widetilde{\Sigma})$. Since X_2 is the stabilizing solution to the CTARE(Σ_2), we get by Theorem 5.2.1

$$N_{\Sigma_2}\widetilde{V}_{\Sigma_2} = M_{\Sigma_2}\widetilde{V}_{\Sigma_2}\, S_{\Sigma_2}, \quad \widetilde{V}_{\Sigma_2} = \begin{bmatrix} I \\ X_2 \\ F_2 \end{bmatrix}, \quad S_{\Sigma_2} \text{ is stable,}$$

where $sM_{\Sigma_2} - N_{\Sigma_2}$ is the EHP(Σ_2). We extend this pencil to get

$$N_{\Sigma_2,e}\widetilde{V}_{\Sigma_2,e} = M_{\Sigma_2,e}\widetilde{V}_{\Sigma_2,e}S_{\Sigma_2}, \quad \widetilde{V}_{\Sigma_2,e} := \begin{bmatrix} I \\ X_2 \\ F_2 \\ O \\ O \end{bmatrix}$$

where $sM_{\Sigma_2,e} - N_{\Sigma_2,e}$ is defined by (8.74), (8.75). Reversing the arguments in the first part of the proof, we arrive at

$$N_{\widetilde{\Sigma},e}\widetilde{V}_{\widetilde{\Sigma},e} = M_{\widetilde{\Sigma},e}\widetilde{V}_{\widetilde{\Sigma},e}S_{\Sigma_2}, \quad \widetilde{V}_{\widetilde{\Sigma},e} = \begin{bmatrix} I - X_1X_2 \\ X_2 \\ \star \\ \star \\ \star \end{bmatrix}.$$

Picking up the first 4×4 entries and using the equivalent form $I - X_1X_2 > 0$ of $\rho(X_1X_2) < 1$ we get

$$N_{\widetilde{\Sigma}}\widetilde{V}_{\widetilde{\Sigma}} = M_{\widetilde{\Sigma}}\widetilde{V}_{\widetilde{\Sigma}}\widetilde{S}_2, \quad \widetilde{V}_{\widetilde{\Sigma}} = \begin{bmatrix} I \\ X_2(I - X_1X_2)^{-1} \\ \star \\ \star \end{bmatrix},$$

$$\widetilde{S}_2 = (I - X_1X_2)S_{\Sigma_2}(I - X_1X_2)^{-1} \text{ is stable.}$$

Hence $\widetilde{X} = X_2(I - X_1X_2)^{-1} \geq 0$ is the stabilizing solution to the CTARE($\widetilde{\Sigma}$) and the proof ends. ■

B. The discrete–time case

8.5 The four block Nehari problem

As in the continuous–time case we focus on the suboptimal version of the discrete–time Nehari problem only and deal directly with the four block case. In discrete–time we formulate the problem for an inverse–time stable system which is the natural discrete–time counterpart of an antistable system in continuous–time.

Consider a discrete inverse–time stable system

$$\begin{aligned} x &= A\sigma x + B_1u_1 + B_2u_2, \\ y_1 &= C_1\sigma x + D_{11}u_1 + D_{12}u_2, \\ y_2 &= C_2\sigma x + D_{21}u_1 + D_{22}u_2, \end{aligned} \tag{8.80}$$

having the TFM given by

$$\mathbf{T} = \left[\begin{array}{c|c} A & B \\ \hline C & D \end{array}\right]^a = \left[\begin{array}{c|cc} A & B_1 & B_2 \\ \hline C_1 & D_{11} & D_{12} \\ C_2 & D_{21} & D_{22} \end{array}\right]^a = \begin{bmatrix} \mathbf{T}_{11} & \mathbf{T}_{12} \\ \mathbf{T}_{21} & \mathbf{T}_{22} \end{bmatrix}, \tag{8.81}$$

$$\mathbf{T}(z) := C(z^{-1}I - A)^{-1}B + D,$$

where $A \in \mathbb{R}^{n\times n}$ is stable, $B_i \in \mathbb{R}^{n\times m_i}$, $C_j \in \mathbb{R}^{p_j\times n}$, $D_{ij} \in \mathbb{R}^{p_i\times m_j}$ $(i, j = 1, 2,$ $m = m_1 + m_2$, $p = p_1 + p_2)$.

We define an *approximant* for the system $\mathbf{T}$ as being a discrete direct–time stable system

$$\begin{aligned} \sigma x_S &= A_S x_S + B_S u_S, \\ y_S &= C_S x_S + D_S u_S, \end{aligned} \tag{8.82}$$

having the TFM given by

$$\begin{aligned} \mathbf{S} &= \left[\begin{array}{c|c} A_S & B_S \\ \hline C_S & D_S \end{array}\right], \\ \mathbf{S}(z) &= C_S(zI - A_S)^{-1}B_S + D_S, \end{aligned} \tag{8.83}$$

where A_S is stable.

Given the inverse–time stable system $\mathbf{T}$ and a threshold $\gamma > 0$, the *suboptimal four block discrete–time Nehari problem (4–NP)* consists in finding all direct–time stable approximants $\mathbf{S}$, when any exists, such that

$$\left\| \begin{bmatrix} \mathbf{T}_{11} + \mathbf{S} & \mathbf{T}_{12} \\ \mathbf{T}_{21} & \mathbf{T}_{22} \end{bmatrix} \right\|_\infty < \gamma. \tag{8.84}$$

As in the continuous–time case, we may assume without loss of generality $D_{11} = 0$ and $\gamma = 1$.

The same scheme of derivation as for the continuous–time counterpart is employed and therefore we highlight the differences in the technical machinery only.

Rewritting (8.84) as

$$\left\| \begin{bmatrix} \mathbf{T}_1 + \overline{\mathbf{S}} \\ \mathbf{T}_2 \end{bmatrix} \right\|_\infty < 1, \tag{8.85}$$

where $\overline{\mathbf{S}} = \begin{bmatrix} \mathbf{S} & O \end{bmatrix}$, $\mathbf{T}_1 := \begin{bmatrix} \mathbf{T}_{11} & \mathbf{T}_{12} \end{bmatrix}$, and $\mathbf{T}_2 := \begin{bmatrix} \mathbf{T}_{21} & \mathbf{T}_{22} \end{bmatrix}$, we obtain further the inequality

$$\begin{bmatrix} I & \overline{\mathbf{S}}^* \end{bmatrix} \mathbf{\Pi}_{\overline{\Sigma}^a} \begin{bmatrix} I \\ \overline{\mathbf{S}} \end{bmatrix} < 0, \text{ on } \partial\mathbb{D} \tag{8.86}$$

for the Popov function

$$\mathbf{\Pi}_{\overline{\Sigma}^a} = \begin{bmatrix} \mathbf{T}^*\mathbf{T} - I & \mathbf{T}_1^* \\ \mathbf{T}_1 & I \end{bmatrix}$$

associated with the Popov triplet

$$\begin{aligned} \overline{\Sigma}^a &= (\overline{A}, \overline{B}; \overline{Q}, \overline{L}, \overline{R}) \\ &:= (A, \begin{bmatrix} B & O \end{bmatrix}; C^TC, \begin{bmatrix} C^TD & C_1^T \end{bmatrix}, \begin{bmatrix} D^TD - I & D_1^T \\ D_1 & I \end{bmatrix}) \end{aligned} \tag{8.87}$$

seen with inverse–time significance. A direct check shows that $\mathbf{\Pi}_{\overline{\Sigma}^a}$ is indeed the Popov function associated with the inverse–time triplet $\overline{\Sigma}^a$. We solve the signature condition (8.86) by employing the general result of Theorem 4.12.9. To fulfill the hypotheses of this theorem we transform first the inverse–time stable triplet $\overline{\Sigma}^a$ into a direct–time stable one as explained below. While doing so we need the following lemma that expresses the contractiveness property of a discrete inverse–time system.

Lemma 8.5.1 *Let* $\mathbf{T}$ *be as in (8.81). Define the Popov triplets*

$$\Sigma_1 = (A, B_2; C^T C, C^T \overline{D}_2, \overline{D}_2^T \overline{D}_2 - I), \quad \overline{D}_2 := \begin{bmatrix} D_{12} \\ D_{22} \end{bmatrix}, \tag{8.88}$$

and

$$\Sigma_2 = (A^T, C_2^T; BB^T, BD_2^T, D_2 D_2^T - I), \quad D_2 := \begin{bmatrix} D_{21} & D_{22} \end{bmatrix}. \tag{8.89}$$

I. *The following assertions are equivalent:*

1. $\left\| \begin{bmatrix} \mathbf{T}_{12} \\ \mathbf{T}_{22} \end{bmatrix} \right\|_\infty < 1.$

2. The KSPYS$(\Sigma_1, -I)$ *given by*

$$\begin{aligned} \overline{D}_2^T \overline{D}_2 - I + B_2^T X B_2 &= -V^T V, \\ C^T \overline{D}_2 + A^T X B_2 &= -W^T V, \\ C^T C + A^T X A - X &= -W^T W, \end{aligned} \tag{8.90}$$

has a stabilizing solution (X_1, V_1, W_1)*, and* $X_1 \geq 0$.

II. *The following assertions are equivalent:*

1. $\| \begin{bmatrix} \mathbf{T}_{21} & \mathbf{T}_{22} \end{bmatrix} \|_\infty < 1.$

2. The KSPYS$(\Sigma_2, -I)$ *given by*

$$\begin{aligned} D_2 D_2^T - I + C_2 X C_2^T &= -V^T V, \\ B D_2^T + A X C_2^T &= -W^T V, \\ BB^T + AXA^T - X &= -W^T W, \end{aligned} \tag{8.91}$$

has a stabilizing solution (X_2, V_2, W_2)*, and* $X_2 \geq 0$.

Proof. We show only part I since part II follows by duality. Let $\overline{\mathbf{T}}_2 := \begin{bmatrix} \mathbf{T}_{12} \\ \mathbf{T}_{22} \end{bmatrix}$. Notice from (3.82) that Σ_1 is a γ–contractiveness Popov triplet (with $\gamma = 1$) associated with the realization of the stable direct–time system $\overline{\mathbf{T}}_2(\frac{1}{z})$ given by

$$\overline{\mathbf{T}}_2(\frac{1}{z}) = \left[\begin{array}{c|c} A & B_2 \\ \hline C_1 & D_{12} \\ C_2 & D_{22} \end{array} \right],$$

and $\|\overline{\mathbf{T}}_2(\frac{1}{z})\|_\infty = \|\overline{\mathbf{T}}_2(z)\|_\infty$. Since A is stable, the equivalence between assertions 1 and 2 is exactly the bounded real lemma 7.7.1 written for the stable system $\overline{\mathbf{T}}_2(\frac{1}{z})$. ■

We use now the stabilizing solution to the KSPYS$(\Sigma_1, -I)$ to define the equivalent Popov triplet $\widehat{\Sigma}^a = (\widehat{A}, \widehat{B}; \widehat{Q}, \widehat{L}, \widehat{R})$ of $\overline{\Sigma}^a$. Notice that both $\widehat{\Sigma}^a$ and $\overline{\Sigma}^a$ are seen with inverse–time significance. More precisely, we construct $\widehat{\Sigma}^a$ to be the $(X_1, \widehat{F})$–equivalent Popov triplet of $\overline{\Sigma}$, where

$$\widehat{F} := \left[\begin{array}{c} O \\ F_1 \\ \hline O \end{array} \right] \begin{array}{l} \}m_1 \\ \\ \}p_1 \end{array} = \left[\begin{array}{c} F_1 \\ \hline O \end{array} \right], \tag{8.92}$$

and $F_1 := -(\overline{D}_2^T\overline{D}_2 + B_2^T X_1 B_2 - I)^{-1}(B_2^T X_1 A + \overline{D}_2^T C)$ is the stabilizing feedback of the KSPYS$(\Sigma_1, -I)$. Substituting the formulas (8.87) and (8.92) in the general expressions that are satisfied by two equivalent Popov triplets (3.100), we obtain for the elements of $\widehat{\Sigma}^a$ the following explicit expressions

$$\begin{aligned}
\widehat{A} &= \overline{A} + \overline{B}\widehat{F} = A + B_2F_1, \\
\widehat{B} &= \overline{B} = \begin{bmatrix} B_1 & B_2 & O \end{bmatrix}, \\
\widehat{L} &= \overline{L} + \widehat{F}^T\overline{R} + \widehat{A}^T X_1 \overline{B} \\
&= \begin{bmatrix} (C_2^T + F_1^T D_{22}^T)D_{21} + \widehat{A}^T X_1 B_1 & O & C_1^T + F_1^T D_{12}^T \end{bmatrix}, \\
\widehat{R} &= \overline{R} + \overline{B}^T X_1 \overline{B} \\
&= \begin{bmatrix} D_{21}^T D_{21} - I + B_1^T X_1 B_1 & D_{21}^T D_{22} + B_1^* X_1 B_2 & O \\ D_{22}^T D_{21} + B_2^T X_1 B_1 & \overline{D}_2^T \overline{D}_2 - I + B_2^T X_1 B_2 & D_{12}^T \\ O & D_{12} & I \end{bmatrix}, \\
\widehat{Q} &= \overline{Q} + \overline{L}\widehat{F} + \widehat{F}^T\overline{L}^T + \widehat{F}^T\overline{R}\widehat{F} + \widehat{A}^T X_1 \widehat{A} - X_1 = 0,
\end{aligned} \tag{8.93}$$

where the last equality is a simple rewriting of the DTARE(Σ_1), with X_1 its stabilizing solution and F_1 its stabilizing Riccati feedback. Consider the dual triplet Popov of $\widehat{\Sigma}^a$ denoted $\widetilde{\Sigma} := (\widetilde{A}, \widetilde{B}; \widetilde{Q}, \widetilde{L}, \widetilde{R})$, and seen with direct–time significance. As $\widehat{Q} = 0$ and $\widehat{A}$ is stable the reducing Lyapunov equation $\widehat{A}^T X A - X + \widehat{Q} = 0$ has a unique solution $X_r = 0$ and we get for the entries of the dual Popov triplet $\widetilde{\Sigma}$ the following expressions (see (3.186))

$$\begin{aligned}
\widetilde{A} &:= \widehat{A}^T = (A + B_2F_1)^T, \\
\widetilde{B} &:= \widehat{L} = \begin{bmatrix} \widetilde{B}_1 & O & \widetilde{B}_3 \end{bmatrix} \\
&\phantom{:= \widehat{L}} = \begin{bmatrix} C_2^T + F_1 D_{22}^T)D_{21} + \widehat{A}^T X_1 B_1 & O & C_1^T + F_1^T D_{12}^T \end{bmatrix}, \\
\widetilde{Q} &:= \widehat{Q} = 0, \\
\widetilde{L} &:= \widehat{B} = \begin{bmatrix} B_1 & B_2 & O \end{bmatrix}, \\
\widetilde{R} &:= \overline{R}.
\end{aligned} \tag{8.94}$$

For the Popov triplet $\widetilde{\Sigma}$ all the hypotheses of Theorem 4.12.9 are fulfilled. Furthermore, if the KSPYS$(\Sigma_1, -I)$ has a stabilizing solution then the signature condition written for $\mathbf{\Pi}_{\widetilde{\Sigma}}$ is solvable if and only if the signature condition written for $\mathbf{\Pi}_{\overline{\Sigma}^a}$ is solvable and the latter one is equivalent to the solvability of the discrete–time 4–NP. Moreover, both signature inequalities and the 4–NP share the same set of solutions. Indeed, since $\widehat{\Sigma}^a$ and $\overline{\Sigma}^a$ are equivalent we have by Proposition 3.6.17

$$\mathbf{\Pi}_{\overline{\Sigma}^a} = \widehat{\mathbf{S}}_F^* \mathbf{\Pi}_{\widehat{\Sigma}^a} \widehat{\mathbf{S}}_F, \tag{8.95}$$

where

$$\widehat{\mathbf{S}}_F := \left[\begin{array}{c|ccc} A & B_1 & B_2 & O \\ \hline O & I & O & O \\ -F_1 & O & I & O \\ O & O & O & I \end{array}\right],$$

and from Lemma 3.8.6 we have

$$\mathbf{\Pi}_{\widetilde{\Sigma}} = \mathbf{\Pi}_{\widehat{\Sigma}^a} \tag{8.96}$$

because $\widetilde{\Sigma}$ is the dual of $\widehat{\Sigma}^a$. From (8.95), (8.96) and the properties of $\widehat{\mathbf{S}}_F$ we conclude that

$$\begin{bmatrix} I & \overline{\mathbf{S}}^* \end{bmatrix} \mathbf{\Pi}_{\overline{\Sigma}^a} \begin{bmatrix} I \\ \overline{\mathbf{S}} \end{bmatrix} < 0 \text{ on } \partial\mathbb{D} \tag{8.97}$$

if and only if

$$\begin{bmatrix} I & \overline{\mathbf{S}}^* \end{bmatrix} \mathbf{\Pi}_{\widetilde{\Sigma}} \begin{bmatrix} I \\ \overline{\mathbf{S}} \end{bmatrix} < 0 \text{ on } \partial\mathbb{D}. \tag{8.98}$$

We get the following result giving necessary and sufficient solvability conditions for the discrete–time 4–NP together with a characterization of the class of solutions to the inequality (8.98). The proof is is based on the discrete–time version of Theorem 8.1.1 whose statement and proof are obtained by simply interchanging the extended imaginary axis $\overline{\mathbb{C}}_0$ with the unit circle $\partial\mathbb{D}$.

Theorem 8.5.2 *Let* $\mathbf{T}$ *be given by (8.81) and let* Σ_1 *and* $\widetilde{\Sigma}$ *be the Popov triplets defined in (8.88) and (8.94), respectively. Then the discrete–time suboptimal 4–NP (8.84) (with* $\gamma = 1$*) has a solution* $\mathbf{S}$ *if and only if the following two conditions hold:*

a. *The KSPYS(*$\Sigma_1, -I$*) given in (8.90) has a stabilizing solution* (X_1, V_1, W_1)*, with* $X_1 \geq 0$.

b. *The KSPYS(*$\widetilde{\Sigma}, J$*) given by*

$$\begin{aligned} \widetilde{R} + \widetilde{B}^T X \widetilde{B} &= V^T J V, \\ \widetilde{L} + \widetilde{A}^T X \widetilde{B} &= W^T J V, \\ \widetilde{Q} + \widetilde{A}^T X \widetilde{A} - X &= W^T J W, \end{aligned} \tag{8.99}$$

with J *given in (8.49), has a stabilizing solution* $(\widetilde{X}, V, W)$*, and* $\widetilde{X} \geq 0$.

If assertions a and b hold, then the Popov function $\mathbf{\Pi}_{\widetilde{\Sigma}}$ *has a* J *spectral factorization*

$$\mathbf{\Pi}_{\widetilde{\Sigma}} = \widetilde{\mathbf{G}}^* J \widetilde{\mathbf{G}}, \tag{8.100}$$

where the J *spectral factor takes the form (8.51) and satisfies all the conditions in the discrete–time version of Theorem 8.1.1 written for* $\mathbf{\Pi}_{\widetilde{\Sigma}}$*. Moreover, the class of all solutions to the discrete–time 4–NP is given by any of the equivalent expressions 1–3 in Corollary 8.1.2. The explicit state–space formulas of the intervening coefficients are given by Corollary 8.3.4 where now* $(\widetilde{X}, V, W)$ *is the stabilizing solution to the KSPYS(*$\widetilde{\Sigma}, J$*) given in (8.99),* $\widetilde{F}$ *is the corresponding stabilizing feedback,* $V^T J V = \widetilde{R} + \widetilde{B}^T \widetilde{X} \widetilde{B}$*, and* V *and* $\overline{V} = V^{-1}$ *have the zero pattern in (8.55) and (8.60), respectively.*

Proof. Conceptually, the proof follows along the same lines as the proof of Theorem 8.3.3. The only minor technical difference is in the "if" part when proving the existence of a J factorization of $\widetilde{R} + \widetilde{B}^T \widetilde{X} \widetilde{B} = V^T J V$ with the special pattern (8.55) of V. To show this, it is enough to check the hypotheses (i)–(iii) of Lemma 8.2.1 for $R := \widetilde{R} + \widetilde{B}^T \widetilde{X} \widetilde{B}$. Indeed, (iii) follows from the first equation (8.99) and it only remains to show

$$\begin{aligned} R_{22} &:= \overline{D}_2^T \overline{D}_2 - I < 0, \\ R_{33} &:= I + \widetilde{B}_3^T \widetilde{X} \widetilde{B}_3 > 0. \end{aligned}$$

Since $X_1 \geq 0$ we get from the first equation in (8.90) written for $X = X_1$ that $R_{22} < 0$ and from $\widetilde{X} \geq 0$ we conclude $R_{33} > 0$. This concludes the proof. ■

Here follows the discrete–time counterpart of Theorem 8.4.1.

Theorem 8.5.3 *Let* $\mathbf{T}$ *be given by (8.81) and let* Σ_1, Σ_2, *and* $\widetilde{\Sigma}$ *be the Popov triplets defined in (8.88), (8.89), and (8.94), respectively.*

I. *The discrete–time suboptimal 4–NP (8.84) (with* $\gamma = 1$*) has a solution* $\mathbf{S}$ *if and only if the following three conditions hold:*

a. *The KSPYS(*$\Sigma_1, -I$*) given in (8.90) has a stabilizing solution* (X_1, V_1, W_1), $X_1 \geq 0$.

b. *The KSPYS(*$\Sigma_2, -I$*) given in (8.91) has a stabilizing solution* (X_2, V_2, W_2), $X_2 \geq 0$.

c. $\rho(X_1 X_2) < 1$.

II. *If conditions a, b and c hold, then the KSPYS(*$\widetilde{\Sigma}, J$*) in (8.99) has a stabilizing solution* $(\widetilde{X}, V, W)$, *and* $\widetilde{X} \geq 0$ *satisfies* $\widetilde{X} = X_2(I - X_1 X_2)^{-1}$.

Therefore, the class of solutions to the 4–NP given in Corollary 8.3.4 can be expressed in terms of the stabilizing solutions X_1 *and* X_2 *of KSPYS(*$\Sigma_1, -I$*) and KSPYS(*$\Sigma_2, -I$*), respectively, whose coefficients are given in terms of the state–space realization (8.81) of the original system* $\mathbf{T}$.

Proof. The proof follows the continuous–time case, by replacing in the proof of Theorem 8.4.1 extended Hamiltonian pencils with extended symplectic pencils. ■

Notes and references

The topic of this chapter has been initiated by Nehari [Neh57] in the framework of complex analysis. Preliminary works in the field are due to Ball and Helton [BH83], and Glover [Glo84]. From the theoretical viewpoint an elegant treatment of the subject has been given by Feintuch and Francis [FF85] where the Szokefalvi–Nagy–Foias commutant lifting theorem [SNF70] played a paramount role. By using an all–pass embedding technique, a complete state–space characterization of all solutions to the four block Nehari problem was given by Glover et. al. [GLD+91]. Other important works in the field are Gohberg, Kaashoek and Woerdeman [GKW91], Alpay, Dewilde and Dym [ADD90, DD92], and Ball, Gohberg and Kaashoek [BGK92]. The content of this chapter is essentially based on the works of Ionescu and Oară [Oar95, IO96a, IO98a]. A singular perturbations based solution to the optimal Nehari problem was given by Ionescu and Oară [IO96b]. The Nehari problem was solved for time–varying systems by Halanay and Ionescu [HI94b] (the two block case), and by Ionescu and Oară [IO96g] (the four block case). The extension to the four block Adamjan–Arov–Krein problem has been recently obtained by Ionescu and Oară [IO98b]. A solution to the two block Nehari problem for infinite–dimensional systems was proposed by Weiss [Wei94a].

9

The Optimal H^2–Control Problem

In this chapter we consider the optimal H^2–control problem which consists in finding a controller, for a given system, that stabilizes the resulting closed–loop system and minimizes its input–output H^2–norm.

In Section 9.1 we formulate the problem and give a useful formula that expresses the H^2–norm of a stable system in terms of the causal impulse response. The main result is given in Section 9.2 and its proof is based on a successive reduction to simpler problems. The discrete–time case is analyzed in Sections 9.3 and 9.4.

A. The continuous–time case

9.1 Problem formulation

Consider a plant that is represented as the linear system

$$\begin{aligned} \dot{x} &= Ax + B_1u_1 + B_2u_2, \\ y_1 &= C_1x + D_{12}u_2, \\ y_2 &= C_2x + D_{21}u_1, \end{aligned} \tag{9.1}$$

having the TFM given by

$$\mathbf{T} = \left[\begin{array}{c|c} A & B \\ \hline C & D \end{array}\right] = \left[\begin{array}{c|cc} A & B_1 & B_2 \\ \hline C_1 & O & D_{12} \\ C_2 & D_{21} & O \end{array}\right] = \begin{bmatrix} \mathbf{T}_{11} & \mathbf{T}_{12} \\ \mathbf{T}_{21} & \mathbf{T}_{22} \end{bmatrix}, \tag{9.2}$$

where $A \in \mathbb{R}^{n\times n}$, $B_i \in \mathbb{R}^{n\times m_i}$, $C_j \in \mathbb{R}^{p_j\times n}$, $D_{ij} \in \mathbb{R}^{p_i\times m_j}$, $(i, j = 1, 2,\ m = m_1 + m_2,\ p = p_1 + p_2)$, and u and y denote the input and output of $\mathbf{T}$, respectively, and are partitioned conformably with $\mathbf{T}$ such that

$$y = \begin{bmatrix} y_1 \\ y_2 \end{bmatrix} = \mathbf{T}u = \begin{bmatrix} \mathbf{T}_{11} & \mathbf{T}_{12} \\ \mathbf{T}_{21} & \mathbf{T}_{22} \end{bmatrix} \begin{bmatrix} u_1 \\ u_2 \end{bmatrix}. \tag{9.3}$$

The vector u_1 is the vector of *external inputs* (disturbance, sensor noise and reference signals), y_1 is the vector of *regulated outputs*, representing in general some *error* signals, u_2 is the vector of *controlled inputs* and y_2 is the *measured outputs* vector.

Controlling the plant (9.1) means using the measurement information y_2 to find an appropriate control input u_2 such that the resulting response y_1 of the plant to the external input u_1 has certain desired features. Roughly speaking, the main concern of this chapter is to find an appropriate control input u_2, as output of a system – called controller – with input y_2, such that the resulting response y_1 has energy (L^2–norm) as small as possible when the external input u_1 is a Dirac impulse.

We define a *controller* for the plant (9.1) as a linear system

$$\begin{aligned} \dot{x}_K &= A_K x_K + B_K u_K, \\ y_K &= C_K x_K, \end{aligned} \tag{9.4}$$

having the TFM given by

$$\mathbf{K} = \left[\begin{array}{c|c} A_K & B_K \\ \hline C_K & O \end{array}\right], \tag{9.5}$$

where $A_K \in \mathbb{R}^{n_K \times n_K}$, $B_K \in \mathbb{R}^{n_K \times p_2}$, $C_K \in \mathbb{R}^{m_2 \times n_K}$, and u_K and y_K are the input and output of $\mathbf{K}$, respectively.

The closed–loop system given in Figure 9.1 is obtained by connecting the controller (9.5) to the plant (9.2) such that $u_2 \equiv y_K$ and $u_K \equiv y_2$. The resulting closed–loop

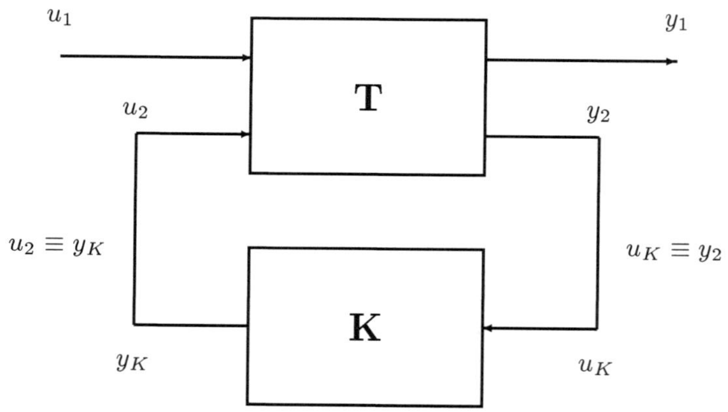

Figure 9.1 $\mathbf{T}_{y_1u_1}$: The closed–loop connection of $\mathbf{T}$ and $\mathbf{K}$

system with input u_1 and output y_1 is well–posed and is given by

$$\begin{aligned} y_1 = \mathbf{T}_{y_1u_1}u_1, \quad \mathbf{T}_{y_1u_1} &= \mathrm{LFT}(\mathbf{T},\mathbf{K}) = \mathbf{T}_{11} + \mathbf{T}_{12}\mathbf{K}(I - \mathbf{T}_{22}\mathbf{K})^{-1}\mathbf{T}_{21}, \\ \mathbf{T}_{y_1u_1} &= \left[\begin{array}{c|c} A_R & B_R \\ \hline C_R & O \end{array}\right] = \left[\begin{array}{cc|c} A & B_2C_K & B_1 \\ B_KC_2 & A_K & B_KD_{21} \\ \hline C_1 & D_{12}C_K & O \end{array}\right]. \end{aligned} \tag{9.6}$$

Given the plant $\mathbf{T}$, the *optimal H^2–control problem* consists in finding a proper controller $\mathbf{K}$ for which the closed–loop system $\mathbf{T}_{y_1u_1}$ is internally stable, i.e.,

$$\Lambda(A_R) \subset \mathbb{C}_- \tag{9.7}$$

and has minimum H^2–norm, i.e., $\|\mathbf{T}_{y_1u_1}\|_2$ attains its minimum over the class of systems $\mathcal{K}=\{\ \mathbf{K} : \mathbf{K}$ strictly proper and $\mathbf{T}_{y_1u_1}$ internally stable$\}$. A controller that satisfies condition (9.7) – denoted (S) – will be called *stabilizing*, and a controller for which the minimum is attained will be called *optimal*.

In the above formulation we have implicitly assumed: (i) $\mathbf{T}_{11}(\infty) = D_{11} = 0$; (ii) $\mathbf{T}_{22}(\infty) = D_{22} = 0$; (iii) $\mathbf{K}(\infty) = D_K = 0$. Assumptions (i) and (iii) assure that $\mathbf{T}_{y_1u_1}(\infty) = D_R = 0$ which is a necessary condition for the H^2–norm of the closed–loop system to be finite. In particular, assumption (iii) automatically ensures that the closed–loop system is well–posed. Assumption (ii) is primarily made for simplifying the formulas, and there is no loss of generality here. Indeed, if $\mathbf{K}$ is a solution to the problem with $D_{22} = 0$, then $\mathbf{K}(I + D_{22}\mathbf{K})^{-1}$ is a solution for the original problem (with arbitrary D_{22}). Anyway, neither (i), nor (iii) are necessary for the solvability of the H^2–control problem and both can be relaxed to a certain extent.

9.1.1 An evaluation of the H^2–norm

We give now a useful evaluation of the H^2–norm of a stable system in terms of the output of the system in the case in which its input is a Dirac impulse.

Let

$$\mathbf{T} = \left[\begin{array}{c|c} A & B \\ \hline C & O \end{array}\right], \qquad y = \mathbf{T}u, \tag{9.8}$$

be an arbitrary system with $A \in \mathbb{R}^{n\times n}$, $B \in \mathbb{R}^{n\times m}$, $C \in \mathbb{R}^{p\times n}$, and A stable. Then the H^2–norm of $\mathbf{T}$ is defined as (see (2.16))

$$\|\mathbf{T}\|_2 := \left\{ \frac{1}{2\pi} \int_{-\infty}^{\infty} \mathrm{Trace}\,[\mathbf{T}^*(j\omega)\mathbf{T}(j\omega)]\mathrm{d}\omega \right\}^{\frac{1}{2}}. \tag{9.9}$$

Let $u^i := \delta e_i$, $(i = 1, \dots, m)$, where δ is the Dirac distribution and $e_1, \dots, e_m$ is the standard basis of $\mathbb{R}^m$. Denote by y^i the output of (9.8) for $u = u^i$, $(i = 1, \dots, m)$. From the stability of A it follows that $y^i \in L_+^{2,p}$ and has the explicit form $y^i(t) = 0$, for $t < 0$, and $y^i(t) = Ce^{At}Be_i$, for $t \geq 0$. Let $\mathcal{G}(t) := 0$, for $t < 0$, and $\mathcal{G}(t) = Ce^{At}B$, for $t \geq 0$, be the *causal impulse matrix* of (9.8). With these, we have successively

$$\begin{aligned} \sum_{i=1}^{m} \|y^i\|_2^2 &= \sum_{i=1}^{m} \int_0^{\infty} e_i^T \mathcal{G}^T(t)\mathcal{G}(t)e_i \mathrm{d}t = \int_0^{\infty} [\sum_{i=1}^{m} e_i^T \mathcal{G}^T(t)\mathcal{G}(t)e_i]\mathrm{d}t \\ &= \int_0^{\infty} \mathrm{Trace}\,[\mathcal{G}^T(t)\mathcal{G}(t)]\mathrm{d}t = \frac{1}{2\pi}\int_{-\infty}^{\infty} \mathrm{Trace}\,[\mathbf{T}^*(j\omega)\mathbf{T}(j\omega)]\mathrm{d}\omega \end{aligned} \tag{9.10}$$

where the last equality follows from Parseval's formula. Hence, with (9.9) and (9.10)

$$\|\mathbf{T}\|_2^2 = \sum_{i=1}^{m} \|y^i\|_2^2. \tag{9.11}$$

9.2 Main result

We give directly the main result and divide its proof into several intermediate steps.

Theorem 9.2.1 *Let the system* $\mathbf{T}$ *be given by (9.2) and associate with it the following two Popov triplets*

$$\Sigma_{12} = (A, B_2; C_1^T C_1, C_1^T D_{12}, D_{12}^T D_{12}), \tag{9.12}$$

$$\Sigma_{21} = (A^T, C_2^T; B_1 B_1^T, B_1 D_{21}^T, D_{21} D_{21}^T). \tag{9.13}$$

Assume the following hypotheses hold:

(H1) *The pair* (A, B_2) *is stabilizable,* D_{12} *has full column rank, and*

$$\operatorname{rank}\begin{bmatrix} A - j\omega I & B_2 \\ C_1 & D_{12} \end{bmatrix} = n + m_2, \quad \forall\, \omega \in \mathbb{R}.$$

(H2) *The pair* (C_2, A) *is detectable,* D_{21} *has full row rank, and*

$$\operatorname{rank}\begin{bmatrix} A - j\omega I & B_1 \\ C_2 & D_{21} \end{bmatrix} = n + p_2, \quad \forall\, \omega \in \mathbb{R}.$$

Then the CTARE(Σ_{12}*) given by*

$$A^T X + XA - (XB_2 + C_1^T D_{12})(D_{12}^T D_{12})^{-1}(B_2^T X + D_{12}^T C_1) + C_1^T C_1 = 0 \tag{9.14}$$

has a stabilizing solution $X_s \geq 0$*, and the CTARE(*Σ_{21}*) given by*

$$AY + YA^T - (YC_2^T + B_1 D_{21}^T)(D_{21} D_{21}^T)^{-1}(C_2 Y + D_{21} B_1^T) + B_1 B_1^T = 0 \tag{9.15}$$

has a stabilizing solution $Y_s \geq 0$*. Moreover, there exists a strictly proper controller*

$$\mathbf{K} = \left[\begin{array}{c|c} A + B_2 F_s + K_s^T C_2 & K_s^T \\ \hline -F_s & O \end{array}\right] \tag{9.16}$$

that solves the optimal H^2*–control problem, where*

$$F_s := -(D_{12}^T D_{12})^{-1}(B_2^T X_s + D_{12}^T C_1), \tag{9.17}$$

$$K_s := -(D_{21} D_{21}^T)^{-1}(C_2 Y_s + D_{21} B_1^T), \tag{9.18}$$

are the corresponding stabilizing Riccati feedbacks of the two Riccati equations. Furthermore, the optimal (minimal) value of the H^2*–norm is*

$$\min_{\mathbf{K}\in\mathcal{K}} \|\mathbf{T}_{y_1 u_1}\|_2 = [\operatorname{Trace}(B_1^T X_s B_1) + \operatorname{Trace}(C_1 Y_s C_1^T)]^{\frac{1}{2}}. \tag{9.19}$$

Before giving the proof we make several remarks. First, (A, B_2) stabilizable and (C_2, A) detectable are necessary for the fulfillment of condition (S) in the statement of the optimal H^2–control problem. The other hypotheses are necessary for the existence of the stabilizing solution to the two Riccati equations but are not in general necessary for the existence of a solution to the H^2–control problem. In case any of these latter assumptions fails we obtain a so-called singular H^2–control problem whose solution involves more complicated arguments.

We proceed now with the proof which is based on the solution of three particular optimal H^2–control problems, each stated as a separate proposition. We call these particular problems by names taken from control jargon: the impulse disturbance estimation problem (IDEP), the impulse disturbance feedforward problem (IDFP), and the impulse output estimation problem (IOEP). The general scheme of derivation runs as follows: we solve first the IDEP which we reduce further to the IDFP. We show that the IOEP is the dual of the IDFP and write down its solution by duality. Finally, we reduce the general optimal H^2–control problem to the IOEP. The main technical tool used for the aforementioned reductions is the Kalman–Popov–Yakubovich system. We start with the IDEP.

Proposition 9.2.2 (IDEP) *Consider the system* $\mathbf{T}$ *given in (9.2) and assume the following hypotheses hold:*

(IDEP–H1) D_{12} *is square and invertible and* $A - B_2 D_{12}^{-1} C_1$ *is stable.*

(IDEP–H2) D_{21} *is square and invertible and* $A - B_1 D_{21}^{-1} C_2$ *is stable.*

Then the controller

$$\mathbf{K} = \left[\begin{array}{c|c} A - B_1 D_{21}^{-1} C_2 - B_2 D_{12}^{-1} C_1 & B_1 D_{21}^{-1} \\ \hline -D_{12}^{-1} C_1 & O \end{array}\right] \tag{9.20}$$

is a solution to the optimal H^2*–control problem and the optimal value of the* H^2*–norm is*

$$\min_{\mathbf{K}\in\mathcal{K}} \|\mathbf{T}_{y_1 u_1}\|_2 = 0.$$

Proof. We substitute the expression of the controller (9.20) in (9.6) and get

$$\begin{aligned} \mathbf{T}_{y_1 u_1} &= \left[\begin{array}{cc|c} A & -B_2 D_{12}^{-1} C_1 & B_1 \\ B_1 D_{21}^{-1} C_2 & A - B_1 D_{21}^{-1} C_2 - B_2 D_{12}^{-1} C_1 & B_1 \\ \hline C_1 & -C_1 & O \end{array}\right] \\ &= \left[\begin{array}{cc|c} A - B_2 D_{12}^{-1} C_1 & -B_2 D_{12}^{-1} C_1 & B_1 \\ O & A - B_1 D_{21}^{-1} C_2 & O \\ \hline O & -C_1 & O \end{array}\right] \end{aligned} \tag{9.21}$$

where, to obtain the second equality, we have performed a similarity transformation. From (9.21) and hypotheses (IDEP–H1) and (IDEP–H2) it follows that (S) holds. Finally, we remove the stable uncontrollable and unobservable parts in (9.21) and get $\mathbf{T}_{y_1 u_1} = 0$ which shows that the minimum of the H^2–norm is attained in zero. ■

Proposition 9.2.3 (IDFP) *Consider the system* $\mathbf{T}$ *given in (9.2). Associate with* $\mathbf{T}$ *the Popov triplet (9.12) and assume the following hypotheses hold:*

(IDFP–H1) *The pair* (A, B_2) *is stabilizable,* D_{12} *has full column rank, and*

$$\operatorname{rank} \begin{bmatrix} A - j\omega I & B_2 \\ C_1 & D_{12} \end{bmatrix} = n + m_2, \quad \forall\, \omega \in \mathbb{R}.$$

(IDFP–H2) D_{21} *is square and invertible and* $A - B_1 D_{21}^{-1} C_2$ *is stable.*

Then the CTARE(Σ_{12}*) given in (9.14) has a stabilizing solution* $X_s \geq 0$*. Moreover, there exists a strictly proper controller*

$$\mathbf{K} = \left[\begin{array}{c|c} A - B_1 D_{21}^{-1} C_2 + B_2 F_s & B_1 D_{21}^{-1} \\ \hline F_s & O \end{array}\right] \tag{9.22}$$

that solves the optimal H^2*–control problem, where* F_s *is the corresponding stabilizing Riccati feedback given in (9.17). Furthermore, the optimal (minimal) value of the* H^2*–norm is given by*

$$\min_{\mathbf{K}\in\mathcal{K}} \|\mathbf{T}_{y_1 u_1}\|_2 = [\operatorname{Trace}(B_1^T X_s B_1)]^{\frac{1}{2}}. \tag{9.23}$$

Proof. Notice from Section 3.4 that Σ_{12} is the positiveness Popov triplet associated with $\mathbf{T}_{12}$. Then with (IDFP–H1) it follows from Proposition 7.2.1 applied to $\mathbf{T}_{12}$ that the CTARE(Σ_{12}) has a stabilizing solution $X_s \geq 0$. Further, the existence of the stabilizing solution X_s to the CTARE(Σ_{12}) is equivalent to the existence of a stabilizing solution (X_s, V_s, W_s), $X_s \geq 0$, to the KPYS(Σ, I) given by

$$\begin{aligned} D_{12}^T D_{12} &= V^T V, \\ C_1^T D_{12} + X B_2 &= W^T V, \\ C_1^T C_1 + A^T X + X A &= W^T W, \end{aligned} \tag{9.24}$$

where $F_s = -V_s^{-1} W_s$ is the stabilizing feedback. Consider the system

$$\begin{aligned} \widetilde{\mathbf{T}} &= \begin{bmatrix} \widetilde{\mathbf{T}}_{11} & \widetilde{\mathbf{T}}_{12} \\ \mathbf{T}_{21} & \mathbf{T}_{22} \end{bmatrix} = \left[\begin{array}{c|cc} A & B_1 & B_2 \\ \hline W_s & O & V_s \\ C_2 & D_{21} & O \end{array}\right], \\ \begin{bmatrix} \widetilde{y}_1 \\ y_2 \end{bmatrix} &= \begin{bmatrix} \widetilde{\mathbf{T}}_{11} & \widetilde{\mathbf{T}}_{12} \\ \mathbf{T}_{21} & \mathbf{T}_{22} \end{bmatrix} \begin{bmatrix} u_1 \\ u_2 \end{bmatrix} \end{aligned} \tag{9.25}$$

which is obtained by replacing in (9.2) the output $y_1 = C_1 x + D_{12} u_2$ with the output

$$\widetilde{y}_1 = W_s x + V_s u_2. \tag{9.26}$$

Since V is invertible, it is straightforward to check that the system $\widetilde{\mathbf{T}}$ satisfies the hypotheses (IDEP–H1) and (IDEP–H2) of Proposition 9.2.2. If we write down the controller (9.20) for $\widetilde{\mathbf{T}}$ given by (9.25) we get (9.22). In other words, the controller $\mathbf{K}$ in (9.22) solves the IDEP formulated for the system (9.25). Thus the resulting closed–loop system in Figure 9.2 given by

$$\widetilde{\mathbf{T}}_{y_1 u_1} = \mathrm{LFT}(\widetilde{\mathbf{T}}, \mathbf{K}) = \widetilde{\mathbf{T}}_{11} + \widetilde{\mathbf{T}}_{12} \mathbf{K} (I - \mathbf{T}_{22} \mathbf{K})^{-1} \mathbf{T}_{21}, \tag{9.27}$$

is internally stable and

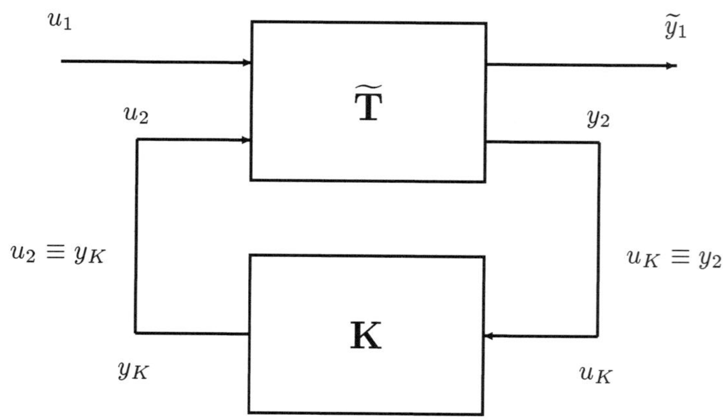

Figure 9.2 $\widetilde{\mathbf{T}}_{y_1 u_1}$: The LFT closed–loop connection of $\widetilde{\mathbf{T}}$ and $\mathbf{K}$

$$\|\widetilde{\mathbf{T}}_{y_1u_1}\|_2^2 = \sum_{i=1}^{m} \|\widetilde{y}_1^i\|_2^2 = 0 \tag{9.28}$$

where $\widetilde{y}_1^i$ is the output of $\widetilde{\mathbf{T}}_{y_1u_1}$ for the input $u_1 = u_1^i = \delta e_i$. Internal stability of (9.27) is equivalent to its state matrix

$$\widetilde{A}_R = \begin{bmatrix} A & B_2F_s \\ B_1D_{21}^{-1}C_2 & A - B_1D_{21}^{-1}C_2 + B_2F_s \end{bmatrix}$$

being stable. However, it is easy to see that $\widetilde{A}_R$ is also the state matrix of

$$\mathbf{T}_{y_1u_1} = \mathrm{LFT}(\mathbf{T},\mathbf{K}) = \mathbf{T}_{11} + \mathbf{T}_{12}\mathbf{K}(I - \mathbf{T}_{22}\mathbf{K})^{-1}\mathbf{T}_{21}, \tag{9.29}$$

and thus $\mathbf{T}_{y_1u_1}$ is internally stable as well. It emerges that $\mathbf{K}$ satisfies the requirement (S) of the H^2–control problem formulated for $\mathbf{T}$ and it remains to show that the optimality requirement is satisfied as well, and to obtain (9.23).

To this end we evaluate successively

$$\begin{aligned}
\|y_1(t)\|^2 &= \|C_1x(t) + D_{12}u_2(t)\|^2 \\
&= x^T(t)C_1^TC_1x(t) + 2x^T(t)C_1^TD_{12}u_2(t) + u_2^T(t)D_{12}^TD_{12}u_2(t) \\
&\overset{(9.24)}{=} x^T(t)W_s^TW_sx(t) - x^T(t)(A^TX_s + X_sA)x(t) + 2x^T(t)W_s^TV_su_2(t) \\
&\quad - 2x^T(t)X_sB_2u_2(t) + u_2^T(t)V_s^TV_su_2(t) \\
&= \|W_sx(t) + V_su_2(t)\|^2 - x^T(t)X_s(Ax(t) + B_2u_2(t)) \\
&\quad - (Ax(t) + B_2u_2(t))^TX_sx(t) \\
&= \|\widetilde{y}_1(t)\|^2 - x^T(t)X_s\dot{x}(t) - \dot{x}^T(t)X_sx(t) \\
&\quad + x^T(t)X_sB_1u_1(t) + u_1^T(t)B_1^TX_sx(t) \\
&= \|\widetilde{y}_1(t)\|^2 - \frac{\mathrm{d}}{\mathrm{d}t}(x^T(t)X_sx(t)) + x^T(t)X_sB_1u_1(t) \\
&\quad + u_1^T(t)B_1^TX_sx(t).
\end{aligned} \tag{9.30}$$

Let now $\mathbf{K}$ denote any controller (9.5) such that $\mathbf{T}_{y_1u_1} = \mathrm{LFT}(\mathbf{T},\mathbf{K})$ is internally stable. As explained earlier, $\widetilde{\mathbf{T}}_{y_1u_1} = \mathrm{LFT}(\widetilde{\mathbf{T}},\mathbf{K})$ is internally stable as well. Let $x_K(t)$ and $x_R(t)$ denote the state vector of $\mathbf{K}$ and of $\mathbf{T}_{y_1u_1}$, respectively. Consider the impulsive input $u_1 = u_1^i := \delta e_i$ and the initial conditions $x(0_-) = 0, x_K(0_-) = 0$. Then we get with (9.6)

$$x_R(0_+) = \begin{bmatrix} x(0_+) \\ x_K(0_+) \end{bmatrix} = B_Re_i = \begin{bmatrix} B_1 \\ B_KD_{21} \end{bmatrix} e_i.$$

Therefore

$$x(0_+) = B_1e_i, \tag{9.31}$$

$x_K(0_+) = B_KD_{21}e_i$, $u_1^i(t) = 0, \forall t > 0$, and all the signals of the closed–loop configurations $\mathbf{T}_{y_1u_1}$ and $\widetilde{\mathbf{T}}_{y_1u_1}$ are exponentially decreasing L_+^2 functions. Denote by

y_1^i and $\widetilde{y}_1^i$ the outputs of $\mathbf{T}_{y_1u_1}$ and $\widetilde{\mathbf{T}}_{y_1u_1}$, respectively, to the input u_1^i. We integrate on $(0,\infty)$ both sides of (9.30) for $u_1(t) = u_1^i := \delta e_i$ and with (9.31) we obtain successively

$$\|y_1^i\|_2^2 = \|\widetilde{y}_1^i\|_2^2 + e_i^T B_1^T X B_1 e_i, \quad \sum_{i=1}^{m_1} \|y_1^i\|_2^2 = \sum_{i=1}^{m_1} \|\widetilde{y}_1^i\|_2^2 + \text{Trace}\,(B_1^T X_s B_1)$$

from where we conclude with (9.11)

$$\|\mathbf{T}_{y_1u_1}\|_2^2 = \|\widetilde{\mathbf{T}}_{y_1u_1}\|_2^2 + \text{Trace}\,(B_1^T X_s B_1). \tag{9.32}$$

The equality (9.32) holds for all controllers $\mathbf{K}$ for which $\mathbf{T}_{y_1u_1}$ (and automatically $\widetilde{\mathbf{T}}_{y_1u_1}$) is internally stable. But (9.32) yields $\|\mathbf{T}_{y_1u_1}\|_2^2 \geq \text{Trace}\,(B_1^T X B_1)$ and equality holds for $\mathbf{K}$ given by (9.22) because of (9.28). From this we conclude that (9.23) holds and the proof ends. ∎

Proposition 9.2.4 (IEOP) *Consider the system* $\mathbf{T}$ *given in (9.2). Associate with* $\mathbf{T}$ *the Popov triplet* Σ_{21} *in (9.13) and assume the following hypotheses hold:*

(IEOP–H1) D_{12} *is square and invertible and* $A - B_2 D_{12}^{-1} C_1$ *is stable.*

(IEOP–H2) *The pair* (C_2, A) *is detectable,* D_{21} *has full row rank and*

$$\text{rank} \begin{bmatrix} A - j\omega I & B_1 \\ C_2 & D_{21} \end{bmatrix} = n + p_2, \quad \forall\, \omega \in \mathbb{R}.$$

Then the CTARE(Σ_{21}*) given in (9.15) has a stabilizing solution* $Y_s \geq 0$*. Moreover, there exists a strictly proper controller*

$$\mathbf{K} = \left[\begin{array}{c|c} A - B_2 D_{12}^{-1} C_1 + K_s^T C_2 & K_s^T \\ \hline D_{12}^{-1} C_1 & O \end{array}\right] \tag{9.33}$$

that solves the optimal H^2*–control problem, where* K_s *is the corresponding stabilizing Riccati feedback given in (9.18). Furthermore, the optimal (minimal) value of the* H^2*–norm is*

$$\min_{\mathbf{K}\in\mathcal{K}} \|\mathbf{T}_{y_1u_1}\|_2 = [\text{Trace}\,(C_1 Y_s C_1^T)]^{\frac{1}{2}}. \tag{9.34}$$

Proof. The proof follows by duality from Proposition 9.2.3. More precisely, if $\mathbf{T}$ satisfies (IEOP–H1) and (IEOP–H2), then $\mathbf{T}^T$ satisfies the hypotheses of the IDFP. We write down the solution $\mathbf{K}$ for the IDFP formulated for $\mathbf{T}^T$ and obtain that $\mathbf{K}^T$ is a solution for the IOEP formulated for $\mathbf{T}$. The rest of the proof is a matter of straightforward substitutions. ∎

We proceed now with the proof of Theorem 9.2.1.

Proof of Theorem 9.2.1. Notice from Section 3.4 that Σ_{12} and Σ_{21} are the positiveness Popov triplets associated with $\mathbf{T}_{12}$ and $\mathbf{T}_{21}$, respectively. Then with (H1) and (H2) it follows from Proposition 7.2.1 applied to $\mathbf{T}_{12}$ and $\mathbf{T}_{21}$ that the CTARE(Σ_{12}) and the CTARE(Σ_{21}) have the stabilizing solution $X_s \geq 0$ and $Y_s \geq 0$, respectively.

Further, we follow the main lines of the proof of Proposition 9.2.3. The existence of the stabilizing solution X_s to the CTARE(Σ_{12}) is equivalent to the existence of a stabilizing solution (X_s, V_s, W_s), $X_s \geq 0$, to the KPYS(Σ_{12}, I) given in (9.24). We consider again the system (9.25) which is obtained by replacing the output y_1 with the output (9.26) in (9.2). Since V_s is invertible, it is straightforward to check that the system $\widetilde{\mathbf{T}}$ satisfies now the hypotheses of Proposition 9.2.3, i.e., the optimal H^2–control problem formulated for $\widetilde{\mathbf{T}}$ is an IEOP. If we write down the controller (9.33) for $\widetilde{\mathbf{T}}$ given by (9.25) we arrive exactly at (9.16). In other words, the controller $\mathbf{K}$ in (9.16) solves the IEOP formulated for the system (9.25) and thus $\widetilde{\mathbf{T}}_{y_1u_1}$ in Figure 9.2 is internally stable and (9.32) holds. From the internal stability of $\widetilde{\mathbf{T}}_{y_1u_1}$ it follows analogously as in the proof of Proposition 9.2.3 that $\mathbf{T}_{y_1u_1}$ in Figure 9.1 is internally stable as well.

It only remains to show that $\mathbf{K}$ is indeed optimal and to get the evaluation (9.19). To this end let $\mathbf{K}$ denote any controller (9.5) such that $\mathbf{T}_{y_1u_1}$ in (9.6) is internally stable. Then from Proposition 9.2.4 applied to (9.25) we get

$$||\widetilde{\mathbf{T}}_{y_1u_1}||_2^2 \geq \text{Trace}\,(C_1 Y_s C_1^T) \tag{9.35}$$

and combining (9.32) with (9.35) this results in

$$||\mathbf{T}_{y_1u_1}||_2^2 \geq \text{Trace}\,(B_1^T X_s B_1) + \text{Trace}\,(C_1 Y_s C_1^T). \tag{9.36}$$

The equality in (9.36) holds if and only if the equality in (9.35) holds. But, as explained above, for $\mathbf{K}$ given by (9.16) the equality (9.35) holds and thus (9.19) follows and the proof of Theorem 9.2.1 ends. ∎

B. The discrete–time case

9.3 Problem formulation

In discrete–time the optimal H^2–control problem is formulated similar to the continuous–time case with the only difference that stability and H^2–norms are taken with respect to the unit disk and the unit circle instead of the left–half plane and the imaginary axis, respectively. We proceed now with the evaluation of the H^2–norm of a stable discrete–time system.

9.3.1 An evaluation of the H^2–norm

Let

$$\begin{aligned} \mathbf{T} &= \left[\begin{array}{c|c} A & B \\ \hline C & O \end{array}\right], \\ y &= \mathbf{T}u, \end{aligned} \tag{9.37}$$

be an arbitrary discrete–time system with $A \in \mathbb{R}^{n\times n}$, $B \in \mathbb{R}^{n\times m}$, $C \in \mathbb{R}^{p\times n}$, and A stable. Then the H^2–norm of $\mathbf{T}$ is defined as (see (2.122))

$$||\mathbf{T}||_2 := \left\{ \frac{1}{2\pi} \int_0^{2\pi} \text{Trace}\,(\mathbf{T}^*(e^{j\theta})\mathbf{T}(e^{j\theta}))d\theta \right\}^{\frac{1}{2}}. \tag{9.38}$$

Let $u = u^i := \delta e_i$ $(i = 1, \ldots, m)$, where $\delta = (\delta_k)_{k\in\mathbb{Z}}$, $\delta(0) = 1$, and $\delta_k = 0$ for $k \neq 0$, and $e_1, \ldots, e_m$ is the standard basis of $\mathbb{R}^m$. Denote by y^i the output of (9.8) for $u = u^i$, $(i = 1, \ldots m)$. From the stability of A it follows that $y^i = (y^i_k)_{k\in\mathbb{Z}} \in \ell_+^{2,p}$ and has the explicit form

$$y^i_k = \begin{cases} 0, & k \leq 0, \\ CA^{k-1}Be_i, & k \geq 1. \end{cases} \tag{9.39}$$

Let $\mathcal{G} = (\mathcal{G}_k)_{k\in\mathbb{Z}}$,

$$\mathcal{G}_k := \begin{cases} 0, & k \leq 0, \\ CA^{k-1}B, & k \geq 1, \end{cases} \tag{9.40}$$

be the *causal impulse matrix* of (9.37). Then using (9.39) and (9.40) one obtains

$$\begin{aligned} \sum_{i=1}^{m} \|y^i\|_2^2 &= \sum_{i=1}^{m}\sum_{k=-\infty}^{\infty} e_i^T\mathcal{G}_k^T\mathcal{G}_k e_i = \sum_{k=-\infty}^{\infty}\sum_{i=1}^{m} e_i^T\mathcal{G}_k^T\mathcal{G}_k e_i = \sum_{k=-\infty}^{\infty} \mathrm{Trace}\,(\mathcal{G}_k^T\mathcal{G}_k) \\ &= \frac{1}{2\pi}\int_0^{2\pi} \mathrm{Trace}\,(\mathbf{T}^*(e^{j\theta})\mathbf{T}(e^{j\theta}))\mathrm{d}\theta \end{aligned} \tag{9.41}$$

where the last equality follows from the discrete–time version of Parseval's formula. Hence, with (9.38) and (9.41), we conclude

$$\|\mathbf{T}\|_2^2 = \sum_{i=1}^{m} \|y^i\|_2^2. \tag{9.42}$$

9.4 Main result

We formulate directly the main result for discrete–time systems. As most of the proofs follow in principle the continuous–time case we shall be brief and concentrate on the nontrivial differences.

Theorem 9.4.1 *Let the discrete–time system* $\mathbf{T}$ *be given by (9.2) and associate with it the Popov triplets given by*

$$\Sigma_{12} = (A, B_2; C_1^TC_1, C_1^TD_{12}, D_{12}^TD_{12}), \tag{9.43}$$

$$\Sigma_{21} = (A^T, C_2^T; B_1B_1^T, B_1D_{21}^T, D_{21}D_{21}^T). \tag{9.44}$$

Assume the following hypotheses hold:

(H1) *The pair* (A, B_2) *is stabilizable and*

$$\mathrm{rank}\begin{bmatrix} A - e^{j\theta}I & B_2 \\ C_1 & D_{12} \end{bmatrix} = n + m_2, \quad \forall\, \theta \in [0,\ 2\pi).$$

(H2) *The pair* (C_2, A) *is detectable and*

$$\mathrm{rank}\begin{bmatrix} A - e^{j\theta}I & B_1 \\ C_2 & D_{21} \end{bmatrix} = n + p_2, \quad \forall\theta \in [0,\ 2\pi).$$

Then the DTARE(Σ_{12}) given by

$$\begin{aligned} &A^T X A - X - (A^T X B_2 + C_1^T D_{12}) \\ &\quad (D_{12}^T D_{12} + B_2^T X B_2)^{-1}(B_2^T X A + D_{12}^T C_1) + C_1^T C_1 = 0 \end{aligned} \tag{9.45}$$

has a stabilizing solution $X_s \geq 0$, and the DTARE(Σ_{21}) given by

$$\begin{aligned} &A Y A^T - Y - (A Y C_2^T + B_1 D_{21}^T) \\ &\quad (D_{21} D_{21}^T + C_2 Y C_2^T)^{-1}(C_2 Y A^T + D_{21} B_1^T) + B_1 B_1^T = 0 \end{aligned} \tag{9.46}$$

has a stabilizing solution $Y_s \geq 0$. Moreover, there exists a strictly proper controller

$$\mathbf{K} = \left[\begin{array}{c|c} A + B_2 F_s + K_s^T C_2 & K_s^T \\ \hline -F_s & O \end{array}\right] \tag{9.47}$$

that solves the discrete–time optimal H^2–control problem, where

$$F_s := -(D_{12}^T D_{12} + B_2^T X B_2)^{-1}(B_2^T X_s A + D_{12}^T C_1), \tag{9.48}$$
$$K_s := -(D_{21} D_{21}^T + C_2 Y C_2^T)^{-1}(C_2 Y A^T + D_{21} B_1^T), \tag{9.49}$$

are the corresponding stabilizing Riccati feedbacks of the two Riccati equations. Furthermore, the optimal (minimal) value of the H^2–norm is

$$\min_{\mathbf{K}\in\mathcal{K}} \|\mathbf{T}_{y_1 u_1}\|_2 = [\text{Trace}\,(B_1^T X_s B_1) + \text{Trace}\,(C_1 Y_s C_1^T)]^{\frac{1}{2}}. \tag{9.50}$$

Proof. Notice that unlike the continuous–time case in discrete–time there is no need to assume that D_{12} and D_{21} have full column and full row ranks, respectively, since these two conditions are not necessary for the existence of the stabilizing solutions to the discrete–time Riccati equations (9.45) and (9.46).

The proof is based on the same intermediate steps as the continuous–time counterpart, and follows almost identically. The existence of the stabilizing solutions to the DTARE(Σ_{12}) in (9.45) and the DTARE(Σ_{21}) in (9.46) follows now from Proposition 7.2.1 and the corresponding hypotheses (H1) and (H2). The only point that needs a different argument is the proof of optimality for the IDFP. We illustrate this aspect only. In doing this we make heavy use of the notation introduced in the continuous–time case.

The existence of the stabilizing solution X_s to the DTARE(Σ_{12}) is equivalent to the existence of a stabilizing solution (X_s, V_s, W_s), $X_s \geq 0$, to the KSPYS(Σ_{12}, I)

$$\begin{aligned} D_{12}^T D_{12} + B_2^T X B_2 &= V^T V, \\ C_1^T D_{12} + A^T X B_2 &= W^T V, \\ C_1^T C_1 + A^T X A - X &= W^T W. \end{aligned} \tag{9.51}$$

We use the KSPYS(Σ_{12}, I) to evaluate $\|y_{1,k}\|$ as follows:

$$\begin{aligned} &\|y_{1,k}\|^2 \\ &= \|C_1 x_k + D_{12} u_{2,k}\|^2 \\ &= x_k^T C_1^T C_1 x_k + 2 x_k^T C_1^T D_{12} u_{2,k} + u_{2,k}^T D_{12}^T D_{12} u_{2,k} \end{aligned}$$

$$\begin{aligned}
&\overset{(9.51)}{=} && x_k^T W_s^T W_s x_k + x_k^T X_s x_k - x_k^T A^T X_s A x_k + 2x_k^T W_s^T V_s u_{2,k} \\
& && - 2x_k^T A^T X_s B_2 u_{2,k} + u_{2,k}^T V_s^T V_s u_{2,k} - u_{2,k}^T B_2^T X_s B_2 u_{2,k} \\
&= && ||W_s x_k + V_s u_{2,k}||^2 + x_k^T X_s x_k - (Ax_k + B_2 u_{2,k})^T X_s (Ax_k + B_2 u_{2,k}) \\
&= && ||\widetilde{y}_{1,k}||^2 + x_k^T X_s x_k - (x_{k+1} - B_1 u_{1,k})^T X_s (x_{k+1} - B_1 u_{1,k}) \\
&= && ||\widetilde{y}_{1,k}||^2 + x_k^T X_s x_k - x_{k+1}^T X_s x_{k+1} \\
& && - u_{1,k}^T B_1^T X_s B_1 u_{1,k} + 2u_{1,k} B_1^T X_s x_{k+1}.
\end{aligned} \tag{9.52}$$

Let now $\mathbf{K}$ be any controller (9.5) such that $\mathbf{T}_{y_1 u_1}$ in (9.6) is internally stable. Then $\widetilde{\mathbf{T}}_{y_1 u_1}$ is internally stable as well. Let $x_{_K}$ and $x_{_R}$ denote the state vector of $\mathbf{K}$ and of $\mathbf{T}_{_R}$, respectively. Consider the impulsive input $u_1 = u_1^i := \delta e_i$ and the initial conditions $x(0) = 0, x_{_K}(0) = 0$. Then we get with (9.6)

$$x_{_R}(1) = \begin{bmatrix} x(1) \\ x_K(1) \end{bmatrix} = B_{_R} u_1^i(0) = \begin{bmatrix} B_1 \\ B_{_K} D_{21} \end{bmatrix} e_i.$$

Hence $x(1) = B_1 e_i$ from where we get

$$u_{1,k} B_1^T X_s x_{k+1} = \begin{cases} e_i B_1^T X_s B_1 e_i, & k = 0, \\ 0, & k \neq 0, \end{cases} \tag{9.53}$$

and

$$u_{1,k}^T B_1^T X_s B_1 u_{1,k} = \begin{cases} e_i B_1^T X_s B_1 e_i, & k = 0, \\ 0, & k \neq 0. \end{cases} \tag{9.54}$$

As $A_{_R}$ is stable, the sequence x is in $\ell^{2,n}$. By summing both sides of (9.52) from $k = 0$ to $k = \infty$ and taking into account (9.53) and (9.54) we get

$$||y_1^i||_2^2 = ||\widetilde{y}_1^i||_2^2 + e_i^T B_1^T X_s B_1 e_i \tag{9.55}$$

and further by summing both sides of (9.55) from $i = 1$ to m_1 we recover (9.32). The rest of the proof is entirely similar to the continuous–time case and is therefore omitted. ■

Notes and references

The solution to the H^2–problem under more restrictive assumptions than those in this chapter appeared in Doyle et. al. [DGKF89]. The presentation in this chapter follows Ionescu and Weiss [IW92a].

10

The H^∞–Control Problem

In this chapter we consider the suboptimal H^∞–control problem (also called in engineering jargon the disturbance attenuation problem) which consists of finding the class of controllers, for a given system, that stabilize the resulting closed–loop system and make its input–output H^∞–norm bounded by a prescribed threshold $\gamma > 0$.

The chapter is organized as follows. In Section 10.1 we formulate the problem, we indicate the assumptions under which we work in Section 10.2, and state the main result in Section 10.3. The statement of the main result consists of necessary and sufficiency solvability conditions for the suboptimal H^∞–control problem, and a construction in terms of realizations of the class of all solutions. A major role in the proof of the main result is played by Redheffer's theorem, which we give in a generalized form in Section 10.4. The proof of the main result is intricate, and we split it along several sections. In Sections 10.5–10.7 we prove the necessity of the conditions, while the proof of sufficiency and the construction of the class of solutions is divided among several subsections in Section 10.8. Finally, in Section 10.9 we give the solution to the H^∞–control problem for the case in which the state is measurable. The discrete–time case is discussed in Sections 10.10–10.18.

A. The continuous–time case

10.1 Problem formulation

In this chapter we work in a framework similar to the one for the H^2–control problem presented in the previous chapter.

Consider a plant that is represented as the linear continuous–time system

$$\begin{aligned} \dot{x} &= Ax + B_1u_1 + B_2u_2, \\ y_1 &= C_1x + D_{11}u_1 + D_{12}u_2, \\ y_2 &= C_2x + D_{21}u_1 + D_{22}u_2, \end{aligned} \tag{10.1}$$

having the TFM given by

$$\mathbf{T} = \left[\begin{array}{c|c} A & B \\ \hline C & D \end{array}\right] = \left[\begin{array}{c|cc} A & B_1 & B_2 \\ \hline C_1 & D_{11} & D_{12} \\ C_2 & D_{21} & D_{22} \end{array}\right] = \left[\begin{array}{cc} \mathbf{T}_{11} & \mathbf{T}_{12} \\ \mathbf{T}_{21} & \mathbf{T}_{22} \end{array}\right], \tag{10.2}$$

where $A \in \mathbb{R}^{n\times n}$, $B_i \in \mathbb{R}^{n\times m_i}$, $C_j \in \mathbb{R}^{p_j\times n}$, $D_{ij} \in \mathbb{R}^{p_i\times m_j}$, ($i, j = 1, 2$, $m = m_1 + m_2$, $p = p_1 + p_2$), u and y denote the input and output of (10.1), respectively, and are partitioned conformably with $\mathbf{T}$ such that

$$y = \begin{bmatrix} y_1 \\ y_2 \end{bmatrix} = \mathbf{T}u = \begin{bmatrix} \mathbf{T}_{11} & \mathbf{T}_{12} \\ \mathbf{T}_{21} & \mathbf{T}_{22} \end{bmatrix} \begin{bmatrix} u_1 \\ u_2 \end{bmatrix}. \tag{10.3}$$

The vector u_1 is the vector of *external inputs* (disturbance, sensor noise and reference signals), y_1 is the vector of *regulated outputs*, representing in general some *error* signals, u_2 is the *controlled inputs* vector and y_2 is the *measured outputs* vector.

Recall form the previous chapter that controlling the plant (10.1) means using the measurement information y_2 to find an appropriate control input u_2 such that the resulting response y_1 of the plant to the external input u_1 has certain desired features. We can cast our problem in the jargon of control engineering as follows: find an appropriate control input u_2, as output of a system – called controller – with input y_2, such that the resulting response y_1 has energy (L^2–norm) as small as possible when the external input u_1 is the worst possible signal of unitary energy (i.e., a function in L^2, of norm 1, chosen such as to maximize y_1). As u_1 is mostly regarded as a disturbance signal, the condition can be regarded as a disturbance attenuation requirement.

We define a controller for the plant (10.1) as a linear system

$$\begin{aligned} \dot{x}_K &= A_K x_K + B_K u_K, \\ y_K &= C_K x_K + D_K u_K, \end{aligned} \tag{10.4}$$

having the TFM given by

$$\mathbf{K} = \left[\begin{array}{c|c} A_K & B_K \\ \hline C_K & D_K \end{array} \right], \tag{10.5}$$

where $A_K \in \mathbb{R}^{n_K \times n_K}$, $B_K \in \mathbb{R}^{n_K \times p_2}$, $C_K \in \mathbb{R}^{m_2 \times n_K}$, $D_K \in \mathbb{R}^{m_2 \times p_2}$, and u_K and y_K are the input and output of (10.4), respectively. The closed–loop system given in Figure 10.1 is obtained by connecting the controller (10.4) to the plant (10.1) such

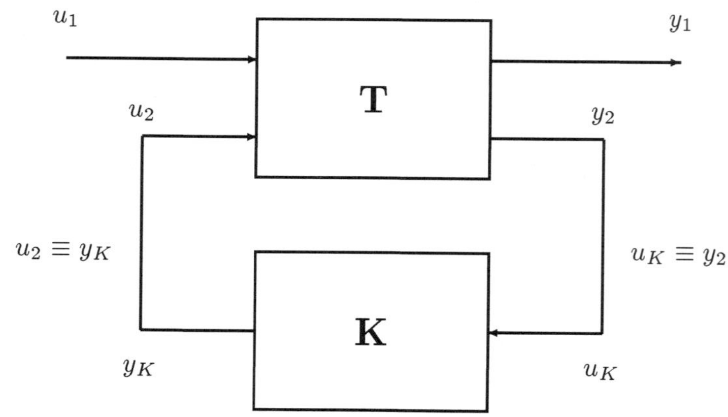

Figure 10.1 The closed–loop connection of $\mathbf{T}$ and $\mathbf{K}$

that $u_2 \equiv y_K$ and $u_K \equiv y_2$ and it is well–posed provided the matrix

$$\begin{bmatrix} I_{p_2} & D_{22} \\ D_K & I_{m_2} \end{bmatrix} \tag{10.6}$$

is nonsingular. Well–posedness can be equivalently expressed by the condition that $S = I_{p_2} - D_{22}D_K$ is nonsingular, or that $\widetilde{S} := I_{m_2} - D_K D_{22}$ is nonsingular.

If the closed–loop system is well–posed, then the resulting system with input u_1 and output y_1 is given by

$$\begin{aligned}
\mathbf{T}_{y_1u_1} &= \mathrm{LFT}(\mathbf{T},\mathbf{K}) = \mathbf{T}_{11} + \mathbf{T}_{12}\mathbf{K}(I - \mathbf{T}_{22}\mathbf{K})^{-1}\mathbf{T}_{21},\\
\mathbf{T}_{y_1u_1} &= \left[\begin{array}{c|c} A_R & B_R \\ \hline C_R & D_R \end{array}\right]\\
&= \left[\begin{array}{cc|c} A + B_2\widetilde{S}^{-1}D_KC_2 & B_2\widetilde{S}^{-1}C_K & B_1 + B_2\widetilde{S}^{-1}D_KD_{21} \\ B_KS^{-1}C_2 & A_K + B_KS^{-1}D_{22}C_K & B_KS^{-1}D_{21} \\ \hline C_1 + D_{12}D_KS^{-1}C_2 & D_{12}\widetilde{S}^{-1}C_K & D_{11} + D_{12}D_KS^{-1}D_{21} \end{array}\right].
\end{aligned} \tag{10.7}$$

With these preliminaries we can state in precise terms the H^∞–control problem. We start with the *suboptimal version.*

Given a system $\mathbf{T}$ and a threshold $\gamma > 0$, the γ–suboptimal H^∞–control problem consists in finding all controllers $\mathbf{K}$ for which the closed–loop system $\mathbf{T}_{y_1u_1}$ is well–posed ($I - D_{22}D_K$ is invertible), internally stable, i.e.,

$$\Lambda(A_R) \subset \mathbb{C}_-, \tag{10.8}$$

and bounded in H^∞–norm by γ, i.e.,

$$\|\mathbf{T}_{y_1u_1}\|_\infty < \gamma. \tag{10.9}$$

Since $\|\mathbf{T}_{y_1u_1}\|_\infty = \sup_{\|u_1\|_2=1}\|y_1\|_2$, it follows that (10.9) is actually imposing a bound on the L_2–norm of the closed–loop response to an arbitrary external input of unitary L_2–norm. This is why we shall refer alternatively, and most often, a γ–suboptimal H^∞–control problem as a γ *disturbance attenuation problem (γ–DAP).* A controller that satisfies (10.8) will be called *stabilizing*, and a controller that satisfies (10.9) will be called *γ–contracting.* If we relax the strict inequality in (10.9) to a nonstrict one, we obtain a so–called *nonstrict γ–DAP.* If $\gamma = 1$, we shall simply talk about a contracting controller, and the corresponding γ–DAP will be simply termed as DAP.

Remark 10.1.1 Simple computations (see Section 2.1.5) show that the realization of $(I - \mathbf{T}_{22}\mathbf{K})^{-1}$, $\mathbf{T}_{y_2u_1} := (I - \mathbf{T}_{22}\mathbf{K})^{-1}\mathbf{T}_{21}$, and $\mathbf{T}_{u_2u_1} = \mathbf{K}(I - \mathbf{T}_{22}\mathbf{K})^{-1}\mathbf{T}_{21}$, written in terms of the realizations (10.2), (10.5), share the same state matrix which is exactly A_R, that is,

$$(I - \mathbf{T}_{22}\mathbf{K})^{-1} = \left[\begin{array}{c|c} A_R & \star \\ \hline \star & \star \end{array}\right], \quad \mathbf{T}_{y_2u_1} = \left[\begin{array}{c|c} A_R & \star \\ \hline \star & \star \end{array}\right], \quad \mathbf{T}_{u_2u_1} = \left[\begin{array}{c|c} A_R & \star \\ \hline \star & \star \end{array}\right].$$

If the controller $\mathbf{K}$ stabilizes $\mathbf{T}$, i.e., the closed–loop is internally stable, then the resulting system is automatically input–output stable, i.e., $\mathbf{T}_{y_1u_1} \in RH^\infty_{+,p_1\times m_1}$, and also $(I - \mathbf{T}_{22}\mathbf{K})^{-1} \in RH^\infty_{+,p_2\times p_2}$, $\mathbf{T}_{y_2u_1} \in RH^\infty_{+,p_2\times m_1}$, and $\mathbf{T}_{u_2u_1} \in RH^\infty_{+,m_2\times m_1}$. This means that a square–integrable input u_1 will generate square–integrable y_1, y_2 and u_2 in the closed–loop scheme in Figure 10.1.

A natural question in this context is also to determine the infimum $\gamma_{\min}$ – called the *least achievable tolerance* – of $\|\mathbf{T}_{y_1u_1}\|_\infty$, over the class of all stabilizing controllers, and to write down formulas for all controllers $\mathbf{K}$ that achieve this minimum (optimum). This is a very tough problem usually known as an *optimal H^∞–control problem* (or optimal disturbance attenuation problem) for which a reasonably short and comprehensive solution is yet not available. Notice that the optimal H^∞–control problem is equivalent to the nonstrict γ–DAP for $\gamma = \gamma_{\min}$.

In this chapter we shall restrict our attention to the γ–DAP whose solution can achieve an arbitrary small (but strictly positive) level $\epsilon := \gamma - \gamma_{\min} > 0$.

10.2 Basic assumptions

Before presenting the solution to the γ–DAP we make a set of additional assumptions on the given plant which either simplify the exposition without restricting the generality, or are of an intrinsic technical nature. These hypotheses which we present below will be in force for the rest of this chapter.

(H1) $\gamma = 1$

Without restricting the generality, we can assume that $\gamma = 1$. Indeed, this is achieved by performing an appropriate data scaling (see Remark 10.3.2 below).

(H2) $D_{22} = 0$

Again, there is no loss of generality in this assumption as if $\mathbf{K}$ is a solution to the problem for the case in which $D_{22} = 0$, then $\mathbf{K}(I+D_{22}\mathbf{K})^{-1}$ is a solution to the original problem. This assumption greatly simplifies the restriction imposed by the well–posedness of the closed–loop since it implies that the matrix (10.6) is automatically invertible. It also leads to simpler formulas for the solution.

(H3) The pair (A, B_2) is stabilizable and the pair (C_2, A) is detectable

These assumptions are necessary for the existence of stabilizing controllers. Indeed, we can write the state matrix A_R of the resulting closed–loop from (10.7) as

$$
\begin{aligned}
A_R &= \begin{bmatrix} A + B_2\widetilde{S}^{-1}D_K C_2 & B_2\widetilde{S}^{-1}C_K \\ B_K S^{-1} C_2 & A_K + B_K S^{-1} D_{22} C_K \end{bmatrix} \qquad (10.10)\\
&= \begin{bmatrix} A & O \\ O & A_K \end{bmatrix} + \begin{bmatrix} B_2 & O \\ O & B_K \end{bmatrix}\begin{bmatrix} \widetilde{S}^{-1}D_K & \widetilde{S}^{-1} \\ S^{-1} & S^{-1}D_{22} \end{bmatrix}\begin{bmatrix} C_2 & O \\ O & C_K \end{bmatrix}.
\end{aligned}
$$

If A_R is stable, we deduce that the matrix pair

$$
\left(\begin{bmatrix} A & O \\ O & A_K \end{bmatrix}, \begin{bmatrix} B_2 & O \\ O & B_K \end{bmatrix}\right)
$$

is stabilizable and the matrix pair

$$
\left(\begin{bmatrix} C_2 & O \\ O & C_K \end{bmatrix}, \begin{bmatrix} A & O \\ O & A_K \end{bmatrix}\right)
$$

is detectable from which it follows that the pair (A, B_2) is stabilizable and the pair (C_2, A) is detectable. Incidentally, it also follows that (A_K, B_K) and (C_K, A_K) must be stabilizable and detectable, respectively.

(H4) Regularity

We make two regularity assumptions (R1) and (R2) below, which are reminiscent from the H^2–control problem.
(R1) The transmission pencil of $\mathbf{T}_{12}$ given by

$$\begin{bmatrix} sI - A & -B_2 \\ -C_1 & -D_{12} \end{bmatrix}$$

has full column rank for all $s \in \overline{\mathbb{C}}_0$. This condition is equivalent to the following two rank conditions:

$$\operatorname{rank} D_{12} = m_2, \quad \operatorname{rank} \begin{bmatrix} j\omega I - A & -B_2 \\ -C_1 & -D_{12} \end{bmatrix} = n + m_2, \quad \forall \omega \in \mathbb{R}.$$

In particular, it follows that $p_1 \geq m_2$, i.e., there are more regulated outputs than controlled inputs, and $\mathbf{T}_{12}$ has no zeros on $\overline{\mathbb{C}}_0$.
(R2) The transmission pencil of $\mathbf{T}_{21}$ given by

$$\begin{bmatrix} sI - A & -B_1 \\ -C_2 & -D_{21} \end{bmatrix}$$

has full row rank for all $s \in \overline{\mathbb{C}}_0$. This condition is equivalent to the following two rank conditions:

$$\operatorname{rank} D_{21} = m_2, \quad \operatorname{rank} \begin{bmatrix} j\omega I - A & -B_1 \\ -C_2 & -D_{21} \end{bmatrix} = n + m_2, \quad \forall \omega \in \mathbb{R}.$$

In particular, it follows that $m_1 \geq p_2$, i.e., there are more external inputs than measured outputs, and $\mathbf{T}_{21}$ has no zeros on $\overline{\mathbb{C}}_0$.

A problem satisfying these two assumptions is called *regular*, otherwise it is called singular.

We make the cautionary remark that these two regularity assumptions are by no means necessary conditions for the existence of a solution to the γ–DAP, and they are primarily made for technical reasons. Singular problems are very difficult to treat, and their complete solution is still missing.

10.3 The solution

With the assumptions made in the previous section, we can state the main result of the chapter that provides the solution to the DAP. This result has two aspects. On one side, there is a solvability criterion, that is a necessary and sufficient condition that a solution to the DAP exists, a condition that can be easily checked numerically. On the other side, there is an explicit formula to generate all the solutions of the problem provided that any exists. Both existence and solution parameterization are expressed in terms of stabilizing solutions to Kalman–Popov–Yakubovich systems. As we shall see, this is better suited to and more handy for our purposes than the traditional Riccati equation formulation.

Theorem 10.3.1 *Assume the system* $\mathbf{T}$ *is given by (10.1)–(10.2) and satisfies hypotheses (H2), (H3) and (H4). Associate with* $\mathbf{T}$ *the following two Popov triplets:*

$$\Sigma_c = (A, \begin{bmatrix} B_1 & B_2 \end{bmatrix}; Q_c, L_c, R_c),$$

and

$$\Sigma_o = (A^T, \begin{bmatrix} C_1^T & C_2^T \end{bmatrix}; Q_o, L_o, R_o),$$

where

$$\begin{aligned} Q_c &= C_1^T C_1, \\ L_c &= C_1^T \begin{bmatrix} D_{11} & D_{12} \end{bmatrix}, \\ R_c &= \begin{bmatrix} D_{11}^T \\ D_{12}^T \end{bmatrix} \begin{bmatrix} D_{11} & D_{12} \end{bmatrix} - \begin{bmatrix} I_{m_1} & O \\ O & O \end{bmatrix}, \end{aligned} \tag{10.11}$$

and

$$\begin{aligned} Q_o &= B_1 B_1^T, \\ L_o &= B_1 \begin{bmatrix} D_{11}^T & D_{21}^T \end{bmatrix}, \\ R_o &= \begin{bmatrix} D_{11} \\ D_{21} \end{bmatrix} \begin{bmatrix} D_{11}^T & D_{21}^T \end{bmatrix} - \begin{bmatrix} I_{p_1} & O \\ O & O \end{bmatrix}. \end{aligned} \tag{10.12}$$

Denote

$$J_c = \begin{bmatrix} -I_{m_1} & O \\ O & I_{m_2} \end{bmatrix}, \quad J_o = \begin{bmatrix} -I_{p_1} & O \\ O & I_{p_2} \end{bmatrix}.$$

Then we have:

(I) *The DAP has a solution if and only if the following conditions are met:*

(C1) *The Kalman–Popov–Yakubovich system KPYS(Σ_c, J_c) has a stabilizing solution* $(X, V_c, W_c) = (X, \begin{bmatrix} V_{c11} & O \\ V_{c21} & V_{c22} \end{bmatrix}, \begin{bmatrix} W_{c1} \\ W_{c2} \end{bmatrix})$, *with* $X \geq 0$.

(C2) *The Kalman–Popov–Yakubovich system KPYS(Σ_o, J_o) has a stabilizing solution* $(Y, V_o, W_o) = (Y, \begin{bmatrix} V_{o11} & O \\ V_{o21} & V_{o22} \end{bmatrix}, \begin{bmatrix} W_{o1} \\ W_{o2} \end{bmatrix})$, *with* $Y \geq 0$.

(C3) *The spectral radius of XY is strictly subunitary, i.e.,*

$$\rho(XY) < 1.$$

(II) *Assume that (C1) is satisfied, let*

$$J_\times = \begin{bmatrix} -I_{m_2} & O \\ O & I_{p_2} \end{bmatrix},$$

and define the Popov triplet

$$\Sigma_\times = (A^T + F_1^T B_1^T, \begin{bmatrix} -F_2^T V_{c22}^T & C_2^T + F_1^T D_{21}^T \end{bmatrix}; Q_\times, L_\times, R_\times),$$

where $\begin{bmatrix} F_1 \\ F_2 \end{bmatrix}$ *is the stabilizing feedback corresponding to the KPYS(Σ_c, J_c) and*

$$\begin{aligned} Q_\times &:= B_1(V_{c11}^T V_{c11})^{-1} B_1^T, \\ L_\times &:= B_1(V_{c11}^T V_{c11})^{-1} \begin{bmatrix} V_{c21}^T & D_{21}^T \end{bmatrix}, \\ R_\times &:= \begin{bmatrix} V_{c21} \\ D_{21} \end{bmatrix} (V_{c11}^T V_{c11})^{-1} \begin{bmatrix} V_{c21}^T & D_{21}^T \end{bmatrix} - \begin{bmatrix} I_{m_2} & O \\ O & O \end{bmatrix}. \end{aligned} \tag{10.13}$$

Then the DAP has a solution if and only if (C1) holds and the following condition is true:

(C4) *The Kalman–Popov–Yakubovich system KPYS($\Sigma_\times, J_\times$) has a stabilizing solution* $(Z, V_\times, W_\times) = (Z, \begin{bmatrix} V_{\times 11} & O \\ V_{\times 21} & V_{\times 22} \end{bmatrix}, \begin{bmatrix} W_{\times 1} \\ W_{\times 2} \end{bmatrix})$, *with* $Z \geq 0$.

Moreover, the following relation among the solutions to the three KPYSs given above holds:

$$Z := Y(I - XY)^{-1}. \tag{10.14}$$

(III) *Assume that (C1) and (C4) both hold and let* $C_{2F_1} := C_2 + D_{21}F_1$, $S_c := (V_{c11}^T V_{c11})^{-1}$, *and* $S_\times := (V_{\times 11}^T V_{\times 11})^{-1}$. *Then the class of all solutions to the DAP can be expressed as* $\mathbf{K} = \mathrm{LFT}(\mathbf{K}_g, \mathbf{Q})$, *where* $\mathbf{Q} \in RH^\infty_{+,m_2 \times p_2}$ *is an arbitrary parameter with* $\|\mathbf{Q}\|_\infty < 1$ *and*

$$\mathbf{K}_g = \left[\begin{array}{c|cc} A_g & B_{g1} & B_{g2} \\ \hline C_{g1} & D_{g11} & D_{g12} \\ C_{g2} & D_{g21} & D_{g22} \end{array}\right] = \begin{bmatrix} \mathbf{K}_{g11} & \mathbf{K}_{g12} \\ \mathbf{K}_{g21} & \mathbf{K}_{g22} \end{bmatrix}, \tag{10.15}$$

where

$$\begin{aligned} A_g &= A + B_1F_1 + B_2F_2 + B_{g1}C_{2F_1}, \\ B_{g1} &= -B_1S_cD_{21}^T(D_{21}S_cD_{21}^T)^{-1} - B_2D_{g11} - ZC_{g2}^TD_{g21}, \\ B_{g2} &= -B_1S_c[(D_{12}^TD_{12})^{-1}D_{12}^TD_{11} + D_{g11}D_{21}]^T(D_{12}^TD_{12})^{\frac{1}{2}}S_\times^{\frac{1}{2}} \\ &\quad -B_2D_{g12} - ZC_{g1}^T(D_{12}^TD_{12})^{\frac{1}{2}}S_\times^{\frac{1}{2}}, \\ C_{g1} &= -F_2 + D_{g11}C_{2F1}, \\ C_{g2} &= (D_{21}S_cD_{21}^T)^{-\frac{1}{2}}C_{2F1}, \\ D_{g11} &= -(D_{12}^TD_{12})^{-1}D_{12}^TD_{11}S_cD_{21}^T(D_{21}S_cD_{21}^T)^{-1}, \\ D_{g12} &= (D_{12}^TD_{12})^{-\frac{1}{2}}S_\times^{-\frac{1}{2}}, \\ D_{g21} &= (D_{21}S_cD_{21}^T)^{-\frac{1}{2}}, \\ D_{g22} &= 0. \end{aligned}$$

10.3.1 The solution under normalizing conditions

It is easy to see that if the following so–called *normalizing conditions* are satisfied

$$D_{11} = 0, \quad D_{12}^T \begin{bmatrix} C_1 & D_{12} \end{bmatrix} = \begin{bmatrix} O & I \end{bmatrix}, \quad \begin{bmatrix} B_1 \\ D_{21} \end{bmatrix} D_{21}^T = \begin{bmatrix} O \\ I \end{bmatrix}, \tag{10.16}$$

then

$$\Sigma_c = (A, \begin{bmatrix} B_1 & B_2 \end{bmatrix}; C_1^T C_1, 0, \begin{bmatrix} -I_{m_1} & \\ & I_{m_2} \end{bmatrix}), \tag{10.17}$$

and

$$\Sigma_o = (A^T, \begin{bmatrix} C_1^T & C_2^T \end{bmatrix}; B_1 B_1^T, 0, \begin{bmatrix} -I_{p_1} & \\ & I_{p_2} \end{bmatrix}). \tag{10.18}$$

Then the CTARE(Σ_c) is

$$A^T X + XA - X(B_2 B_2^T - B_1 B_1^T)X + C_1^T C_1 = 0 \tag{10.19}$$

and the CTARE(Σ_o) is

$$AY + YA^T - Y(C_2^T C_2 - C_1^T C_1)Y + B_1 B_1^T = 0, \tag{10.20}$$

and Theorem 10.3.1 can be formulated in terms of the stabilizing solutions to these Riccati equations. Indeed, it is easy to see that

$$V_{c22} = I, \quad V_{c21} = 0, \quad V_{c11} = I, \quad W_{c1} = -B_1^T X, \quad W_{c2} = B_2^T X. \tag{10.21}$$

Hence

$$F_1 = B_1^T X, F_2 = -B_2^T X. \tag{10.22}$$

With (10.21) and (10.22), the Popov triplet $\Sigma_\times$ becomes

$$\Sigma_\times = (A^T + XB_1B_1^T, \begin{bmatrix} XB_2 & C_2^T \end{bmatrix}; B_1B_1^T, \begin{bmatrix} O & O \end{bmatrix}, \begin{bmatrix} -I_{m_2} & O \\ O & I_{p_2} \end{bmatrix}). \tag{10.23}$$

The CTARE($\Sigma_\times$) can be written explicitly as

$$(A + B_1B_1^T X)Z + Z(A + B_1B_1^T X)^T - Z(C_2^T C_2 - XB_2B_2^T X)Z + B_1B_1^T = 0. \tag{10.24}$$

Finally, the realization for $\mathbf{K}_g$ becomes

$$\begin{aligned}
A_g &= A + B_1B_1^T X - B_2B_2^T X - ZC_2^T C_2, \\
B_{g1} &= -ZC_2^T, \\
B_{g2} &= -\{B_2 - [Z(-XB_2)]\} = -(I + ZX)B_2 = -(I - YX)^{-1}B_2, \\
C_{g1} &= B_2^T X, \\
C_{g2} &= C_2, \\
D_{g11} &= 0, \\
D_{g12} &= I_{m2}, \\
D_{g21} &= I_{p2}, \\
D_{g22} &= 0.
\end{aligned} \tag{10.25}$$

If we take the parameter $\mathbf{Q} = 0$ we obtain the *central controller*. If the normalizing conditions (10.16) hold then the central controller has the realization

$$\mathbf{K} = \left[\begin{array}{c|c} A + B_1B_1^T X - B_2B_2^T X - (I - YX)^{-1}YC_2^T C_2 & -(I - YX)^{-1}YC_2^T \\ \hline B_2^T X & O \end{array}\right]. \tag{10.26}$$

Remark 10.3.2 The conversion of a γ–DAP with arbitrary $\gamma > 0$ to a DAP (with $\gamma = 1$) can be achieved by the following scaling transformation on the initial problem

$$\mathbf{T}_{scaled} = \left[\begin{array}{c|cc} A & \gamma^{-1/2}B_1 & \gamma^{1/2}B_2 \\ \hline \gamma^{-1/2}C_1 & \gamma^{-1}D_{11} & D_{12} \\ \gamma^{1/2}C_2 & D_{21} & \gamma D_{22} \end{array}\right]. \tag{10.27}$$

The solution $\mathbf{K}_{scaled}$ to the DAP (with $\gamma = 1$) formulated for (10.27) can be expressed in terms of the solution (10.15) to the original γ–DAP as

$$\mathbf{K}_{g,scaled} = \left[\begin{array}{c|cc} A_g & \gamma^{-1/2}B_{g1} & \gamma^{1/2}B_{g2} \\ \hline \gamma^{-1/2}C_{g1} & \gamma^{-1}D_{g11} & D_{g12} \\ \gamma^{1/2}C_{g2} & D_{g21} & \gamma D_{g22} \end{array}\right]. \tag{10.28}$$

The above expressions for the scaling can be obtained from

$$\mathbf{T}_{y_1u_1,scaled} := \gamma^{-1}\mathbf{T}_{y_1u_1} = \gamma^{-1}\mathbf{T}_{11} + \mathbf{T}_{12}\gamma^{-1}\mathbf{K}(I - \gamma\mathbf{T}_{22}\gamma^{-1}\mathbf{K})\mathbf{T}_{21}$$

and

$$\mathbf{K}_{scaled} = \gamma^{-1}\mathbf{K} = \gamma^{-1}\mathbf{K}_{g,11} + \mathbf{K}_{g,12}\mathbf{Q}_{scaled}(I - \gamma\mathbf{K}_{g,22}\mathbf{Q}_{scaled})\mathbf{K}_{g,21}$$

where $\mathbf{Q}_{scaled} = \gamma^{-1}\mathbf{Q}$ with $\|\mathbf{Q}\|_\infty < \gamma$ and $\|\mathbf{Q}_{scaled}\|_\infty < 1$. The solutions to the KPYSs intervening in Theorem 10.3.1 written for the original and the scaled data are connected by

$$(X, \begin{bmatrix} V_{c11} & O \\ V_{c21} & V_{c22} \end{bmatrix}, \begin{bmatrix} W_{c1} \\ W_{c2} \end{bmatrix})_{scaled} = (\gamma^{-1}X, \begin{bmatrix} \gamma^{-1}V_{c11} & O \\ \gamma^{-1}V_{c21} & V_{c22} \end{bmatrix}, \begin{bmatrix} \gamma^{-\frac{1}{2}}W_{c1} \\ \gamma^{-\frac{1}{2}}W_{c2} \end{bmatrix}),$$

$$(Y, \begin{bmatrix} V_{o11} & O \\ V_{o21} & V_{o22} \end{bmatrix}, \begin{bmatrix} W_{o1} \\ W_{o2} \end{bmatrix})_{scaled} = (\gamma^{-1}Y, \begin{bmatrix} \gamma^{-1}V_{o11} & O \\ \gamma^{-1}V_{o21} & V_{o22} \end{bmatrix}, \begin{bmatrix} \gamma^{-\frac{1}{2}}W_{o1} \\ \gamma^{-\frac{1}{2}}W_{o2} \end{bmatrix}),$$

$$(Z, \begin{bmatrix} V_{\times 11} & O \\ V_{\times 21} & V_{\times 22} \end{bmatrix}, \begin{bmatrix} W_{\times 1} \\ W_{\times 2} \end{bmatrix})_{scaled} = (\gamma^{-1}Z, \begin{bmatrix} \gamma^{-1}V_{\times 11} & O \\ \gamma^{-1}V_{\times 21} & V_{\times 22} \end{bmatrix}, \begin{bmatrix} \gamma^{-\frac{1}{2}}W_{\times 1} \\ \gamma^{-\frac{1}{2}}W_{\times 2} \end{bmatrix}).$$

The above formulas allow a handy conversion between the solutions to the γ–DAP and the scaled DAP (with $\gamma = 1$).

10.3.2 The plan of the proof

The plan for the proof of Theorem 10.3.1 is as follows. First we prove in Section 10.5 the necessity of (C1), and thereafter the necessity of (C2) follows by duality. In order to prove the necessity of (C3), we need a more tricky reasoning. We show first in Section 10.6 that if (C1) holds then (C4) is necessary for the existence of a solution to the DAP. Further, we show in Section 10.7 that (C4) implies (C2) together with (C3) which ends the proof of the necessity of (C3). In fact we prove a stronger result that will be used for the sufficiency part as well, namely, if (C1) holds, then (C4) is equivalent to (C2) together with (C3). This incidentally proves also the other assertions of the theorem except for the parameterization of all the solutions. For the parameterization formula, we successively consider three simpler problems and use the solution of the KPYS(Σ_c, J_c) to reduce the general problem to a simpler problem that turns out to be solvable by using the solution of the KPYS($\Sigma_\times, J_\times$). Before pursuing this plan we need some technical preliminaries given in the next section.

10.4 Redheffer's theorem

In this section we prove Redheffer's theorem under relaxed assumptions. Redheffer's theorem is a special case of closed–loop strict subunitary disturbance attenuation.

We need first the following definitions. Recall that a system with TFM $\mathbf{T} \in RL^{\infty}$ is called *isometric* if $\mathbf{T}^*\mathbf{T} = I$ and is called inner if it is isometric and $\mathbf{T} \in RH^{\infty}_{+}$.

Assume $\mathbf{T}$ is given by a realization

$$\mathbf{T} = \left[\begin{array}{c|c} A & B \\ \hline C & D \end{array}\right]. \tag{10.29}$$

If A is dichotomic (stable), $D^T D = I$, and there exists an $X = X^T$ such that

$$\begin{array}{rcl} C^T D + XB & = & 0, \\ C^T C + A^T X + XA & = & 0, \end{array} \tag{10.30}$$

then we say that the realization (10.29) is isometric (inner). It is easy to check that if a system $\mathbf{T}$ has an isometric (inner) realization then it is isometric (inner), i.e., $\mathbf{T}^*\mathbf{T} = I$ and $\mathbf{T} \in RL^{\infty}$ ($\mathbf{T} \in RH^{\infty}_{+}$). However, not every realization of an isometric (inner) system $\mathbf{T}$ is isometric (inner). Since in this section we shall work with systems given by realizations we shall make the abuse to call a system isometric (inner) if and only if its realization is isometric (inner).

Theorem 10.4.1 (Generalized Redheffer's theorem) *Assume that the system* $\mathbf{T}$ *in (10.1)–(10.2) is isometric,* $p_2 = m_1$, D_{21} *is nonsingular and* $A - B_1 D_{21}^{-1} C_2$ *is dichotomic, and let* $\mathbf{K}$ *given in (10.4) be a controller for* $\mathbf{T}$. *Then the following two assertions are equivalent:*

1. The closed–loop system $\mathbf{T}_{y_1 u_1}$ *given in (10.7) is well–posed, and the controller* $\mathbf{K}$ *is a stabilizing and contracting controller for* $\mathbf{T}$.

2. $\mathbf{T}$ *is inner,* A_K *is stable, and* $\|\mathbf{K}\|_{\infty} < 1$.

Proof. 2. $\Rightarrow$ 1. Since $\mathbf{T}$ is inner, it follows from $\mathbf{T}(\infty) = D$ that $\|D\| = 1$, and consequently $\|D_{22}\| \leq 1$, whereas since $\|\mathbf{K}\|_{\infty} < 1$, we deduce that $\|D_K\| < 1$. It follows that $S = I - D_{22} D_K$ is nonsingular, and the closed–loop system is well–posed.

On the other hand, since A_K is stable and $\|\mathbf{K}\|_{\infty} < 1$, it follows from part I of Lemma 7.1.1 that the KPYS

$$\begin{array}{rcl} -I + D_K^T D_K & = & -V_K^T V_K, \\ C_K^T D_K + X_K B_K & = & -W_K^T V_K, \\ C_K^T C_K + A_K^T X_K + X_K A_K & = & -W_K^T W_K, \end{array} \tag{10.31}$$

has a stabilizing solution (X_K, V_K, W_K), with $X_K \geq 0$. Since $\mathbf{T}$ is inner, there exists an $X = X^T$ such that the equations in (10.30) are satisfied. Moreover, $X \geq 0$ as it follows from the second equation (10.30) because A is stable.

If we take

$$X_R := \begin{bmatrix} X & O \\ O & X_K \end{bmatrix}, V_R := V_K S^{-1} D_{21}, W_R := \begin{bmatrix} V_K S^{-1} C_2 & W_K + V_K S^{-1} D_{22} C_K \end{bmatrix}, \tag{10.32}$$

then $X_R \geq 0$ and it can be checked by direct computation that

$$\begin{aligned} -I + D_R^T D_R &= -V_R^T V_R, \\ C_R^T D_R + X_R B_R &= -W_R^T V_R, \\ C_R^T C_R + A_R^T X_R + X_R A_R &= -W_R^T W_R. \end{aligned} \tag{10.33}$$

On the other hand, for $K_R := \begin{bmatrix} O & -B_2 D_K V_K^{-1} \\ O & -B_K V_K^{-1} \end{bmatrix}$ one gets

$$A_R + K_R \begin{bmatrix} C_R \\ W_R \end{bmatrix} = \begin{bmatrix} A & \star \\ O & A_K - B_K V_K^{-1} W_K \end{bmatrix}. \tag{10.34}$$

Since (X_K, V_K, W_K) is a stabilizing solution to (10.31) and A is stable, it follows that the upper left and lower right corners in the right–hand side of (10.34) are stable. Hence so is the left–hand side of the same equality, from which one deduces the detectability of the pair $(\begin{bmatrix} C_R \\ W_R \end{bmatrix}, A_R)$. By combining this conclusion with the fact that the last equation in (10.33) is fulfilled for $X_R \geq 0$, we conclude with Theorem 1.5.5 that A_R is stable. Hence $\mathbf{K}$ is a stabilizing controller for $\mathbf{T}$.

Let us prove now that $\mathbf{K}$ is a contracting controller for $\mathbf{T}$, i.e., $\|\mathbf{T}_{y_1u_1}\|_\infty < 1$. Notice that (10.33) is the KPYS$(\Sigma_R, -I)$ associated with the Popov triplet $\Sigma_R = (A_R, B_R; C_R^T C_R, C_R^T D_R, -I + D_R^T D_R)$ which, in turn, is the contractiveness Popov triplet associated with the system $\mathbf{T}_{y_1u_1}$ (see Section 3.4). Thus, the Popov function associated with Σ_R is $\mathbf{\Pi}_{\Sigma_R} = -I + \mathbf{T}_{y_1u_1}^* \mathbf{T}_{y_1u_1}$. Since A_R is stable and the KPYS$(\Sigma_R, -I)$ has a solution (X_R, V_R, W_R) given by (10.32), it follows from the spectral factorization identity (see (3.39))

$$\mathbf{\Pi}_{\Sigma_R} = -\left[\begin{array}{c|c} A_R & B_R \\ \hline W_R & V_R \end{array}\right]^* \left[\begin{array}{c|c} A_R & B_R \\ \hline W_R & V_R \end{array}\right] \leq 0$$

from where $\|\mathbf{T}_{y_1u_1}\|_\infty \leq 1$. Let us show now that the inequality is strict. Assume the contrary, i.e., $\|\mathbf{T}_{y_1u_1}\|_\infty = 1$. Then there exists a sequence $(u_1^k)_{k\geq 0}$ with $u_1^k \in L^{2,m_1}$ and $\|u_1^k\|_2 = 1$ such that for $(y_1^k)_{k\geq 0}$ with $y_1^k \in L^{2,p_1}$ obtained as $y_1^k(j\omega) = \mathbf{T}_{y_1u_1}(j\omega)u_1^k(j\omega)$ we have

$$\|y_1^k\|_2^2 - \|u_1^k\|_2^2 \to 0 \text{ as } \quad k \to \infty. \tag{10.35}$$

Since both A and A_K are stable it follows that $I - \mathbf{T}_{22}\mathbf{K} \in RH^\infty_{+,p_2\times p_2}$. But $(I - \mathbf{T}_{22}\mathbf{K})^{-1} \in RH^\infty_{+,p_2\times p_2}$ since $\mathbf{K}$ is a stabilizing controller (see Remark 10.1.1). Hence $I - \mathbf{T}_{22}\mathbf{K}$ is a unit in $RH^\infty_{+,p_2\times p_2}$. Further, $\mathbf{T}_{21} \in RH^\infty_{+,p_2\times p_2}$ and

$$\mathbf{T}_{21}^{-1} = \left[\begin{array}{c|c} A - B_1 D_{21}^{-1} C_2 & B_1 D_{21}^{-1} \\ \hline -D_{21}^{-1} C_2 & D_{21}^{-1} \end{array}\right] \in RL^\infty_{p_2\times p_2}$$

as follows from the assumption that $A - B_1 D_{21}^{-1} C_2$ is dichotomic. Thus $\mathbf{T}_{21}$ is a unit in $RL^\infty_{p_2\times p_2}$. Since $\mathbf{T}_{21}$ and $I - \mathbf{T}_{22}\mathbf{K}$ are both units in $RL^\infty_{p_2\times p_2}$ it follows that $\mathbf{T}_{y_2u_1} := (I - \mathbf{T}_{22}\mathbf{K})^{-1}\mathbf{T}_{21}$ is a unit in $RL^\infty_{p_2\times p_2}$ as well. Hence there exists $\rho_0 > 0$ such that

$$\|y_2\|_2 \geq \rho_o \|u_1\|_2, \quad \forall u_1 \in L^{2,p_2}, \tag{10.36}$$

for $y_2 \in L^{2,p_2}$ defined by $y_2(j\omega) = \mathbf{T}_{y_2u_1}(j\omega)u_1(j\omega)$. Since $\mathbf{T}$ is isometric it follows that

$$||y_1^k||_2^2 - ||u_1^k||_2^2 = ||u_2^k||_2^2 - ||y_2^k||_2^2, \tag{10.37}$$

where u_1^k and y_1^k have been introduced before, and $y_2^k \in L^{2,p_2}$ and $u_2^k \in L^{2,m_2}$ are defined by $y_2^k(j\omega) = \mathbf{T}_{y_2u_1}(j\omega)u_1^k(j\omega)$ and $u_2^k(j\omega) = \mathbf{K}(j\omega)y_2^k(j\omega) = \mathbf{T}_{u_2u_1}(j\omega)u_1^k(j\omega)$, respectively. In the last relation we have denoted $\mathbf{T}_{u_2u_1} := \mathbf{K}\mathbf{T}_{y_2u_1} \in RH^{\infty}_{+,p_2\times m_1}$. By combining (10.35) with (10.37) one deduces that

$$||y_2^k||_2^2 - ||u_2^k||_2^2 \to 0, \text{ as } \quad k \to \infty. \tag{10.38}$$

But $||u_2^k||_2^2 \le ||\mathbf{K}||_\infty^2 ||y_2^k||_2^2$. Hence by taking into account (10.36) one gets

$$\begin{aligned} ||y_2^k||_2^2 - ||u_2^k||_2^2 &\ge (1 - ||\mathbf{K}||_\infty^2)||y_2^k||_2^2 \ge (1 - ||\mathbf{K}||_\infty^2)\rho_0^2||u_1^k||_\infty^2 \\ &= (1 - ||\mathbf{K}||_\infty^2)\rho_0^2 =: c_0 > 0. \end{aligned} \tag{10.39}$$

But (10.39) contradicts (10.38), and the strict contractiveness of $\mathbf{T}_{y_1u_1}$ follows.

1. $\Rightarrow$ 2. Assume that $\mathbf{K}$ is a stabilizing and contracting controller for $\mathbf{T}$.

Let us show first that $\mathbf{K} \in RL^{\infty}_{m_2\times p_2}$. It is trivial to check that the following identity holds

$$\begin{bmatrix} \mathbf{T}_{y_1u_1} \\ \mathbf{T}_{y_2u_1} \end{bmatrix} = \begin{bmatrix} \mathbf{T}_{11} & \mathbf{T}_{12} \\ \mathbf{T}_{21} & \mathbf{T}_{22} \end{bmatrix} \begin{bmatrix} I \\ \mathbf{T}_{u_2u_1} \end{bmatrix}. \tag{10.40}$$

If the both sides of (10.40) are premultiplied by $\mathbf{T}^*$ then, by taking into account that $\mathbf{T}$ is isometric, one obtains $\mathbf{T}_{11}^*\mathbf{T}_{y_1u_1} + \mathbf{T}_{21}^*\mathbf{T}_{y_2u_1} = I$ from which

$$\mathbf{T}_{y_2u_1} = \mathbf{T}_{21}^{-*}(I - \mathbf{T}_{11}^*\mathbf{T}_{y_1u_1}) \in RL^{\infty}_{p_2\times p_2}. \tag{10.41}$$

But $\mathbf{T}_{21}$ is a unit in $RL^{\infty}_{p_2\times m_1}$ because of the dichotomy of both A and $A - B_1D_{21}^{-1}C_2$. On the other hand $||\mathbf{T}||_\infty = 1$ implies $||\mathbf{T}_{11}|| \le 1$. Hence $||\mathbf{T}_{11}^*\mathbf{T}_{y_1u_1}||_\infty \le ||\mathbf{T}_{11}||_\infty ||\mathbf{T}_{y_1u_1}||_\infty < 1$ from which it follows that $I - \mathbf{T}_{11}^*\mathbf{T}_{y_1u_1}$ is a unit in $RL^{\infty}_{m_1\times m_1}$. Hence $\mathbf{T}_{y_2u_1}$ is a unit in $RL^{\infty}_{p_2\times p_2}$ as follows from (10.41). Using this conclusion we have that

$$\mathbf{K} = \mathbf{T}_{u_2u_1}\mathbf{T}_{y_2u_1}^{-1} \in RL^{\infty}_{m_2\times p_2}. \tag{10.42}$$

Moreover, $||\mathbf{K}||_\infty < 1$. Assume the contrary, i.e. $||\mathbf{K}||_\infty \ge 1$. Then there exists a sequence $(y_2^k)_{k\ge 0}$ with $y_2^k \in L^{2,p_2}$ and $||y_2^k||_2 = 1$, such that for $(u_2^k)_{k\ge 0}$ with $u_2^k \in L^{2,m_2}$ defined by $u_2^k(j\omega) = \mathbf{K}(j\omega)y_2^k(j\omega)$, we have

$$\lim_{k\to\infty}(||u_2^k||_2^2 - ||y_2^k||_2^2) \ge 0. \tag{10.43}$$

Since $\mathbf{T}_{y_2u_1}$ is a unit in $RL^{\infty}_{p_2\times p_2}$, one can define the sequence $(u_1^k)_{k\ge 0}, u_1^k \in L^{2,m_1}$ by $u_1^k(j\omega) = \mathbf{T}_{y_2u_1}^{-1}(j\omega)y_2^k(j\omega) \ne 0, \forall k$. Let further $(y_1^k)_{k\ge 0}, y_1^k \in L^{2,p_1}$ be defined by $y_1^k(j\omega) = \mathbf{T}_{y_1u_1}(j\omega)u_1^k(j\omega)$. Since $\mathbf{T}$ is isometric

$$||y_1^k||_2^2 - ||u_1^k||_2^2 = ||u_2^k||_2^2 - ||y_2^k||_2^2. \tag{10.44}$$

This combined with (10.43) yields $\lim_{k\to\infty}(||y_1^k||_2^2 - ||u_1^k||_2^2) \ge 0$ which contradicts the strict contractiveness of $\mathbf{T}_{y_1u_1}$. Thus $||\mathbf{K}||_\infty < 1$.

Let us show now that A_K is stable. Notice first that $\mathbf{K}$ stabilizes $\mathbf{T}$. Hence the pairs (A_K, B_K) and (A, B_2) are stabilizable and the pairs (C_K, A_K) and (C_2, A) are detectable. Using this conclusion in conjunction with (10.42) it follows that A_K must be dichotomic. Therefore, since $||\mathbf{K}||_\infty < 1$ it follows from part II of Lemma 7.1.1 that the KPYS (10.31) has a stabilizing solution (X_K, V_K, W_K). By combining (10.30) and (10.31), (10.33) will hold for (X_R, V_R, W_R) defined by (10.32). But A_R in the last equation (10.33) is stable. Hence $X_R \geq 0$ and consequently $X \geq 0$ and $X_K \geq 0$. Further, since (C_K, A_K) is detectable and $X_K \geq 0$, it follows from the last equation (10.31) with Theorem 1.5.5 that A_K is stable.

Finally, A is stable as follows from the last equation (10.30) where $X \geq 0$ and (C, A) is detectable since (C_2, A) is detectable. Thus $\mathbf{T}$ is inner and the theorem is completely proved. ∎

Remark 10.4.2 Assume that hypotheses of Theorem 10.4.1 hold. Assume also that (10.33) is fulfilled for (10.32) where (X_K, V_K, W_K) is a stabilizing solution to (10.31). Then from (10.7) and (10.32) one gets

$$A_R - B_R V_R^{-1} W_R = \begin{bmatrix} A - B_1 D_{21}^{-1} C_2 & \star \\ O & A_K - B_K V_K^{-1} W_K \end{bmatrix}. \tag{10.45}$$

By inspecting both sides of the equality (10.45) one concludes that (X_R, V_R, W_R) given by (10.32) is a stabilizing solution to (10.33) if and only if $A - B_1 D_{21}^{-1} C_2$ is stable. In the original statement of (standard) Redheffer's theorem $A - B_1 D_{21}^{-1} C_2$ is assumed stable and in this case the implication 2. ⇒ 1. is a trivial consequence of Lemma 7.1.1.

10.5 Necessity of (C1) and (C2)

We start in this section with the proof of Theorem 10.3.1. In this section we assume that a stabilizing controller exists such that $||\mathbf{T}_{y_1 u_1}||_\infty < 1$ and we prove that (C1) and (C2) are fulfilled. Notice that (C2) is simply the dual of (C1) in the sense that it states that the dual system

$$\mathbf{T}_{dual} = \left[\begin{array}{c|cc} A^T & C_1^T & C_2^T \\ \hline B_1^T & D_{11}^T & D_{21}^T \\ B_2^T & D_{12}^T & O \end{array}\right]$$

satisfies (C1). Since for every stabilizing and contracting controller $\mathbf{K}$, its dual

$$\mathbf{K}_{dual} = \left[\begin{array}{c|c} A_K^T & C_K^T \\ \hline B_K^T & D_K^T \end{array}\right],$$

is stabilizing and contracting for $\mathbf{T}_{dual}$, it follows that it is sufficient to prove (C1).

The basic idea of the proof of (C1) is to use the implication 1. ⇒ 2. in part I of Theorem 4.5.9 for the Popov triplet Σ_c. However, we see that A is not stable. To remedy this, let $\widetilde{F}_2$ be such that $A + B_2 \widetilde{F}_2$ is stable and denote by $\widetilde{\Sigma}_c = (\widetilde{A}, B; \widetilde{Q}_c, \widetilde{L}_c, \widetilde{R}_c)$ the $\widetilde{F} = \begin{bmatrix} O \\ \widetilde{F}_2 \end{bmatrix}$-equivalent Popov triplet of Σ_c, with the entries given by $\widetilde{A} := A + B\widetilde{F} = A + B_2F_2$, $\widetilde{Q}_c := \widetilde{C}_1^T\widetilde{C}_1$, $\widetilde{L}_c := \widetilde{C}_1^T \begin{bmatrix} D_{11} & D_{12} \end{bmatrix}$, where $\widetilde{C}_1 := C_1 + D_{12}\widetilde{F}_2$. We show that the Popov triplet $\widetilde{\Sigma}_c$ satisfies the hypotheses of Theorem 4.5.9.

It is easy to check that the Popov function associated with $\widetilde{\Sigma}_c$ is given by

$$\mathbf{\Pi}_{\widetilde{\Sigma}_c} = \begin{bmatrix} \widetilde{\mathbf{T}}_{11}^*\widetilde{\mathbf{T}}_{11} - I & \widetilde{\mathbf{T}}_{11}^*\widetilde{\mathbf{T}}_{12} \\ \widetilde{\mathbf{T}}_{12}^*\widetilde{\mathbf{T}}_{11} & \widetilde{\mathbf{T}}_{12}^*\widetilde{\mathbf{T}}_{12} \end{bmatrix}, \qquad \begin{bmatrix} \widetilde{\mathbf{T}}_{11} & \widetilde{\mathbf{T}}_{12} \end{bmatrix} := \left[\begin{array}{c|cc} \widetilde{A} & B_1 & B_2 \\ \hline \widetilde{C}_1 & D_{11} & D_{12} \end{array}\right].$$

From the regularity assumption (R1) we have

$$\mathbf{\Pi}_{\widetilde{\Sigma}_c,22} = \widetilde{\mathbf{T}}_{12}^*\widetilde{\mathbf{T}}_{12} > 0 \tag{10.46}$$

on $\overline{\mathbb{C}}_0$ and clearly $\begin{bmatrix} \widetilde{C}_1^T \\ D_{12}^T \end{bmatrix} \begin{bmatrix} \widetilde{C}_1 & D_{12} \end{bmatrix} \geq 0$. Since $\widetilde{A}$ is stable it follows that $\widetilde{\Sigma}_c$ satisfies all the hypotheses of Theorem 4.5.9.

We show now that assertion 1 in part I of this theorem holds for $\widetilde{\Sigma}_c$. The Popov function associated with Σ_c can be readily computed to be

$$\mathbf{\Pi}_{\Sigma_c} = \begin{bmatrix} \mathbf{T}_{11}^*\mathbf{T}_{11} - I & \mathbf{T}_{11}^*\mathbf{T}_{12} \\ \mathbf{T}_{12}^*\mathbf{T}_{11} & \mathbf{T}_{12}^*\mathbf{T}_{12} \end{bmatrix} = \begin{bmatrix} \mathbf{T}_{11}^* \\ \mathbf{T}_{12}^* \end{bmatrix} \begin{bmatrix} \mathbf{T}_{11} & \mathbf{T}_{12} \end{bmatrix} - \begin{bmatrix} I & O \\ O & O \end{bmatrix}. \tag{10.47}$$

If we take $\mathbf{S} := \mathbf{T}_{u_2u_1} = \mathbf{K}(I - \mathbf{T}_{22}\mathbf{K})^{-1}\mathbf{T}_{21}$ (the closed–loop TFM from u_1 to u_2), then clearly $\mathbf{S} \in RH_+^\infty$ (see Remark 10.1.1) and

$$\begin{bmatrix} I & \mathbf{S}^* \end{bmatrix} \mathbf{\Pi}_{\Sigma_c} \begin{bmatrix} I \\ \mathbf{S} \end{bmatrix} = -I + (\mathbf{T}_{11} + \mathbf{T}_{12}\mathbf{S})^*(\mathbf{T}_{11} + \mathbf{T}_{12}\mathbf{S}).$$

But we have $\mathbf{T}_{11} + \mathbf{T}_{12}\mathbf{S} = \mathbf{T}_{11} + \mathbf{T}_{12}\mathbf{T}_{u_2u_1} = \mathbf{T}_{y_1u_1}$. Since we assumed that the controller is contracting, we have $-I + \mathbf{T}_{y_1u_1}^*\mathbf{T}_{y_1u_1} < 0$ on $\overline{\mathbb{C}}_0$, and it follows that

$$\begin{bmatrix} I & \mathbf{S}^* \end{bmatrix} \mathbf{\Pi}_{\Sigma_c} \begin{bmatrix} I \\ \mathbf{S} \end{bmatrix} < 0 \tag{10.48}$$

on $\overline{\mathbb{C}}_0$. According to Proposition 3.2.16 we have

$$\mathbf{\Pi}_{\Sigma_c} = \mathbf{S}_{\widetilde{F}}^*\mathbf{\Pi}_{\widetilde{\Sigma}_c}\mathbf{S}_{\widetilde{F}}, \tag{10.49}$$

where

$$\mathbf{S}_{\widetilde{F}} = \left[\begin{array}{c|c} A & B \\ \hline -\widetilde{F} & I \end{array}\right] = \left[\begin{array}{c|cc} A & B_1 & B_2 \\ \hline O & I & O \\ -\widetilde{F}_2 & O & I \end{array}\right] = \begin{bmatrix} I & O \\ \mathbf{S}_{\widetilde{F},21} & \mathbf{S}_{\widetilde{F},22} \end{bmatrix}.$$

We also have

$$\begin{bmatrix} I & O \\ \mathbf{S}_{\widetilde{F},21} & \mathbf{S}_{\widetilde{F},22} \end{bmatrix} \begin{bmatrix} I \\ \mathbf{S} \end{bmatrix} = \begin{bmatrix} I \\ \widetilde{\mathbf{S}} \end{bmatrix} \tag{10.50}$$

where $\widetilde{\mathbf{S}} := \mathbf{S}_{\widetilde{F},21} + \mathbf{S}_{\widetilde{F},22}\mathbf{S}$. From (10.48) it follows

$$\begin{bmatrix} I & \mathbf{S}^* \end{bmatrix} \mathbf{\Pi}_{\Sigma_c} \begin{bmatrix} I \\ \mathbf{S} \end{bmatrix} = \begin{bmatrix} I & \widetilde{\mathbf{S}}^* \end{bmatrix} \mathbf{\Pi}_{\widetilde{\Sigma}_c} \begin{bmatrix} I \\ \widetilde{\mathbf{S}} \end{bmatrix} < 0 \tag{10.51}$$

on $\mathbb{C}_0$. Thus assertion 1 in part I of Theorem 4.5.9 holds for $\mathbf{\Pi}_{\widetilde{\Sigma}_c}$ provided we show that $\widetilde{\mathbf{S}} \in RH^\infty_{+m_2\times m_1}$. Indeed,

$$\widetilde{\mathbf{S}} \ := \ \mathbf{S}_{\widetilde{F},21} + \mathbf{S}_{\widetilde{F},22}\mathbf{S} = \mathbf{S}_{\widetilde{F},21} + \mathbf{S}_{\widetilde{F},22}\mathbf{K}(I - \mathbf{T}_{22}\mathbf{K})^{-1}\mathbf{T}_{21} = \mathrm{LFT}(\mathbf{T}_S, \mathbf{K})$$

where $\mathbf{T}_S := \begin{bmatrix} \mathbf{S}_{\widetilde{F},21} & \mathbf{S}_{\widetilde{F},22} \\ \mathbf{T}_{21} & \mathbf{T}_{22} \end{bmatrix}$. But $\mathbf{K}$ is a stabilizing controller for $\mathbf{T}$ if and only if it is a stabilizing controller for $\mathbf{T}_S$ from where it follows that $\widetilde{\mathbf{S}} \in RH^\infty_{+m_2\times m_1}$.

Applying Theorem 4.5.9 to $\widetilde{\Sigma}_c$, it follows that the the KPYS$(\widetilde{\Sigma}_c, J_c)$ has a stabilizing solution

$$(X, V_c, \widetilde{W}_c) = (X, \begin{bmatrix} V_{c11} & O \\ V_{c21} & V_{c22} \end{bmatrix}, \begin{bmatrix} \widetilde{W}_{c1} \\ \widetilde{W}_{c2} \end{bmatrix})$$

with $X \geq 0$. Finally, we conclude with Theorem 3.2.13 that the KPYS(Σ_c, J_c) has a stabilizing solution (X, V_c, W_c), with $X \geq 0$, which is exactly (C1).

10.6 An auxiliary necessary condition

In order to prove the necessity of (C3) we need to follow a rather more intricate route in which we prove that the existence of a stabilizing and contracting controller implies the existence of the stabilizing solution to the KPYS$(\Sigma_\times, J_\times)$ (condition (C4) in Theorem 10.3.1). In the next section we show that if (C1) holds then (C4) is equivalent to (C2) together with (C3). Incidentally, we also prove relation (10.14) and the results in this and the next section will be very useful in pursuing with the rest of the proof of Theorem 10.3.1.

With the solution (X, V_c, W_c) of the KPYS(Σ_c, J_c), let us introduce the systems

$$\begin{aligned} \dot{x}_I &= (A - B_2V_{c22}^{-1}W_{c2})x_I + (B_1 - B_2V_{c22}^{-1}V_{c21})u_1 + B_2V_{c22}^{-1}\widetilde{y}_1, \\ y_1 &= (C_1 - D_{12}V_{c22}^{-1}W_{c2})x_I + (D_{11} - D_{12}V_{c22}^{-1}V_{c21})u_1 + D_{12}V_{c22}^{-1}\widetilde{y}_1, \\ \widetilde{u}_1 &= W_{c1}x_I + V_{c11}u_1, \end{aligned} \tag{10.52}$$

and

$$\begin{aligned} \dot{x}_O &= (A + B_1F_1)x_O + B_1V_{c11}^{-1}\widetilde{u}_1 + B_2u_2, \\ \widetilde{y}_1 &= -V_{c22}F_2x_O + V_{c21}V_{c11}^{-1}\widetilde{u}_1 + V_{c22}u_2, \\ y_2 &= (C_2 + D_{21}F_1)x_O + D_{21}V_{c11}^{-1}\widetilde{u}_1, \end{aligned} \tag{10.53}$$

whose genesis will be explained later on (see Remark 10.6.4). We denote by $\mathbf{T}_I$ the transfer function of the system (10.52) and by $\mathbf{T}_O$ the transfer function of the system (10.53), and we have

$$\begin{aligned} \mathbf{T}_I &= \left[\begin{array}{c|c} A_I & B_I \\ \hline C_I & D_I \end{array}\right] = \left[\begin{array}{c|cc} A_I & B_{I1} & B_{I2} \\ \hline C_{I1} & D_{I11} & D_{I12} \\ C_{I2} & D_{I21} & D_{I22} \end{array}\right] \\ &:= \left[\begin{array}{c|cc} A - B_2V_{c22}^{-1}W_{c2} & B_1 - B_2V_{c22}^{-1}V_{c21} & B_2V_{c22}^{-1} \\ \hline C_1 - D_{12}V_{c22}^{-1}W_{c2} & D_{11} - D_{12}V_{c22}^{-1}V_{c21} & D_{12}V_{c22}^{-1} \\ W_{c1} & V_{c11} & O \end{array}\right] \\ \begin{bmatrix} y_1 \\ \widetilde{u}_1 \end{bmatrix} &= \mathbf{T}_I \begin{bmatrix} u_1 \\ \widetilde{y}_1 \end{bmatrix}, \end{aligned} \tag{10.54}$$

$$
\begin{aligned}
\mathbf{T}_O &= \left[\begin{array}{c|c} A_O & B_O \\ \hline C_O & D_O \end{array}\right] = \left[\begin{array}{c|cc} A_O & B_{O1} & B_{O2} \\ \hline C_{O1} & D_{O11} & D_{O12} \\ C_{O2} & D_{O21} & D_{O22} \end{array}\right] \\
&:= \left[\begin{array}{c|cc} A + B_1F_1 & B_1V_{c11}^{-1} & B_2 \\ \hline -V_{c22}F_2 & V_{c21}V_{c11}^{-1} & V_{c22} \\ C_2 + D_{21}F_1 & D_{21}V_{c11}^{-1} & O \end{array}\right] \\
\begin{bmatrix} \widetilde{y}_1 \\ y_2 \end{bmatrix} &= \mathbf{T}_O \begin{bmatrix} \widetilde{u}_1 \\ u_2 \end{bmatrix}.
\end{aligned}
\tag{10.55}
$$

The two systems are connected such that the system with inputs $\begin{bmatrix} u_1 \\ u_2 \end{bmatrix}$ and outputs $\begin{bmatrix} y_1 \\ y_2 \end{bmatrix}$ is the Redheffer product $\mathbf{T}_I \otimes \mathbf{T}_O$ (see Figure 10.2). Computing the realization

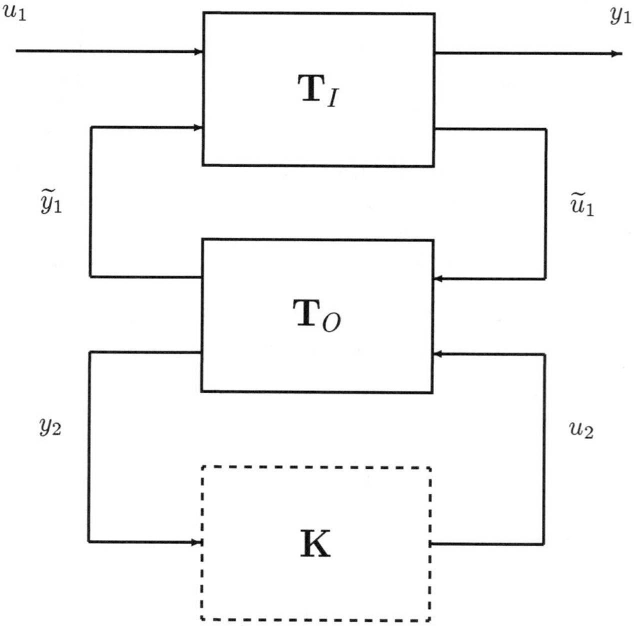

Figure 10.2 The connection of the systems $\mathbf{T}_I$ and $\mathbf{T}_O$

of the Redheffer product $\mathbf{T}_I \otimes \mathbf{T}_O$ (see Section 2.1.5), we obtain

$$
\begin{aligned}
\dot{x}_I &= (A + B_2F_2)x_I - B_2F_2x_O + B_1u_1 + B_2u_2, \\
\dot{x}_O &= -B_1F_2x_I + (A + B_1F_1)x_O + B_1u_1 + B_2u_2, \\
y_1 &= (C_1 + D_{12}F_2)x_I - D_{12}F_2x_O + D_{11}u_1 + D_{12}u_2, \\
y_2 &= (C_2 + D_{21}F_1)x_O - D_{21}F_1x_I + D_{21}u_1,
\end{aligned}
\tag{10.56}
$$

or in the new state coordinates

$$\begin{bmatrix} x_I \\ \xi \end{bmatrix} = \begin{bmatrix} I & O \\ -I & I \end{bmatrix} \begin{bmatrix} x_I \\ x_O \end{bmatrix} = \begin{bmatrix} x_I \\ x_O - x_I \end{bmatrix},$$

the same system becomes

$$\begin{aligned} \dot{x}_I &= Ax_I - B_2F_2\xi + B_1u_1 + B_2u_2, \\ \dot{\xi} &= (A + BF)\xi, \\ y_1 &= C_1x_I - D_{12}F_2\xi + D_{11}u_1 + D_{12}u_2, \\ y_2 &= C_2x_I + (C_2 + D_{21}F_1)\xi + D_{21}u_1. \end{aligned} \tag{10.57}$$

Having the system in this form, it is easy to prove the following result.

Lemma 10.6.1 *The controller* **K** *given by the realization (10.4) is a stabilizing and contracting controller for* **T** *if and only if it is a stabilizing and contracting controller for the system given by the Redheffer product* $\mathbf{T}_I \otimes \mathbf{T}_O$.

Proof. It is easy to check that the state matrix of the closed–loop connection of $\mathbf{T}_I \otimes \mathbf{T}_O$ with **K** (see Figure 10.2), in the case that the realization is written in the state variables

$$\begin{bmatrix} x_I \\ x_K \\ \xi \end{bmatrix},$$

is of the form

$$A'_R = \begin{bmatrix} A_R & \star \\ O & A + BF \end{bmatrix}.$$

Since $A + BF$ is stable, it follows that A'_R is stable if and only if A_R is stable. This means that **K** is stabilizing for **T** if and only if it is stabilizing for $\mathbf{T}_I \otimes \mathbf{T}_O$.

Further, it is easy to check that the system (10.57) with inputs $\begin{bmatrix} u_1 \\ u_2 \end{bmatrix}$ and outputs $\begin{bmatrix} y_1 \\ y_2 \end{bmatrix}$ has the TFM equal to **T**. In fact, ξ represents an uncontrollable stable dynamics that does not affect the TFM and, if $\xi = 0$, then (10.57) becomes identical to the realization of **T**. Thus, **K** is contracting for **T** if and only if it is contracting for $\mathbf{T}_I \otimes \mathbf{T}_O$. ■

Now we are in position to prove the main result of this section. The aim of this section is a mere consequence of this result.

Lemma 10.6.2 *The controller* **K** *given by the realization (10.4) is a stabilizing and contracting controller for* **T** *if and only if it is a stabilizing and contracting controller for* $\mathbf{T}_O$.

Proof. Assume that **K** is stabilizing and contracting for **T**. Then according to Lemma 10.6.1, **K** is stabilizing and contracting for $\mathbf{T}_I \otimes \mathbf{T}_O$. We deduce that the system with input $\widetilde{u}_1$ and output $\widetilde{y}_1$ obtained from the feedback connection of $\mathbf{T}_O$ and **K** and having the TFM equal to LFT($\mathbf{T}_O$, **K**) is stabilizing and contracting for $\mathbf{T}_I$ (see Figure 10.2). We shall show that $\mathbf{T}_I$ is inner and satisfies all other conditions of

Redheffer's theorem 10.4.1 and, according to this result, we deduce that **K** is stabilizing and contracting for $\mathbf{T}_O$.

Conversely, assume that **K** is stabilizing and contracting for $\mathbf{T}_O$. As we shall see in the sequel, $\mathbf{T}_I$ is inner. Then applying again Redheffer's theorem 10.4.1, we deduce that the closed–loop system LFT$(\mathbf{T}_O, \mathbf{K})$ is stabilizing and contracting for $\mathbf{T}_I$. We conclude that **K** is stabilizing and contracting for $\mathbf{T}_I \otimes \mathbf{T}_O$ and according to Lemma 10.6.1, it is stabilizing and contracting for **T**.

It remains to prove that $\mathbf{T}_I$ is inner and satisfies all other conditions of Redheffer's theorem. Indeed, it is a matter of rather lengthy, but straightforward computations to show that the realization (10.52) of $\mathbf{T}_I$ satisfies $D_I^T D_I = I$, and

$$\begin{array}{rcl} C_I^T D_I + XB_I & = & 0, \\ C_I^T C_I + A_I^T X + XA_I & = & 0, \end{array} \tag{10.58}$$

where $X \geq 0$ comes from the solution of the KPYS(Σ_c, J_c). The matrix $D_{I21} = V_{c11}$ is nonsingular and

$$A_I - B_{I1} D_{I21}^{-1} C_{I2} = A - B_2 V_{c22}^{-1} W_{c2} - (B_1 - B_2 V_{c22}^{-1} V_{c21}) V_{c11}^{-1} W_{c1} = A + BF$$

is stable. In particular, $(C_{I2}, A_I) = (W_{c1}, A - B_2 V_{c22}^{-1} W_{c2})$ is detectable and thus (C_I, A_I) is detectable. Since $X \geq 0$, it follows from the last equation in (10.58) and Theorem 1.5.5 that A_I is stable. Thus we checked that $\mathbf{T}_I$ is inner. We can therefore apply Redheffer's theorem to $\mathbf{T}_I$ as claimed and the proof is completed. ■

The necessity of (C4) follows now from the next result.

Corollary 10.6.3 *If the DAP has a solution, then the KPYS$(\Sigma_\times, J_\times)$, where $\Sigma_\times$ and $J_\times$ were introduced in Theorem 10.3.1, has a stabilizing solution of the form*

$$(Z, \begin{bmatrix} V_{\times 11} & O \\ V_{\times 21} & V_{\times 22} \end{bmatrix}, \begin{bmatrix} W_{\times 1} \\ W_{\times 2} \end{bmatrix}), \qquad Z \geq 0.$$

Proof. If there exists a stabilizing and contracting controller for **T**, then there exists a stabilizing and contracting controller for $\mathbf{T}_O$. The corollary states merely the fact that $\mathbf{T}_O$ satisfies (C1) which, as we have shown, is necessary for the existence of the stabilizing and contracting controller. We only need to show that $\mathbf{T}_O$ satisfies (R1) and (R2).

For (R1) notice that $D_{O12} = V_{c22}$ is nonsingular and for each $\omega \in \mathbb{R}$

$$\text{rank} \begin{bmatrix} j\omega I - A - B_1 F_1 & -B_2 \\ -V_{c22} F_2 & V_{c22} \end{bmatrix} = \text{rank} \begin{bmatrix} j\omega I - A - BF & -B_2 \\ O & V_{c22} \end{bmatrix} = n + m_2.$$

For (R2), we have that $D_{21} V_{c11}^{-1}$ has full row rank, since D_{21} has, and

$$\begin{aligned} \text{rank} \begin{bmatrix} j\omega I - A - B_1 F_1 & -B_1 V_{c11}^{-1} \\ C_2 + D_{21} F_1 & D_{21} V_{c11}^{-1} \end{bmatrix} &= \text{rank} \begin{bmatrix} j\omega I - A - B_1 F_1 & -B_1 \\ C_2 + D_{21} F_1 & D_{21} \end{bmatrix} \\ &= \text{rank} \begin{bmatrix} j\omega I - A & -B_1 \\ C_2 & D_{21} \end{bmatrix} = n + p_2 \end{aligned}$$

for each $\omega \in \mathbb{R}$ because of the original assumption. ■

Remark 10.6.4 The systems $\mathbf{T}_I$ and $\mathbf{T}_O$ and their properties have a simple genesis as explained now. Elementary computations lead to (see (3.7))

$$J_{\Sigma_c,e}(u_1,u_2) = -||u_1||_2^2 + ||y_1||_2^2.$$

On the other hand, by using Proposition 3.2.12, one gets

$$\begin{aligned} &J_{\Sigma_c,e}(u_1,u_2)\\ &= \langle V_c u + W_c x, J_c(V_c u + W_c x)\rangle_{L^{2,m}\times L^{2,n}}\\ &= \left\langle \begin{bmatrix} V_{c1}u_1 + W_{c1}x \\ V_{c21}u_1 + V_{c22}u_2 + W_{c2}x \end{bmatrix}, \begin{bmatrix} -I_{m1} & O \\ O & I_{m2} \end{bmatrix} \begin{bmatrix} V_{c1}u_1 + W_{c1}x \\ V_{c21}u_1 + V_{c22}u_2 + W_{c2}x \end{bmatrix} \right\rangle\\ &= -||\widetilde{u}_1||_2^2 + ||\widetilde{y}_1||_2^2 \end{aligned}$$

for

$$\begin{aligned} \widetilde{u}_1 &:= V_{c11}u_1 + W_{c1}x, \\ \widetilde{y}_1 &:= V_{c21}u_1 + V_{c22}u_2 + W_{c2}x. \end{aligned} \tag{10.59}$$

Thus, we have the identity

$$-||u_1||_2^2 + ||y_1||_2^2 = -||\widetilde{u}_1||_2^2 + ||\widetilde{y}_1||_2^2. \tag{10.60}$$

With (10.59) in (10.1) one gets that $\mathbf{T}_I$ is the system with inputs $\begin{bmatrix} u_1 \\ \widetilde{y}_1 \end{bmatrix}$ and outputs $\begin{bmatrix} y_1 \\ \widetilde{u}_1 \end{bmatrix}$, and $\mathbf{T}_O$ is the system with inputs $\begin{bmatrix} \widetilde{u}_1 \\ u_2 \end{bmatrix}$ and outputs $\begin{bmatrix} \widetilde{y}_1 \\ y_2 \end{bmatrix}$. Now the meaning of (10.60) is twofold. Firstly, one can remark that if $-||u_1||_2^2 + ||y_1||_2^2 < 0$ then $-||\widetilde{u}_1||_2^2 + ||\widetilde{y}_1||_2^2 < 0$, which in turn suggests the statement of Lemma 10.6.2. Further, if we read (10.60) as $||u_1||_2^2 + ||\widetilde{y}_1||_2^2 = ||\widetilde{u}_1||_2^2 + ||y_1||_2^2$, this suggests that $\mathbf{T}_I$ is inner.

10.7 Necessity of (C3)

The main result of this section not only establishes the necessity of (C3), but it will also serve later to prove the sufficiency part of Theorem 10.3.1. The main result given below actually states that if (C1) holds, then (C2) together with (C3) are equivalent to (C4).

Lemma 10.7.1 *Assume that (C1) in Theorem 10.3.1 holds. Then the following statements are equivalent.*

1. *The KPYS$(\Sigma_\times, J_\times)$ has a stabilizing solution*

$$\left(Z, \begin{bmatrix} V_{\times 11} & O \\ V_{\times 21} & V_{\times 22} \end{bmatrix}, \begin{bmatrix} W_{\times 1} \\ W_{\times 2} \end{bmatrix}\right), \qquad Z \geq 0.$$

2. *The KPYS(Σ_o, J_o) has a stabilizing solution* $\left(Y, \begin{bmatrix} V_{o11} & O \\ V_{o21} & V_{o22} \end{bmatrix}, \begin{bmatrix} W_{o1} \\ W_{o2} \end{bmatrix}\right)$, *with* $Y \geq 0$, *and* $\rho(XY) < 1$.

In this case the two solutions Y and Z are related through (10.14).

Proof. We prove first that under the given hypotheses the following statements are equivalent:

(S1) The CTARE($\Sigma_\times$) has a stabilizing solution, $Z \geq 0$.

(S2) The CTARE(Σ_o) has a stabilizing solution $Y \geq 0$ and $\rho(XY) < 1$.

The idea of the proof is to relate the EHP($\Sigma_\times$) and the EHP(Σ_o) in order to find a corresponding relation between the stabilizing solutions to CTARE($\Sigma_\times$) and CTARE(Σ_o). According to (3.32) and (10.13), the EHP($\Sigma_\times$), denoted $sM_{\Sigma_\times} - N_{\Sigma_\times}$, is given by

$$M_{\Sigma_\times} := \operatorname{diag}(I_n, I_n, O_{m_2}, O_{p_2}), \tag{10.61}$$

$$N_{\Sigma_\times} := \left[\begin{array}{c|c} N_{11}^\times & N_{12}^\times \\ \hline N_{21}^\times & N_{22}^\times \end{array}\right]$$
$$= \left[\begin{array}{cc|cc} A^T - W_{c1}^T V_{c11}^{-T} B_1^T & O & -W_{c1}^T V_{c11}^{-T} V_{c21}^T + W_{c2}^T & C_2^T - W_{c1}^T V_{c11}^{-T} D_{21}^T \\ -B_1 V_{c11}^{-1} V_{c11}^{-T} B_1^T & -A + B_1 V_{c11}^{-1} W_{c1} & -B_1 V_{c11}^{-1} V_{c11}^{-T} V_{c21}^T & -B_1 V_{c11}^{-1} V_{c11}^{-T} D_{21}^T \\ \hline V_{c21} V_{c11}^{-1} V_{c11}^{-T} B_1^T & -V_{c21} V_{c11}^{-1} W_{c1} + W_{c2} & V_{c21} V_{c11}^{-1} V_{c11}^{T} V_{c21}^T - I_{m_2} & V_{c21} V_{c11}^{-1} V_{c11}^{-T} D_{21}^T \\ D_{21} V_{c11}^{-1} V_{c11}^{-T} B_1^T & C_2 - D_{21} V_{c11}^{-1} W_{c1} & D_{21} V_{c11}^{-1} V_{c11}^{-T} V_{c21}^T & D_{21} V_{c11}^{-1} V_{c11}^{-T} D_{21}^T \end{array}\right] \tag{10.62}$$

and with (3.32) and (10.12) we have that the EHP(Σ_o), denoted $sM_{\Sigma_o} - N_{\Sigma_o}$, is given by

$$M_{\Sigma_o} := \operatorname{diag}(I_n, I_n, O_{p_1}, O_{p_2}), \tag{10.63}$$

$$N_{\Sigma_o} := \left[\begin{array}{c|c} N_{11}^o & N_{12}^o \\ \hline N_{21}^o & N_{22}^o \end{array}\right] = \left[\begin{array}{cc|cc} A^T & O & C_1^T & C_2^T \\ -B_1 B_1^T & -A & -B_1 D_{11}^T & -B_1 D_{21}^T \\ \hline D_{11} B_1^T & C_1 & D_{11} D_{11}^T - I_{p_1} & D_{11} D_{21}^T \\ D_{21} B_1^T & C_2 & D_{21} D_{11}^T & D_{21} D_{21}^T \end{array}\right]. \tag{10.64}$$

Since the above EHPs are not of the same dimension we shall work with their augmented versions $\lambda M_a^\times - N_a^\times$ and $\lambda \widehat{M}_a - \widehat{N}_a$ defined by

$$M_a^\times := \operatorname{diag}(M_{\Sigma_\times}, O_{m_1}, O_{p_1}), \quad N_a^\times := \operatorname{diag}(N_{\Sigma_\times}, -V_{c11}^{-1} V_{c11}^{-T}, -I_{p_1}) \tag{10.65}$$

and

$$\widehat{M}_a := \operatorname{diag}(M_{\Sigma_o}, O_{m_1}, O_{m_2}), \quad \widehat{N}_a := \operatorname{diag}(N_{\Sigma_o}, -I_{m_1}, -I_{m_2}), \tag{10.66}$$

respectively. Thus $M_a^\times$, $N_a^\times$, $\widehat{M}_a$ and $\widehat{N}_a$ are all $(2n+m+p) \times (2n+m+p)$ matrices.

Introduce the following nonsingular matrices

$$U_a = \left[\begin{array}{c|c} U_{a11} & U_{a12} \\ \hline O & U_{a22} \end{array}\right]$$
$$:= \left[\begin{array}{cc|cccc} I_n & X & W_{c2}^T & O & C_1^T D_{11} & -C_1^T \\ O & I_n & -B_1 V_{c21}^T & O & B_1(-I + \Xi) & O \\ \hline O & O & D_{11} V_{c21}^T & O & -D_{11}\Xi & I_{p_1} - D_{11} D_{11}^T \\ O & O & D_{21} V_{c21}^T & I_{p_2} & D_{21}(I - \Xi) & -D_{21} D_{11}^T \\ O & O & -V_{c21}^T & O & \Xi & D_{11}^T \\ O & O & I_{m_2} & O & V_{c21} & O \end{array}\right], \tag{10.67}$$

and

$$W_a = \left[\begin{array}{c|c} W_{a11} & O \\ \hline W_{a21} & W_{a22} \end{array}\right]$$

$$:= \left[\begin{array}{cc|cccc} I_n & -X & O & O & O & O \\ O & I_n & O & O & O & O \\ \hline V_{c21}B_1^T & W_{c2} & V_{c21}D_{11}^T & V_{c21}D_{21}^T & -V_{c21} & I_{m_2} \\ O & O & O & I_{p_2} & O & O \\ (I-\Xi)B_1^T & D_{11}^T C_1 & -\Xi D_{11}^T & (I-\Xi)D_{21}^T & \Xi & V_{c21}^T \\ -D_{11}B_1^T & -C_1 & I_{p_1} - D_{11}D_{11}^T & -D_{11}D_{21}^T & D_{11} & O \end{array}\right] \tag{10.68}$$

where we have denoted $\Xi := V_{c11}^T V_{c11} - V_{c21}^T V_{c21}$. Then by using the KPYS$(\Sigma_c, J_c)$ one can check (the computations are omitted) that for U_a and W_a introduced above we have

$$U_a M_a^\times W_a = \widehat{M}_a, \qquad U_a N_a^\times W_a = \widehat{N}_a. \tag{10.69}$$

The above two equalities show that the augmented EHPs $sM_a^\times - N_a^\times$ and $s\widehat{M}_a - \widehat{N}_a$ are strictly equivalent. The matrices U_a and W_a have been obtained by storing the elementary row and column operations performed on $sM_a^\times - N_a^\times$ in order to bring it to the strict equivalent form $s\widehat{M}_a - \widehat{N}_a$. We are ready now for the proof.

(S1) $\Rightarrow$ (S2). Since the CTARE$(\Sigma_\times)$ has a stabilizing solution $Z \geq 0$, it follows in accordance with Theorem 5.2.1 that

$$N_{\Sigma_\times} V^\times = M_{\Sigma_\times} V^\times S^\times \tag{10.70}$$

with

$$V^\times := \begin{bmatrix} I \\ Z \\ F^\times \end{bmatrix} \tag{10.71}$$

where $F^\times$ is the stabilizing feedback and the $n \times n$ matrix $S^\times$ is stable.

Let

$$V_a^\times := \begin{bmatrix} V^\times \\ O \\ O \end{bmatrix} \begin{matrix} \\ \}m_1 \\ \}p_1. \end{matrix} \tag{10.72}$$

With (10.65), (10.72), and (10.70) we obtain

$$N_a^\times V_a^\times = M_a^\times V_a^\times S^\times. \tag{10.73}$$

Using now (10.69), equality (10.73) yields

$$\widehat{N}_a W_a^{-1} V_a^\times = \widehat{M}_a W_a^{-1} V_a^\times S^\times. \tag{10.74}$$

But, in accordance with the structure of W_a, we have that

$$W_a^{-1} V_a^\times = \begin{bmatrix} I + XZ \\ Z \\ \star \end{bmatrix} \begin{matrix} \\ \\ \}m+p \end{matrix} . \tag{10.75}$$

Since both X and Z are positive semidefinite, $I + XZ$ is nonsingular. Hence, by combining (10.74) and (10.75) one gets

$$\widehat{N}_a \begin{bmatrix} I \\ Y \\ \star \end{bmatrix} = \widehat{M}_a \begin{bmatrix} I \\ Y \\ \star \end{bmatrix} \widehat{S} \tag{10.76}$$

where we have denoted

$$Y := Z(I + XZ)^{-1} \tag{10.77}$$

and

$$\widehat{S} := (I + XZ)S^{\times}(I + XZ)^{-1}, \tag{10.78}$$

with $\widehat{S}$ stable. Now it is easy to see that (10.76) reveals that

$$N_{\Sigma_o}\begin{bmatrix} I \\ Y \\ \star \end{bmatrix} = M_{\Sigma_o}\begin{bmatrix} I \\ Y \\ \star \end{bmatrix}\widehat{S} \tag{10.79}$$

where $\star$ stands now for an irrelevant p–rows block. But (10.79) shows by Theorem 5.2.1 that Y is the stabilizing solution to the CTARE(Σ_o) and in addition $Y \geq 0$, as (10.77) shows. Furthermore, (10.77) yields

$$I - XY = I - XZ(I + XZ)^{-1} = (I + XZ)^{-1}. \tag{10.80}$$

As all the eigenvalues of $I + XZ$ are strictly positive, (10.80) implies

$$\rho(XY) < 1. \tag{10.81}$$

Using (10.80) one gets

$$Z = Y(I - XY)^{-1} \tag{10.82}$$

and the first implication is proved.

(S2) $\Rightarrow$ (S1). This part follows by simply reversing the above line of reasoning. Thus, we have (S1) $\Leftrightarrow$ (S2).

To complete the proof of the theorem we have to switch from the CTAREs to the corresponding KPYSs. Using Remark 3.2.9, all it remains to prove is that the first equation of the KPYS($\Sigma_\times, J_\times$), which is $R_\times = V_\times^T J_\times V_\times$, implies and is implied by the first equation of the KPYS(Σ_o, J_o) in (C2), which is $R_o = V_o^T J_o V_o$. Equivalently, we have to prove

$$\operatorname{sgn} R_\times = J_\times \qquad \Leftrightarrow \qquad \operatorname{sgn} R_o = J_o. \tag{10.83}$$

Indeed, we notice from (10.67) and (10.68) that $W_{a22}^T = U_{a22}$. Then, (10.69) shows that

$$U_{a22}\operatorname{diag}(N_{22}^{\times}, -V_{c11}^{-1}V_{c11}^{-T}, -I_{p_1})U_{a22}^T = \operatorname{diag}(N_{22}^{o}, -I_{m_1}, -I_{m_2}),$$

from where (10.83) follows by recalling the notation $N_{22}^{\times} := R_\times$ and $N_{22}^{o} := R_o$ made in (10.62) and (10.64), respectively. ■

Now (C3) follows from Corollary 10.6.3 and the implication 1. $\Rightarrow$ 2. of Lemma 10.7.1.

10.8 Proof of sufficiency

The proof of sufficiency of (C1), (C2), and (C3) in Theorem 10.3.1 is constructive. Namely, we show that if these conditions are fulfilled we can construct, in terms of explicit formulas, a stabilizing and contracting controller. In fact, we give the class of all those controllers, completing in this way the proof of Theorem 10.3.1.

To this end we shall successively solve the DAP for some simpler cases, and in the end we reduce the general problem to these particular ones.

10.8.1 The disturbance estimation (DE) case

The plant (10.1) in the DE case satisfies the following assumptions:

(DE1) D_{12} is square and nonsingular.

(DE2) $A - B_2 D_{12}^{-1} C_1$ is stable.

(DE3) D_{21} is square and nonsingular.

(DE4) $A - B_1 D_{21}^{-1} C_2$ is stable.

Theorem 10.8.1 *The class of all stabilizing and contracting controllers for the plant (10.1) satisfying the DE assumptions is given by*

$$\mathbf{K} = \mathrm{LFT}(\mathbf{K}_g, \mathbf{Q}) \tag{10.84}$$

where $\mathbf{Q} \in RH^{\infty}_{+,p_1 \times m_1}$ *is an arbitrary parameter with* $\|\mathbf{Q}\|_{\infty} < 1$ *and*

$$\mathbf{K}_g = \left[\begin{array}{c|cc} A_g & B_{g1} & B_{g2} \\ \hline C_{g1} & D_{g11} & D_{g12} \\ C_{g2} & D_{g21} & D_{g22} \end{array}\right], \tag{10.85}$$

with

$$\begin{array}{rcl} A_g & = & A - B_1 D_{21}^{-1} C_2 - B_2 D_{12}^{-1} C_1 + B_2 D_{12}^{-1} D_{11} D_{21}^{-1} C_2, \\ B_{g1} & = & (B_1 - B_2 D_{12}^{-1} D_{11}) D_{21}^{-1}, \\ B_{g2} & = & B_2 D_{12}^{-1}, \\ C_{g1} & = & -D_{12}^{-1}(C_1 - D_{11} D_{21}^{-1} C_2), \\ C_{g2} & = & -D_{21}^{-1} C_2, \\ D_{g11} & = & -D_{12}^{-1} D_{11} D_{21}^{-1}, \\ D_{g12} & = & D_{12}^{-1}, \\ D_{g21} & = & D_{21}^{-1}, \\ D_{g22} & = & 0. \end{array} \tag{10.86}$$

Proof. We prove first that (10.84) with (10.85) and (10.86) give a solution to the DAP. Indeed, if $\mathbf{K}_g$ is coupled to $\mathbf{T}$ such that $\begin{bmatrix} u_2 \\ u_3 \end{bmatrix} = \mathbf{K}_g \begin{bmatrix} y_2 \\ y_3 \end{bmatrix}$ then the resultant closed–loop system

$$\begin{bmatrix} y_1 \\ u_3 \end{bmatrix} = \mathbf{T}_R \begin{bmatrix} u_1 \\ y_3 \end{bmatrix} \tag{10.87}$$

is given by the Redheffer product

$$\mathbf{T}_R = \mathbf{T} \otimes \mathbf{K}_g$$

$$= \left[\begin{array}{cc|cc} \widehat{A} & -B_2 D_{12}^{-1}(C_1 - D_{11} D_{21}^{-1} C_2) & B_1 - B_2 D_{12}^{-1} D_{11} & B_2 D_{12}^{-1} \\ (B_1 - B_2 D_{12}^{-1} D_{11}) D_{21}^{-1} C_2 & \widehat{A} - B_1 D_{21}^{-1} C_2 - B_2 D_{12}^{-1} C_1 & B_1 - B_2 D_{12}^{-1} D_{11} & B_2 D_{12}^{-1} \\ \hline C_1 - D_{11} D_{21}^{-1} C_2 & -(C_1 - D_{11} D_{21}^{-1} C_2) & O & I_{p_1} \\ D_{21}^{-1} C_2 & -D_{21}^{-1} C_2 & I_{p_2} & O \end{array}\right] \tag{10.88}$$

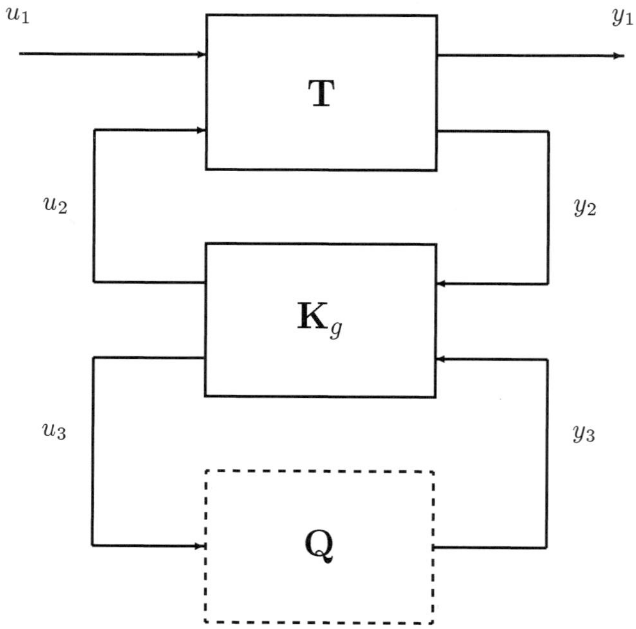

Figure 10.3 The connection of the systems $\mathbf{T}$ and $\mathbf{K}_g$

(see Figure 10.3) which becomes, after a similarity transformation,

$$\begin{aligned}\mathbf{T}_R &= \left[\begin{array}{cc|cc} A - B_2D_{12}^{-1}C_1 & -B_2D_{12}^{-1}(C_1 - D_{11}D_{21}^{-1}C_2) & B_1 - B_2D_{12}^{-1}D_{11} & B_2D_{12}^{-1} \\ O & A - B_1D_{21}^{-1}C_2 & O & O \\ \hline O & -(C_1 - D_{11}D_{21}^{-1}C_2) & O & I_{p_1} \\ O & -D_{21}^{-1}C_2 & I_{p_2} & O \end{array}\right] \\ &= \begin{bmatrix} O & I_{p_1} \\ I_{p_2} & O \end{bmatrix} \end{aligned} \tag{10.89}$$

which clearly is internally stable due to the assumption (DE2) and (DE4), and where we have denoted $\widehat{A} := A - B_2D_{12}^{-1}D_{11}D_{21}^{-1}C_2$. By coupling $y_3 = \mathbf{Q}u_3$ to (10.87) (see Figure 10.3) we obtain $y_1 = y_3, u_3 = u_1, y_3 = \mathbf{Q}u_3$ or, eventually,

$$y_1 = \mathbf{Q}u_1 \tag{10.90}$$

where the *internal* dynamics of (10.90) consists of the internal dynamics of (10.89) and of the stable system $\mathbf{Q}$. Hence the resultant system $\mathbf{T}_{y_1u_1}$ (see (10.7)) is stable and the attenuation property (10.9) holds since $||\mathbf{Q}||_\infty < 1$. So we proved that (10.84) is stabilizing and contracting.

Conversely, assume now that $\mathbf{K}$ is stabilizing and contracting. That means $\mathbf{T}_{y_1u_1} \in RH^\infty_{+,p_1\times m_1}$ and $||\mathbf{T}_{y_1u_1}||_\infty < 1$, and consequently $\mathbf{T}_{y_1u_1}$ may play the role of the parameter $\mathbf{Q}$. Set $\mathbf{Q} := \mathbf{T}_{y_1u_1}$, and $y_3 = y_1, u_3 = u_1$, and evaluate $\text{LFT}(\mathbf{K}_g, \mathbf{Q}) = \text{LFT}(\mathbf{K}_g, \text{LFT}(\mathbf{T}, \mathbf{K})) = \text{LFT}(\widehat{\mathbf{T}}_R, \mathbf{K})$ where $\widehat{\mathbf{T}}_R$ is the Redheffer product $\widehat{\mathbf{T}}_R :=$

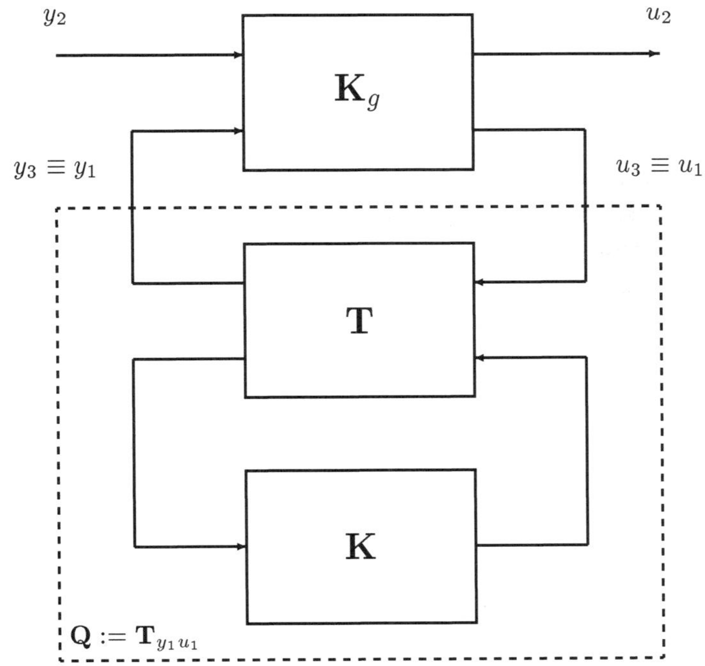

Figure 10.4 The connection of the systems $\mathbf{K}_g$ and $\mathbf{Q} := \mathbf{T}_{y_1 u_1}$

$\mathbf{K}_g \otimes \mathbf{T}$ as it is shown in Figure 10.4. Using (10.85), (10.86), and (10.2) one gets

$$\widehat{\mathbf{T}}_R = \left[\begin{array}{cc|cc} A & -B_1 D_{21}^{-1} C_2 & B_1 D_{21}^{-1} & B_2 \\ B_2 D_{12}^{-1} C_1 & A - B_1 D_{21}^{-1} C_2 - B_2 D_{12}^{-1} C_1 & B_1 D_{21}^{-1} & B_2 \\ \hline D_{12}^{-1} C_1 & -D_{12}^{-1} C_1 & O & I_{p_1} \\ C_2 & -C_2 & I_{p_2} & O \end{array}\right]. \tag{10.91}$$

Performing a similarity transformation on (10.91), we get

$$\widehat{\mathbf{T}}_R = \left[\begin{array}{cc|cc} A - B_1 D_{21}^{-1} C_2 & -B_1 D_{21}^{-1} C_2 & B_1 D_{21}^{-1} & B_2 \\ O & A - B_2 D_{12}^{-1} C_1 & O & O \\ \hline O & -D_{12}^{-1} C_1 & O & I_{p_1} \\ O & -C_2 & I_{p_2} & O \end{array}\right] = \left[\begin{array}{cc} O & I_{p_1} \\ I_{p_2} & O \end{array}\right]. \tag{10.92}$$

With (10.92) we have

$$\mathrm{LFT}(\widehat{\mathbf{T}}_R, \mathbf{K}) = \mathbf{K}, u_2 = \mathbf{K} y_2.$$

Hence we conclude that for any solution $\mathbf{K}$ there exists $\mathbf{Q} \in RH^{\infty}_{+,p_1 \times m_1}$ with $\|\mathbf{Q}\|_{\infty} < 1$ such that $\mathrm{LFT}(\mathbf{K}_g, \mathbf{Q}) = \mathbf{K}$. Thus the theorem is completely proved. ■

10.8.2 The disturbance feedforward (DF) case

The plant (10.1) in the DF case satisfies the following assumptions:

(DF1) D_{21} is square and nonsingular.

(DF2) $A - B_1 D_{21}^{-1} C_2$ is stable.

(DF3) Condition (C1) in Theorem 10.3.1 holds.

Theorem 10.8.2 *The class of all stabilizing and contracting controllers for the plant (10.1) satisfying the DF assumptions is given by*

$$\mathbf{K} = \mathrm{LFT}(\mathbf{K}_g, \mathbf{Q}), \tag{10.93}$$

where $\mathbf{Q} \in RH^{\infty}_{+,m_2 \times m_1}$ *is an arbitrary parameter with* $\|\mathbf{Q}\|_\infty < 1$ *and* $\mathbf{K}_g$ *has the realization*

$$\begin{aligned}
A_g &= A - B_1 D_{21}^{-1} C_2 - B_2 (D_{12}^T D_{12})^{-1} [D_{12}^T (C_1 - D_{11} D_{21}^{-1} C_2) + B_2^T X], \\
B_{g1} &= (B_1 - B_2 (D_{12}^T D_{12})^{-1} D_{12}^T D_{11}) D_{21}^{-1}, \\
B_{g2} &= B_2 (D_{12}^T D_{12})^{-1/2}, \\
C_{g1} &= -(D_{12}^T D_{12})^{-1} [D_{12}^T (C_1 - D_{11} D_{21}^{-1} C_2) + B_2^T X], \\
C_{g2} &= -S_c^{-1/2} [D_{21}^{-1} C_2 + S_c D_{11}^T (I - D_{12} (D_{12}^T D_{12})^{-1} D_{12}^T) C_1 \\
&\quad + S_c (B_1^T X - D_{11}^T D_{12} (D_{12}^T D_{12})^{-1} B_2^T X)], \\
D_{g11} &= -(D_{12}^T D_{12})^{-1} D_{12}^T D_{11} D_{21}^{-1} \\
D_{g12} &= (D_{12}^T D_{12})^{-1/2}, \\
D_{g21} &= S_c^{-1/2} D_{21}^{-1}, \\
D_{g22} &= 0,
\end{aligned} \tag{10.94}$$

and

$$S_c := (V_{c11}^T V_{c11})^{-1} = \{I - D_{11}^T [I - D_{12} (D_{12}^T D_{12})^{-1} D_{12}^T] D_{11}\}^{-1}. \tag{10.95}$$

Proof. Assumption (DF3) allows us to consider again the systems $\mathbf{T}_I$ (10.52) and $\mathbf{T}_O$ (10.53) that we introduced in Section 10.6. According to Lemma 10.6.2, it is enough to find all the stabilizing and contracting controllers for $\mathbf{T}_O$. We show that

$$\mathbf{T}_O = \left[\begin{array}{c|cc} A_O & B_{O1} & B_{O2} \\ \hline C_{O1} & D_{O11} & D_{O12} \\ C_{O2} & D_{O21} & D_{O22} \end{array}\right] = \left[\begin{array}{c|cc} A + B_1 F_1 & B_1 V_{c11}^{-1} & B_2 \\ \hline -V_{c22} F_2 & V_{c21} V_{c11}^{-1} & V_{c22} \\ C_2 + D_{21} F_1 & D_{21} V_{c11}^{-1} & O \end{array}\right]$$

satisfies the assumptions in the DE case and we apply Theorem 10.8.1 to obtain the parameterization of all the stabilizing and contracting controllers.

Since $D_{O12} = V_{c22}$, (DE1) is trivial. Also $D_{O21} = D_{21} V_{c11}^{-1}$ is invertible by (DF1). For (DE2) we compute

$$A_O - B_{O2} D_{O12}^{-1} C_{O1} = A + B_1 F_1 - B_2 V_{c22}^{-1} (-V_{c22} F_2) = A + BF$$

which is stable. For (DE4), we have

$$A_O - B_{O1} D_{O21}^{-1} C_{O2} = A + B_1 F_1 - B_1 V_{c11}^{-1} V_{c11} D_{21}^{-1} (C_2 + D_{21} F_1) = A - B_1 D_{21}^{-1} C_2$$

which is stable because of (DF2).

The rest of the proof follows from Theorem 10.8.1 and some simple manipulations using the KPYS(Σ_c, J_c). ■

If the normalizing conditions (10.16) hold then the formulas for the parameterization of all the stabilizing and contracting controllers in the DF case are given by

$$\mathbf{K}_g = \left[\begin{array}{c|cc} A - B_1D_{21}^{-1}C_2 - B_2B_2^TX & B_1D_{21}^{-1} & B_2 \\ \hline -B_2^TX & O & I_{m_2} \\ -(D_{21}^{-1}C_2 + B_1^TX) & D_{21}^{-1} & O \end{array}\right]. \tag{10.96}$$

Thus the central controller is

$$\mathbf{K} = \left[\begin{array}{c|c} A - B_1D_{21}^{-1}C_2 - B_2B_2^TX & B_1D_{21}^{-1} \\ \hline -B_2^TX & O \end{array}\right]. \tag{10.97}$$

10.8.3 The output estimation (OE) case

The plant (10.1) in the OE case satisfies the following assumptions:

(OE1) D_{12} is square and nonsingular.

(OE2) $A - B_2D_{12}^{-1}C_1$ is stable.

(OE3) Condition (C2) in Theorem 10.3.1 holds.

Theorem 10.8.3 *The class of all stabilizing and contracting controllers for the plant (10.1) satisfying the OE assumptions is given by*

$$\mathbf{K} = \mathrm{LFT}(\mathbf{K}_g, \mathbf{Q}), \tag{10.98}$$

where $\mathbf{Q} \in RH^\infty_{+,p_1\times p_2}$ *is an arbitrary parameter with* $\|\mathbf{Q}\|_\infty < 1$ *and* $\mathbf{K}_g$ *has the realization*

$$\begin{aligned}
A_g &= A - B_2D_{12}^{-1}C_1 - [(B_1 - B_2D_{12}^{-1}D_{11})D_{21}^T + YC_2^T](D_{21}D_{21}^T)^{-1}C_2, \\
B_{g1} &= -[(B_1 - B_2D_{12}^{-1}D_{11})D_{21}^T + YC_2^T](D_{21}D_{21}^T)^{-1}, \\
B_{g2} &= -[B_2D_{12}^{-1} + B_1(I - D_{21}^T(D_{21}D_{21}^T)^{-1}D_{21})D_{11}^TS_o \\
&\quad + (YC_1^T - YC_2^T(D_{21}D_{21}^T)^{-1}D_{21}D_{11}^T)S_o]S_o^{-1/2} \\
C_{g1} &= D_{12}^{-1}(C_1 - D_{11}D_{21}^T(D_{21}D_{21}^T)^{-1}C_2), \\
C_{g2} &= (D_{21}D_{21}^T)^{-1/2}C_2, \\
D_{g11} &= -D_{12}^{-1}D_{11}D_{21}^T(D_{21}D_{21}^T)^{-1}, \\
D_{g12} &= D_{12}^{-1}S_o^{-1/2}, \\
D_{g21} &= (D_{21}D_{21}^T)^{-1/2}, \\
D_{g22} &= 0,
\end{aligned} \tag{10.99}$$

and

$$S_o := (V_{o11}^TV_{o11})^{\frac{1}{2}} = \{I - D_{11}[I - D_{21}^T(D_{21}D_{21}^T)^{-1}D_{21}]D_{11}^T\}^{-1}. \tag{10.100}$$

This theorem is the dual of Theorem 10.8.2.

If the normalizing conditions (10.16) hold then (10.99) and (10.100) give

$$\mathbf{K}_g = \left[\begin{array}{c|cc} A - B_2D_{12}^{-1}C_1 - YC_2^TC_2 & -YC_2^T & -(B_2D_{12}^{-1} + YC_1^T) \\ \hline D_{12}^{-1}C_1 & O & D_{12}^{-1} \\ (D_{21}^TD_{21})^{-1}C_2 & I_{p_2} & O \end{array}\right], \tag{10.101}$$

whereas the central controller is

$$\mathbf{K} = \left[\begin{array}{c|c} A - B_2D_{12}^{-1}C_1 - YC_2^TC_2 & -YC_2^T \\ \hline D_{12}^{-1}C_1 & O \end{array}\right]. \tag{10.102}$$

10.8.4 The general case

We turn now to the remains of the proof of Theorem 10.3.1. We assume that (C1), (C2), and (C3) hold and we find the class of all stabilizing and contracting controllers.

In fact, we repeat the scheme in the proof of Theorem 10.8.2. We consider the systems $\mathbf{T}_I$ (10.52) and $\mathbf{T}_O$ (10.53) that we introduced in Section 10.6. According to Lemma 10.6.2, it is enough to find all the stabilizing and contracting controllers for $\mathbf{T}_O$. We show that

$$\mathbf{T}_O = \left[\begin{array}{c|cc} A + B_1F_1 & B_1V_{c11}^{-1} & B_2 \\ \hline -V_{c22}F_2 & V_{c21}V_{c11}^{-1} & V_{c22} \\ C_2 + D_{21}F_1 & D_{21}V_{c11}^{-1} & O \end{array}\right]$$

satisfies the assumptions in the OE case and we apply Theorem 10.8.3 to obtain the parameterization of all the stabilizing and contracting controllers.

Assumptions (OE1) and (OE2) are obvious as they are the same as in the proof of Theorem 10.8.2. Condition (OE3) with respect to $\mathbf{T}_O$ requires the existence of a stabilizing solution for the KPYS($\Sigma_\times, J_\times$). As we know from Lemma 10.7.1, this is true if and only if (C2) and (C3) are satisfied. Thus the sufficiency part of part I and II of Theorem 10.3.1 is completed.

The parameterization of all the stabilizing and contracting controllers in Theorem 10.3.1 is now a consequence of Theorem 10.8.3 and some manipulations using KPYSs.

Remark 10.8.4 Invoking Lemma 10.6.2 we get that (DF3) and (OE3) are also necessary solvability conditions for the respective particular DAPs.

10.9 State feedback solution

When the state is accessible for measurement the solution to the γ–DAP can be expressed in simple state–space feedback form. In this case the system (10.1) reduces to

$$\begin{aligned} \dot{x} &= Ax + B_1u_1 + B_2u_2, \\ y_1 &= C_1x + D_{11}u_1 + D_{12}u_2, \end{aligned} \tag{10.103}$$

or, equivalently, (10.2) reduces to

$$\mathbf{T} = \left[\begin{array}{c|c} A & B \\ \hline C_1 & D \end{array}\right] = \left[\begin{array}{c|cc} A & B_1 & B_2 \\ \hline C_1 & D_{11} & D_{12} \end{array}\right] = \left[\begin{array}{cc} \mathbf{T}_{11} & \mathbf{T}_{12} \end{array}\right]. \tag{10.104}$$

The γ–DAP consists, in this case, in finding a feedback $F_2 \in \mathbb{R}^{m_2 \times n}$ such that the state–space feedback

$$u_2 = F_2 x \tag{10.105}$$

stabilizes the system (10.103), which means that the resulting closed–loop system

$$\begin{array}{rcl} \dot{x} & = & (A + B_2F_2)x + B_1u_1, \\ y_1 & = & (C_1 + D_{12}F_2)x + D_{11}u_1, \end{array} \tag{10.106}$$

is stable, i.e.,

$$\Lambda(A + B_2F_2) \subset \mathbb{C}^- \tag{10.107}$$

and its TFM satisfies

$$||\mathbf{T}_{y_1u_1}||_\infty < \gamma.$$

As before, we can reduce the problem to the case $\gamma = 1$, and we shall call a stabilizing feedback that solves the γ-DAP, a stabilizing and contracting state feedback.

We assume further that the regularity assumption (R1) holds. We have the following theorem.

Theorem 10.9.1 *There exists a stabilizing and contracting state feedback if and only if*

$$D_{11}^T D_{11} < I_{m_1} \tag{10.108}$$

and (C1) in Theorem 10.3.1 is true.

Moreover, in this case, if

$$F = \begin{bmatrix} F_1 \\ F_2 \end{bmatrix}$$

is the stabilizing feedback of the KPYS(Σ_c, J_c), then F_2 is a stabilizing and contracting state feedback.

Proof. "Sufficiency". We assume that conditions (10.108) and (C1) hold. The idea is to prove that F_2 is a stabilizing and contracting state feedback using Theorem 4.5.10. We see that

$$R_{c11} = -I + D_{11}^T D_{11} < 0$$

and $R_{c22} = D_{12}^T D_{12} > 0$. The stabilizing solution to the Riccati associated with Σ_c exists and is positive semidefinite according to (C1). Then by using Theorem 4.5.10 we deduce that $A + B_2F_2$ is stabilizing and there exists $\rho > 0$ such that

$$J_{e,\Sigma_c}(u_1, u_2) \leq -\rho^2 ||u_1||_2^2, \quad \forall u_1 \in L^{2,m_1}, \tag{10.109}$$

for $u_2 = F_2x$. Using (10.11) we can easily deduce that

$$J_{e,\Sigma_c}(u_1, u_2) = -||u_1||_2^2 + ||y_1||_2^2. \tag{10.110}$$

By combining (10.109) and (10.110),

$$||y_1||_2^2 \leq (1 - \rho^2)||u||_1^2, \quad \forall u_1 \in L^{2,m_1},$$

which in fact is equivalent to the contracting condition.

"Necessity". Assume that a stabilizing and contracting state feedback exists. Then $\mathbf{T}_{y_1u_1}(\infty) = D_{11}$ and therefore $\|D_{11}\| < 1$. Condition (10.108) follows.

The rest of the proof is quite similar to the argument used in Section 10.5 to prove (C1). Indeed, we assume for the moment that A is stable and we prove that Σ_c satisfies the hypotheses of Theorem 4.5.9. The Popov function associated with Σ_c is

$$\mathbf{\Pi}_{\Sigma_c} = \begin{bmatrix} \mathbf{T}_{11}^*\mathbf{T}_{11} - I & \mathbf{T}_{11}^*\mathbf{T}_{12} \\ \mathbf{T}_{12}^*\mathbf{T}_{11} & \mathbf{T}_{12}^*\mathbf{T}_{12} \end{bmatrix} = \begin{bmatrix} \mathbf{T}_{11}^* \\ \mathbf{T}_{12}^* \end{bmatrix} \begin{bmatrix} \mathbf{T}_{11} & \mathbf{T}_{12} \end{bmatrix} - \begin{bmatrix} I & O \\ O & O \end{bmatrix}. \tag{10.111}$$

From the regularity assumption (R1) we have

$$\mathbf{\Pi}_{\Sigma_c,22} = \mathbf{T}_{12}^*\mathbf{T}_{12} > 0 \tag{10.112}$$

on $\overline{\mathbb{C}}_0$, and clearly

$$\begin{bmatrix} C_1^T \\ D_{12}^T \end{bmatrix} \begin{bmatrix} C_1 & D_{12} \end{bmatrix} \geq 0. \tag{10.113}$$

Since A was assumed stable, it follows that the hypotheses of Theorem Theorem 4.5.9 are fulfilled.

We show further that assertion 1 in part I of this theorem is satisfied. If we define $\mathbf{S} := \mathbf{T}_{u_2u_1}$, the closed–loop TFM from u_1 to u_2, then

$$\mathbf{S}(s) := F_2(sI - A - B_2F_2)^{-1}B_1. \tag{10.114}$$

Clearly $\mathbf{S} \in RH^\infty_{+,m_2\times n}$ because F_2 is stabilizing. Since

$$y_1 = \mathbf{T}_{11}u_1 + \mathbf{T}_{12}u_2, \tag{10.115}$$

it is clear that

$$\mathbf{T}_{y_1u_1} = \mathbf{T}_{11} + \mathbf{T}_{12}\mathbf{S}. \tag{10.116}$$

On the other hand,

$$\begin{bmatrix} I & \mathbf{S}^* \end{bmatrix} \mathbf{\Pi}_{\Sigma_c} \begin{bmatrix} I \\ \mathbf{S} \end{bmatrix} = -I + (\mathbf{T}_{11} + \mathbf{T}_{12}\mathbf{S})^*(\mathbf{T}_{11} + \mathbf{T}_{12}\mathbf{S}) = -I + \mathbf{T}_{y_1u_1}^*\mathbf{T}_{y_1u_1} < 0$$

on $\overline{\mathbb{C}}_0$. Thus assertion 1 in part I of Theorem Theorem 4.5.9 is checked and (C1) follows from part I of the same theorem.

In the case that A is not stable, we take $\widetilde{F}_2$ such that $A + B_2\widetilde{F}_2$ is stable. Further, we can repeat the above reasoning for $\widetilde{\Sigma}_c$ which is the $\begin{bmatrix} O \\ \widetilde{F}_2 \end{bmatrix}$–equivalent Popov triplet of Σ_c. Finally, we return to Σ_c by using the Proposition below that can be proved by straightforward substitutions, and this concludes the proof. ■

Proposition 10.9.2 *The closed–loop system obtained by applying $u_2 = F_2x$ to the system*

$$\mathbf{T} = \left[\begin{array}{c|cc} A & B_1 & B_2 \\ \hline C_1 & D_{11} & D_{12} \end{array}\right]$$

coincides with the closed–loop system obtained by applying $u_2 = (F_2 - \widetilde{F}_2)x$ to the system

$$\widetilde{\mathbf{T}} = \left[\begin{array}{c|cc} A + B_2\widetilde{F}_2 & B_1 & B_2 \\ \hline C_1 + D_{12}\widetilde{F}_2 & D_{11} & D_{12} \end{array}\right].$$

B. The discrete–time case

10.10 Problem formulation

The framework in the discrete–time case is formally very similar to the continuous–time case, and we recast it only for convenience.

Consider a plant that is represented as the discrete–time linear system

$$\begin{aligned} \sigma x &= Ax + B_1 u_1 + B_2 u_2, \\ y_1 &= C_1 x + D_{11} u_1 + D_{12} u_2, \\ y_2 &= C_2 x + D_{21} u_1 + D_{22} u_2, \end{aligned} \tag{10.117}$$

having the TFM given by

$$\mathbf{T} = \left[\begin{array}{c|c} A & B \\ \hline C & D \end{array}\right] = \left[\begin{array}{c|cc} A & B_1 & B_2 \\ \hline C_1 & D_{11} & D_{12} \\ C_2 & D_{21} & D_{22} \end{array}\right] = \begin{bmatrix} \mathbf{T}_{11} & \mathbf{T}_{12} \\ \mathbf{T}_{21} & \mathbf{T}_{22} \end{bmatrix}, \tag{10.118}$$

where $A \in \mathbb{R}^{n\times n}$, $B_i \in \mathbb{R}^{n\times m_i}$, $C_j \in \mathbb{R}^{p_j\times n}$, $D_{ij} \in \mathbb{R}^{p_i\times m_j}$, $(i, j = 1, 2,\ m = m_1 + m_2,\ p = p_1 + p_2)$, u and y denote the input and output of $\mathbf{T}$, respectively, and are partitioned conformably with $\mathbf{T}$ such that

$$y = \begin{bmatrix} y_1 \\ y_2 \end{bmatrix} = \mathbf{T}u = \begin{bmatrix} \mathbf{T}_{11} & \mathbf{T}_{12} \\ \mathbf{T}_{21} & \mathbf{T}_{22} \end{bmatrix} \begin{bmatrix} u_1 \\ u_2 \end{bmatrix}. \tag{10.119}$$

The vectors u_1, y_1, u_2, and y_2 have the same significance as in the continuous–time case.

We define a controller for the plant (10.117) as a linear discrete–time system

$$\begin{aligned} \sigma x_K &= A_K x_K + B_K u_K, \\ y_K &= C_K x_K + D_K u_K, \end{aligned} \tag{10.120}$$

with the TFM given by

$$\mathbf{K} = \left[\begin{array}{c|c} A_K & B_K \\ \hline C_K & D_K \end{array}\right], \tag{10.121}$$

where $A_K \in \mathbb{R}^{n_K\times n_K}$, $B_K \in \mathbb{R}^{n_K\times p_2}$, $C_K \in \mathbb{R}^{m_2\times n_K}$, $D_K \in \mathbb{R}^{m_2\times p_2}$, and u_K and y_K are the input and output of (10.120), respectively.

The closed–loop system that is obtained by connecting the controller (10.120) to the plant (10.117) (see Figure 10.1) such that $u_2 \equiv y_K$ and $u_K \equiv y_2$ is well–posed provided the matrix

$$\begin{bmatrix} I_{p_2} & D_{22} \\ D_K & I_{m_2} \end{bmatrix}$$

is nonsingular. Well–posedness can be equivalently expressed by the condition that $\widetilde{S} := I_{m_2} - D_K D_{22}$ is nonsingular, or that $S = I_{p_2} - D_{22} D_K$ is nonsingular.

If the closed–loop system is well–posed, then the resulting system with input u_1 and output y_1 is given by

$$\begin{aligned}
\mathbf{T}_{y_1u_1} &= \mathrm{LFT}(\mathbf{T},\mathbf{K}) = \mathbf{T}_{11} + \mathbf{T}_{12}\mathbf{K}(I - \mathbf{T}_{22}\mathbf{K})^{-1}\mathbf{T}_{21}, \\
\mathbf{T}_{y_1u_1} &= \left[\begin{array}{c|c} A_R & B_R \\ \hline C_R & D_R \end{array}\right] \\
&= \left[\begin{array}{cc|c} A + B_2\widetilde{S}^{-1}D_KC_2 & B_2\widetilde{S}^{-1}C_K & B_1 + B_2\widetilde{S}^{-1}D_KD_{21} \\ B_KS^{-1}C_2 & A_K + B_KS^{-1}D_{22}C_K & B_KS^{-1}D_{21} \\ \hline C_1 + D_{12}D_KS^{-1}C_2 & D_{12}\widetilde{S}^{-1}C_K & D_{11} + D_{12}D_KS^{-1}D_{21} \end{array}\right].
\end{aligned} \tag{10.122}$$

With these preliminaries we can state in precise terms the suboptimal H^∞–control problem (or the γ disturbance attenuation problem) for discrete–time systems.

Given a discrete–time system $\mathbf{T}$ and a threshold $\gamma > 0$, the γ–suboptimal H^∞–control problem (or the γ–DAP) consists in finding all discrete–time controllers $\mathbf{K}$ for which the closed–loop system $\mathbf{T}_{y_1u_1}$ is well–posed ($I - D_{22}D_K$ is invertible), internally stable, i.e.,

$$\Lambda(A_R) \subset \mathbb{D}, \tag{10.123}$$

and bounded in H^∞–norm by γ, i.e.,

$$\|\mathbf{T}_{y_1u_1}\|_\infty < \gamma. \tag{10.124}$$

A controller that satisfies (10.123) will be called *stabilizing*, and a controller that satisfies (10.124) will be called *γ–contracting*. If $\gamma = 1$, we shall simply talk about a contracting controller, and the corresponding γ–DAP will be simply termed as DAP.

10.11 Basic assumptions

As in the continuous–time case, we shall make some basic assumptions.

(HD1) $\gamma = 1$

(HD2) $D_{22} = 0$

The arguments that allow these simplifications are identical to those in the continuous–time case and were given in Section 10.2. We also make the assumption that:

(HD3) The pair (A, B_2) is stabilizable and the pair (C_2, A) is detectable

Again, these assumptions are necessary for the existence of stabilizing controllers, and the argument is identical to that in Section 10.2.

(HD4) Regularity

In discrete–time the regularity assumptions are the following:
(RD1) The transmission pencil of $\mathbf{T}_{12}$ given by

$$\begin{bmatrix} zI - A & -B_2 \\ -C_1 & -D_{12} \end{bmatrix}$$

has full column rank for all $z \in \partial\mathbb{D}$. This condition is equivalent to the following rank condition:

$$\operatorname{rank}\begin{bmatrix} e^{j\theta}I - A & -B_2 \\ -C_1 & -D_{12} \end{bmatrix} = n + m_2, \quad \forall\theta \in [0, 2\pi).$$

In particular, it follows that $p_1 \geq m_2$, i.e., there are more regulated outputs than controlled inputs, and $\mathbf{T}_{12}$ has no zeros on $\partial\mathbb{D}$.

(RD2) The transmission pencil of $\mathbf{T}_{21}$ given by

$$\begin{bmatrix} zI - A & -B_1 \\ -C_2 & -D_{21} \end{bmatrix}$$

has full row rank for all $z \in \partial\mathbb{D}$. This condition is equivalent to the following rank condition:

$$\operatorname{rank}\begin{bmatrix} e^{j\theta}I - A & -B_1 \\ -C_2 & -D_{21} \end{bmatrix} = n + m_2, \quad \forall\theta \in [0, 2\pi).$$

In particular, it follows that $m_1 \geq p_2$, i.e., there are more external inputs than measured outputs, and $\mathbf{T}_{21}$ has no zeros on $\partial\mathbb{D}$.

Notice the difference with respect to the continuous–time case, where we assumed additionally that D_{12} and D_{21} have full column and row ranks, respectively.

A problem that satisfies (RD1) and (RD2) is called *regular*, otherwise it is called singular. Again, the two regularity assumptions are not necessary for the existence of a solution to the γ–DAP, and they are made solely for technical reasons.

10.12 The solution

With the assumptions introduced in the previous section, we can state the main result of this chapter that provides the solution to the DAP. Both existence and solution parameterization are expressed in terms of stabilizing solutions to Kalman–Szego–Popov–Yakubovich systems. In the discrete–time case the advantage of this form over the traditional Riccati equation formulation is even more obvious than in the continuous–time case. One of the obvious advantages is that it allows for a formulation very similar to the continuous–time case, although it should be remembered that the respective systems of equations have a quite different and more complicated look.

Theorem 10.12.1 *Assume the system* $\mathbf{T}$ *given by (10.117)–(10.118) satisfies assumptions (HD2), (HD3), and (HD4). Associate with* $\mathbf{T}$ *the following two Popov triplets:*

$$\Sigma_c = (A, \begin{bmatrix} B_1 & B_2 \end{bmatrix}; Q_c, L_c, R_c),$$

and

$$\Sigma_o = (A^T, \begin{bmatrix} C_1^T & C_2^T \end{bmatrix}; Q_o, L_o, R_o),$$

where

$$\begin{aligned} Q_c &= C_1^T C_1, \\ L_c &= C_1^T \begin{bmatrix} D_{11} & D_{12} \end{bmatrix}, \\ R_c &= \begin{bmatrix} D_{11}^T \\ D_{12}^T \end{bmatrix} \begin{bmatrix} D_{11} & D_{12} \end{bmatrix} - \begin{bmatrix} I_{m_1} & O \\ O & O \end{bmatrix}, \end{aligned} \tag{10.125}$$

and

$$\begin{aligned} Q_o &= B_1B_1^T, \\ L_o &= B_1\begin{bmatrix} D_{11}^T & D_{21}^T \end{bmatrix}, \\ R_o &= \begin{bmatrix} D_{11} \\ D_{21} \end{bmatrix}\begin{bmatrix} D_{11}^T & D_{21}^T \end{bmatrix} - \begin{bmatrix} I_{p_1} & O \\ O & O \end{bmatrix}. \end{aligned} \tag{10.126}$$

Denote

$$J_c = \begin{bmatrix} -I_{m_1} & O \\ O & I_{m_2} \end{bmatrix}, \quad J_o = \begin{bmatrix} -I_{p_1} & O \\ O & I_{p_2} \end{bmatrix}.$$

Then we have:

(I) *The DAP in discrete–time has a solution if and only if the following conditions are met:*

(CD1) *The Kalman–Szego–Popov–Yakubovich system KSPYS(Σ_c, J_c) has a stabilizing solution* $(X, V_c, W_c) = (X, \begin{bmatrix} V_{c11} & O \\ V_{c21} & V_{c22} \end{bmatrix}, \begin{bmatrix} W_{c1} \\ W_{c2} \end{bmatrix})$, *with* $X \geq 0$.

(CD2) *The Kalman–Szego–Popov–Yakubovich system KSPYS(Σ_o, J_o) has a stabilizing solution* $(Y, V_o, W_o) = (Y, \begin{bmatrix} V_{o11} & O \\ V_{o21} & V_{o22} \end{bmatrix}, \begin{bmatrix} W_{o1} \\ W_{o2} \end{bmatrix})$, *with* $Y \geq 0$.

(CD3) *The spectral radius $\rho(XY)$ is strictly subunitary.*

(II) *Assume that (CD1) is satisfied, let*

$$J_\times = \begin{bmatrix} -I_{m_2} & O \\ O & I_{p_2} \end{bmatrix},$$

and define the Popov triplet

$$\Sigma_\times = (A^T + F_1^TB_1^T, \begin{bmatrix} -F_2^TV_{c22}^T & C_2^T + F_1^TD_{21}^T \end{bmatrix}; Q_\times, L_\times, R_\times),$$

where $\begin{bmatrix} F_1 \\ F_2 \end{bmatrix}$ *is the stabilizing feedback corresponding to the KSPYS(Σ_c, J_c) and*

$$\begin{aligned} Q_\times &:= B_1(V_{c11}^TV_{c11})^{-1}B_1^T, \\ L_\times &:= B_1(V_{c11}^TV_{c11})^{-1}\begin{bmatrix} V_{c21}^T & D_{21}^T \end{bmatrix}, \\ R_\times &:= \begin{bmatrix} V_{c21} \\ D_{21} \end{bmatrix}(V_{c11}^TV_{c11})^{-1}\begin{bmatrix} V_{c21}^T & D_{21}^T \end{bmatrix} - \begin{bmatrix} I_{m_2} & O \\ O & O \end{bmatrix}. \end{aligned} \tag{10.127}$$

Assume that (CD1), (CD2), and (CD3) hold. Then we have that the following assertion is true:

(CD4) *The Kalman–Szego–Popov–Yakubovich system KSPYS($\Sigma_\times, J_\times$) has a stabilizing solution* $(Z, V_\times, W_\times) = (Z, \begin{bmatrix} V_{\times 11} & O \\ V_{\times 21} & V_{\times 22} \end{bmatrix}, \begin{bmatrix} W_{\times 1} \\ W_{\times 2} \end{bmatrix})$, *with* $Z \geq 0$.

Moreover,

$$Z := Y(I - XY)^{-1} \geq 0. \tag{10.128}$$

(III) *Let* $C_{2F_1} := C_2 + D_{21}F_1$, $S_c := (V_{c11}^TV_{c11})^{-1}$, *and* $S_\times := (V_{\times 11}^TV_{\times 11})^{-1}$. *Then the class of all solutions to the DAP can be expressed as* $\mathbf{K} = \mathrm{LFT}(\mathbf{K}_g, \mathbf{Q})$ *where*

$\mathbf{Q} \in RH^\infty_{+,m_2 \times p_2}$ *is an arbitrary parameter with* $\|\mathbf{Q}\|_\infty < 1$ *and*

$$\mathbf{K}_g = \left[\begin{array}{c|cc} A_g & B_{g1} & B_{g2} \\ \hline C_{g1} & D_{g11} & D_{g12} \\ C_{g2} & D_{g21} & D_{g22} \end{array}\right],$$

where

$$\begin{aligned}
A_g &= A + B_1F_1 + B_2F_2 + B_{g1}C_{2F_1}, \\
B_{g1} &= -B_1S_cD_{21}^T(D_{21}S_cD_{21}^T + C_{2F_1}ZC_{2F_1}^T)^{-1} - B_2D_{g11} \\
&\quad -(A + B_1F_1)ZC_{g2}^TD_{g21}, \\
B_{g2} &= -B_2D_{g12} - B_1S_c[V_{c22}^{-1}V_{c21} + D_{g11}D_{21}]^TV_{c22}^TS_\times^{\frac{1}{2}} \\
&\quad -(A + B_1F_1)ZC_{g1}^TV_{c2}^TS_\times^{\frac{1}{2}}, \\
C_{g1} &= -F_2 + D_{g11}C_{2F1}, \\
C_{g2} &= (D_{21}S_cD_{21}^T + C_{2F_1}ZC_{2F_1}^T)^{-\frac{1}{2}}C_{2F1}, \\
D_{g11} &= -(V_{c22}^{-1}V_{c21}S_cD_{21}^T - F_2ZC_{2F_1}^T)(D_{21}S_cD_{21}^T + C_{2F_1}ZC_{2F_1}^T)^{-1}, \\
D_{g12} &= V_{c22}^{-1}S_\times^{-\frac{1}{2}}, \\
D_{g21} &= (D_{21}S_cD_{21}^T + C_{2F_1}ZC_{2F_1}^T)^{-\frac{1}{2}}, \\
D_{g22} &= 0,
\end{aligned}$$

and

$$\begin{aligned}
V_{c22}^{-1}V_{c21} &= (D_{12}^TD_{12} + B_2^TXB_2)^{-1}(D_{12}^TD_{11} + B_2^TXB_1), \\
V_{c22} &= V_{c22}^T = (D_{12}^TD_{12} + B_2^TXB_2)^{\frac{1}{2}}
\end{aligned} \tag{10.129}$$

have to be substituted in D_{g12}, D_{g11} *and* B_{g2}.

Notice that in the continuous–time case the substitutions corresponding to (10.129) (see Theorem 10.3.1) were already performed as their expressions are much simpler. The continuous–time counterparts of (10.129) can formally be obtained by vanishing the terms containing X.

The plan for the proof of Theorem 10.12.1 follows closely that for the continuous–time counterpart.

10.13 Redheffer's theorem

Here we give the discrete–time counterpart of the theory developed in Section 10.4.

We need first the following definitions. Recall that a discrete–time system with TFM $\mathbf{T} \in RL^\infty$ is called *isometric* if $\mathbf{T}^*\mathbf{T} = I$ and is called inner if it is isometric and $\mathbf{T} \in RH^\infty_+$.

Assume $\mathbf{T}$ is given by a realization

$$\mathbf{T} = \left[\begin{array}{c|c} A & B \\ \hline C & D \end{array}\right]. \tag{10.130}$$

If A is dichotomic (stable), and there exists an $X = X^T$ such that

$$\begin{array}{rcl} -I + D^T D + B^T X B & = & 0, \\ C^T D + A^T X B & = & 0, \\ C^T C + A^T X A - X & = & 0, \end{array} \tag{10.131}$$

then we say that the realization (10.130) is isometric (inner). It is easy to check that if a system $\mathbf{T}$ has an isometric (inner) realization then it is isometric (inner), i.e., $\mathbf{T}^*\mathbf{T} = I$ and $\mathbf{T} \in RL^\infty$ ($\mathbf{T} \in RH_+^\infty$). However, not every realization of an isometric (inner) system $\mathbf{T}$ is isometric (inner). Since in this section we shall work with systems given by realizations we shall make the abuse to call a system isometric (inner) if and only if its realization is isometric (inner).

Theorem 10.13.1 (Generalized Redheffer's theorem in discrete–time) *Assume that the discrete–time system* $\mathbf{T}$ *in (10.117)–(10.118) is isometric,* $p_2 = m_1$*,* D_{21} *is nonsingular and* $A - B_1 D_{21}^{-1} C_2$ *is dichotomic, and let* $\mathbf{K}$ *given by (10.120) be a controller for* $\mathbf{T}$*. Then the following two assertions are equivalent:*

1. The closed–loop system $\mathbf{T}_{y_1 u_1}$ *given in (10.122) is well–posed, and the controller* $\mathbf{K}$ *is a stabilizing and contracting controller for* $\mathbf{T}$*.*

2. $\mathbf{T}$ *is inner,* A_K *is stable, and* $\|\mathbf{K}\|_\infty < 1$*.*

Proof. 2 $\Rightarrow$ 1. Since $\mathbf{T}$ is inner, it follows that there exists some symmetric matrix X that satisfies (10.131). However, as A is stable, $X \geq 0$, as follows from the last equation (10.131) with Theorem 1.5.5.

Using the first equation (10.131) we have that $\|D\| \leq 1$, hence

$$\|D_{22}\| \leq 1. \tag{10.132}$$

On the other hand, since A_K is stable and $\|\mathbf{K}\|_\infty < 1$, it follows from part I of Lemma 7.7.1 that the KSPYS

$$\begin{array}{rcl} -I + D_K^T D_K + B_K^T X_K B_K & = & -V_K^T V_K, \\ C_K^T D_K + A_K^T X_K B_K & = & -W_K^T V_K, \\ C_K^T C_K + A_K^T X_K A_K - X_K & = & -W_K^T W_K, \end{array} \tag{10.133}$$

has a stabilizing solution (X_K, V_K, W_K), with V_K invertible and $X_K \geq 0$. It follows from the first equation (10.133) that $\|D_K\| < 1$ which combined with (10.132) leads to the invertibility of $S = I - D_{22} D_K$ which in turn implies the well–posedness of the closed–loop system (10.122).

If we take

$$X_R := \begin{bmatrix} X & O \\ O & X_K \end{bmatrix}, V_R := V_K S^{-1} D_{21}, W_R := \begin{bmatrix} V_K S^{-1} C_2 & W_K + V_K S^{-1} D_{22} C_K \end{bmatrix} \tag{10.134}$$

then $X_R \geq 0$ and it can be checked by direct computation that

$$\begin{array}{rcl} -I + D_R^T D_R + B_R^T X_R B_R & = & -V_R^T V_R, \\ C_R^T D_R + A_R^T X_R B_R & = & -W_R^T V_R, \\ C_R^T C_R + A_R^T X_R A_R - X_R & = & -W_R^T W_R. \end{array} \tag{10.135}$$

On the other hand, for

$$K_R := \begin{bmatrix} O & -B_2 D_K V_K^{-1} \\ O & -B_K V_K^{-1} \end{bmatrix}$$

one gets

$$A_R + K_R \begin{bmatrix} C_R \\ W_R \end{bmatrix} = \begin{bmatrix} A & \star \\ O & A_K - B_K V_K^{-1} W_K \end{bmatrix}. \qquad (10.136)$$

Since (X_K, V_K, W_K) is a stabilizing solution to (10.133) and A is stable, it follows that the upper left and lower right corners in the right–hand side of (10.136) are stable. Hence the pair $(\begin{bmatrix} C_R \\ W_R \end{bmatrix}, A_R)$ is detectable. Combining this conclusion with the fact that the last equation in (10.135) is fulfilled for $X_R \geq 0$ we deduce with Theorem 1.5.5 that A_R is stable. Hence $\mathbf{K}$ is a stabilizing controller for $\mathbf{T}$.

Let us prove now that $\mathbf{K}$ is a contracting controller for $\mathbf{T}$, i.e., $\|\mathbf{T}_{y_1u_1}\|_\infty < 1$. Since A_R in (10.135) is stable, we can check, similarly to the continuous case, that

$$\|\mathbf{T}_{y_1u_1}\|_\infty \leq 1. \qquad (10.137)$$

We shall prove now that the assumption that $\|\mathbf{T}_{y_1u_1}\|_\infty = 1$ leads to a contradiction, from where we conclude that the inequality in (10.137) is strict.

Indeed, if $\|\mathbf{T}_{y_1u_1}\|_\infty = 1$ then there exists a sequence of input signals $(u_1^k)_{k\geq 0}$ with $u_1^k \in L^{2,m_1}$ (recall that here $L^{2,m_1} = L^{2,m_1}(\partial\mathbb{D})$) and $\|u_1^k\|_2 = 1$ such that for $(y_1^k)_{k\geq 0}$ with $y_1^k \in L^{2,p_1}$ obtained as $y_1^k(e^{j\theta}) = \mathbf{T}_{y_1u_1}(e^{j\theta})u_1^k(e^{j\theta})$ we have

$$\|y_1^k\|_2^2 - \|u_1^k\|_2^2 \to 0 \text{ as k} \to \infty. \qquad (10.138)$$

Since both A and A_K are stable it follows that $I - \mathbf{T}_{22}\mathbf{K}$ and $\mathbf{T}_{21}$ are in $RH^\infty_{+,p_2\times p_2}$. Due to the closed–loop stability and to the fact that $A - B_1 D_{21}^{-1} C_2$ is dichotomic, we have that $(I - \mathbf{T}_{22}\mathbf{K})^{-1} \in RH^\infty_{+,p_2\times p_2}$ and $\mathbf{T}_{21}^{-1} \in RL^\infty_{p_2\times p_2}$, respectively. Thus both $I - \mathbf{T}_{22}\mathbf{K})^{-1}$ and $\mathbf{T}_{21}$ are units in $RL^\infty_{p_2\times p_2}$ and consequently $\mathbf{T}_{y_2u_1} := (I - \mathbf{T}_{22}\mathbf{K})^{-1}\mathbf{T}_{21}$ (the closed–loop TFM from u_1 to y_2) is a unit in $RL^\infty_{p_2\times p_2}$ as well. Hence there exists $\rho_0 > 0$ such that

$$\|y_2\|_2 \geq \rho_o \|u_1\|_2, \qquad (10.139)$$

for each $u_1 \in L^{2,p_1}$ and $y_2 \in L^{2,p_2}$ defined by $y_2(e^{j\theta}) = \mathbf{T}_{y_2u_1}(e^{j\theta})u_1(e^{j\theta})$. Since $\mathbf{T}$ is isometric it follows that

$$\|y_1^k\|_2^2 - \|u_1^k\|_2^2 = \|u_2^k\|_2^2 - \|y_2^k\|_2^2, \qquad (10.140)$$

with u_1^k and y_1^k as introduced above, and with $y_2^k \in L^{2,p_2}$ and $u_2^k \in L^{2,m_2}$ defined by $y_2^k(e^{j\theta}) = \mathbf{T}_{y_2u_1}(e^{j\theta})u_1^k(e^{j\theta})$ and by $u_2^k(e^{j\theta}) = \mathbf{K}(e^{j\theta})y_2^k(e^{j\theta}) = \mathbf{T}_{u_2u_1}(e^{j\theta})u_1^k(e^{j\theta})$, respectively (where $\mathbf{T}_{u_2u_1} := \mathbf{K}\mathbf{T}_{y_2u_1} \in RH^\infty_{+,p_2\times m_1}$ is the closed–loop TFM from u_1 to u_2). By combining (10.138) with (10.140)

$$\|y_2^k\|_2^2 - \|u_2^k\|_2^2 \to 0, \quad k \to \infty. \qquad (10.141)$$

But $\|u_2^k\|_2^2 \leq \|\mathbf{K}\|_\infty^2 \|y_2^k\|_2^2$. Hence by taking into account (10.139)

$$\|y_2^k\|_2^2 - \|u_2^k\|_2^2 \geq (1 - \|\mathbf{K}\|_\infty^2)\rho_0^2 \|u_1^k\|_\infty^2 = (1 - \|\mathbf{K}\|_\infty^2)\rho_0^2 =: c_0 > 0. \qquad (10.142)$$

But (10.142) contradicts (10.141) and the strict contractiveness of $\mathbf{T}_{y_1u_1}$ follows.

1. $\Rightarrow$ 2. Assume that $\mathbf{K}$ is a stabilizing and contracting controller for $\mathbf{T}$.

Let us show first that $\mathbf{K} \in RL^{\infty}_{m_2\times p_2}$. To this end we repeat the same scheme of proof as in the continuous–time case. It is simple to check that

$$\begin{bmatrix} \mathbf{T}_{y_1u_1} \\ \mathbf{T}_{y_2u_1} \end{bmatrix} = \begin{bmatrix} \mathbf{T}_{11} & \mathbf{T}_{12} \\ \mathbf{T}_{21} & \mathbf{T}_{22} \end{bmatrix} \begin{bmatrix} I \\ \mathbf{T}_{u_2u_1} \end{bmatrix}. \tag{10.143}$$

Premultiplying both sides of (10.143) by $\mathbf{T}$ and taking into account that $\mathbf{T}$ is isometric one gets

$$\mathbf{T}^*_{11}\mathbf{T}_{y_1u_1} + \mathbf{T}^*_{21}\mathbf{T}_{y_2u_1} = I$$

from which

$$\mathbf{T}_{y_2u_1} = \mathbf{T}^{-*}_{21}(I - \mathbf{T}^*_{11}\mathbf{T}_{y_1u_1}) \in RL^{\infty}_{p_2\times p_2}. \tag{10.144}$$

Since $\|\mathbf{T}_{y_1u_1}\|_{\infty} < 1$ and $\mathbf{T}$ isometric implies $\|\mathbf{T}_{11}\| \leq 1$, it follows that $\|\mathbf{T}^*_{11}\mathbf{T}_{y_1u_1}\|_{\infty} < 1$. Hence it follows that $I - \mathbf{T}^*_{11}\mathbf{T}_{y_1u_1}$ is a unit in $RL^{\infty}_{m_1\times m_1}$. Using this conclusion, (10.144) reveals that $\mathbf{T}_{y_2u_1}$ is a unit in $RL^{\infty}_{m_2\times p_2}$ and therefore

$$\mathbf{K} = \mathbf{T}_{u_2u_1}\mathbf{T}^{-1}_{y_2u_1} \in RL^{\infty}_{m_2\times p_2}.$$

Moreover $\|\mathbf{K}\|_{\infty} < 1$. Assume the contrary, i.e. $\|\mathbf{K}\|_{\infty} \geq 1$. Then there exists a sequence $(y_2^k)_{k\geq 0}$ with $y_2^k \in L^{2,p_2}$ and $\|y_2^k\|_2 = 1$, such that for $(u_2^k)_{k\geq 0}$ with $u_2^k \in L^{2,m_2}$ defined by $u_2^k(e^{j\theta}) = \mathbf{K}(e^{j\theta})y_2^k(e^{j\theta})$ we have

$$\lim_{k\to\infty}(\|u_2^k\|_2^2 - \|y_2^k\|_2^2) \geq 0. \tag{10.145}$$

Since $\mathbf{T}_{y_2u_1}$ is a unit in $RL^{\infty}_{m_2\times p_2}$ one can define the sequence $(u_1^k)_{k\geq 0}, u_1^k \in L^{2,m_1}$ by $u_1^k(e^{j\theta}) = \mathbf{T}^{-1}_{y_2u_1}(e^{j\theta})y_2^k(e^{j\theta}) \neq 0, \forall k$. Let $(y_1^k)_{k\geq 0}, y_1^k \in L^{2,p_1}$ be defined by $y_1^k(e^{j\theta}) = \mathbf{T}_{y_1u_1}(e^{j\theta})u_1^k(e^{j\theta})$. Since $\mathbf{T}$ is isometric we can write

$$\|y_1^k\|_2^2 - \|u_1^k\|_2^2 = \|u_2^k\|_2^2 - \|y_2^k\|_2^2. \tag{10.146}$$

Hence (10.145), (10.146) yields

$$\lim_{k\to\infty}(\|y_1^k\|_2^2 - \|u_1^k\|_2^2) \geq 0$$

which contradicts the strict contractiveness of $\mathbf{T}_{y_1u_1}$. Thus $\|\mathbf{K}\|_{\infty} < 1$.

Let us show now that A_K is stable. Notice first that $\mathbf{K}$ stabilizes $\mathbf{T}$. Hence the pairs (A_K, B_K) and (A, B_2) are stabilizable and the pairs (C_K, A_K) and (C_2, A) are detectable. Using this conclusion in conjunction with the fact that $\mathbf{K} \in RL^{\infty}_{m_2\times p_2}$ we deduce that A_K is dichotomic. Therefore, since $\|\mathbf{K}\|_{\infty} < 1$ it follows from part II of Lemma 7.7.1 that the KSYPS (10.133) has a stabilizing solution (X_K, V_K, W_K). Hence, as above, by combining (10.131) and (10.133), the KSPYS (10.135) will be fulfilled for (10.134). But A_R in the last equation (10.135) is stable. Hence $X_R \geq 0$ and (10.134) reveals that $X \geq 0$ and $X_K \geq 0$. Since $X_K \geq 0$ and the pair (C_K, A_K) is detectable, it follows from the last equation (10.133) with Theorem 1.5.5 that A_K is stable.

Finally, A is stable as follows from the last equation (10.131) where $X \geq 0$ and (C, A) is detectable since (C_2, A) is detectable. Thus $\mathbf{T}$ is inner and the theorem is completely proved. ∎

10.14 Necessity of (CD1) and (CD2)

We start in this section with the proof of Theorem 10.12.1. We assume that a stabilizing controller exists such that $\|\mathbf{T}_{y_1 u_1}\|_\infty < 1$ and we prove that (CD1) and (CD2) are fulfilled. Notice that (CD2) is simply the dual of (CD1), as in the continuous–time case, and hence it is sufficient to prove (CD1).

The basic idea of the proof of (CD1) is to use the implication 1. $\Rightarrow$ 2. in part I of Theorem 4.12.9 for the Popov triplet Σ_c. However, we see that A is not stable. To remedy this, we proceed similarly as in the continuous–time case, and conclude finally that (CD1) holds.

10.15 An auxiliary necessary condition

Just as we did in the continuous–time case, we prove that the existence of a stabilizing and contracting controller imply the existence of the stabilizing solution to the KSPYS($\Sigma_\times, J_\times$) (i.e., (CD4) in Theorem 10.12.1 is fulfilled). In the next section we show that (CD4) together with (CD2) imply (CD3). Incidentally we also prove relation (10.128) and the results in this and the next section will be very useful in pursuing with the rest of the proof of Theorem 10.12.1.

With the solution (X, V_c, W_c) of the KSPYS(Σ_c, J_c), let us introduce the systems

$$\begin{aligned} \sigma x_I &= (A - B_2 V_{c22}^{-1} W_{c2}) x_I + (B_1 - B_2 V_{c22}^{-1} V_{c21}) u_1 + B_2 V_{c22}^{-1} \widetilde{y}_1, \\ y_1 &= (C_1 - D_{12} V_{c22}^{-1} W_{c2}) x_I + (D_{11} - D_{12} V_{c22}^{-1} V_{c21}) u_1 + D_{12} V_{c22}^{-1} \widetilde{y}_1, \\ \widetilde{u_1} &= W_{c1} x_I + V_{c11} u_1, \end{aligned} \tag{10.147}$$

and

$$\begin{aligned} \sigma x_O &= (A + B_1 F_1) x_O + B_1 V_{c11}^{-1} \widetilde{u_1} + B_2 u_2, \\ \widetilde{y}_1 &= -V_{c22} F_2 x_O + V_{c21} V_{c11}^{-1} \widetilde{u_1} + V_{c22} u_2, \\ y_2 &= (C_2 + D_{21} F_1) x_O + D_{21} V_{c11}^{-1} \widetilde{u_1}. \end{aligned} \tag{10.148}$$

We denote by $\mathbf{T}_I$ the transfer function of the system (10.147) and by $\mathbf{T}_O$ the transfer function of the system (10.148), and we have

$$\begin{aligned} \mathbf{T}_I &= \left[\begin{array}{c|c} A_I & B_I \\ \hline C_I & D_I \end{array}\right] = \left[\begin{array}{c|cc} A_I & B_{I1} & B_{I2} \\ \hline C_{I1} & D_{I11} & D_{I12} \\ C_{I2} & D_{I21} & D_{I22} \end{array}\right] \\ &= \left[\begin{array}{c|cc} A - B_2 V_{c22}^{-1} W_{c2} & B_1 - B_2 V_{c22}^{-1} V_{c21} & B_2 V_{c22}^{-1} \\ \hline C_1 - D_{12} V_{c22}^{-1} W_{c2} & D_{11} - D_{12} V_{c22}^{-1} V_{c21} & D_{12} V_{c22}^{-1} \\ W_{c1} & V_{c11} & O \end{array}\right] \\ \begin{bmatrix} y_1 \\ \widetilde{u}_1 \end{bmatrix} &= \mathbf{T}_I \begin{bmatrix} u_1 \\ \widetilde{y}_1 \end{bmatrix} \end{aligned} \tag{10.149}$$

and

$$
\begin{aligned}
\mathbf{T}_O &= \left[\begin{array}{c|c} A_O & B_O \\ \hline C_O & D_O \end{array}\right] = \left[\begin{array}{c|cc} A_O & B_{O1} & B_{O2} \\ \hline C_{O1} & D_{O11} & D_{O12} \\ C_{O2} & D_{O21} & D_{O22} \end{array}\right] \\
&= \left[\begin{array}{c|cc} A + B_1 F_1 & B_1 V_{c11}^{-1} & B_2 \\ \hline -V_{c22} F_2 & V_{c21} V_{c11}^{-1} & V_{c22} \\ C_2 + D_{21} F_1 & D_{21} V_{c11}^{-1} & O \end{array}\right] \qquad (10.150) \\
\left[\begin{array}{c} \widetilde{y}_1 \\ y_2 \end{array}\right] &= \mathbf{T}_O \left[\begin{array}{c} \widetilde{u}_1 \\ u_2 \end{array}\right].
\end{aligned}
$$

As the computations are identical to those in the continuous–time case, we can readily state the following result.

Lemma 10.15.1 *The controller* $\mathbf{K}$ *given by the realization (10.120) is a stabilizing and contracting controller for* $\mathbf{T}$ *if and only if it is a stabilizing and contracting controller for the system given by the Redheffer product* $\mathbf{T}_I \otimes \mathbf{T}_O$.

Now we are in a position to give the discrete–time counterpart of Lemma 10.6.2.

Lemma 10.15.2 *The controller* $\mathbf{K}$ *given by the realization (10.120) is a stabilizing and contracting controller for* $\mathbf{T}$ *if and only if it is a stabilizing and contracting controller for* $\mathbf{T}_O$.

Proof. The single part of the proof which is slightly different from the continuous–time case is that to prove that $\mathbf{T}_I$ is inner and satisfies all the other conditions of Redheffer's theorem 10.13.1. Indeed, we can show through direct substitutions that

$$
\begin{aligned}
D_I^T D_I + B_I^T X B_I &= I, \\
C_I^T D_I + A_I^T X B_I &= 0, \qquad (10.151) \\
C_I^T C_I + A_I^T X A_I - X &= 0,
\end{aligned}
$$

where $X \geq 0$ comes from the solution of the KSPYS(Σ_c, J_c). The matrix $D_{I21} = V_{c11}$ is nonsingular and

$$
A_I - B_{I1} D_{I21}^{-1} C_{I2} = A - B_2 V_{c22}^{-1} W_{c2} - (B_1 - B_2 V_{c22}^{-1} V_{c21}) V_{c11}^{-1} W_{c1} = A + BF
$$

is stable. In particular, $(C_{I2}, A_I) = (W_{c1}, A - B_2 V_{c22}^{-1} W_{c2})$ is detectable and thus (C_I, A_I) is detectable. Since $X \geq 0$, it follows from the last equation in (10.151) and Theorem 1.5.5 that A_I is stable. Thus we checked that $\mathbf{T}_I$ is inner. We can therefore apply Redheffer's theorem to $\mathbf{T}_I$ as claimed and the proof is completed. ■

Since the matrices in the realization (10.148) of $\mathbf{T}_O$ are identical to those in the continuous–time case (10.53), we can readily state the auxiliary necessary condition (CD4) that we were looking for.

Corollary 10.15.3 *If the discrete DAP has a solution, then the KSPYS($\Sigma_\times, J_\times$), where $\Sigma_\times$ and $J_\times$ were introduced in Theorem 10.12.1, has a stabilizing solution of the form*

$$
(Z, \left[\begin{array}{cc} V_{\times 11} & O \\ V_{\times 21} & V_{\times 22} \end{array}\right], \left[\begin{array}{c} W_{\times 1} \\ W_{\times 2} \end{array}\right]), \quad Z \geq 0.
$$

To explain the origin of the systems (10.147) and (10.148) we can employ similar arguments as in Remark 10.6.4.

10.16 Necessity of (CD3)

The main result of this section is the discrete–time version of Lemma 10.7.1. Due to the inherent technical difficulties, we shall prove only a weaker result than in the continuous–time case. Namely, instead of proving the genuine discrete–time counterpart of Theorem 10.7.1 formulated in terms of the KSPYS$(\Sigma_\times, J_\times)$ and the KSPYS(Σ_o, J_o), we prove the result formulated in terms of the corresponding Riccati equations which are DTARE$(\Sigma_\times)$ and DTARE(Σ_o). It turns out that this weaker result, although enough for proving the necessity part of Theorem 10.12.1, does not suffice for the sufficiency part which will require an auxiliary result stated in Lemma 10.17.4.

Lemma 10.16.1 *Assume that (CD1) in Theorem 10.12.1 holds. Then the following statements are equivalent:*

1. *The DTARE$(\Sigma_\times)$ has a stabilizing solution $Z \geq 0$.*
2. *The DTARE(Σ_o) has a stabilizing solution $Y \geq 0$, and $\rho(XY) < 1$.*

In this case the two solutions Y and Z are related through (10.128).

Proof. The idea of the proof is to relate the ESP$(\Sigma_\times)$ and the ESP(Σ_o) in order to find a corresponding relation between the stabilizing solutions to DTARE$(\Sigma_\times)$ and DTARE(Σ_o). According to (3.131) and (10.127), the ESP$(\Sigma_\times)$ denoted $zM_{\Sigma_\times} - N_{\Sigma_\times}$, is given by

$$M_{\Sigma_\times} := \begin{bmatrix} I_n & O & O & O \\ O & -A + B_1 V_{c11}^{-1} W_{c1} & O & O \\ O & V_{c21} V_{c11}^{-1} W_{c1} - W_{c2} & O & O \\ O & -C_2 + D_{21} V_{c11}^{-1} W_{c1} & O & O \end{bmatrix}, \tag{10.152}$$

$$N_{\Sigma_\times} := \begin{bmatrix} A^T - W_{c1}^T V_{c11}^{-T} B_1^T & O & -W_{c1}^T V_{c11}^{-T} V_{c21}^T + W_{c2}^T & C_2^T - W_{c1}^T V_{c11}^{-T} D_{21}^T \\ B_1 V_{c11}^{-1} V_{c11}^{-T} B_1^T & -I_n & B_1 V_{c11}^{-1} V_{c11}^{-T} V_{c21}^T & B_1 V_{c11}^{-1} V_{c11}^{-T} D_{21}^T \\ V_{c21} V_{c11}^{-1} V_{c11}^{-T} B_1^T & O & V_{c21} V_{c11}^{-1} V_{c11}^{-T} V_{c21}^T - I_{m_2} & V_{c21} V_{c11}^{-1} V_{c11}^{-T} D_{21}^T \\ D_{21} V_{c11}^{-1} V_{c11}^{-T} B_1^T & O & D_{21} V_{c11}^{-1} V_{c11}^{-T} V_{c21}^T & D_{21} V_{c11}^{-1} V_{c11}^{-T} D_{21}^T \end{bmatrix}, \tag{10.153}$$

and with (3.131) and (10.126) we have that the ESP(Σ_o), denoted $zM_{\Sigma_o} - N_{\Sigma_o}$, is given by

$$M_{\Sigma_o} := \begin{bmatrix} I_n & O & O & O \\ O & -A & O & O \\ O & -C_1 & O & O \\ O & -C_2 & O & O \end{bmatrix}, \tag{10.154}$$

$$N_{\Sigma_o} := \begin{bmatrix} A^T & O & C_1^T & C_2^T \\ B_1 B_1^T & -I_n & B_1 D_{11}^T & B_1 D_{21}^T \\ D_{11} B_1^T & O & D_{11} D_{11}^T - I_{p_1} & D_{11} D_{21}^T \\ D_{21} B_1^T & O & D_{21} D_{11}^T & D_{21} D_{21}^T \end{bmatrix}. \tag{10.155}$$

Notice that $M_{\Sigma_\times}$ and $N_{\Sigma_\times}$ are $(2n+m_2+p_2) \times (2n+m_2+p_2)$ matrices while M_{Σ_o} and N_{Σ_o} are $(2n+p_1+p_2) \times (2n+p_1+p_2)$ matrices. Since the ESPs are not of the same dimension we shall work with their augmented versions $zM_a^\times - N_a^\times$ and $z\widehat{M}_a - \widehat{N}_a$

defined by

$$M_a^\times := \begin{bmatrix} M_{\Sigma_\times} & O & O \\ O & O_{m_1} & O \\ \overline{\Theta} & O & O_{p_1} \end{bmatrix}, \tag{10.156}$$

$$N_a^\times := \begin{bmatrix} N_{\Sigma_\times} & & \\ & V_{c11}^{-1}V_{c11}^{-T} & \\ & & -I_{p_1} \end{bmatrix}, \tag{10.157}$$

and

$$\widehat{M}_a := \begin{bmatrix} M_{\Sigma_o} & & \\ X_1 & O_{m_1} & \\ X_2 & & O_{m_2} \end{bmatrix}, \widehat{N}_a := \begin{bmatrix} N_{\Sigma_o} & & \\ & I_{m_1} & \\ & & -I_{m_2} \end{bmatrix}, \tag{10.158}$$

where

$\overline{\Theta} := [\ O_{p_1\times n} \quad \Theta \quad O_{p_1\times m_2} \quad O_{p_1\times p_2}\ ]$, $\Theta := -C_1 - D_{11}S_cM_1^T + D_{11}S_cV_{c21}^TV_{c22}^{-T}M_2^T$,

$S_c := V_{c11}^{-1}V_{c11}^{-T}$, and X_1 and X_2 are elements to be specified later on. Introduce the following notation:

$$\begin{aligned} M_1 &:= C_1^TD_{11} + A^TXB_1, \\ M_2 &:= C_1^TD_{12} + A^TXB_2. \end{aligned} \tag{10.159}$$

Using the KSPYS(Σ_c, J_c) one gets

$$W_{c1}^T = -M_1V_{c11}^{-1} + M_2V_{c22}^{-1}V_{c21}V_{c11}^{-1}, \quad W_{c2}^T = M_2V_{c22}^{-1} \tag{10.160}$$

and the DTARE(Σ_c) can be written into the equivalent form

$$\begin{aligned} &A^TXA - X \\ &-[\ M_1\ \ M_2\ ]\begin{bmatrix} S_c & S_cV_{c21}V_{c22}^{-1} \\ V_{c22}^{-1}V_{c21}S_c & -V_{c22}^{-1}V_{c21}S_cV_{c21}^TV_{c22}^{-T} + V_{c22}^{-1}V_{c22}^{-T} \end{bmatrix}\begin{bmatrix} M_1^T \\ M_2^T \end{bmatrix} + C_1^TC_1 = 0. \end{aligned} \tag{10.161}$$

Let the $(2n+m+p)\times(2n+m+p)$ nonsingular matrices U_a and W_a be introduced by

$$U_a := \begin{bmatrix} I_n & -A^TX & W_2^T & O & C_1^TD_{11} & -C_1^T \\ O & I_n - B_1B_1^TX & B_1V_{c21}^T & O & B_1(I - \Xi - B_1^TXB_1) & -B_1D_{11}^T \\ O & -D_{11}B_1^TX & D_{11}V_{c21}^T & O & -D_{11}(\Xi + B_1^TXB_1) & I_{p_1} - D_{11}D_{11}^T \\ O & -D_{21}B_1^TX & D_{21}V_{c21}^T & I_{p_2} & -D_{21}(\Xi + B_1^TXB_1) & -D_{21}D_{11}^T \\ O & B_1^TX & -V_{c21}^T & O & \Xi + B_1^TXB_1 & D_{11}^T \\ O & O & I_{m_2} & O & V_{c21} & O \end{bmatrix} \tag{10.162}$$

and

$$W_a := \begin{bmatrix} I_n & -X & O & O & O & O \\ O & I_n & O & O & O & O \\ V_{c21}B_1^T & W_{c2} - V_{c22}^{-T}M_2^T & V_{c21}D_{11}^T & V_{c21}D_{21}^T & V_{c21} & I_{m_2} \\ O & O & O & I_{p_2} & O & O \\ (-I+\Xi)B_1^T & B_1^TX & \Xi D_{11}^T & (-I+\Xi)D_{21}^T & \Xi & -V_{c21}^T \\ -D_{11}B_1^T & O & I_{p_1} - D_{11}D_{11}^T & -D_{11}D_{21}^T & -D_{11} & O \end{bmatrix} \tag{10.163}$$

with $\Xi := V_{c11}^T V_{c11} - V_{c21}^T V_{c21}$. Then using (10.160) and (10.161) we obtain after some tedious but routine computations

$$U_a M_a^\times W_a = \widehat{M}_a, \tag{10.164}$$

$$U_a N_a^\times W_a = \widehat{N}_a. \tag{10.165}$$

Notice that the equality (10.165) automatically specifies also the elements X_1 and X_2 introduced in (10.158). The equalities (10.164), (10.165) show that the augmented ESPs $sM_a^\times - N_a^\times$ and $s\widehat{M}_a - \widehat{N}_a$ are strictly equivalent. The matrices U_a and W_a have been obtained by storing the elementary row and column operations performed on $zM_a^\times - N_a^\times$ in order to bring it into the strict equivalent form $z\widehat{M}_a - \widehat{N}_a$. We are ready now for the proof.

1. $\Rightarrow$ 2. Since the DTARE($\Sigma_\times$) has a stabilizing solution $Z \geq 0$, it follows in accordance with Theorem 5.5.1 that

$$N_{\Sigma_\times} V^\times = M_{\Sigma_\times} V^\times S^\times \tag{10.166}$$

where

$$V^\times := \begin{bmatrix} I \\ Z \\ F^\times \end{bmatrix}, \tag{10.167}$$

and $F^\times$ is the stabilizing feedback and the $n \times n$ matrix $S^\times$ is stable.

Let

$$V_a^\times := \begin{bmatrix} V^\times \\ O \\ -\Theta Z S^\times \end{bmatrix}. \tag{10.168}$$

With (10.156), (10.157) and (10.166) one can check

$$N_a^\times V_a^\times = M_a^\times V_a^\times S^\times. \tag{10.169}$$

Using now (10.164) and (10.165), (10.169) yields

$$\widehat{N}_a W_a^{-1} V_a^\times = \widehat{M}_a W_a^{-1} V_a^\times S^\times. \tag{10.170}$$

But, in accordance with the structure of W_a in (10.163), we deduce that

$$W_a^{-1} V_a^\times = \begin{bmatrix} I + XZ \\ Z \\ \star \end{bmatrix}. \tag{10.171}$$

As both X and Z are positive definite, $I + XZ$ is nonsingular with all its eigenvalues strictly positive. Hence, by combining (10.170) and (10.171) one gets

$$\widehat{N}_a \begin{bmatrix} I \\ Y \\ \star \end{bmatrix} = \widehat{M}_a \begin{bmatrix} I \\ Y \\ \star \end{bmatrix} \widehat{S} \tag{10.172}$$

where

$$Y := Z(I + XZ)^{-1} \geq 0 \tag{10.173}$$

and

$$\widehat{S} = (I + XZ)S^{\times}(I + XZ)^{-1} \tag{10.174}$$

is stable. Using now (10.158) we deduce from (10.169) that

$$N_{\Sigma_o} \begin{bmatrix} I \\ Y \\ \star \end{bmatrix} = M_{\Sigma_o} \begin{bmatrix} I \\ Y \\ \star \end{bmatrix} \widehat{S}. \tag{10.175}$$

But (10.175) in conjunction with (10.173) and (10.174), shows by Theorem 5.5.1 that Y is the positive semidefinite stabilizing solution to the DTARE(Σ_o). Further we have

$$I - XY = (I + XZ)^{-1} \tag{10.176}$$

and therefore

$$\rho(XY) < 1. \tag{10.177}$$

Using (10.173) in conjunction with (10.176) one gets

$$Z = Y(I - XY)^{-1}. \tag{10.178}$$

2. $\Rightarrow$ 1. This implication follows by simply reversing the above arguments. ■

Now the necessity of (CD3) follows by combining Corollary 10.15.3 with the implication 1. $\Rightarrow$ 2. of Lemma 10.16.1.

10.17 Proof of sufficiency

As in the continuous–time case, we show that if (CD1), (CD2), and (CD3) are fulfilled then we can construct in terms of explicit formulas, a stabilizing and contracting controller. In fact, we give the class of all those controllers, completing in this way the proof of Theorem 10.12.1.

To this purpose we shall successively solve the discrete–time DAP for some simpler cases, and in the end we reduce the general problem to these particular ones.

10.17.1 The disturbance estimation (DE) case

This case goes identically as for continuous–time. Even the formulas for the solution have the same form. We repeat them here only for convenience.

The plant (10.117) in the DE case satisfies the following assumptions:

(DE1) D_{12} is square and nonsingular.

(DE2) $A - B_2 D_{12}^{-1} C_1$ is stable.

(DE3) D_{21} is square and nonsingular.

(DE4) $A - B_1 D_{21}^{-1} C_2$ is stable.

Theorem 10.17.1 *The class of all stabilizing and contracting controllers for the discrete–time plant (10.117) satisfying the DE assumptions is given by*

$$\mathbf{K} = \mathrm{LFT}(\mathbf{K}_g, \mathbf{Q}) \tag{10.179}$$

where $\mathbf{Q} \in RH^\infty_{+,p_1\times m_1}$ *is an arbitrary parameter with* $\|\mathbf{Q}\|_\infty < 1$ *and*

$$\mathbf{K}_g = \left[\begin{array}{c|cc} A_g & B_{g1} & B_{g2} \\ \hline C_{g1} & D_{g11} & D_{g12} \\ C_{g2} & D_{g21} & D_{g22} \end{array}\right], \tag{10.180}$$

with

$$\begin{aligned} A_g &= A - B_1 D_{21}^{-1} C_2 - B_2 D_{12}^{-1} C_1 + B_2 D_{12}^{-1} D_{11} D_{21}^{-1} C_2, \\ B_{g1} &= (B_1 - B_2 D_{12}^{-1} D_{11}) D_{21}^{-1}, \\ B_{g2} &= B_2 D_{12}^{-1}, \\ C_{g1} &= -D_{12}^{-1}(C_1 - D_{11} D_{21}^{-1} C_2), \\ C_{g2} &= -D_{21}^{-1} C_2, \\ D_{g11} &= -D_{12}^{-1} D_{11} D_{21}^{-1}, \\ D_{g12} &= D_{12}^{-1}, \\ D_{g21} &= D_{21}^{-1}, \\ D_{g22} &= 0. \end{aligned} \tag{10.181}$$

10.17.2 The disturbance feedforward (DF) case

The plant (10.117) in the DF case satisfies the following assumptions:

(DF1) D_{21} is square and nonsingular.

(DF2) $A - B_1 D_{21}^{-1} C_2$ is stable.

(DF3) Condition (CD1) in Theorem 10.12.1 holds.

Theorem 10.17.2 *The class of all stabilizing and contracting controllers for the plant (10.117) satisfying the DF assumptions is given by*

$$\mathbf{K} = \mathrm{LFT}(\mathbf{K}_g, \mathbf{Q}), \tag{10.182}$$

where $\mathbf{Q} \in RH^\infty_{+,m_2\times m_1}$ *is an arbitrary parameter with* $\|\mathbf{Q}\|_\infty < 1$ *and* $\mathbf{K}_g$ *has the realization*

$$\begin{aligned} A_g &= A - B_1 D_{21}^{-1} C_2 - B_2 (D_{12}^T D_{12} + B_2^T X B_2)^{-1} [D_{12}^T C_1 + B_2^T X A \\ &\quad - (D_{12}^T D_{11} + B_2^T X B_1) D_{21}^{-1} C_2], \\ B_{g1} &= B_1 D_{21}^{-1} - B_2 (D_{12}^T D_{12} + B_2^T X B_2)^{-1} (D_{12}^T D_{11} + B_2^T X B_1) D_{21}^{-1}, \\ B_{g2} &= B_2 (D_{12}^T D_{12} + B_2^T X B_2)^{-1/2}, \\ C_{g1} &= -(D_{12}^T D_{12} + B_2^T X B_2)^{-1} [D_{12}^T C_1 + B_2^T X A - (D_{12}^T D_{11} + B_2^T X B_1) D_{21}^{-1} C_2)], \\ C_{g2} &= -S_c^{-1/2} [D_{21}^{-1} C_2 + S_c (D_{11}^T C_1 + B_1^T X A) \\ &\quad - S_c (D_{11}^T D_{12} + B_1^T X B_2)(D_{12}^T D_{12} + B_2^T X B_2)^{-1} (D_{12}^T C_1 + B_2^T X A)], \\ D_{g11} &= -(D_{12}^T D_{12} + B_2^T X B_2)^{-1} (D_{12}^T D_{11} + B_2^T X B_1) D_{21}^{-1}, \\ D_{g12} &= (D_{12}^T D_{12} + B_2^T X B_2)^{-1/2}, \\ D_{g21} &= S_c^{-1/2} D_{21}^{-1}, \\ D_{g22} &= 0, \end{aligned} \tag{10.183}$$

where

$$\begin{aligned} S_c := (V_{c11}^T V_{c11})^{-1} &= [I - D_{11}^T D_{11} - B_1^T X B_1 - (D_{11}^T D_{12} + B_1^T X B_2) \\ &\quad \times (D_{12}^T D_{12} + B_2^T X B_2)^{-1} (D_{12}^T D_{11} + B_2^T X B_1)]^{-1}. \end{aligned} \tag{10.184}$$

The proof is entirely similar to that of Theorem 10.8.2 except that it uses the KSPYS(Σ_c, J_c) to compute the formulas for the parameterization.

10.17.3 The output estimation (OE) case

The plant (10.117) in the OE case satisfies the following assumptions:

(OE1) D_{12} is square and nonsingular.

(OE2) $A - B_2 D_{12}^{-1} C_1$ is stable.

(OE3) Condition (CD2) in Theorem 10.12.1 holds.

Theorem 10.17.3 *The class of all stabilizing and contracting controllers for the plant (10.117) satisfying the OE assumptions is given by*

$$\mathbf{K} = \mathrm{LFT}(\mathbf{K}_g, \mathbf{Q}), \tag{10.185}$$

where $\mathbf{Q} \in RH^{\infty}_{+,p_1 \times p_2}$ *is an arbitrary parameter with* $\|\mathbf{Q}\|_\infty < 1$ *and* $\mathbf{K}_g$ *has the realization*

$$\begin{array}{rcl} A_g & = & A - B_2 D_{12}^{-1} C_1 - [B_1 D_{21}^T \\ & & + AYC_2^T - B_2 D_{12}^{-1}(D_{11} D_{21}^T + C_1 Y C_2^T)](D_{21} D_{21}^T + C_2 Y C_2^T)^{-1} C_2, \\ B_{g1} & = & -[B_1 D_{21}^T + AYC_2^T - B_2 D_{12}^{-1}(D_{11} D_{21}^T + C_1 Y C_2^T)](D_{21} D_{21}^T + C_2 Y C_2^T)^{-1}, \\ B_{g2} & = & -[B_2 D_{12}^{-1} + (B_1 D_{11}^T + AYC_1^T) S_o \\ & & -(AYC_2^T + B_1 D_{21}^T)(D_{21} D_{21}^T + C_2 Y C_2^T)^{-1}(D_{21} D_{11}^T + C_2 Y C_1^T) S_o] S_o^{-\frac{1}{2}}, \\ C_{g1} & = & D_{12}^{-1} C_1 - D_{12}^{-1}(D_{11} D_{21}^T + C_1 Y C_2^T)(D_{21} D_{21}^T + C_2 Y C_2^T)^{-1} C_2, \\ C_{g2} & = & (D_{21} D_{21}^T + C_2 Y C_2^T)^{-1/2} C_2, \\ D_{g11} & = & -D_{12}^{-1}(D_{11} D_{21}^T + C_1 Y C_2^T)(D_{21} D_{21}^T + C_2 Y C_2^T)^{-1}, \\ D_{g12} & = & D_{12}^{-1} S_o^{-1/2}, \\ D_{g21} & = & (D_{21} D_{21}^T + C_2 Y C_2^T)^{-1/2}, \\ D_{g22} & = & 0, \end{array} \tag{10.186}$$

and

$$\begin{array}{rcl} S_o & := & (V_{o11}^T V_{o11})^{-1} = [I - D_{11} D_{11}^T - C_1 Y C_1^T - (D_{11} D_{21}^T + C_1 Y C_2^T) \\ & & \times (D_{21} D_{21}^T + C_2 Y C_2^T)^{-1}(D_{21} D_{11}^T + C_2 Y C_1^T)]^{-1}. \end{array} \tag{10.187}$$

This theorem is the dual of Theorem 10.17.2.

10.17.4 The general case

We turn now to the remains of the proof of Theorem 10.12.1. We assume that (CD1), (CD2) and (CD3) hold together with (RD1) and (RD2) and we find the class of all stabilizing and contracting controllers.

As in the continuous–time case, we repeat the scheme in the proof of Theorem 10.17.2. We consider the systems $\mathbf{T}_I$ (10.147) and $\mathbf{T}_O$ (10.148) that we introduced in Section 10.15. According to Lemma 10.15.2, it is enough to find all the stabilizing and contracting controllers for $\mathbf{T}_O$. We need to show that

$$\mathbf{T}_O = \left[\begin{array}{c|c} A_O & B_O \\ \hline C_O & D_O \end{array}\right] = \left[\begin{array}{c|cc} A_O & B_{O1} & B_{O2} \\ \hline C_{O1} & D_{O11} & D_{O12} \\ C_{O2} & D_{O21} & O \end{array}\right] := \left[\begin{array}{c|cc} A+B_1F_1 & B_1V_{c11}^{-1} & B_2 \\ \hline -V_{c22}F_2 & V_{c21}V_{c11}^{-1} & V_{c22} \\ C_2+D_{21}F_1 & D_{21}V_{c11}^{-1} & O \end{array}\right]$$

satisfies the OE assumptions and we apply Theorem 10.17.3 to obtain the parameterization of all the stabilizing and contracting controllers.

Assumption (OE1) is obvious and for (OE2) we have

$$A_O - B_{O2}D_{O12}^{-1}C_{O1} = A + B_1F_1 - B_2V_{c22}^{-1}(-V_{c22}F_2) = A + BF,$$

which is stable.

The triplet Σ_o for the system $\mathbf{T}_O$ coincides with $\Sigma_\times$ and according to Lemma 10.16.1, we deduce that the Riccati equation associated to $\Sigma_\times$ has the positive semidefinite stabilizing solution Z that is given by $Z = Y(I - XY)^{-1}$. In order to prove that (OE3) holds for $\mathbf{T}_O$ (or, equivalently, (CD2) holds), we still need to prove the following result.

Lemma 10.17.4 *If (CD1), (CD2), and (CD3) in Theorem 10.12.1 hold, then there exists*

$$V_\times = \begin{bmatrix} V_{\times 11} & O \\ V_{\times 21} & V_{\times 22} \end{bmatrix},$$

with $V_{\times 11}$ and $V_{\times 22}$ invertible, such that

$$R_\times + C_O Z C_O^T = V_\times^T J_\times V_\times. \tag{10.188}$$

Proof. Notice that (10.188) is simply the first equation in the KSPYS$(\Sigma_\times, J_\times)$.

First of all, let us introduce some notation that will help with later computations. We denote

$$\overline{D}_{11} = \begin{bmatrix} D_{11} & C_1Y^{1/2} \\ X^{1/2}B_1 & X^{1/2}AY^{1/2} \end{bmatrix}, \tag{10.189}$$

$$\overline{D}_{12} = \begin{bmatrix} D_{12} \\ X^{1/2}B_2 \end{bmatrix}, \tag{10.190}$$

$$\overline{D}_{21} = \begin{bmatrix} D_{21} & C_2Y^{1/2} \end{bmatrix}. \tag{10.191}$$

Let us write

$$Z = Y^{1/2}(I - Y^{1/2}XY^{1/2})^{-1}Y^{1/2}$$

and, if we denote

$$T := (I - Y^{1/2}XY^{1/2})^{-1}, \tag{10.192}$$

we have

$$Z = Y^{1/2}TY^{1/2}. \tag{10.193}$$

Recall also the notation

$$S_c := (V_{c11}^T V_{c11})^{-1}$$

introduced in the statement of Theorem 10.12.1.

With this notation, we compute the (2,2)–block of $R_\times + C_O Z C_O^T$ as

$$\begin{aligned} & D_{21} V_{c11}^{-1} V_{c11}^{-T} D_{21}^T + (C_2 + D_{21} F_1) Z (C_2 + D_{21} F_1)^T \\ &= D_{21} S_c D_{21}^T + (C_2 - D_{21} S_c V_{c11}^T W_{c1}) Y^{1/2} T Y^{1/2} (C_2 - D_{21} S_c V_{c11}^T W_{c1})^T \\ &= \overline{D}_{21} \begin{bmatrix} S_c + S_c V_{c11}^T W_{c1} Y^{1/2} T Y^{1/2} W_{c1}^T V_{c11} S_c & -S_c V_{c11}^T W_{c1} Y^{1/2} T \\ -T Y^{1/2} W_{c1}^T V_{c11} S_c & T \end{bmatrix} \overline{D}_{21}^T . \end{aligned}$$

An essential observation for the rest of this proof is that the matrix

$$\begin{bmatrix} S_c + S_c V_{c11}^T W_{c1} Y^{1/2} T Y^{1/2} W_{c1}^T V_{c11} S_c & -S_c V_{c11}^T W_{c1} Y^{1/2} T \\ -T Y^{1/2} W_{c1}^T V_{c11} S_c & T \end{bmatrix}$$

is the inverse of the matrix

$$\mathcal{V} := \begin{bmatrix} V_{c11}^T V_{c11} & V_{c11}^T W_{c1} Y^{1/2} \\ Y^{1/2} W_{c1}^T V_{c11} & I - Y^{1/2} X Y^{1/2} + Y^{1/2} W_{c1}^T W_{c1} Y^{1/2} \end{bmatrix} . \tag{10.194}$$

This fact can be checked directly.

We can write

$$D_{21} V_{c11}^{-1} V_{c11}^{-T} D_{21}^T + (C_2 + D_{21} F_1) Z (C_2 + D_{21} F_1)^T = \overline{D}_{21} \mathcal{V}^{-1} \overline{D}_{21}^T . \tag{10.195}$$

Now $\mathcal{V} > 0$ since $V_{c11}^T V_{c11} > 0$, and its Schur complement is

$$\begin{aligned} I - Y^{1/2} X Y^{1/2} + Y^{1/2} W_{c1}^T W_{c1} Y^{1/2} - Y^{1/2} W_{c1}^T V_{c11} (V_{c11}^T V_{c11})^{-1} V_{c11}^T W_{c1} Y^{1/2} \\ = I - Y^{1/2} X Y^{1/2} > 0, \end{aligned}$$

because of (CD3).

We show now that $\overline{D}_{21}$ has full row rank. To this end, consider the Popov triplet $\Sigma_{o2} = (A^T, C_2^T; B_1 B_1^T, B_1 D_{21}^T, D_{21} D_{21}^T)$ which is the positiveness Popov triplet associated with $\mathbf{T}_{21}^T$. It follows from the regularity assumption (RD2) that the hypotheses of Proposition 7.8.1 are fulfilled for $\mathbf{T}_{21}^T$, and thus the DTARE(Σ_{o2}) has a stabilizing solution $Y_2 \geq 0$. We have $Y \geq Y_2$. This follows by using similar arguments to those used in the proof of the implication 1. $\Rightarrow$ 2. in part II of Theorem 4.12.9 (see the proof of Theorem 4.5.9). However, in our case A is not stable as required by Theorem 4.12.9. Nevertheless, we can make the reasoning in terms of the feedback equivalent Popov triplets $\widetilde{\Sigma}_o$ and $\widetilde{\Sigma}_{o2}$ which have the "A" matrices stable, where $\widetilde{\Sigma}_o$ is the $\begin{bmatrix} O \\ \widetilde{F}_2 \end{bmatrix}$–equivalent of Σ_o and $\widetilde{\Sigma}_{o2}$ is the $\widetilde{F}_2$–equivalent of Σ_{o2}, for some matrix $\widetilde{F}_2$ for which $A^T + C_2^T \widetilde{F}_2$ is stable. As the DTARE($\widetilde{\Sigma}_o$) and DTARE($\widetilde{\Sigma}_{o2}$) have the same stabilizing solutions Y and Y_2 as DTARE(Σ_o) and DTARE(Σ_{o2}), respectively, it follows that $Y \geq Y_2$.

From this we get $\overline{D}_{21} \overline{D}_{21}^T = D_{21} D_{21}^T + C_2 Y C_2^T \geq D_{21} D_{21}^T + C_2 Y_2 C_2^T > 0$ which means that $\overline{D}_{21}$ has full row rank.

Hence we deduce that the left–hand side of (10.195) is positive definite and there exists some invertible matrix $V_{\times 22}$ such that

$$[R_\times + C_O Z C_O^T]_{22} = V_{\times 22}^T V_{\times 22} . \tag{10.196}$$

By $[M]_{ij}$ we denote the (i,j) block element of the matrix M. We choose now $V_{\times 21}$ to be

$$V_{\times 21} = V_{\times 22}^{-T}[R_\times + C_O Z C_O^T]_{12}^T \tag{10.197}$$

and, in order to introduce $V_{\times 11}$ (see (10.127)), we need to show that

$$-[R_\times + C_O Z C_O^T]_{11} + V_{\times 21}^T V_{\times 21} > 0,$$

or, equivalently,

$$\begin{aligned} &-[R_\times + C_O Z C_O^T]_{11} \\ &+[R_\times + C_O Z C_O^T]_{12}[R_\times + C_O Z C_O^T]_{22}^{-1}[R_\times + C_O Z C_O^T]_{12}^T > 0. \end{aligned} \tag{10.198}$$

Indeed, if (10.198) is satisfied, then we can choose an invertible $V_{\times 11}$ such that

$$V_{\times 11}^T V_{\times 11} = -[R_\times + C_O Z C_O^T]_{11} + V_{\times 21}^T V_{\times 21} \tag{10.199}$$

and, with this choice, relation (10.188) is satisfied.

It only remains to prove the inequality (10.198). To this end we shall need to bring $\mathcal{V}$ into a more amenable form.

But let us first notice that from the KSPYS(Σ_c, J_c) we have that

$$\overline{D}_{12}^T \overline{D}_{12} = V_{c22}^T V_{c22}.$$

We deduce that there exists some unitary matrix U_c such that

$$\overline{D}_{12} = U_c \begin{bmatrix} I_{m_2} \\ O \end{bmatrix} V_{c22}. \tag{10.200}$$

Similarly, from the KSPYS(Σ_o, J_o), we have that

$$\overline{D}_{21} \overline{D}_{21}^T = V_{o22}^T V_{o22},$$

such that there exists some unitary matrix U_o such that

$$\overline{D}_{21}^T = U_o \begin{bmatrix} I_{p_2} \\ O \end{bmatrix} V_{o22}. \tag{10.201}$$

Let us also introduce the following partition of $U_c^T \overline{D}_{11} U_o$ compatible with relations (10.200) and (10.201),

$$U_c^T \overline{D}_{11} U_o = \begin{bmatrix} \overline{D}_{11}^{11} & \overline{D}_{11}^{12} \\ \overline{D}_{11}^{21} & \overline{D}_{11}^{22} \end{bmatrix}. \tag{10.202}$$

From the KSPYS(Σ_c, J_c) we have that

$$\begin{aligned} &I - \overline{D}_{11}^T \overline{D}_{11} \\ &= \begin{bmatrix} I - D_{11}^T D_{11} - B_1^T X B_1 & (-D_{11}^T C_1 - B_1^T X A) Y^{1/2} \\ Y^{1/2}(-C_1^T D_{11} - A^T X B_1) & I + Y^{1/2}(-A^T X A - C_1^T C_1) Y^{1/2} \end{bmatrix} \\ &= \begin{bmatrix} V_{c11}^T V_{c11} - V_{c21}^T V_{c21} & (V_{c11}^T W_{c1} - V_{c21}^T W_{c2}) Y^{1/2} \\ Y^{1/2}(W_{c1}^T V_{c11} - W_{c2}^T V_{c21}) & I + Y^{1/2}(-X + W_{c1}^T W_{c1} - W_{c2}^T W_{c2}) Y^{1/2} \end{bmatrix}. \end{aligned}$$

Accordingly, with (10.194), we have

$$\mathcal{V} = I - \overline{D}_{11}^T \overline{D}_{11} + \begin{bmatrix} V_{c21}^T V_{c21} & V_{c21}^T W_{c2} Y^{1/2} \\ Y^{1/2} W_{c2}^T V_{c21} & Y^{1/2} W_{c2}^T W_{c2} Y^{1/2} \end{bmatrix}.$$

If we introduce

$$\overline{V}_{21} := \begin{bmatrix} V_{c21} & W_{c2} Y^{1/2} \end{bmatrix}, \tag{10.203}$$

then we get

$$\mathcal{V} = I - \overline{D}_{11}^T \overline{D}_{11} + \overline{V}_{21}^T \overline{V}_{21}. \tag{10.204}$$

On the other hand, using again the KSPYS(Σ_c, J_c), we have

$$\overline{D}_{11}^T \overline{D}_{12} = \begin{bmatrix} D_{11}^T D_{12} + B_1^T X B_2 \\ Y^{1/2}(C_1^T D_{12} + A^T X B_2) \end{bmatrix} = \begin{bmatrix} V_{c21}^T V_{c22} \\ Y^{1/2} W_{c2} V_{c22} \end{bmatrix} = \overline{V}_{21}^T V_{c22}.$$

With relation (10.200), we deduce that

$$\overline{V}_{21} = \begin{bmatrix} I_{m_2} & O \end{bmatrix} U_c^T \overline{D}_{11}. \tag{10.205}$$

Using this expression and (10.202) in (10.204), we compute

$$\begin{aligned} U_o^T \mathcal{V} U_o &= I - U_o^T \overline{D}_{11}^T U_c U_c^T \overline{D}_{11} U_o + U_o^T \overline{D}_{11}^T U_c \begin{bmatrix} I_{m_2} \\ O \end{bmatrix} \begin{bmatrix} I_{m_2} & O \end{bmatrix} U_c^T \overline{D}_{11} U_o \\ &= I - \begin{bmatrix} \overline{D}_{11}^{11} & \overline{D}_{11}^{12} \\ \overline{D}_{11}^{21} & \overline{D}_{11}^{22} \end{bmatrix}^T \begin{bmatrix} O & O \\ O & I \end{bmatrix} \begin{bmatrix} \overline{D}_{11}^{11} & \overline{D}_{11}^{12} \\ \overline{D}_{11}^{21} & \overline{D}_{11}^{22} \end{bmatrix} \\ &= I - \begin{bmatrix} \overline{D}_{11}^{21} & \overline{D}_{11}^{22} \end{bmatrix}^T \begin{bmatrix} \overline{D}_{11}^{21} & \overline{D}_{11}^{22} \end{bmatrix}. \end{aligned} \tag{10.206}$$

Since $\mathcal{V} > 0$, we deduce that

$$I > \begin{bmatrix} \overline{D}_{11}^{21} & \overline{D}_{11}^{22} \end{bmatrix}^T \begin{bmatrix} \overline{D}_{11}^{21} & \overline{D}_{11}^{22} \end{bmatrix}. \tag{10.207}$$

By duality, we can also write the inequality

$$I > \begin{bmatrix} \overline{D}_{11}^{12} \\ \overline{D}_{11}^{22} \end{bmatrix}^T \begin{bmatrix} \overline{D}_{11}^{12} \\ \overline{D}_{11}^{22} \end{bmatrix}$$

and, as a consequence, we shall use further in the proof that

$$I > (\overline{D}_{11}^{12})^T \overline{D}_{11}^{12} + (\overline{D}_{11}^{22})^T \overline{D}_{11}^{22}. \tag{10.208}$$

As it will appear frequently in our further computations we introduce

$$E = (U_o^T \mathcal{V} U_o)^{-1} = \begin{bmatrix} E_{11} & E_{12} \\ E_{12}^T & E_{22} \end{bmatrix} \tag{10.209}$$

where the partition is conformal to the expression (10.206).

The next step in proving (10.198) is to express the terms in the left–hand side in terms of $\mathcal{V}$ and then in terms of the matrices in (10.202) and E in (10.209). We have seen from (10.195) that

$$[R_\times + C_O Z C_O^T]_{22} = \overline{D}_{21}\mathcal{V}^{-1}\overline{D}_{21}^T.$$

With (10.206) and (10.201) we have

$$\begin{aligned}[R_\times + C_O Z C_O^T]_{22} &= V_{o22}^T \begin{bmatrix} I_{p_2} & O \end{bmatrix} E \begin{bmatrix} I_{p2} \\ O \end{bmatrix} V_{o22} \\ &= V_{o22}^T E_{11} V_{o22}. \end{aligned} \tag{10.210}$$

Further,

$$\begin{aligned}[R_\times + C_O Z C_O^T]_{12} &= V_{c21} S_c D_{21}^T - V_{c22} F_2 Z (C_2^T + F_1^T D_{21}^T) \\ &= [V_{c21} S_c - V_{c22} F_2 Z F_1^T] D_{21}^T - V_{c22} F_2 Y^{1/2} T Y^{1/2} C_2^T \\ &= \begin{bmatrix} V_{c21} S_c - V_{c22} F_2 Z F_1^T & -V_{c22} F_2 Y^{1/2} T \end{bmatrix} \overline{D}_{21}^T \end{aligned}$$

and, substituting

$$V_{c22} F_2 = V_{c21} S_c V_{c11}^T W_{c1} - W_{c2}\ , \quad F_1 = W_{c1}^T V_{c11} S_c, \tag{10.211}$$

simple manipulations lead to

$$[R_\times + C_O Z C_O^T]_{12} = \overline{V}_{21}\mathcal{V}^{-1}\overline{D}_{21}^T. \tag{10.212}$$

Using here (10.205) and (10.201), we compute

$$\begin{aligned}[R_\times + C_O Z C_O^T]_{12} &= \begin{bmatrix} I_{m_2} & O \end{bmatrix} U_c^T \overline{D}_{11} U_o (U_o^T \mathcal{V} U_o)^{-1} U_o^T \overline{D}_{21}^T \\ &= \begin{bmatrix} \overline{D}_{11}^{11} & \overline{D}_{11}^{12} \end{bmatrix} E \begin{bmatrix} I_{p2} \\ O \end{bmatrix} V_{o22}. \end{aligned} \tag{10.213}$$

To finish with the terms of (10.198), we have to compute

$$-[R_\times + C_O Z C_O^T]_{11} = I_{m_2} - V_{c21}(V_{c11}^T V_{c11})^{-1} V_{c21}^T - V_{c22} F_2 Z F_2^T V_{c22}^T.$$

With (10.211), we arrive at

$$-[R_\times + C_O Z C_O^T]_{11} = I_{m_2} - \overline{V}_{21}\mathcal{V}^{-1}\overline{V}_{21}^T.$$

With (10.205),

$$\begin{aligned}&-[R_\times + C_O Z C_O^T]_{11} && (10.214) \\ &= I_{m_2} - \begin{bmatrix} I_{m_2} & O \end{bmatrix} U_c^T \overline{D}_{11} U_o (U_o^T \mathcal{V} U_o)^{-1} U_o^T \overline{D}_{11}^T U_c \begin{bmatrix} I_{m_2} & O \end{bmatrix}^T \\ &= I_{m_2} - \begin{bmatrix} \overline{D}_{11}^{11} & \overline{D}_{11}^{12} \end{bmatrix} E \begin{bmatrix} \overline{D}_{11}^{11} & \overline{D}_{11}^{12} \end{bmatrix}^T. && (10.215)\end{aligned}$$

The left–hand side of (10.198) can now be readily computed from (10.210), (10.213) and (10.215) and we obtain that it equals

$$I - \overline{D}_{11}^{12}(E_{22} - E_{12}^T E_{11}^{-1} E_{12})(\overline{D}_{11}^{12})^T.$$

It is a matter of simple computations using the definition of E in (10.209) to prove that

$$E_{22} - E_{12}^T E_{11}^{-1} E_{12} = (I - (\overline{D}_{11}^{22})^T \overline{D}_{11}^{22})^{-1}.$$

It remains to prove that

$$I - \overline{D}_{11}^{12}[I - (\overline{D}_{11}^{22})^T \overline{D}_{11}^{22}]^{-1} (\overline{D}_{11}^{12})^T > 0.$$

However, this last inequality is equivalent to the fact that the eigenvalues of the matrix

$$I - [I - (\overline{D}_{11}^{22})^T \overline{D}_{11}^{22}]^{-1} (\overline{D}_{11}^{12})^T \overline{D}_{11}^{12} = [I - (\overline{D}_{11}^{22})^T \overline{D}_{11}^{22}]^{-1} [I - (\overline{D}_{11}^{22})^T \overline{D}_{11}^{22} - (\overline{D}_{11}^{12})^T \overline{D}_{11}^{12}]$$

are all positive, which in turn follows immediately from the matrix inequality (10.208). ■

10.18 State feedback solution

When the state is accessible for measurements the solution to the discrete–time γ–DAP can be expressed in simple state–space feedback form. In this case the system (10.117) reduces to

$$\begin{aligned} \sigma x &= Ax + B_1 u_1 + B_2 u_2, \\ y_1 &= C_1 x + D_{11} u_1 + D_{12} u_2, \end{aligned} \tag{10.216}$$

or, equivalently, (10.118) reduces to

$$\mathbf{T} = \left[\begin{array}{c|c} A & B \\ \hline C_1 & D \end{array}\right] = \left[\begin{array}{c|cc} A & B_1 & B_2 \\ \hline C_1 & D_{11} & D_{12} \end{array}\right] = \left[\begin{array}{cc} \mathbf{T}_{11} & \mathbf{T}_{12} \end{array}\right]. \tag{10.217}$$

The discrete–time γ–DAP consists, in this case, in finding a feedback $F_2 \in \mathbb{R}^{m_2 \times n}$ such that the state feedback law

$$u_2 = F_2 x \tag{10.218}$$

stabilizes the system (10.216), which means that the resulting closed–loop system

$$\begin{aligned} \sigma x &= (A + B_2 F_2) x + B_1 u_1, \\ y_1 &= (C_1 + D_{12} F_2) x + D_{11} u_1, \end{aligned} \tag{10.219}$$

is stable, i.e.,

$$\Lambda(A + B_2 F_2) \subset \mathbb{D} \tag{10.220}$$

and its TFM satisfies

$$||\mathbf{T}_{y_1 u_1}||_\infty < \gamma.$$

As before, we can reduce the problem to the case $\gamma = 1$, and we shall call a stabilizing feedback that solves the γ–DAP, a stabilizing and contracting state feedback.

We assume in the sequel that the regularity assumption (RD1) holds.

Theorem 10.18.1 *There exists a stabilizing and contracting state feedback if and only if (CD1) in Theorem 10.12.1 is true and, additionally,*

$$D_{11}^T D_{11} + B_1^T X B_1 < I_{m_1}. \tag{10.221}$$

Moreover, in this case, if

$$F = \begin{bmatrix} F_1 \\ F_2 \end{bmatrix}$$

is the stabilizing feedback given by the solution of the KSPYS(Σ_c, J_c), then F_2 is a stabilizing and contracting state feedback.

Proof. "Necessity". Assume that a stabilizing and contracting state feedback exists.

An argument similar to the one used in Section 10.14 can be used to prove (CD1). As it follows along the same lines as the continuous–time counterpart we omit it.

To prove (10.221) we need to prove first that

$$D_{11}^T D_{11} - I < 0, \tag{10.222}$$

which, since $X \geq 0$, is a weaker inequality.

Take an arbitrary $v \in \mathbb{R}^{m_1}$ and let $u_1 = (u_{1,k})_{k\in\mathbb{Z}}$ be defined as $u_{1,0} = v$ and $u_{1,k} = 0$ for $k \neq 0$. Clearly $u_1 \in \ell^{2,m_1}$. Since $A + B_2F_2$ is stable, the output $y_1 = (y_{1,k})_{k\in\mathbb{Z}}$ belongs to ℓ^{2,p_1} and

$$\begin{aligned} y_{1,k} &= \sum_{i=-\infty}^{k-1} C_{1F_2}(A + B_2F_2)^{k-i-1}B_1u_{1,i} + D_{11}u_{1,k} \\ &= \begin{cases} 0, & k < 0, \\ D_{11}v, & k = 0, \\ C_{1F_2}(A + B_2F_2)^{k-1}B_1v, & k > 0. \end{cases} \end{aligned} \tag{10.223}$$

Here $C_{1F_2} = C_1 + D_{12}F_2$. Since F_2 is contracting, we can write

$$||D_{11}v||^2 = ||y_{1,0}||^2 \leq ||y_1||_2^2 < ||u_1||_2^2 = ||v||_2^2,$$

and (10.222) is proved.

In order to prove (10.221), we use again the construction in Section 10.15. Indeed, a stabilizing and contracting feedback can be regarded as a stabilizing and contracting controller for the system

$$\mathbf{T}_{aux} = \left[\begin{array}{c|cc} A & B_1 & B_2 \\ \hline C_1 & D_{11} & D_{12} \\ I & O & O \end{array}\right].$$

This system does not satisfy the regularity assumption (RD2). Nonetheless, that assumption was not used in Section 10.15 and therefore the entire argument carries on to the new situation. In particular, we deduce from Lemma 10.15.2 that F_2 is a stabilizing and contracting feedback for the system

$$\mathbf{T}_O = \left[\begin{array}{c|cc} A + B_1F_1 & B_1V_{c11}^{-1} & B_2 \\ \hline -V_{c22}F_2 & V_{c21}V_{c11}^{-1} & V_{c22} \end{array}\right]. \tag{10.224}$$

Therefore this system must also satisfy (10.222), that is,

$$V_{c11}^{-T}V_{c21}^TV_{c21}V_{c11}^{-1} - I < 0$$

or, equivalently,

$$-V_{c11}^T V_{c11} + V_{c21}^T V_{c21} < 0. \tag{10.225}$$

From the KSPYS(Σ_c, J_c) we have that

$$-I + D_{11}^T D_{11} + B_1^T X B_1 = -V_{c11}^T V_{c11} + V_{c21}^T V_{c21} \tag{10.226}$$

and this proves (10.221).

"Sufficiency". We assume now that (RD1), (10.221) and (CD1) hold. The idea is to prove that F_2 is a stabilizing and contracting state feedback using Theorem 4.12.10. We see that

$$R_{c11} + B_1^T X B_1 = -I + D_{11}^T D_{11} + B_1^T X B_1 < 0,$$

and $R_{c22} + B_2^T X B_2 = D_{12}^T D_{12} + B_2^T X B_2 > 0$. The stabilizing solution to the DTARE associated with Σ_c exists and is positive semidefinite according to (CD1). The hypotheses of Theorem 4.12.10 are fulfilled and we deduce that $A + B_2 F_2$ is stabilizing and there exists $\rho > 0$ such that

$$J_{e,\Sigma_c}(u_1, u_2) \leq -\rho^2 \|u_1\|_2^2, \quad \forall u_1 \in \ell^{2,m_1}, \tag{10.227}$$

for $u_2 = F_2 x$. Using (10.125) we can easily deduce that

$$J_{e,\Sigma_c}(u_1, u_2) = -\|u_1\|_2^2 + \|y_1\|_2^2. \tag{10.228}$$

By combining (10.227) and (10.228),

$$\|y_1\|_2^2 \leq (1 - \rho^2)\|u\|_1^2, \quad \forall u_1 \in \ell^{2,m_1},$$

which in fact is equivalent to the contracting condition. ■

Notes and References

The idea to design a feedback control system as to reduce the sensitivity in closed–loop is due to Zames [Zam81] who showed also that this design problem is equivalent to an optimization problem having as performance criterion the L^∞ operator norm. The fundamental works of Sarason [Sar67], and Adamjan, Arov, and Krein [AAK71], have established the connection between operator theory and complex function theory, in particular L^∞–norms. For good surveys on the classical topics of the field we refer to Francis and Doyle [FD87] and Francis [Fra87]. Remarkable contributions to the state–space solution to the H^∞–control problem are due to Francis [Fra87], Doyle, Glover, Khargonekar, and Francis [DGKF89], Tadmor [Tad90], Stoorvogel [Sto92], and Iglesias [Igl91]. A parallel technique of solving the regular H^∞–control problem under normalizing conditions was developed by Ionescu and Weiss [IW93c] in a discrete–time setting. For both continuous and discrete–time systems, the H^∞–control problem is solved in this chapter for the most general class of systems which satisfy the regularity assumptions, and the formulas for solutions are the most general available and appear here for the first time. The extension of this approach to more general classes of systems is due to Drăgan, Halanay and Ionescu [DHI94], and Barb, Ionescu, and de Koning [BIdK95, BIdK94a, BIdK94b]. The proof of Redheffer's theorem as presented in this chapter is due to Ionescu [Ion98a].

11

Robust Stabilization

In this chapter we solve robust stabilization problems for various types of uncertainties by employing the theory developed in Chapter 10. The key argument for this approach is that a system with uncertainties modeled additive, multiplicative, or on coprime factors is equivalent to a disturbance attenuation problem formulated for an adequately chosen generalized system.

The chapter is organized as follows. In Section 11.1 we introduce different models of uncertainties and explain the aforementioned equivalence. We give an evaluation of the least achievable tolerance $\gamma_{\min}$ for which the disturbance feedforward (DF) problem has an optimal solution, and construct this solution by using singular perturbation techniques in Section 11.2. The main goal of this section is to put ground for Section 11.3 in which we solve the robust stabilization problem with normalized coprime factors uncertainties, which is nothing but a particular case of the disturbance feedforward problem. The breakthrough here is that we obtain an optimal (as opposed to suboptimal) solution which is amenable to numerical algorithms. The numerical features include the reduced order of the solution and the removal of the large pole–zeros that usually lead to an ill–conditioned computation of the solution near the optimum. The problem of robust stabilization with multiplicative uncertainties which turns out to be a singular disturbance attenuation problem, is solved again by a singular perturbations approach in Section 11.4. The corresponding discrete–time cases are treated in detail in Sections 11.5–11.8.

A. The continuous–time case

11.1 Problem formulation and some prerequisites

Consider the $p \times m$ continuous–time system $\mathbf{G}$, $y = \mathbf{G}u$, and let $(\widetilde{\mathbf{N}}, \widetilde{\mathbf{M}})$ be a left coprime factorization over RH_{+}^{∞} (see Section 7.3) such that $\mathbf{G} = \widetilde{\mathbf{M}}^{-1}\widetilde{\mathbf{N}}$. We define a *stabilizing controller* for $\mathbf{G}$ as an $m \times p$ system $\mathbf{K}$ for which the feedback closed–loop of $\mathbf{G}$ with $\mathbf{K}$ is internally stable (see Section 2.1.5).

For a $\delta > 0$, introduce the following classes of systems

$$\mathcal{D}_{\delta}^{a} := \{\mathbf{G}_{\Delta}^{a} : \mathbf{G}_{\Delta}^{a} = \mathbf{G} + \mathbf{\Delta}, \mathbf{\Delta} \in RH_{+,p\times m}^{\infty} \text{ and } \|\mathbf{\Delta}\|_{\infty} < \delta\}, \tag{11.1}$$

$$\mathcal{D}_{\delta}^{m} := \{\mathbf{G}_{\Delta}^{m} : \mathbf{G}_{\Delta}^{m} = (I + \mathbf{\Delta})\mathbf{G}, \mathbf{\Delta} \in RH_{+,p\times p}^{\infty} \text{ and } \|\mathbf{\Delta}\|_{\infty} < \delta\}, \tag{11.2}$$

$$\begin{aligned} \mathcal{D}_{\delta}^{cf} := \{\mathbf{G}_{\Delta}^{cf} : \mathbf{G}_{\Delta}^{cf} &= (\widetilde{\mathbf{M}} + \mathbf{\Delta}_M)^{-1}(\widetilde{\mathbf{N}} + \mathbf{\Delta}_N), \\ \mathbf{\Delta} &= \left[\begin{array}{c|c} \mathbf{\Delta}_N & -\mathbf{\Delta}_M \end{array} \right] \in RH_{+,p\times(p+m)}^{\infty} \text{ and } \|\mathbf{\Delta}\|_{\infty} < \delta\}. \end{aligned} \tag{11.3}$$

Each class $\mathcal{D}_\delta^a$, $\mathcal{D}_\delta^m$ and $\mathcal{D}_\delta^{cf}$ represents a class of perturbed (uncertain) systems obtained from the $\mathbf{G}$ by *additive, multiplicative* and *coprime factors uncertainties.* Denote by $\mathcal{D}_\delta$ any of the three classes of systems introduced above, and by $\mathbf{G}_\Delta$ a generic element of $\mathcal{D}_\delta$.

The largest $\delta = \delta_{\max}$ for which there exists a single stabilizing controller for all systems in $\mathcal{D}_{\delta_{\max}}$ is called the *maximum stability margin* for the respective class.

Given a nominal system $\mathbf{G}$, a bound δ and a class of systems $\mathcal{D}_\delta$, where $\delta < \delta_{\max}$, the *suboptimal robust stabilization problem* consists in constructing a single stabilizing controller $\mathbf{K}$, $u = \mathbf{K}y$, for all the systems $\mathbf{G}_\Delta \in \mathcal{D}_\delta$ (see Figure 11.1). The controller

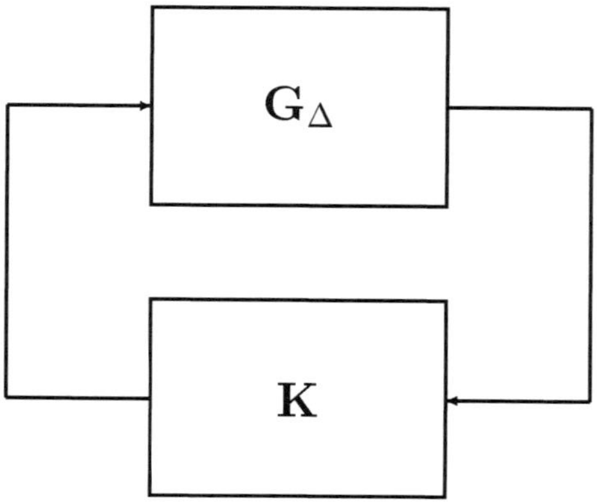

Figure 11.1 Robustly stabilizing controller $\mathbf{K}$ for the systems $\mathbf{G}_\Delta \in \mathcal{D}_\delta$

$\mathbf{K}$ is called a *robustly stabilizing controller.* The *optimal robust stabilization problem* consists of constructing a single stabilizing controller $\mathbf{K}$ for all systems in $\mathcal{D}_{\delta_{\max}}$.

We show now how the robust stabilization problem can be converted into an equivalent disturbance attenuation problem (DAP). Each class of uncertain systems $\mathcal{D}_\delta$ can be represented as a ULFT of a nominal generalized system and a perturbation system $\boldsymbol{\Delta}$. Indeed, consider the three generalized systems

$$\mathbf{T}^a = \left[\begin{array}{c|c} \mathbf{T}_{11}^a & \mathbf{T}_{12}^a \\ \hline \mathbf{T}_{21}^a & \mathbf{T}_{22}^a \end{array}\right] = \left[\begin{array}{c|c} O & I_m \\ \hline I_p & \mathbf{G} \end{array}\right], \tag{11.4}$$

$$\mathbf{T}^m = \left[\begin{array}{c|c} \mathbf{T}_{11}^m & \mathbf{T}_{12}^m \\ \hline \mathbf{T}_{21}^m & \mathbf{T}_{22}^m \end{array}\right] = \left[\begin{array}{c|c} O & \mathbf{G} \\ \hline I_p & \mathbf{G} \end{array}\right], \tag{11.5}$$

$$\mathbf{T}^{cf} = \left[\begin{array}{c|c} \mathbf{T}_{11}^{cf} & \mathbf{T}_{12}^{cf} \\ \hline \mathbf{T}_{21}^{cf} & \mathbf{T}_{22}^{cf} \end{array}\right] = \left[\begin{array}{c|c} O & I_m \\ \widetilde{\mathbf{M}}^{-1} & \mathbf{G} \\ \hline \widetilde{\mathbf{M}}^{-1} & \mathbf{G} \end{array}\right]. \tag{11.6}$$

A routine computation shows that

$$\begin{aligned} \mathrm{ULFT}(\mathbf{T}^a, \boldsymbol{\Delta}) &= \mathbf{G} + \boldsymbol{\Delta} = \mathbf{G}_\Delta^a, & (11.7)\\ \mathrm{ULFT}(\mathbf{T}^m, \boldsymbol{\Delta}) &= (I + \boldsymbol{\Delta})\mathbf{G} = \mathbf{G}_\Delta^m, & (11.8)\\ \mathrm{ULFT}(\mathbf{T}^{cf}, \boldsymbol{\Delta}) &= (\widetilde{\mathbf{M}} + \boldsymbol{\Delta}_M)^{-1}(\mathbf{N} + \boldsymbol{\Delta}_N) = \mathbf{G}_\Delta^{cf}, \quad \boldsymbol{\Delta} = \left[\begin{array}{c|c} \boldsymbol{\Delta}_N & -\boldsymbol{\Delta}_M \end{array}\right] & (11.9) \end{aligned}$$

Comparing (11.7), (11.8) and (11.9) with (11.1), (11.2) and (11.3), respectively, reveals that the class of uncertain systems can, in each case, be described as ULFT($\mathbf{T}, \mathbf{\Delta}$) as shown in Figure 11.2. Thus we get the equivalent configuration in Figure 11.3 in

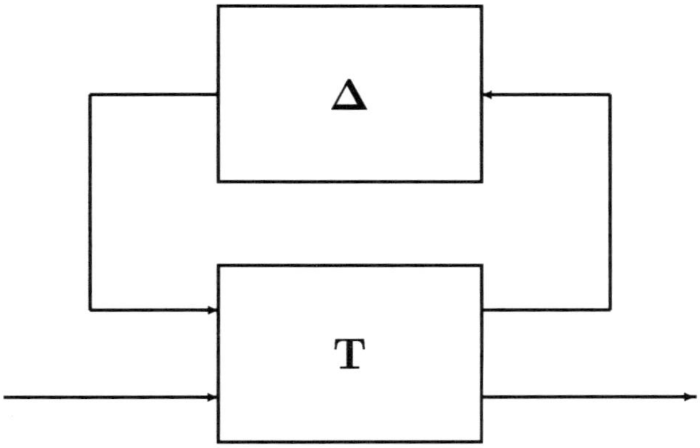

Figure 11.2 Model of uncertain systems $\mathbf{G}_\Delta$ = ULFT($\mathbf{T}, \mathbf{\Delta}$)

which *both* $\mathbf{T}_{y_1u_1}$ and $\mathbf{\Delta}$ are stable and $\mathbf{T}_{y_1u_1}$ = LLFT($\mathbf{T}, \mathbf{K}$), where $\mathbf{T}$ is given by either (11.4), (11.5), or (11.6). Returning to our original problem of converting the robust stabilization problem into a DAP, we have the following result.

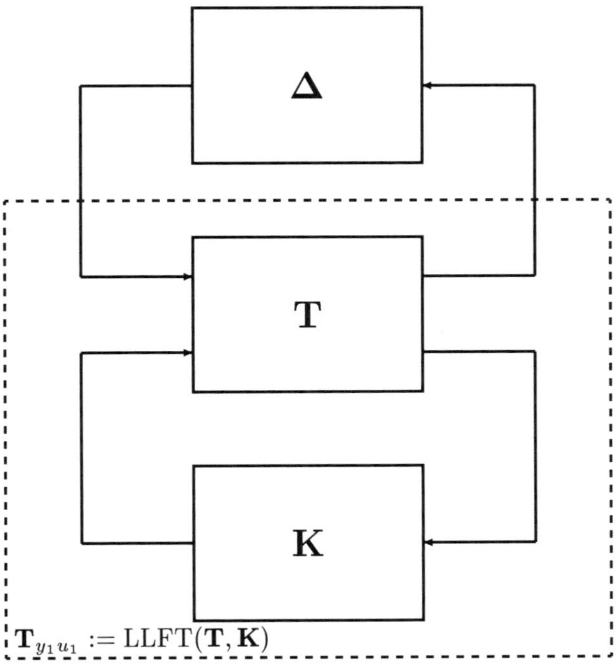

Figure 11.3 Model of uncertain systems with feedback controller

Theorem 11.1.1 *A controller* $\mathbf{K}$ *is a solution to the robust stabilization problem with respect to any of the class of systems* $\mathcal{D}_\delta$, $\delta \leq \delta_{\max}$, *given in either (11.1), (11.2), or (11.3), if and only if* $\mathbf{K}$ *is a solution to the corresponding nonstrict* $\frac{1}{\delta}$*-DAP for* $\mathbf{T}$ *given by either (11.4), (11.5), or (11.6).*

Proof. "If". From the Small Gain Theorem 7.4.1 we see that the closed–loop system in Figure 11.3 is internally stable for all stable $\mathbf{\Delta}$ with $\|\mathbf{\Delta}\|_\infty < \delta$, provided $\mathbf{T}_{y_1u_1}$ is stable and $\|\mathbf{T}_{y_1u_1}\|_\infty \leq \frac{1}{\delta}$, which is equivalent to saying that $\mathbf{K}$ is a solution to the nonstrict $\frac{1}{\delta}$–DAP.

"Only if". If the closed–loop system in Figure 11.3 is stable for all stable $\mathbf{\Delta}$ with $\|\mathbf{\Delta}\|_\infty < \delta$, then, in particular, it is stable for $\mathbf{\Delta} = 0$ from which it follows that $\mathbf{T}_{y_1u_1}$ is stable. To prove that $\|\mathbf{T}_{y_1u_1}\|_\infty \leq \frac{1}{\delta}$ we need a more sophisticated operatorial argument. Assume the contrary, i.e., $\|\mathbf{T}_{y_1u_1}\|_\infty > \frac{1}{\delta}$. Then there exists a stable $\mathbf{\Delta}$, with $\|\mathbf{\Delta}\|_\infty < \delta$, such that $I - \mathbf{T}_{y_1u_1}\mathbf{\Delta}$ is not a unit in RH_+^∞. This follows for example from the Spectral Mapping Theorem in Douglas [Dou72]. Since $I - \mathbf{T}_{y_1u_1}\mathbf{\Delta}$ is not a unit in RH_+^∞, the closed–loop feedback system in Figure 11.3 is not internally stable which contradicts the hypotheses. Thus $\|\mathbf{T}_{y_1u_1}\|_\infty \leq \frac{1}{\delta}$ and it follows that $\mathbf{K}$ is a solution to the nonstrict $\frac{1}{\delta}$–DAP. ■

In the next sections we shall solve successively the robust stabilization problems with normalized coprime factors and multiplicative uncertainties. As we shall see, the DAP problem corresponding to the robust stabilization problem with normalized coprime factors uncertainties is in fact a particular DF problem. Thus, we can obtain an expression for the corresponding maximum stability margin and a solution to the optimal robust stabilization problem by evaluating first the least achievable tolerance $\gamma_{\min} = \frac{1}{\delta_{\max}}$ and constructing an optimal controller (solution) for the DF problem. These two tasks are the topic of the next section.

11.2 Optimal solution to the DF problem

In this section we give an evaluation of the least achievable tolerance $\gamma_{\min}$ for which the DF problem has an optimal solution, and give a construction of that solution under certain hypotheses. Further, we shall show in Section 11.3 that these hypotheses are all satisfied in the case of the DF problem arising from the robust stabilization problem with normalized coprime factors uncertainties, and we shall provide a solution for the latter.

Consider the continuous–time generalized system

$$\mathbf{T} = \left[\begin{array}{c|c} A & B \\ \hline C & D \end{array}\right] = \left[\begin{array}{c|cc} A & B_1 & B_2 \\ \hline C_1 & D_{11} & D_{12} \\ C_2 & D_{21} & O \end{array}\right] = \left[\begin{array}{cc} \mathbf{T}_{11} & \mathbf{T}_{12} \\ \mathbf{T}_{21} & \mathbf{T}_{22} \end{array}\right], \tag{11.10}$$

$$\left[\begin{array}{c} y_1 \\ y_2 \end{array}\right] = \mathbf{T}\left[\begin{array}{c} u_1 \\ u_2 \end{array}\right],$$

where $A \in \mathbb{R}^{n\times n}$, $B_i \in \mathbb{R}^{n\times m_i}$, $C_j \in \mathbb{R}^{p_j\times n}$, $D_{ij} \in \mathbb{R}^{p_i\times m_j}$, $(i,j = 1,2$, $m = m_1 + m_2$, $p = p_1 + p_2)$. Assume the following conditions are fulfilled:

(A1) D_{21} is square and invertible.

(A2) $A - B_1 D_{21}^{-1} C_2$ is stable.

(A3) (A, B_2) is stabilizable.

(A4) rank $\begin{bmatrix} j\omega I - A & -B_2 \\ C_1 & D_{12} \end{bmatrix} = n + m_2, \quad \forall \omega \in \mathbb{R}, \qquad \operatorname{rank} D_{12} = m_2.$

Notice that (A1) and (A2) are exactly the first two assumptions made for the DF problem in Section 10.8.2, while (A3) and (A4) are among the general hypotheses made in Section 10.2 for the DAP. The result below is a particular case of Theorem 10.3.1 (see also Remarks 10.3.2 and 10.8.4).

Theorem 11.2.1 *Consider the system* $\mathbf{T}$ *in (11.10), and assume it satisfies conditions (A1), (A2), (A3), and (A4). Let* $J_c := \operatorname{diag}(-I_{m_1}, I_{m_2})$ *and consider the Popov triplet* $\Sigma_c = (A, B; Q_c, L_c, R_c)$ *where*

$$\begin{aligned} Q_c &:= C_1^T C_1, \\ L_c &= \begin{bmatrix} L_{c1} & L_{c2} \end{bmatrix} := C_1^T \begin{bmatrix} D_{11} & D_{12} \end{bmatrix}, \\ R_c &= \begin{bmatrix} R_{c11} & R_{c12} \\ R_{c12}^T & R_{c22} \end{bmatrix} := \begin{bmatrix} D_{11}^T \\ D_{12}^T \end{bmatrix} \begin{bmatrix} D_{11} & D_{12} \end{bmatrix} - \begin{bmatrix} \gamma^2 I_{m_1} & O \\ O & O \end{bmatrix}. \end{aligned} \tag{11.11}$$

Then the γ*-DF problem for* $\mathbf{T}$ *has a solution if and only if the KPYS(*Σ_c, J_c*) has a stabilizing solution*

$$(X, V_c, W_c) = (X, \begin{bmatrix} V_{c11} & O \\ V_{c21} & V_{c22} \end{bmatrix}, \begin{bmatrix} W_{c1} \\ W_{c2} \end{bmatrix}), \qquad X \geq 0. \tag{11.12}$$

Using this theorem, we give now an evaluation of the least achievable tolerance $\gamma_{\min}$.

11.2.1 An evaluation of $\gamma_{\min}$

Consider the generalized system (11.10) satisfying assumptions (A1)–(A4) and the Popov triplet $\Sigma_c = (A, B; Q_c, L_c, R_c)$, with the entries given by (11.11). Before stating the main result of the section we make some preparations.

Since (A3) and (A4) hold, it follows with Proposition 7.2.1 that the KPYS(Σ_{c2}, I_{m_2}), where $\Sigma_{c2} = (A, B_2; Q_c, L_{c2}, R_{c22})$, has a stabilizing solution (X_2, V_2, W_2), with $X_2 \geq 0$. Let $\widetilde{\Sigma}_c = (\widetilde{A}, B; \widetilde{Q}_c, \widetilde{L}_c, R_c)$ be the $\begin{bmatrix} O \\ \widetilde{F}_2 \end{bmatrix}$-equivalent Popov triplet of Σ_c, where $\widetilde{F}_2$ is the stabilizing feedback associated with the KPYS(Σ_{c2}, I_{m_2}), with the entries given explicitly by

$$\begin{aligned} \widetilde{A} &:= A + B_2 \widetilde{F}_2, \\ \widetilde{Q}_c &:= \widetilde{C}_1^T \widetilde{C}_1, \\ \widetilde{L}_c &:= \widetilde{C}_1 \begin{bmatrix} D_{11} & D_{12} \end{bmatrix}, \end{aligned} \tag{11.13}$$

and $\widetilde{C}_1 := C_1 + D_{12} \widetilde{F}_2$. The Popov function associated with $\widetilde{\Sigma}_c$ is given by

$$\mathbf{\Pi}_{\widetilde{\Sigma}_c} = \begin{bmatrix} \mathbf{\Pi}_{\widetilde{\Sigma}_c,11} & \mathbf{\Pi}_{\widetilde{\Sigma}_c,12} \\ \mathbf{\Pi}_{\widetilde{\Sigma}_c,12}^* & \mathbf{\Pi}_{\widetilde{\Sigma}_c,22} \end{bmatrix} = \begin{bmatrix} \widetilde{\mathbf{T}}_{11}^* \\ \widetilde{\mathbf{T}}_{12}^* \end{bmatrix} \begin{bmatrix} \widetilde{\mathbf{T}}_{11} & \widetilde{\mathbf{T}}_{12} \end{bmatrix} - \begin{bmatrix} \gamma^2 I & O \\ O & O \end{bmatrix} \tag{11.14}$$

where

$$\left[\begin{array}{cc} \widetilde{\mathbf{T}}_{11} & \widetilde{\mathbf{T}}_{12} \end{array} \right] = \left[\begin{array}{c|cc} \widetilde{A} & B_1 & B_2 \\ \hline \widetilde{C}_1 & D_{11} & D_{12} \end{array} \right]. \tag{11.15}$$

We are ready now for the main result.

Theorem 11.2.2 *Consider the system* $\mathbf{T}$ *in (11.10), and assume it satisfies conditions (A1), (A2), (A3), and (A4). Let*

$$\mathcal{R}_{\widetilde{\Sigma}_c} = \left[\begin{array}{cc} \mathcal{R}_{11} & \mathcal{R}_{12} \\ \mathcal{R}_{12}^* & \mathcal{R}_{22} \end{array} \right] \tag{11.16}$$

be the causal Toeplitz operator with symbol $\mathbf{\Pi}_{\widetilde{\Sigma}_c}$ *(the partition is taken in accordance with* $\mathbf{\Pi}_{\widetilde{\Sigma}_c}$ *given in (11.14)) and let*

$$\mathcal{R}_{11}^{\times} := \mathcal{R}_{11} - \mathcal{R}_{12}\mathcal{R}_{22}^{-1}\mathcal{R}_{12}^* \tag{11.17}$$

be the Schur complement of $\mathcal{R}_{22}$ *in (11.16). Then we have:*

1. The (suboptimal) γ*–DF problem for* $\mathbf{T}$ *has a solution if and only if*

$$\mathcal{R}_{11}^{\times} << 0. \tag{11.18}$$

2. The least achievable $\gamma = \gamma_{\min}$ *(the optimal value) for the DF problem is given by*

$$\gamma_{\min} = \rho^{\frac{1}{2}}(\mathbb{H}_{\widetilde{\mathbf{T}}_{11}^*\widehat{\mathbf{T}}_{12}}\mathbb{H}_{\widetilde{\mathbf{T}}_{11}^*\widehat{\mathbf{T}}_{12}}^* + \mathbb{T}_{(\widetilde{\mathbf{T}}_{11}^*\widehat{\mathbf{T}}_{12}^{\perp})(\widetilde{\mathbf{T}}_{11}^*\widehat{\mathbf{T}}_{12}^{\perp})^*}), \tag{11.19}$$

where $\widehat{\mathbf{T}}_{12}^{\perp} \in RL^{\infty}_{p_1\times(p_1-m_2)}$ *is an orthogonal completion of the inner TFM*

$$\widehat{\mathbf{T}}_{12} = \widetilde{\mathbf{T}}_{12}V_2^{-1} \tag{11.20}$$

such that $\widehat{\mathbf{T}}_{12}^a := \left[\begin{array}{cc} \widehat{\mathbf{T}}_{12} & \widehat{\mathbf{T}}_{12}^{\perp} \end{array} \right]$ *is* all–pass, *i.e.,* $(\widehat{\mathbf{T}}_{12}^a)^*(\widehat{\mathbf{T}}_{12}^a) = (\widehat{\mathbf{T}}_{12}^a)(\widehat{\mathbf{T}}_{12}^a)^* = I$.

Proof. Notice first that $\mathbf{\Pi}_{\widetilde{\Sigma}_c,22} = \widetilde{\mathbf{T}}_{12}^*\widetilde{\mathbf{T}}_{12} = V_2^TV_2$, which is a consequence of the spectral factorization identity (3.39). Hence $\widehat{\mathbf{T}}_{12}$ defined in (11.20) is inner and $\mathcal{R}_{22} >> 0$. It is easy to see by some standard arguments in Hilbert spaces that we always can construct an orthogonal completion of $\widehat{\mathbf{T}}_{12}$ such that $\widehat{\mathbf{T}}_{12}^a$ is all–pass.

From Theorem 11.2.1 it follows that the γ–DF problem has a solution if and only if the KPYS(Σ_c, J_c) has a stabilizing solution (11.12). Further, from the equivalence of the Popov triplets $\widetilde{\Sigma}_c$ and Σ_c and Proposition 3.2.13 we infer that the KPYS(Σ_c, J_c) has a stabilizing solution (11.12) if and only if the KPYS$(\widetilde{\Sigma}_c, J_c)$ has a stabilizing solution

$$(X, V_c, \widetilde{W}_c) = (X, \left[\begin{array}{cc} V_{c11} & O \\ V_{c21} & V_{c22} \end{array} \right], \left[\begin{array}{c} W_{c1} \\ W_{c2} + V_{c22}\widetilde{F}_2 \end{array} \right]). \tag{11.21}$$

Thus, all it remains to prove is that the KPYS$(\widetilde{\Sigma}_c, J_c)$ has a stabilizing solution $(X, \left[\begin{array}{cc} V_{c11} & O \\ V_{c21} & V_{c22} \end{array} \right], \widetilde{W}_c)$ if and only if (11.18) holds.

1. "Necessity". We show first that the condition (11.18) is necessary. We check that the hypotheses of Theorem 4.5.9 are fulfilled for the triplet $\widetilde{\Sigma}_c$. Indeed, $\widetilde{A}$ given in (11.13) is stable because $\widetilde{F}_2$ is the stabilizing feedback associated with the KPYS(Σ_{c2}, I_{m_2}), $\mathbf{\Pi}_{\widetilde{\Sigma}_c,22} = V_2^T V_2 > 0$, and

$$\left[\begin{array}{c} \widetilde{C}_1^T \\ D_{12}^T \end{array}\right]\left[\begin{array}{cc} \widetilde{C}_1 & D_{12} \end{array}\right] \geq 0.$$

From the implication 2. ⇒ 1. in part I of Theorem 4.5.9 it follows that there exists $\mathbf{S} \in RH_+^\infty$ which satisfies the signature inequality

$$\left[\begin{array}{cc} I & \mathbf{S}^* \end{array}\right] \mathbf{\Pi}_{\widetilde{\Sigma}_c} \left[\begin{array}{c} I \\ \mathbf{S} \end{array}\right] < 0 \text{ on } \overline{\mathbb{C}}_0. \tag{11.22}$$

Finally, from (11.22) it follows similarly as in the proof of Theorem 4.5.8 that $\mathcal{R}_{11}^\times << 0$.

"Sufficiency". We assume now that (11.18) holds. With $\mathcal{R}_{22} >> 0$ and (11.18) we get from Theorem 4.4.1 that the CTARE$(\widetilde{\Sigma}_c)$ has a stabilizing solution $X \geq 0$. We also have

$$\operatorname{sgn}(\mathbf{\Pi}_{\widetilde{\Sigma}_c}) = J_c \tag{11.23}$$

on $\overline{\mathbb{C}}_0$ because $\mathbf{\Pi}_{\widetilde{\Sigma}_c,22} > 0$ and its Schur complement satisfies

$$\mathbf{\Pi}_{\widetilde{\Sigma}_c,11}^\times := \widetilde{\mathbf{T}}_{11}^* \widetilde{\mathbf{T}}_{11} - \gamma^2 I - \widetilde{\mathbf{T}}_{11}^* \widehat{\mathbf{T}}_{12}^* \widehat{\mathbf{T}}_{12} \widetilde{\mathbf{T}}_{11} = -\gamma^2 I < 0.$$

From the spectral factorization identity (3.39) and (11.23) we get also $\operatorname{sgn}(R_c) = J_c$, and since $R_{c22} > 0$ there exists $V_c = \left[\begin{array}{cc} V_{c11} & O \\ V_{c21} & V_{c22} \end{array}\right]$ such that

$$V_c^T J_c V_c = R_c. \tag{11.24}$$

Since we already proved that the CTARE$(\widetilde{\Sigma}_c)$ has a stabilizing solution $X \geq 0$, we get from (11.24) with Remark 3.2.9 that the KPYS$(\widetilde{\Sigma}_c, J_c)$ has a stabilizing solution $(X, V_c, \widetilde{W}_c) = (X, \left[\begin{array}{cc} V_{c11} & O \\ V_{c21} & V_{c22} \end{array}\right], \widetilde{W}_c)$, and the proof ends.

2. To obtain the expression (11.19) of $\gamma_{\min}$ we evaluate the operator $\mathcal{R}_{11}^\times$. With (11.14) one gets

$$\mathcal{R}_{11}^\times = \mathbb{T}_{\widetilde{\mathbf{T}}_{11}^* \widetilde{\mathbf{T}}_{11}} - \gamma^2 I - \mathbb{T}_{\widetilde{\mathbf{T}}_{11}^* \widehat{\mathbf{T}}_{12}} \mathbb{T}_{\widetilde{\mathbf{T}}_{11}^* \widehat{\mathbf{T}}_{12}}^*. \tag{11.25}$$

Since $\widehat{\mathbf{T}}_{12}^a$ is all–pass, we have

$$\left[\begin{array}{cc} \widehat{\mathbf{T}}_{12} & \widehat{\mathbf{T}}_{12}^\perp \end{array}\right]^* \left[\begin{array}{cc} \widehat{\mathbf{T}}_{12} & \widehat{\mathbf{T}}_{12}^\perp \end{array}\right] = \left[\begin{array}{cc} \widehat{\mathbf{T}}_{12} & \widehat{\mathbf{T}}_{12}^\perp \end{array}\right] \left[\begin{array}{cc} \widehat{\mathbf{T}}_{12} & \widehat{\mathbf{T}}_{12}^\perp \end{array}\right]^* = I_{p_1}. \tag{11.26}$$

Using (11.26), (11.25) becomes

$$\mathcal{R}_{11}^\times = \mathbb{T}_{\widetilde{\mathbf{T}}_{11}^* [\widehat{\mathbf{T}}_{12}\ \widehat{\mathbf{T}}_{12}^\perp][\widehat{\mathbf{T}}_{12} \widehat{\mathbf{T}}_{12}^\perp]^* \widetilde{\mathbf{T}}_{11}} - \mathbb{T}_{\widetilde{\mathbf{T}}_{11}^* \widehat{\mathbf{T}}_{12}} \mathbb{T}_{\widetilde{\mathbf{T}}_{11}^* \widehat{\mathbf{T}}_{12}}^* - \gamma^2 I. \tag{11.27}$$

With the identity

$$\mathbb{T}_{\mathbf{G}\mathbf{G}^*} = \mathbb{T}_{\mathbf{G}}\mathbb{T}_{\mathbf{G}}^* + \mathbb{H}_{\mathbf{G}}\mathbb{H}_{\mathbf{G}}^* \tag{11.28}$$

which holds for any $\mathbf{G}$, and where $\mathbb{T}_{\mathbf{G}}$ and $\mathbb{H}_{\mathbf{G}}$ are the causal Toeplitz and Hankel operators associated with $\mathbf{G}$, we get from (11.27)

$$\begin{aligned}\mathcal{R}_{11}^{\times} &= \mathbb{T}_{\widetilde{\mathbf{T}}_{11}^*\widehat{\mathbf{T}}_{12}\widehat{\mathbf{T}}_{12}^*\widetilde{\mathbf{T}}_{11}} + \mathbb{T}_{\widetilde{\mathbf{T}}_{11}^*\widehat{\mathbf{T}}_{12}^{\perp}(\widehat{\mathbf{T}}_{12}^{\perp})^*\widetilde{\mathbf{T}}_{11}} - \mathbb{T}_{\widetilde{\mathbf{T}}_{11}^*\widehat{\mathbf{T}}_{12}\widehat{\mathbf{T}}_{12}^*\widetilde{\mathbf{T}}_{11}} + \mathbb{H}_{\widetilde{\mathbf{T}}_{11}^*\widehat{\mathbf{T}}_{12}}\mathbb{H}_{\widetilde{\mathbf{T}}_{11}^*\widehat{\mathbf{T}}_{12}}^* - \gamma^2 I \\ &= \mathbb{H}_{\widetilde{\mathbf{T}}_{11}^*\widehat{\mathbf{T}}_{12}}\mathbb{H}_{\widetilde{\mathbf{T}}_{11}^*\widehat{\mathbf{T}}_{12}}^* + \mathbb{T}_{(\widetilde{\mathbf{T}}_{11}^*\widehat{\mathbf{T}}_{12}^{\perp})(\widetilde{\mathbf{T}}_{11}^*\widehat{\mathbf{T}}_{12}^{\perp})^*} - \gamma^2 I.\end{aligned} \tag{11.29}$$

Hence condition (11.18) is equivalent in this case to

$$\rho^{\frac{1}{2}}(\mathbb{H}_{\widetilde{\mathbf{T}}_{11}^*\widehat{\mathbf{T}}_{12}}\mathbb{H}_{\widetilde{\mathbf{T}}_{11}^*\widehat{\mathbf{T}}_{12}}^* + \mathbb{T}_{(\widetilde{\mathbf{T}}_{11}^*\widehat{\mathbf{T}}_{12}^{\perp})(\widetilde{\mathbf{T}}_{11}^*\widehat{\mathbf{T}}_{12}^{\perp})^*}) < \gamma, \tag{11.30}$$

where ρ denotes the spectral radius. The conclusion follows immediately from (11.30) and point 1 of the theorem. ∎

11.2.2 An optimal solution

In this section we construct an optimal solution to a DF problem formulated for a particular generalized system $\mathbf{T}$. As already explained, our final goal is to obtain in Section 11.3 an optimal solution to the robust stabilization problem with normalized coprime factors uncertainties. We take here a singular perturbations approach as explained further. We take $\gamma > \gamma_{\min}$ for which the suboptimal γ–DF problem has a solution, say $\mathbf{K}_\gamma$, which we know how to construct from Chapter 10. Further, we investigate the behavior of the state–space realization of $\mathbf{K}_\gamma$ as $\gamma \downarrow \gamma_{\min}$, and find the limit which is a state–space system, say $\mathbf{K}_{\gamma_{\min}}$. Finally, we prove that $\mathbf{K}_{\gamma_{\min}}$ is a stabilizing and optimal controller for the DF problem at hand by investigating the behavior of the closed–loop system $\mathrm{LFT}(\mathbf{T}, \mathbf{K}_\gamma)$ as $\gamma \downarrow \gamma_{\min}$. We shall divide this reasoning among several subsections.

Before pursuing our plan, we make some notation and additional assumptions which will considerably simplify the exposition and which, as we shall see, are all fulfilled by the robust stabilization problem with normalized coprime factors uncertainties.

Consider the generalized system (11.10) satisfying assumptions (A1)–(A4). We introduce a number of additional hypotheses (A5)–(A8) as follows.

(A5) $D_{12}^T \begin{bmatrix} C_1 & D_{11} & D_{12} \end{bmatrix} = \begin{bmatrix} O & O & I_{m_2} \end{bmatrix}$.

Consider also the Popov triplet $\Sigma_c = (A, B; Q_c, L_c, R_c)$, with the entries given by (11.11). Due to (A5) the entries of Σ_c have the following simpler forms:

$$Q_c = C_1^T C_1, \quad L_c = \begin{bmatrix} C_1^T D_{11} & O \end{bmatrix}, \quad R_c = \begin{bmatrix} -\gamma^2 I + D_{11}^T D_{11} & O \\ O & I \end{bmatrix}. \tag{11.31}$$

Take $\gamma > \gamma_{\min}$ such that according to Theorem 11.2.2 the KPYS(Σ_c, J_c) has a stabilizing solution (X, V_c, W_c) and the γ–DF problem has a suboptimal solution

$$\mathbf{K} = \left[\begin{array}{c|c} A_K & B_K \\ \hline C_K & D_K \end{array}\right] = \left[\begin{array}{c|c} A - B_1 D_{21}^{-1} C_2 - B_2 B_2^T X_\gamma & B_1 D_{21}^{-1} \\ \hline -B_2^T X_\gamma & O \end{array}\right], \quad u_2 = \mathbf{K} y_2. \tag{11.32}$$

The explicit expression for the realization of **K** was obtained from Theorem 10.8.2 by taking in the formulas (10.94) the central controller ($\mathbf{Q} = 0$), removing the scaling $\gamma = 1$, and denoting $X = X_\gamma$ to stress the dependence on γ. Define

$$\epsilon := \gamma^2 - \gamma_{\min}^2, \tag{11.33}$$

and assume **T** satisfies the following additional hypotheses:

(A6) $\gamma_{\min} > \overline{\sigma}(D_{11})$.

(A7) There exists $r \in \{1, \ldots, n\}$ such that

$$X_\gamma = \begin{bmatrix} \epsilon^{-1} X_1 & \\ & X_2 \end{bmatrix} \tag{11.34}$$

where

$$\epsilon^{-1} X_1 = \operatorname{diag}(\nu_1, \ldots, \nu_r), \quad X_2 = \operatorname{diag}(\nu_{r+1}, \ldots, \nu_n),$$

X_1 and X_2 depend continuously on ϵ and there exist the (finite) limits

$$\lim_{\epsilon \downarrow 0} X_i = \overline{X}_i, \quad \text{with} \quad \overline{X}_1 > 0.$$

(A8) $B_{21} B_{21}^T > 0$ where B_{21} is defined by partitioning (11.10) conformably with the partition of X_γ in (11.34) as

$$\mathbf{T} = \left[\begin{array}{c|c:c} A & B_1 & B_2 \\ \hline C_1 & D_{11} & D_{12} \\ \hdashline C_2 & D_{21} & O \end{array}\right] = \left[\begin{array}{cc|c:c} A_{11} & A_{12} & B_{11} & B_{21} \\ A_{21} & A_{22} & B_{12} & B_{22} \\ \hline C_{11} & C_{12} & D_{11} & D_{12} \\ \hdashline C_{21} & C_{22} & D_{21} & O \end{array}\right]. \tag{11.35}$$

Following our plan, we investigate now the behavior of **K** and of the closed-loop LFT(**T**, **K**) when $\epsilon \downarrow 0$.

The behavior of the controller when $\epsilon \downarrow 0$

We study the expression of the controller as $\epsilon \downarrow 0$. Using (11.34) and (11.35) we get for the realization of the controller **K** in (11.32) the expressions

$$\begin{aligned} A_K &= A - B_1 D_{21}^{-1} C_2 - \begin{bmatrix} \epsilon^{-1} B_{21} B_{21}^T X_1 & B_{21} B_{22}^T X_2 \\ \epsilon^{-1} B_{22} B_{21}^T X_1 & B_{22} B_{22}^T X_2 \end{bmatrix} =: \begin{bmatrix} A_{11}^o - \epsilon^{-1} M_{11} & A_{12}^o \\ A_{21}^o - \epsilon^{-1} M_{21} & A_{22}^o \end{bmatrix}, \\ B_K &= \begin{bmatrix} B_{11} \\ B_{12} \end{bmatrix} D_{21}^{-1} =: \begin{bmatrix} B_{K1} \\ B_{K2} \end{bmatrix}, \\ C_K &= \begin{bmatrix} -\epsilon^{-1} B_{21}^T X_1 & -B_{22}^T X_2 \end{bmatrix} =: \begin{bmatrix} \epsilon^{-1} C_{K1} & C_{K2} \end{bmatrix}, \\ D_K &= 0. \end{aligned} \tag{11.36}$$

If $x_K = \begin{bmatrix} x_{K1} \\ x_{K2} \end{bmatrix}$ is the state of the controller partitioned conformably, then let

$$\xi_K = \begin{bmatrix} \xi_{K1} \\ \xi_{K2} \end{bmatrix} := \begin{bmatrix} \epsilon^{-1} x_{K1} \\ x_{K2} \end{bmatrix}. \tag{11.37}$$

With (11.37), (11.36) leads to the two time–scale system

$$\begin{aligned} \epsilon\dot{\xi}_{K1} &= (\epsilon A^o_{11} - M_{11})\xi_{K1} + A^o_{12}\xi_{K2} + B_{K1}y_2, \\ \dot{\xi}_{K2} &= (\epsilon A^o_{21} - M_{21})\xi_{K1} + A^o_{22}\xi_{K2} + B_{K2}y_2, \\ u_2 &= C_{K1}\xi_{K1} + C_{K2}\xi_{K2}. \end{aligned} \tag{11.38}$$

Making $\epsilon \downarrow 0$ in (11.38) gives

$$\begin{aligned} 0 &= -\overline{M}_{11}\xi_{K1} + \overline{A}^o_{12}\xi_{K2} + B_{K1}y_2, \\ \dot{\xi}_{K2} &= -\overline{M}_{21}\xi_{K1} + \overline{A}^o_{22}\xi_{K2} + B_{K2}y_2, \\ u_2 &= \overline{C}_{K1}\xi_{K1} + \overline{C}_{K2}\xi_{K2}, \end{aligned} \tag{11.39}$$

where $\overline{M}_{11}$, $\overline{A}^o_{12}$, $\overline{M}_{21}$, $\overline{A}^o_{22}$, $\overline{C}_{K1}$, $\overline{C}_{K2}$ are the limits for the corresponding entries when $\epsilon \downarrow 0$. Since $\overline{M}_{11} = B_{21}B^T_{21}X_1$ is nonsingular due to (A7) and (A8), we can eliminate ξ_{K1} in (11.39) and get

$$\begin{aligned} \dot{\xi}_{K2} &= (\overline{A}^o_{22} - \overline{M}_{21}\overline{M}^{-1}_{11}\overline{A}^o_{12})\xi_{K2} + (B_{K2} - \overline{M}_{21}\overline{M}^{-1}_{11}B_{K1})y_2, \\ u_2 &= (\overline{C}_{K2} + \overline{C}_{K1}\overline{M}^{-1}_{11}\overline{A}^o_{12})\xi_{K2} + \overline{C}_{K1}\overline{M}^{-1}_{11}B_{K1}y_2, \end{aligned}$$

which are the state–space equations for the reduced order controller

$$\begin{aligned} \overline{\mathbf{K}} &= \left[\begin{array}{c|c} \overline{A}_K & \overline{B}_K \\ \hline \overline{C}_K & \overline{D}_K \end{array}\right] = \left[\begin{array}{c|c} \overline{A}^o_{22} - \overline{M}_{21}\overline{M}^{-1}_{11}\overline{A}^o_{12} & B_{K2} - \overline{M}_{21}\overline{M}^{-1}_{11}B_{K1} \\ \hline \overline{C}_{K2} + \overline{C}_{K1}\overline{M}^{-1}_{11}\overline{A}^o_{12} & \overline{C}_{K1}\overline{M}^{-1}_{11}B_{K1} \end{array}\right], \\ u_2 &= \overline{\mathbf{K}}y_2. \end{aligned} \tag{11.40}$$

We are ready for the main result of this section.

Theorem 11.2.3 *Assume that the generalized system* $\mathbf{T}$ *given in (11.35) satisifies (A1)–(A8). Then the controller* $\overline{\mathbf{K}}$ *given in (11.40), of reduced order* $n-r$*, is a solution to the optimal DF problem formulated for* $\mathbf{T}$.

Proof. The proof of the theorem follows immediately once we show that the limit of the closed–loop system $\mathbf{T}_{y_1u_1}$, when $\gamma \downarrow \gamma_{\min}$, is a system, say $\overline{\mathbf{T}}_{y_1u_1}$, which is internally stable and for which $||\overline{\mathbf{T}}_{y_1u_1}|| = \gamma_{\min}$. The proof of these latter facts implies some tedious manipulations and several auxilliary results on the behavior of the closed–loop system which we defer to the next subsection. ■

The behavior of the closed–loop system when $\epsilon \downarrow 0$

Let $\gamma > \gamma_{\min}$. The resultant closed–loop system of $\mathbf{T}$ and $\mathbf{K}$ has the realization

$$\mathbf{T}_{y_1u_1} = \mathrm{LFT}(\mathbf{T},\mathbf{K}) = \left[\begin{array}{cc|c} A & -B_2B^T_2X_\gamma & B_1 \\ B_1D^{-1}_{21}C_2 & A - B_1D^{-1}_{21}C_2 - B_2B^T_2X_\gamma & B_1 \\ \hline C_1 & -D_{12}B^T_2X_\gamma & D_{11} \end{array}\right] \tag{11.41}$$

where we have used (11.32) in (2.44). After a similarity transformation and removal of the uncontrollable stable part, we get

$$\mathbf{T}_{y_1u_1} = \left[\begin{array}{c|c} A - B_2B_2^TX_\gamma & B_1 \\ \hline C_1 - D_{12}B_2^TX_\gamma & D_{11} \end{array}\right] =: \left[\begin{array}{c|c} A_R & B_R \\ \hline C_R & D_R \end{array}\right]. \tag{11.42}$$

Clearly A_R is stable and $\|\mathbf{T}_{y_1u_1}\|_\infty < \gamma$ since $\mathbf{K}$ is a solution to the suboptimal γ–DF problem. Consider the γ–contractiveness Popov triplet of $\mathbf{T}_{y_1u_1}$ given explicitly by

$$\begin{aligned} \Sigma_R &= (A_R, B_R; C_R^TC_R, C_R^TD_R, -\gamma^2I + D_R^TD_R) \\ &= (A - B_2B_2^TX_\gamma, B_1; C_1^TC_1 + X_\gamma B_2B_2^TX_\gamma, C_1^TD_{11}, -\gamma^2I + D_{11}^TD_{11}) \end{aligned} \tag{11.43}$$

where for the last equality we have used (A5). By applying Lemma 7.1.1 it follows that the KPYS$(\Sigma_R, -I)$

$$\begin{aligned} -\gamma^2I + D_R^TD_R &= -V_R^TV_R, \\ C_R^TD_R + X_RB_R &= -W_R^TV_R, \\ C_R^TC_R + A_R^TX_R + X_RA_R &= -W_R^TW_R, \end{aligned} \tag{11.44}$$

has a stabilizing solution (X_R, V_R, W_R), with $X_R \geq 0$.

The result below connects X_R to X_γ and is crucial to understand the behavior of the closed–loop system $\mathbf{T}_{y_1u_1}$.

Proposition 11.2.4 *Assume that the generalized system* $\mathbf{T}$ *given in (11.10) satisfies (A1)–(A5), let* $\gamma > \gamma_{\min}$ *and* $\mathbf{K}$ *be the central controller (11.32) that solves the* γ*–DF problem for* $\mathbf{T}$*. Then the following equality holds*

$$X_\gamma = X_R. \tag{11.45}$$

Proof. The idea of the proof is to relate the EHPs associated with the triplets Σ_R and Σ_c. The EHP(Σ_c), denoted $sM_{\Sigma_c} - N_{\Sigma_c}$, is given by

$$M_{\Sigma_c} = \begin{bmatrix} I & O & O & O \\ O & I & O & O \\ O & O & O & O \\ O & O & O & O \end{bmatrix}, \quad N_{\Sigma_c} = \begin{bmatrix} A & O & B_1 & B_2 \\ -C_1^TC_1 & -A^T & -C_1^TD_{11} & O \\ D_{11}^TC_1 & B_1^T & -\gamma^2I + D_{11}^TD_{11} & O \\ O & B_2^T & O & I \end{bmatrix}, \tag{11.46}$$

and the EHP(Σ_R), denoted $sM_{\Sigma_R} - N_{\Sigma_R}$, is given by

$$M_{\Sigma_R} = \begin{bmatrix} I & O & O \\ O & I & O \\ O & O & O \end{bmatrix}, N_{\Sigma_R} = \begin{bmatrix} A_R & O & B_1 \\ -C_1^TC_1 - X_\gamma B_2B_2^TX_\gamma & -A_R^T & -C_1^TD_{11} \\ D_{11}^TC_1 & B_1 & -\gamma^2I + D_{11}^TD_{11} \end{bmatrix}, \tag{11.47}$$

where $A_R = A - B_2B_2^TX_\gamma$. Since the KPYS$(\Sigma_c, J_c)$ and the KPYS$(\Sigma_R, -I)$ have stabilizing solutions, we can switch to the corresponding CTAREs and get by Theorem 5.2.1,

$$N_{\Sigma_c}V_{\Sigma_c} = M_{\Sigma_c}V_{\Sigma_c}S_{\Sigma_c}, \quad V_{\Sigma_c} := \begin{bmatrix} I \\ X_\gamma \\ F_1 \\ F_2 \end{bmatrix}, \quad \Lambda(S_{\Sigma_c}) \subset \mathbb{C}_-, \tag{11.48}$$

where $S_{\Sigma_c} := A + BF = A + B_1 F_1 + B_2 F_2$, and

$$N_{\Sigma_R} V_{\Sigma_R} = M_{\Sigma_R} V_{\Sigma_R} S_{\Sigma_R}, \quad V_{\Sigma_R} := \begin{bmatrix} I \\ X_R \\ F_R \end{bmatrix}, \quad \Lambda(S_{\Sigma_R}) \subset \mathbb{C}_-, \tag{11.49}$$

where $S_{\Sigma_R} := A_R + B_R F_R = A - B_2 B_2^T X_\gamma + B_1 F_R$. Here $F = \begin{bmatrix} F_1 \\ F_2 \end{bmatrix}$ and F_R are the stabilizing Riccati feedbacks of the CTARE(Σ_c) and CTARE(Σ_R), respectively. Eliminating $F_2 = -B_2^T X_\gamma$ from the first equality in (11.48) and using (11.46) yields

$$N_{\Sigma_R} \begin{bmatrix} I \\ X_\gamma \\ F_1 \end{bmatrix} = M_{\Sigma_R} \begin{bmatrix} I \\ X_\gamma \\ F_1 \end{bmatrix} S_{\Sigma_c}. \tag{11.50}$$

Comparing (11.50) with (11.49) and taking into account the uniqueness of the stabilizing solution to the CTARE, it follows from Theorem 5.2.1 that $X_R = X_\gamma$, $F_R = F_1$, $S_{\Sigma_R} = S_{\Sigma_c}$, and the proof ends. ■

Finally, we look at the behavior of the closed–loop as $\epsilon \downarrow 0$. Using (11.34) and (11.35) in (11.42) we get

$$\begin{aligned} A_R &= \begin{bmatrix} A_{11} - \epsilon^{-1} B_{21} B_{21}^T X_1 & A_{12} - B_{21} B_{22}^T X_2 \\ A_{21} - \epsilon^{-1} B_{22} B_{21}^T X_1 & A_{22} - B_{22} B_{22}^T X_2 \end{bmatrix}, \\ B_R &= \begin{bmatrix} B_{11} \\ B_{12} \end{bmatrix}, \\ C_R &= \begin{bmatrix} C_{11} - D_{12} B_{21}^T \epsilon^{-1} X_1 & C_{12} - D_{12} B_{22}^T X_2 \end{bmatrix}, \\ D_R &= D_{11}. \end{aligned} \tag{11.51}$$

If $x_R = \begin{bmatrix} x_{R1} \\ x_{R2} \end{bmatrix}$ is the state of the closed–loop partitioned conformably, then let

$$\xi_R = \begin{bmatrix} \xi_{R1} \\ \xi_{R2} \end{bmatrix} := \begin{bmatrix} \epsilon^{-1} x_{R1} \\ x_{R2} \end{bmatrix}. \tag{11.52}$$

With (11.52), (11.51) leads to the two time–scale system

$$\begin{aligned} \epsilon \dot{\xi}_{R1} &= (\epsilon A_{11} - B_{21} B_{21}^T X_1) \xi_{R1} + (A_{12} - B_{21} B_{22}^T X_2) \xi_{R2} + B_{11} u_1, \\ \dot{\xi}_{R2} &= (\epsilon A_{21} - B_{22} B_{21}^T X_1) \xi_{R1} + (A_{22} - B_{22} B_{22}^T X_2) \xi_{R2} + B_{12} u_1, \\ y_1 &= (\epsilon C_{11} - D_{12} B_{21}^T X_1) \xi_{R1} + (C_{12} - D_{12} B_{22}^T X_2) \xi_{R2} + D_{11} u_1. \end{aligned} \tag{11.53}$$

By taking $\epsilon \downarrow 0$ in (11.53) gives

$$\begin{aligned} 0 &= -B_{21} B_{21}^T \overline{X}_1 \xi_{R1} + (A_{12} - B_{21} B_{22}^T \overline{X}_2) \xi_{R2} + B_{11} u_1, \\ \dot{\xi}_{R2} &= -B_{22} B_{21}^T \overline{X}_1 \xi_{R1} + (A_{22} - B_{22} B_{22}^T \overline{X}_2) \xi_{R2} + B_{12} u_1, \\ y_1 &= -D_{12} B_{21}^T \overline{X}_1 \xi_{R1} + (C_{12} - D_{12} B_{22}^T \overline{X}_2) \xi_{R2} + D_{11} u_1. \end{aligned} \tag{11.54}$$

Using (A8), ξ_{R1} can be eliminated and (11.54) generates the reduced system

$$\overline{\mathbf{T}}_{y_1u_1} = \left[\begin{array}{c|c} \overline{A}_R & \overline{B}_R \\ \hline \overline{C}_R & \overline{D}_R \end{array}\right]$$

where

$$\begin{aligned}
\overline{A}_R &:= A_{22} - B_{22}B_{22}^T\overline{X}_2 - B_{22}B_{21}^T(B_{21}B_{21}^T)^{-1}(A_{12} - B_{21}B_{22}^T\overline{X}_2), \quad (11.55)\\
\overline{B}_R &:= B_{12} - B_{22}B_{21}^T(B_{21}B_{21}^T)^{-1}B_{11},\\
\overline{C}_R &:= C_{12} - D_{12}B_{22}^T\overline{X}_2 - D_{12}B_{21}^T(B_{21}B_{21}^T)^{-1}(A_{12} - B_{21}B_{22}^T\overline{X}_2),\\
\overline{D}_R &:= D_{11} - D_{12}B_{21}^T(B_{21}B_{21}^T)^{-1}B_{11},
\end{aligned}$$

For the system (11.54) we can prove the following key result.

Theorem 11.2.5 *1. The system (11.54) is internally stable, i.e.,*

$$\Lambda(-B_{21}B_{21}^T\overline{X}_1) \subset \mathbb{C}_-, \qquad \Lambda(\overline{A}_R) \subset \mathbb{C}_-.$$

2. $\|\overline{\mathbf{T}}_{y_1u_1}\|_\infty = \gamma_{\min}$.

Proof. 1. Indeed, as $B_{21}B_{21}^T > 0$ and $\overline{X}_1 > 0$, the stability of $-B_{21}B_{21}^T\overline{X}_1$ follows from inertia arguments.

To prove that $\overline{A}_R$ is stable we need some lengthy computations. We write first the KPYS($\Sigma_R, -I$) given in (11.44) in explicit form by using (11.43), (11.45), (11.33) and (11.34), and get

$$\begin{aligned}
-\epsilon I - \gamma_{\min}^2 I + D_{11}^T D_{11} &= -V_R^T V_R, \quad (11.56)\\
C_{11}^T D_{11} + \epsilon^{-1}X_1 B_{11} &= -W_{R1}^T V_R, \quad (11.57)\\
C_{12}^T D_{11} + X_2 B_{12} &= -W_{R2}^T V_R, \quad (11.58)\\
(A_{11} - \epsilon^{-1}B_{21}B_{21}^T X_1)^T\epsilon^{-1}X_1 + \epsilon^{-1}X_1(A_{11} - \epsilon^{-1}B_{21}B_{21}^T X_1) &\\
+\epsilon^{-2}X_1B_{21}B_{21}^T X_1 + C_{11}^T C_{11} &= -W_{R1}^T W_{R1}, (11.59)\\
(A_{21} - \epsilon^{-1}B_{22}B_{21}^T X_1)^T X_2 + \epsilon^{-1}X_1(A_{12} - B_{21}B_{22}^T X_2) &\\
+\epsilon^{-1}X_1B_{21}B_{22}^T X_2 + C_{11}^T C_{12} &= -W_{R1}^T W_{R2}, (11.60)\\
(A_{22} - B_{22}B_{22}^T X_2)^T X_2 + X_2(A_{22} - B_{22}B_{22}^T X_2) &\\
+X_2B_{22}B_{22}^T X_2 + C_{12}^T C_{12} &= -W_{R2}^T W_{R2}, (11.61)
\end{aligned}$$

where $W_R = \begin{bmatrix} W_{R1} & W_{R2} \end{bmatrix}$ has been partitioned conformably with X_γ in (11.34).

Using (11.56), W_{R1} and W_{R2} can be made explicit from (11.57) and (11.58). By performing this computation (11.59), (11.60) and (11.61) give

$$\begin{aligned}
&(A_{11}-\epsilon^{-1}B_{21}B_{21}^T X_1)^T\epsilon^{-1}X_1+\epsilon^{-1}X_1(A_{11}-\epsilon^{-1}B_{21}B_{21}^T X_1)+\epsilon^{-2}X_1B_{21}B_{21}^T X_1+C_{11}^T C_{11}\\
&\quad + (C_{11}^T D_{11} + \epsilon^{-1}X_1B_{11})(\epsilon I + \gamma_{\min}^2 I - D_{11}^T D_{11})^{-1}(C_{11}^T D_{11} + \epsilon^{-1}X_1B_{11})^T = 0,\\
&(A_{21} - \epsilon^{-1}B_{22}B_{21}^T X_1)^T X_2 + \epsilon^{-1}X_1(A_{12} - B_{21}B_{22}^T X_2) + \epsilon^{-1}X_1B_{21}B_{22}^T X_2 + C_{11}^T C_{12}\\
&\quad + (C_{11}^T D_{11} + \epsilon^{-1}X_1B_{11})(\epsilon I + \gamma_{\min}^2 I - D_{11}^T D_{11})^{-1}(C_{12}^T D_{11} + X_2B_{12})^T = 0,
\end{aligned}$$

$$(A_{22} - B_{22}B_{22}^T X_2)^T X_2 + X_2(A_{22} - B_{22}B_{22}^T X_2) + X_2 B_{22} B_{22}^T X_2 + C_{12}^T C_{12}$$
$$+ (C_{12}^T D_{11} + X_2 B_{12})(\epsilon I + \gamma_{\min}^2 I - D_{11}^T D_{11})^{-1}(C_{12}^T D_{11} + X_2 B_{12})^T = 0.$$

By taking $\epsilon \downarrow 0$ in the above three equations we get

$$-\overline{X}_1 B_{21} B_{21}^T \overline{X}_1 + \overline{X}_1 B_{11}(\gamma_{\min}^2 I - D_{11}^T D_{11})^{-1} B_{11}^T \overline{X}_1 = 0, \tag{11.62}$$

$$\overline{X}_1(A_{12} - B_{21}B_{22}^T \overline{X}_2) + \overline{X}_1 B_{11}(\gamma_{\min}^2 I - D_{11}^T D_{11})^{-1}(B_{12}^T \overline{X}_2 + D_{11}^T C_{12}) = 0, \tag{11.63}$$

$$(A_{22} - B_{22}B_{22}^T \overline{X}_2)^T \overline{X}_2 + \overline{X}_2(A_{22} - B_{22}B_{22}^T \overline{X}_2) + \overline{X}_2 B_{22} B_{22}^T \overline{X}_2$$
$$+ C_{12}^T C_{12} + (C_{12}^T D_{11} + \overline{X}_2 B_{12})(\gamma_{\min}^2 I - D_{11}^T D_{11})^{-1}(D_{11}^T C_{12} + B_{12}^T \overline{X}_2) = 0, \tag{11.64}$$

where $\gamma_{\min}^2 I > D_{11}^T D_{11}$ according to (A6). Introduce now

$$\widetilde{B}_{11} := B_{11}(\gamma_{\min}^2 I - D_{11}^T D_{11})^{\frac{1}{2}},$$
$$\widetilde{B}_{12} := B_{12}(\gamma_{\min}^2 I - D_{11}^T D_{11})^{\frac{1}{2}},$$
$$\widetilde{D}_{11} := D_{11}(\gamma_{\min}^2 I - D_{11}^T D_{11})^{\frac{1}{2}}.$$

With this notation (11.62)–(11.64) become

$$-\overline{X}_1 B_{21} B_{21}^T \overline{X}_1 + \overline{X}_1 \widetilde{B}_{11} \widetilde{B}_{11}^T \overline{X}_1 = 0, \tag{11.65}$$

$$\overline{X}_1(A_{12} - B_{21}B_{22}^T \overline{X}_2) + \overline{X}_1 \widetilde{B}_{11}(\widetilde{B}_{12}^T \overline{X}_2 + \widetilde{D}_{11}^T C_{12}) = 0, \tag{11.66}$$

$$(A_{22} - B_{22}B_{22}^T \overline{X}_2)^T \overline{X}_2 + \overline{X}_2(A_{22} - B_{22}B_{22}^T \overline{X}_2) + \overline{X}_2 B_{22} B_{22}^T \overline{X}_2 + C_{12}^T C_{12}$$
$$+ (C_{12}^T \widetilde{D}_{11} + \overline{X}_2 \widetilde{B}_{12})(\widetilde{B}_{12}^T \overline{X}_2 + \widetilde{D}_{11}^T C_{12}) = 0. \tag{11.67}$$

Now (11.65) and (11.66) give

$$B_{21} B_{21}^T = \widetilde{B}_{11} \widetilde{B}_{11}^T \tag{11.68}$$

and

$$A_{12} - B_{21} B_{22}^T \overline{X}_2 = -\widetilde{B}_{11}(\widetilde{B}_{12}^T \overline{X}_2 + \widetilde{D}_{11}^T C_{12}). \tag{11.69}$$

Using (11.68) and (11.69),

$$B_{22}B_{21}^T(B_{21}B_{21}^T)^{-1}(A_{12} - B_{21}B_{22}^T \overline{X}_2) = -B_{22}B_{21}^T(\widetilde{B}_{11}\widetilde{B}_{11}^T)^{-1}\widetilde{B}_{11}(\widetilde{B}_{12}^T \overline{X}_2 + \widetilde{D}_{11}^T C_{12}). \tag{11.70}$$

Using (11.70) and (11.55) we rewrite (11.67) as

$$\overline{A}_R^T \overline{X}_2 + \overline{X}_2 \overline{A}_R + C_{12}^T C_{12} - (\overline{X}_2 \widetilde{B}_{12} + C_{12}^T \widetilde{D}_{11})\widetilde{B}_{11}^T(\widetilde{B}_{11}\widetilde{B}_{11}^T)^{-1} B_{21} B_{22}^T \overline{X}_2$$
$$- \overline{X}_2 B_{22} B_{21}^T (\widetilde{B}_{11}\widetilde{B}_{11}^T)^{-1}\widetilde{B}_{11}(\widetilde{B}_{12}^T \overline{X}_2 + \widetilde{D}_{11}^T C_{12}) + \overline{X}_2 B_{22} B_{22}^T \overline{X}_2$$
$$+ (C_{12}^T \widetilde{D}_{11} + \overline{X}_2 \widetilde{B}_{12})(\widetilde{B}_{12}^T \overline{X}_2 + \widetilde{D}_{11}^T C_{12}) = 0. \tag{11.71}$$

Let

$$\overline{B}_{12} := \widetilde{B}_{12} + \overline{X}_2^{-1} C_{12}^T \widetilde{D}_{11}.$$

Then (11.71) becomes

$$\overline{A}_R^T \overline{X}_2 + \overline{X}_2 \overline{A}_R + C_{12}^T C_{12} + \overline{X}_2[B_{22}B_{22}^T + \overline{B}_{12}\overline{B}_{12}^T$$

$$- \overline{B}_{12}\widetilde{B}_{11}^T(\widetilde{B}_{11}\widetilde{B}_{11}^T)^{-1}B_{21}B_{22}^T - B_{22}B_{21}^T(\widetilde{B}_{11}\widetilde{B}_{11}^T)^{-1}\widetilde{B}_{11}\overline{B}_{12}^T]\overline{X}_2 = 0. \tag{11.72}$$

Let

$$E := B_{22}B_{22}^T + \overline{B}_{12}\overline{B}_{12}^T - \overline{B}_{12}\widetilde{B}_{11}^T(\widetilde{B}_{11}\widetilde{B}_{11}^T)^{-1}B_{21}B_{22}^T - B_{22}B_{21}^T(\widetilde{B}_{11}\widetilde{B}_{11}^T)^{-1}\widetilde{B}_{11}\overline{B}_{12}^T. \tag{11.73}$$

With this, (11.72) becomes

$$\overline{A}_R^T\overline{X}_2 + \overline{X}_2\overline{A}_R + C_{12}^TC_{12} + \overline{X}_2E\overline{X}_2 = 0. \tag{11.74}$$

Provided $E \geq 0$ and the following condition holds:

(A9) $\left[\begin{array}{c} C_{12} \\ \sqrt{E}\ \overline{X}_2 \end{array}\right], \overline{A}_R)$ is detectable,

we deduce from (11.74) with Theorem 1.5.5 that $\overline{A}_R$ is stable. Notice that (A9) is generically fulfilled. Thus, we only prove $E \geq 0$. From (A8) and (11.68) we get that both $\widetilde{B}_{11}$ and B_{21} have full row rank. Thus, there exist orthogonal matrices U and V such that $\widetilde{B}_{11}U = \left[B_{11}^{(1)}\ O\right]$ and $B_{21}V = \left[B_{21}^{(1)}\ O\right]$, with $B_{11}^{(1)}$ and $B_{21}^{(1)}$ both invertible. Partition conformably $\overline{B}_{12}U = \left[B_{12}^{(1)}\ B_{12}^{(2)}\right]$ and $B_{22}V = \left[B_{22}^{(1)}\ B_{22}^{(2)}\right]$ and get

$$E = [B_{22}^{(1)} - B_{12}^{(1)}(B_{11}^{(1)})^{-1}B_{21}^{(1)}][B_{22}^{(1)} - B_{12}^{(1)}(B_{11}^{(1)})^{-1}B_{21}^{(1)}]^T + B_{12}^{(2)}(B_{12}^{(2)})^T + B_{22}^{(2)}(B_{22}^{(2)})^T \tag{11.75}$$

where we have used $B_{11}^{(1)}(B_{11}^{(1)})^T = B_{21}^{(1)}(B_{21}^{(1)})^T$ which is a consequence of (11.68). From (11.75) it follows that $E \geq 0$ and the proof of this part ends.

2. According to point 1 of the theorem, the system $\overline{\mathbf{T}}_{y_1u_1}$ is stable. As $\mathbf{T}_{y_1u_1} \to \overline{\mathbf{T}}_{y_1u_1}$, we invoke Theorem 2.1.3 to deduce that $\|\overline{\mathbf{T}}_{y_1u_1}\|_\infty \leq \gamma_{\min}$ where only equality can take place. Hence $\|\overline{\mathbf{T}}_{y_1u_1}\|_\infty = \gamma_{\min}$ which concludes the proof. ■

11.3 Robust stabilization for normalized coprime factors uncertainties

In this section we solve the optimal robust stabilization problem for the case in which the uncertainties act on normalized coprime factors. Consider the continuous–time system

$$\mathbf{G} = \left[\begin{array}{c|c} A & B \\ \hline C & O \end{array}\right], \quad y = \mathbf{G}u, \tag{11.76}$$

where $A \in \mathbb{R}^{n\times n}$, $B \in \mathbb{R}^{n\times m}$, $C \in \mathbb{R}^{p\times n}$, the pair (A, B) is stabilizable and the pair (C, A) is detectable. Let $(\widetilde{\mathbf{N}}, \widetilde{\mathbf{M}})$ be a normalized left coprime factorization over RH_+^∞ (see Section 7.3) such that $\mathbf{G} = \widetilde{\mathbf{M}}^{-1}\widetilde{\mathbf{N}}$. For a $\delta > 0$, introduce the class of systems

$$\begin{aligned} \mathcal{D}_\delta^{cf} \quad := \quad & \{\mathbf{G}_\Delta^{cf} : \mathbf{G}_\Delta^{cf} = (\widetilde{\mathbf{M}} + \mathbf{\Delta}_M)^{-1}(\widetilde{\mathbf{N}} + \mathbf{\Delta}_N), \\ & \mathbf{\Delta} = \left[\begin{array}{c|c} \mathbf{\Delta}_N & -\mathbf{\Delta}_M \end{array}\right] \in RH_{+,p\times(p+m)}^\infty \text{ and } \|\mathbf{\Delta}\|_\infty < \delta\}. \end{aligned} \tag{11.77}$$

The main goal of this section is to evaluate the maximum stability margin $\delta = \delta_{\max}$ for which there exists a single stabilizing controller $\mathbf{K}$ for all systems in the class $\mathcal{D}_{\delta_{\max}}^{cf}$, and to construct such an optimal controller $\mathbf{K}$.

Consider the standard CTARE for control

$$A^T X + XA - XBB^T X + C^T C = 0 \tag{11.78}$$

and the standard CTARE for estimation

$$AY + YA^T - YC^T CY + BB^T = 0, \tag{11.79}$$

both associated with the system (11.76). Let X, Y be their positive semidefinite stabilizing solutions which exist according to Corollary 4.2.2. Let $F := -B^T X$ and $K := -YC^T$.

As we have proved in Section 11.1, the robust stabilization problem for the class of systems (11.77) is equivalent to a DAP formulated for the generalized system $\mathbf{T}^{cf}$ given in (11.6). In the particular case at hand in which the coprime factors $\widetilde{\mathbf{M}}$ and $\widetilde{\mathbf{N}}$ are normalized, we use the formulas for the normalizing coprime factors given in Theorem 7.3.1, and obtain an explicit realization in the form

$$\mathbf{T}^{cf} = \left[\begin{array}{c|c:c} A & B_1 & B_2 \\ \hline C_1 & D_{11} & D_{12} \\ \hdashline C_2 & D_{21} & O \end{array}\right] = \left[\begin{array}{c|c:c} A & -K & B \\ \hline O & O & I \\ C & I & O \\ \hdashline C & I & O \end{array}\right]. \tag{11.80}$$

It is easy to check with (11.80) that (A1)–(A5) are fulfilled and the Popov triplet Σ_c given in (11.11) is

$$\Sigma_c = (A, \begin{bmatrix} -K & B \end{bmatrix}; C^T C, \begin{bmatrix} C^T & O \end{bmatrix}, \begin{bmatrix} (1-\gamma^2)I_p & O \\ O & I_m \end{bmatrix}). \tag{11.81}$$

Our approach to solving the robust stabilization problem at hand is to apply the results in Section 11.2 for the particular case of $\mathbf{T}^{cf}$.

11.3.1 An evaluation of the maximum stability margin

The main result of the section is encapsulated in the following theorem.

Theorem 11.3.1 *The maximum stability margin for the robust stabilization problem with normalized coprime factors uncertainties formulated for the stabilizable and detectable system (11.76) is given by*

$$\delta_{\max} = (1 + \rho(XY))^{-\frac{1}{2}} \tag{11.82}$$

where X and Y are the unique stabilizing solutions to the CTAREs given in (11.78) and (11.79), respectively.

Proof. From (11.15) we have for the actual data

$$\begin{bmatrix} \widetilde{\mathbf{T}}_{11} & \widetilde{\mathbf{T}}_{12} \end{bmatrix} = \left[\begin{array}{c|cc} A_F & -K & B \\ \hline F & O & I \\ C & I & O \end{array}\right] = \begin{bmatrix} -\widetilde{\mathbf{W}} & \mathbf{M} \\ \widetilde{\mathbf{U}} & \mathbf{N} \end{bmatrix} \tag{11.83}$$

where for the last relation we have used the notation (7.27) introduced in Theorem 7.3.1. Notice also that we can take $V_2 = I$ in (11.20) due to (A5). Hence (11.83) and (11.20) give

$$\widetilde{\mathbf{T}}_{11} = \begin{bmatrix} -\widetilde{\mathbf{W}} \\ \widetilde{\mathbf{U}} \end{bmatrix}, \quad \widehat{\mathbf{T}}_{12} = \begin{bmatrix} \mathbf{M} \\ \mathbf{N} \end{bmatrix}. \tag{11.84}$$

Further, since the pairs $(\mathbf{N}, \mathbf{M})$ and $(\widetilde{\mathbf{N}}, \widetilde{\mathbf{M}})$ define the normalized right and left coprime factorizations of $\mathbf{G}$, respectively, $\begin{bmatrix} \mathbf{M} & -\widetilde{\mathbf{N}}^* \\ \mathbf{N} & \widetilde{\mathbf{M}}^* \end{bmatrix}$ is all–pass. Hence

$$\widehat{\mathbf{T}}_{12}^{\perp} = \begin{bmatrix} -\widetilde{\mathbf{N}}^* \\ \widetilde{\mathbf{M}}^* \end{bmatrix} \tag{11.85}$$

is an orthogonal completion of $\widehat{\mathbf{T}}_{12}$. Then we have

$$\widetilde{\mathbf{T}}_{11}^* \widehat{\mathbf{T}}_{12} = \begin{bmatrix} -\widetilde{\mathbf{W}}^* & \widetilde{\mathbf{U}}^* \end{bmatrix} \begin{bmatrix} \mathbf{M} \\ \mathbf{N} \end{bmatrix} = -\widetilde{\mathbf{W}}^*\mathbf{M} + \widetilde{\mathbf{U}}^*\mathbf{N} =: \mathbf{\Omega} \tag{11.86}$$

and

$$\widetilde{\mathbf{T}}_{11}^* \widehat{\mathbf{T}}_{12}^{\perp} = \begin{bmatrix} -\widetilde{\mathbf{W}}^* & \widetilde{\mathbf{U}}^* \end{bmatrix} \begin{bmatrix} -\widetilde{\mathbf{N}}^* \\ \widetilde{\mathbf{M}}^* \end{bmatrix} = I \tag{11.87}$$

as follows from (7.28). Using (11.86) and (11.87), (11.19) yields

$$\gamma_{\min} = (1 + \rho(\mathbb{H}_{\mathbf{\Omega}} \mathbb{H}_{\mathbf{\Omega}}^*))^{\frac{1}{2}}. \tag{11.88}$$

To get an expression for $\gamma_{\min}$ in terms of the original data, we first evaluate $\mathbf{\Omega}$. Using (11.86) and Theorem 7.3.1,

$$\begin{aligned} \mathbf{\Omega} &= \left[\begin{array}{c|cc} -A_F^T & F^T & C^T \\ \hline K^T & O & I \end{array}\right] \left[\begin{array}{c|c} A_F & B \\ \hline F & I \\ C & O \end{array}\right] = \left[\begin{array}{cc|c} A_F & O & B \\ F^TF + C^TC & -A_F^T & F^T \\ \hline C & K^T & O \end{array}\right] \\ &= \left[\begin{array}{cc|c} A_F & O & B \\ O & -A_F^T & O \\ \hline C - K^TX & K^T & O \end{array}\right] = \left[\begin{array}{c|c} A_F & B \\ \hline C - K^TX & O \end{array}\right] \end{aligned} \tag{11.89}$$

where we have used (11.78) written in the form $A_F^TX + XA_F + C^TC + F^TF = 0$. Now it is a matter of simple computation to show that the controllability and observability Gramians associated with (11.89) are

$$P_c = Y(I + XY)^{-1}, \quad P_o = (I + XY)X. \tag{11.90}$$

Further, we have

$$\rho(\mathbb{H}_{\mathbf{\Omega}} \mathbb{H}_{\mathbf{\Omega}}^*) = \rho(P_cP_o) = \rho(XY) \tag{11.91}$$

as follows from (11.90) and Corollary 2.4.9. Combining (11.91) with (11.88) and taking into account that $\delta_{\max} = 1/\gamma_{\min}$ the conclusion is immediate. ■

11.3.2 An optimal robustly stabilizing controller

In order to find an optimal robustly stabilizing controller we apply Theorem 11.2.3 in the particular case of the generalized system $\mathbf{T}^{cf}$. But we need first to check the hypotheses (A1)–(A8) for $\mathbf{T}^{cf}$. Assumptions (A1)–(A5) are obvious, while (A6) follows from (11.82) and (11.80). To check (A7) we need first the following result.

Proposition 11.3.2 *Let $\gamma > \gamma_{\min}$ and Σ_c be the Popov triplet given in (11.81). Then the CTARE(Σ_c) has a stabilizing solution $X_\gamma \geq 0$ which is given by*

$$X_\gamma = -\gamma^2 X[(1-\gamma^2)I + YX]^{-1} \tag{11.92}$$

where $X \geq 0$ and $Y \geq 0$ are the stabilizing solutions to the CTAREs (11.78) and (11.79), respectively.

Proof. Since $\gamma > \gamma_{\min}$, the existence of the stabilizing solution X_γ follows directly from Theorem 11.2.1.

The rest of the proof makes intense use of manipulations on EHPs. Let $sM_{\Sigma_c} - N_{\Sigma_c}$ be the EHP(Σ_c), where

$$M_{\Sigma_c} = \begin{bmatrix} I & O & O & O \\ O & I & O & O \\ O & O & O & O \\ O & O & O & O \end{bmatrix}, \quad N_{\Sigma_c} = \begin{bmatrix} A & O & -K & B \\ -C^TC & -A^T & -C^T & O \\ C & -K^T & (1-\gamma^2)I & O \\ O & B^T & O & I \end{bmatrix}. \tag{11.93}$$

Introduce the following two nonsingular matrices

$$U_0 := \begin{bmatrix} I & \gamma^{-2}Y & -\gamma^{-2}K & \gamma^{-2}B \\ O & -\gamma^{-2}(1-\gamma^2)I & -\gamma^{-2}C^T & O \\ O & O & O & -\gamma^{-2}(1-\gamma^2)I \\ O & O & (1-\gamma^2)^{-1}I & O \end{bmatrix} \tag{11.94}$$

and

$$W_0 := \begin{bmatrix} I & (1-\gamma^2)^{-1}Y & O & O \\ O & -\gamma^2(1-\gamma^2)^{-1}I & O & O \\ -(1-\gamma^2)^{-1}C & (1-\gamma^2)^{-1}K^T & O & I \\ O & O & -\gamma^2(1-\gamma^2)^{-1}I & O \end{bmatrix}. \tag{11.95}$$

Then M_0 and N_0 introduced by

$$M_0 := U_0 M_{\Sigma_c} W_0, \quad N_0 := U_0 N_{\Sigma_c} W_0, \tag{11.96}$$

have the explicit expressions

$$M_0 = \begin{bmatrix} I & O & O & O \\ O & I & O & O \\ O & O & O & O \\ O & O & O & O \end{bmatrix}, \quad N_0 = \begin{bmatrix} A & O & B & O \\ -C^TC & -A^T & O & O \\ O & B^T & I & O \\ O & O & O & I \end{bmatrix}. \tag{11.97}$$

Since X_γ is a stabilizing solution to the CTARE(Σ_c), one can write

$$N_{\Sigma_c} V_{\Sigma_c} = M_{\Sigma_c} V_{\Sigma_c} S_{\Sigma_c}, \quad V_{\Sigma_c} = \begin{bmatrix} I \\ X_\gamma \\ F_1 \\ F_2 \end{bmatrix}, \quad \Lambda(S_{\Sigma_c}) \subset \mathbb{C}_-, \tag{11.98}$$

where $\begin{bmatrix} F_1 \\ F_2 \end{bmatrix}$ is the stabilizing Riccati feedback of CTARE(Σ_c). Using (11.96), (11.98) yields

$$N_0 W_0^{-1} V_{\Sigma_c} = M_0 W_0^{-1} V_{\Sigma_c} S_{\Sigma_c}. \tag{11.99}$$

From (11.95) we depict for W_0^{-1} the following structure

$$W_0^{-1} = \begin{bmatrix} I & \gamma^{-2}Y & O & O \\ O & -\gamma^{-2}(1-\gamma^2)I & O & O \\ \star & \star & O & I \\ \star & \star & I & O \end{bmatrix}. \tag{11.100}$$

Hence (11.98) and (11.100) give

$$W_0^{-1} V_{\Sigma_c} = \begin{bmatrix} I + \gamma^{-2} Y X_\gamma \\ -\gamma^{-2}(1-\gamma^2) X_\gamma \\ \star \\ \star \end{bmatrix} = \begin{bmatrix} I \\ \widehat{X}_\gamma \\ \star \\ \star \end{bmatrix} (I + \gamma^{-2} Y X_\gamma), \tag{11.101}$$

where

$$\widehat{X}_\gamma := -\gamma^{-2}(1-\gamma^2) X_\gamma (I + \gamma^{-2} Y X_\gamma)^{-1}. \tag{11.102}$$

Inspecting now the structure of M_0 and N_0 and taking into account (11.101), (11.99) yields

$$\begin{bmatrix} A & O & B \\ -C^T C & -A^T & O \\ O & B^T & I \end{bmatrix} \begin{bmatrix} I \\ \widehat{X}_\gamma \\ \star \end{bmatrix} = \begin{bmatrix} I & O & O \\ O & I & O \\ O & O & O \end{bmatrix} \begin{bmatrix} I \\ \widehat{X}_\gamma \\ \star \end{bmatrix} S_0, \tag{11.103}$$

where $S_0 := (I + \gamma^{-2} Y X_\gamma) S_{\Sigma_c} (I + \gamma^{-2} Y X_\gamma)^{-1}$ is stable, too. It follows from (11.103) and Theorem 5.2.1 that the CTARE in (11.78) has the unique stabilizing solution

$$X = \widehat{X}_\gamma. \tag{11.104}$$

Finally, (11.92) is a mere rewriting of (11.102) with $\widehat{X}_\gamma$ substituted by X. ■

We return now to the proof of (A7). Assume that the realization of **G** is such that both the CTAREs (11.78) and (11.79) are already in *balanced* form,

$$X = Y = \begin{bmatrix} \overline{\sigma} I_r & O \\ O & \text{diag}\,(\sigma_{r+1}, \ldots, \sigma_n) \end{bmatrix}, \tag{11.105}$$

where $\overline{\sigma} > \sigma_{r+1} \geq \ldots \geq \sigma_n$, and r is the multiplicity of the maximum singular value $\overline{\sigma}$. Such a balanced realization of **G** can be obtained similarly as in Theorem 2.1.4 where instead of the controllability and observability Gramians we use the solutions X and Y to the standard CTAREs (11.78) and (11.79), respectively. We have $\rho(XY) = \overline{\sigma}^2$ and (11.82) becomes

$$\gamma_{\min}^2 = 1 + \overline{\sigma}^2 \tag{11.106}$$

from which

$$\gamma^2 = \epsilon + 1 + \overline{\sigma}^2. \tag{11.107}$$

With (11.105) substituted in (11.92), X_γ takes the form (11.34) where

$$X_1 := (\epsilon + 1 + \overline{\sigma}^2)\overline{\sigma} I_r, \quad X_2 := \operatorname{diag}\left(\frac{\epsilon + 1 + \overline{\sigma}^2}{\overline{\sigma}^2 - \sigma_i^2 + \epsilon}\sigma_i\right)_{i=r+1}^{n} \tag{11.108}$$

and

$$\overline{X}_1 := (1 + \overline{\sigma}^2)\overline{\sigma} I_r > 0, \quad \overline{X}_2 := \operatorname{diag}\left(\frac{1 + \overline{\sigma}^2}{\overline{\sigma}^2 - \sigma_i^2}\sigma_i\right)_{i=r+1}^{n}.$$

Here $\overline{X}_i$ stands for the limit when $\epsilon \downarrow 0$ of X_i, for $i = 1, 2$. It follows that (A7) holds. Let now

$$A = \begin{bmatrix} A_{11} & A_{12} \\ A_{21} & A_{22} \end{bmatrix}, \; B = \begin{bmatrix} B_{21} \\ B_{22} \end{bmatrix}, \; K = \begin{bmatrix} K_{11} \\ K_{12} \end{bmatrix}, \; C = \begin{bmatrix} C_{11} & C_{12} \end{bmatrix}, \tag{11.109}$$

be partitioned conformably with $X = Y$ in (11.105). We are now ready for the main result of this section.

Theorem 11.3.3 *If (A8) and (A9) hold for $\mathbf{T}^{cf}$, and $\mathbf{G}$ in (11.76) is in balanced form with respect to the standard CTAREs (11.78) and (11.79) then*

$$\overline{\mathbf{K}} = \begin{bmatrix} \overline{A}_K & \overline{B}_K \\ \overline{C}_K & \overline{D}_K \end{bmatrix},$$

where

$$\begin{aligned}
\overline{A}_K &= A_{22} + K_{12}C_{12} - B_{22}B_{22}^T\overline{X}_2 - B_{22}B_{21}^T(B_{21}B_{21}^T)^{-1}(A_{12} + K_{11}C_{12} - B_{21}B_{22}^T\overline{X}_2),\\
\overline{B}_K &= -K_{12} + B_{22}B_{21}^T(B_{21}B_{21}^T)^{-1}K_{11},\\
\overline{C}_K &= -B_{22}^T\overline{X}_2 - B_{21}^T(B_{21}B_{21}^T)^{-1}(A_{12} + K_{11}C_{12} - B_{21}B_{22}^T\overline{X}_2),\\
\overline{D}_K &= B_{21}^T(B_{21}B_{21}^T)^{-1}K_{11},
\end{aligned} \tag{11.110}$$

is an optimal robustly stabilizing controller of (reduced) order $n - r$ for $\mathbf{G}$, with respect to normalized coprime factors uncertainties. Moreover, the maximum stability margin is given by

$$\delta_{\max} = (1 + \overline{\sigma}^2)^{-\frac{1}{2}}. \tag{11.111}$$

Proof. We have already proved that (A1)–(A7) hold for $\mathbf{T}^{cf}$, and together with (A8) and (A9) assumed in the statement it follows that Theorem 11.2.3 is applicable for $\mathbf{T}^{cf}$. The explicit formulas (11.110) are a matter of straightforward substitutions in (11.40) with data given by (11.80), while formula (11.111) is a simple rewriting of (11.82) with (11.105). ■

Notice that since generically $r = 1$, (A8) reduces to the condition that the first row of B is nonzero.

11.4 Robust stabilization for multiplicative uncertainties

In this section we solve the suboptimal robust stabilization problem for the case in which the uncertainties act multiplicative. Consider the continuous–time system

$$\mathbf{G} = \left[\begin{array}{c|c} A & B \\ \hline C & D \end{array}\right], \tag{11.112}$$

where $A \in \mathbb{R}^{n\times n}$, $B \in \mathbb{R}^{n\times m}$, $C \in \mathbb{R}^{p\times n}$, the pair (A, B) is stabilizable and the pair (C, A) is detectable. For a $\delta > 0$, introduce the class of systems

$$\mathcal{D}_{\delta}^{m} := \{\mathbf{G}_{\Delta}^{m} : \mathbf{G}_{\Delta}^{m} = (I + \mathbf{\Delta})\mathbf{G}, \mathbf{\Delta} \in RH_{+,p\times p}^{\infty} \text{ and } \|\mathbf{\Delta}\|_{\infty} < \delta\}. \tag{11.113}$$

The main goal of this section is to evaluate the maximum stability margin $\delta = \delta_{\max}$ for which there exists a single stabilizing controller $\mathbf{K}$ for all systems in the class $\mathcal{D}_{\delta_{\max}}^{m}$, and to construct a robustly stabilizing controller $\mathbf{K}$ in the suboptimal case in which $\delta < \delta_{\max}$.

Notice that if $\mathbf{G} \in RH_{+,p\times m}^{\infty}$, then $\mathbf{G}_{\Delta} \in RH_{+,p\times m}^{\infty}$ for all $\mathbf{\Delta} \in RH_{+,p\times p}^{\infty}$, and there is no need to stabilize the system. Hence this case is trivial and the interesting problem is when $\mathbf{G} \notin RH_{+,p\times m}^{\infty}$.

As we have proved in Section 11.1, the suboptimal robust stabilization problem for the class of perturbed systems (11.113) is equivalent to a suboptimal γ–DAP formulated for the generalized system $\mathbf{T}^m$ given in (11.5), and some $\gamma > 0$. A realization of the generalized system in (11.5) is given by

$$\begin{aligned} \mathbf{T}^{m} &= \left[\begin{array}{c|cc} A & O & B \\ \hline C & O & D \\ C & I & O \end{array}\right] =: \left[\begin{array}{c|c} A & \widehat{B} \\ \hline \widehat{C} & \widehat{D} \end{array}\right], \\ \left[\begin{array}{c} y_1 \\ y_2 \end{array}\right] &= \mathbf{T}\left[\begin{array}{c} u_1 \\ u_2 \end{array}\right] \end{aligned} \tag{11.114}$$

and we take $\gamma > \gamma_{\min} := \frac{1}{\delta_{\max}}$.

To solve this particular γ–DAP, we intend to use Theorem 10.3.1. To this end, consider the Popov triplets $\Sigma_c = (A, \widehat{B}; Q_c, L_c, R_c)$ and $\Sigma_o = (A^T, \widehat{C}^T; Q_o, L_o, R_o)$ introduced in (10.11) and (10.12), respectively. For the actual data these triplets are given explicitly by

$$\begin{aligned} Q_c &:= C^T C, \\ L_c &:= \left[\begin{array}{cc} O & C^T D \end{array}\right], \\ R_c &= \left[\begin{array}{cc} R_{c11} & R_{c12} \\ R_{c12}^T & R_{c22} \end{array}\right] := \left[\begin{array}{cc} -\gamma^2 I_p & O \\ O & D^T D \end{array}\right], \end{aligned} \tag{11.115}$$

and

$$\begin{aligned} Q_o &:= 0, \\ L_o &:= \left[\begin{array}{cc} O & O \end{array}\right], \\ R_o &= \left[\begin{array}{cc} R_{o11} & R_{o12} \\ R_{o12}^T & R_{o22} \end{array}\right] = \left[\begin{array}{cc} -\gamma^2 I_p & O \\ O & I \end{array}\right]. \end{aligned} \tag{11.116}$$

Expressing the KPYS(Σ_c, J_c) and KPYS(Σ_o, J_o) in Theorem 10.3.1 in terms of the corresponding CTAREs and adequate sign constraints, we obtain that the γ–DAP for $\mathbf{T}^m$ has a solution if the following conditions (C1)–(C3) below hold.

(C1) The CTARE(Σ_c) given by

$$A^T X + XA - (L_c + X\widehat{B})R_c^{-1}(\widehat{B}^T X + L_c^T) + Q_c = 0 \tag{11.117}$$

has a positive semidefinite stabilizing solution X and the following two sign constraints

$$\begin{aligned} R_{c22} &> 0, \\ \operatorname{sgn} R_c = J_c &:= \begin{bmatrix} -I_p & O \\ O & I_m \end{bmatrix}, \end{aligned} \tag{11.118}$$

are fulfilled.

(C2) The CTARE(Σ_o) given by

$$AY + YA^T - (L_o + Y\widehat{C}^T)R_o^{-1}(\widehat{C}Y + L_o^T) + Q_o = 0 \tag{11.119}$$

has a positive semidefinite stabilizing solution Y and the following two sign constraints

$$\begin{aligned} R_{o22} &> 0, \\ \operatorname{sgn}(R_o) = J_o &:= \begin{bmatrix} -I_p & O \\ O & I_p \end{bmatrix}, \end{aligned} \tag{11.120}$$

are fulfilled.

(C3) The solutions X and Y are linked by

$$\rho(XY) < \gamma^2. \tag{11.121}$$

Notice that since $Q_o = 0$ and $L_o = 0$, the CTARE (11.119) is actually a Bernoulli equation of the form

$$AY + YA - Y\widehat{C}R_o^{-1}\widehat{C}Y = 0, \tag{11.122}$$

and the sign constraints (11.120) are automatically fulfilled for any $\gamma > 0$. What concerns (C1), if D is not full column rank, then R_c is singular and the CTARE (11.117) does not make sense. To circumvent this difficulty we shall employ a singular perturbation technique as explained further.

We introduce a new regulated output in (11.114), and get an augmented generalized system $\mathbf{T}_\epsilon^m$ of the form

$$\begin{aligned} \mathbf{T}_\epsilon^m &= \left[\begin{array}{c|cc} A & O & B \\ \hline C & O & D \\ O & O & \epsilon I_m \\ C & I & O \end{array}\right] \\ \begin{bmatrix} y_1 \\ y_{1\epsilon} \\ y_2 \end{bmatrix} &= \mathbf{T}_\epsilon^m \begin{bmatrix} u_1 \\ u_2 \end{bmatrix}, \end{aligned} \tag{11.123}$$

where clearly $y_{1\epsilon} = \epsilon u_2$. If $\mathbf{K}$ is a solution to the γ–DAP for $\mathbf{T}_\epsilon^m$ then $\mathbf{K}$ is a solution to the γ–DAP for $\mathbf{T}^m$. Indeed, this follows from $\mathrm{LFT}(\mathbf{T}_\epsilon^m, \mathbf{K}) = \begin{bmatrix} \mathrm{LFT}(\mathbf{T}^m, \mathbf{K}) \\ \epsilon \mathbf{T}_{u_2 u_1}^m \end{bmatrix}$, where $\mathbf{T}_{u_2 u_1}^m$ is the closed–loop TFM from u_1 to u_2 for $\mathrm{LFT}(\mathbf{T}^m, \mathbf{K})$.

We investigate now conditions (C1)–(C3) for $\mathbf{T}_\epsilon^m$. If we write down the corresponding Popov triplets Σ_c^ϵ and Σ_o^ϵ, we see that the corresponding sign constraints (11.118) and (11.120) are automatically fulfilled, and the CTAREs (11.117) and (11.119) expressed in the original data are

$$A^T X_\epsilon + X_\epsilon A - (C^T D + X_\epsilon B)(\epsilon^2 I + D^T D)^{-1}(B^T X_\epsilon + D^T C) + C^T C = 0 \quad (11.124)$$

and

$$AY + YA^T - (1 - \gamma^{-2})YC^T CY = 0. \quad (11.125)$$

Let $\gamma \in (0,1)$ and assume that (11.125) has a positive semidefinite stabilizing solution Y. Since $1-\gamma^{-2} < 0$ it follows from Proposition 5.3.2 that A is stable, and as explained earlier this case is trivial. Hence we take $\gamma > 1$ and assume that A is dichotomic. From Proposition 5.3.1 we see that this assumption is necessary for the existence of the stabilizing solution to (11.125) and therefore for the solvability of the problem at hand. Since $1 - \gamma^{-2} > 0$ and the pair (C, A) is detectable, it follows, according to Theorem 5.3.3, that (11.125) has a positive semidefinite stabilizing solution Y.

We focus further on the Riccati equation (11.124). Since the pair (A, B) is stabilizable and the pair (C, A) is detectable, Proposition 7.2.1 shows that the equation (11.124) always has a positive semidefinite stabilizing solution X_ϵ. Assume further that $\mathbf{G}$ has a stable inverse to the right (not necessarily proper). Then by combining 2. in Theorem 7.2.3 with Remark 7.2.4 it follows that

$$\lim_{\epsilon \downarrow 0} X_\epsilon = 0. \quad (11.126)$$

Hence by taking ϵ small enough, condition (11.121), written explicitly

$$\rho(X_\epsilon Y) < \gamma^2,$$

can be achieved. Thus we conclude that all three conditions mentioned at the beginning of this section are fulfilled for ϵ small enough. Meanwhile we have proved the following theorem.

Theorem 11.4.1 *Assume for the system* $\mathbf{G}$ *given in (11.112) that the pair* (A, B) *is stabilizable, the pair* (C, A) *is detectable,* A *is dichotomic, and* $\mathbf{G}$ *has a stable inverse to the right. Then we have:*

1. For each $\delta \in (0,1)$ *the controller*

$$\mathbf{K} = \left[\begin{array}{c|c} A + BF_\epsilon - Z_\epsilon C^T C & Z_\epsilon C^T \\ \hline F_\epsilon & O \end{array}\right] \quad (11.127)$$

is a suboptimal robustly stabilizing controller for $\mathbf{G}$*, with respect to multiplicative uncertainties, where*

$$Z_\epsilon := Y(I - \delta^2 X_\epsilon Y)^{-1}, \quad F_\epsilon := -(\epsilon^2 I + D^T D)^{-1}(B^T X_\epsilon + D^T C),$$

with X_ϵ *and* Y *the unique stabilizing solutions to the CTAREs (11.124) and (11.125), respectively, and* ϵ *chosen such that* $\rho(X_\epsilon Y) < \frac{1}{\delta^2}$.

2. The maximum stability margin for the problem of robust stabilization with multiplicative uncertainties is $\delta_{\max} = 1$.

Proof. 1. We have already shown that under the assumptions of the theorem, conditions (C1)–(C3) written for $\mathbf{T}_\epsilon^m$ are fulfilled. Thus the $\frac{1}{\delta}$–DAP problem for $\mathbf{T}_\epsilon^m$ has a solution $\mathbf{K}$ which, as we have proved, is a solution to the $\frac{1}{\delta}$–DAP for the original $\mathbf{T}^m$ and, of course, to the corresponding robust stabilization problem under multiplicative uncertainties for $\mathbf{G}$. Finally, the explicit formulas for the controller are obtained from the general formulas in Theorem 10.3.1 written for the actual data.

2. This is a trivial consequence of $\delta = \frac{1}{\gamma}$ where, as we have shown, $\gamma \in (1, \infty)$. ■

B. The discrete–time case

11.5 Problem formulation and some prerequisites

All of the developments in Section 11.1 are valid irrespective of the continuous or discrete–time setting, and there is no need to repeat them here. The reader is referred to Section 11.1 for the details. Again, the robust stabilization problems formulated for the class of systems (11.1), (11.2) and (11.3) are equivalent to DAPs formulated for the discrete–time generalized system (11.7), (11.8), and (11.9), respectively. Theorem 11.1.1 holds in a discrete–time setting as well. Nevertheless, the discrete–time counterparts of Sections 11.2 and 11.3 differ drastically from the continuous case, and in no way can they be achieved by simple comparison. The explanation originates in the completely different way in which the singular perturbation theory works in the discrete–time framework.

11.6 Optimal solution to the DF problem

In this section we give an evaluation of the least achievable tolerance $\gamma_{\min}$ for which the discrete–time DF problem has an optimal solution, and give a construction of that solution under certain hypotheses. Further, we shall show in Section 11.7 that these hypotheses are all satisfied in the case of the DF problem arising from the discrete–time robust stabilization problem with normalized coprime factors uncertainties, and we shall provide a solution for the latter.

Consider the discrete–time generalized system

$$\mathbf{T} = \left[\begin{array}{c|c} A & B \\ \hline C & D \end{array}\right] = \left[\begin{array}{c|cc} A & B_1 & B_2 \\ \hline C_1 & D_{11} & D_{12} \\ C_2 & D_{21} & O \end{array}\right] = \left[\begin{array}{cc} \mathbf{T}_{11} & \mathbf{T}_{12} \\ \mathbf{T}_{21} & \mathbf{T}_{22} \end{array}\right], \qquad (11.128)$$

$$\left[\begin{array}{c} y_1 \\ y_2 \end{array}\right] = \mathbf{T}\left[\begin{array}{c} u_1 \\ u_2 \end{array}\right]$$

where $A \in \mathbb{R}^{n\times n}$, $B_i \in \mathbb{R}^{n\times m_i}$, $C_j \in \mathbb{R}^{p_j\times n}$, $D_{ij} \in \mathbb{R}^{p_i\times m_j}$, $(i, j = 1, 2,\ m = m_1 + m_2,\ p = p_1 + p_2)$.

Assume that the following conditions are fulfilled:

(AD1) D_{21} is square and invertible.

(AD2) $A - B_1 D_{21}^{-1} C_2$ is stable.

(AD3) (A, B_2) is stabilizable.

(AD4) rank $\begin{bmatrix} e^{j\theta}I - A & -B_2 \\ C_1 & D_{12} \end{bmatrix} = n + m_2, \quad \forall \theta \in [0,\ 2\pi).$

Notice that (AD1) and (AD2) are exactly the first two assumptions made for the discrete–time DF problem in Section 10.17.2, while (AD3) and (AD4) are among the general hypotheses made in Section 10.11 for the discrete–time DAP.

11.6.1 An evaluation of $\gamma_{\min}$

Consider the system $\mathbf{T}$ in (11.128), and assume it satisfies conditions (AD1), (AD2), (AD3), and (AD4). Consider the Popov triplet $\Sigma_2 = (A, B_2; C_1^T C_1, C_1^T D_{12}, D_{12}^T D_{12})$. With (AD3) and (AD4), Proposition 7.8.1 shows that the stabilizing solution X_2 to the DTARE(Σ_2) does exist, is positive semidefinite, and $D_{12}^T D_{12} + B_2^T X_2 B_2 > 0$. Let $\widetilde{F}_2$ be the stabilizing Riccati feedback of the DTARE(Σ_2), and let

$$\begin{bmatrix} \widetilde{\mathbf{T}}_{11} & \widetilde{\mathbf{T}}_{12} \end{bmatrix} = \left[\begin{array}{c|cc} \widetilde{A} & B_1 & B_2 \\ \hline \widetilde{C}_1 & D_{11} & D_{12} \end{array}\right],$$

where $\widetilde{A} = A + B\widetilde{F}_2$, $\widetilde{C}_1 = C_1 + D_{12}\widetilde{F}_2$. For $D_{12}^T D_{12} + B_2^T X_2 B_2 = V_2^T V_2$, where V_2 is invertible, let $\widehat{\mathbf{T}}_{12} := \widetilde{\mathbf{T}}_{12} V_2^{-1}$. From the spectral factorization identity (3.142) it follows that $\widehat{\mathbf{T}}_{12}$ is inner, and let $\widehat{\mathbf{T}}_{12}^{\perp}$ be the orthogonal completion of $\widehat{\mathbf{T}}_{12}$ such that $\begin{bmatrix} \widehat{\mathbf{T}}_{12} & \widehat{\mathbf{T}}_{12}^{\perp} \end{bmatrix}$ is *all–pass*, i.e., $(\widehat{\mathbf{T}}_{12}^a)^*(\widehat{\mathbf{T}}_{12}^a) = (\widehat{\mathbf{T}}_{12}^a)(\widehat{\mathbf{T}}_{12}^a)^* = I$. Following the same route as in the continuous–time case, we arrive at the following result.

Theorem 11.6.1 *Consider the system* $\mathbf{T}$ *in (11.128), and assume it satisfies conditions (AD1), (AD2), (AD3), and (AD4). The least achievable tolerance (the optimal value)* $\gamma = \gamma_{\min}$ *for the DF problem is given by*

$$\gamma_{\min} = \rho^{\frac{1}{2}}\left(\mathbb{H}_{\widetilde{\mathbf{T}}_{11}^*\widehat{\mathbf{T}}_{12}}\mathbb{H}_{\widetilde{\mathbf{T}}_{11}^*\widehat{\mathbf{T}}_{12}}^* + \mathbb{T}_{(\widetilde{\mathbf{T}}_{11}^*\widehat{\mathbf{T}}_{12}^{\perp})(\widetilde{\mathbf{T}}_{11}^*\widehat{\mathbf{T}}_{12}^{\perp})^*}\right). \tag{11.129}$$

11.6.2 An optimal solution

In this section we construct an optimal solution to a DF problem formulated for a particular discrete–time generalized system $\mathbf{T}$. Our final goal is to obtain in Section 11.7 an optimal solution to the discrete–time robust stabilization problem with normalized coprime factors uncertainties. We follow closely the same approach taken in the continuous–time case.

Consider the generalized system (11.128) satisfying assumptions (AD1)–(AD4). We introduce a number of additional hypotheses (AD5)–(AD7) as follows.

(AD5) $D_{12}^T \begin{bmatrix} C_1 & D_{11} & D_{12} \end{bmatrix} = \begin{bmatrix} O & O & I_{m_2} \end{bmatrix}$.

Consider also the Popov triplet $\Sigma_c = (A, B; Q_c, L_c, R_c)$ in Theorem 10.12.1, which due to (AD5) has the following explicit expression

$$Q_c = C_1^T C_1, \quad L_c = \begin{bmatrix} C_1^T D_{11} & O \end{bmatrix}, \quad R_c = \begin{bmatrix} -\gamma^2 I + D_{11}^T D_{11} & O \\ O & I \end{bmatrix}. \tag{11.130}$$

Take $\gamma > \gamma_{\min}$ such that according to the discrete version of Theorem 11.2.2 the KSPYS(Σ_c, J_c) has a stabilizing solution (X, V_c, W_c) and the (suboptimal) γ–DF problem has a solution. Using Theorem 10.12.1 (after removing the scaling) and (AD5), we obtain the realization of the central controller ($\mathbf{Q} = 0$) in the form

$$\mathbf{K} = \left[\begin{array}{c|c} A_K & B_K \\ \hline C_K & D_K \end{array}\right] = \left[\begin{array}{c|c} M_\gamma^{-1}(A - B_1 D_{21}^{-1} C_2) & M_\gamma^{-1} B_1 D_{21}^{-1} \\ \hline -B_2^T X_\gamma M_\gamma^{-1}(A - B_1 D_{21}^{-1} C_2) & -B_2^T X_\gamma M_\gamma^{-1} B_1 D_{21}^{-1} \end{array}\right] \tag{11.131}$$

where

$$M_\gamma := I + B_2 B_2^T X_\gamma, \tag{11.132}$$

and the solution X to the KSPYS(Σ_c, J_c) in Theorem 10.12.1 has been denoted by X_γ to stress the dependence on γ. Define

$$\epsilon := \gamma^2 - \gamma_{\min}^2 \tag{11.133}$$

and assume $\mathbf{T}$ satisfies the following additional hypotheses:

(AD6) There exists $r \in \{1, \ldots, n\}$ such that

$$X_\gamma = \begin{bmatrix} \epsilon^{-1} X_1 & \\ & X_2 \end{bmatrix} \tag{11.134}$$

where

$$\epsilon^{-1} X_1 = \operatorname{diag}(\nu_1, \ldots, \nu_r), \quad X_2 = \operatorname{diag}(\nu_{r+1}, \ldots, \nu_n),$$

X_1 and X_2 depend continuously on ϵ and there exist the (finite) limits

$$\lim_{\epsilon \downarrow 0} X_i = \overline{X}_i, \quad \text{with} \quad \overline{X}_1 > 0.$$

(AD7) $B_{21} B_{21}^T > 0$ where B_{21} is defined by partitioning (11.128) in accordance with the partition of X_γ in (11.134) as

$$\mathbf{T} = \left[\begin{array}{c|c:c} A & B_1 & B_2 \\ \hline C_1 & D_{11} & D_{12} \\ \hdashline C_2 & D_{21} & O \end{array}\right] = \left[\begin{array}{cc|c:c} A_{11} & A_{12} & B_{11} & B_{21} \\ A_{21} & A_{22} & B_{12} & B_{22} \\ \hline C_{11} & C_{12} & D_{11} & D_{12} \\ \hdashline C_{21} & C_{22} & D_{21} & O \end{array}\right]. \tag{11.135}$$

Following the same plan as in the continuous–time case, we investigate now the behavior of $\mathbf{K}$ and of the closed–loop LFT$(\mathbf{T}, \mathbf{K})$ when $\epsilon \downarrow 0$.

The behavior of the controller when $\epsilon \downarrow 0$

We study the expression of the controller (11.131) as $\epsilon \downarrow 0$. We get successively

$$M_\gamma = \begin{bmatrix} I + \epsilon^{-1} B_{21} B_{21}^T X_1 & B_{21} B_{22}^T X_2 \\ \epsilon^{-1} B_{22} B_{21}^T X_1 & I + B_{22} B_{22}^T X_2 \end{bmatrix},$$

$$M_\gamma^{-1} = \begin{bmatrix} \epsilon I & O \\ O & I \end{bmatrix} \begin{bmatrix} \epsilon I + B_{21} B_{21}^T X_1 & B_{21} B_{22}^T X_2 \\ B_{22} B_{21}^T X_1 & I + B_{22} B_{22}^T X_2 \end{bmatrix}^{-1}, \tag{11.136}$$

$$X_\gamma M_\gamma^{-1} = \begin{bmatrix} X_1 & O \\ O & X_2 \end{bmatrix} \begin{bmatrix} \epsilon I + B_{21}B_{21}^T X_1 & B_{21}B_{22}^T X_2 \\ B_{22}B_{21}^T X_1 & I + B_{22}B_{22}^T X_2 \end{bmatrix}^{-1}. \tag{11.137}$$

But according to (AD6) and (AD7), the matrix

$$\widehat{M} := \begin{bmatrix} B_{21}B_{21}^T \overline{X}_1 & B_{21}B_{22}^T \overline{X}_2 \\ B_{22}B_{21}^T \overline{X}_1 & I + B_{22}B_{22}^T \overline{X}_2 \end{bmatrix}$$

is nonsingular. Indeed, since $B_{21}B_{21}^T\overline{X}_1$ is nonsingular, the Schur complement of the (1,1) entry of the matrix above is

$$\begin{aligned} I + B_{22}B_{22}^T\overline{X}_2 - B_{22}B_{21}^T(B_{21}B_{21}^T\overline{X}_1)^{-1}B_{21}B_{22}^T\overline{X}_2 \\ = I + B_{22}[I - B_{21}^T(B_{21}B_{21}^T)^{-1}B_{21}]B_{22}^T\overline{X}_2 \end{aligned}$$

which is clearly nonsingular because of $I - B_{21}^T(B_{21}B_{21}^T)^{-1}B_{21} \geq 0$ and $\overline{X}_2 \geq 0$. Let

$$\overline{M} = \begin{bmatrix} \overline{M}_{11} & \overline{M}_{12} \\ \overline{M}_{21} & \overline{M}_{22} \end{bmatrix} := \widehat{M}^{-1}, \quad \overline{N} := \begin{bmatrix} O & O \\ \overline{M}_{21} & \overline{M}_{22} \end{bmatrix}. \tag{11.138}$$

From (11.136) and (11.137) we get

$$M_\gamma^{-1} \to \overline{N} \text{ as } \epsilon \downarrow 0, \tag{11.139}$$

$$X_\gamma M_\gamma^{-1} \to \operatorname{diag}(\overline{X}_1, \overline{X}_2)\overline{M} \text{ as } \epsilon \downarrow 0. \tag{11.140}$$

Using (11.139) and (11.140), (11.131) gives

$$\mathbf{K} = \left[\begin{array}{c|c} A_K & B_K \\ \hline C_K & D_K \end{array}\right] = \left[\begin{array}{cc|c} \epsilon A_{K11} & \epsilon A_{K12} & \epsilon B_{K1} \\ A_{K21} & A_{K22} & B_{K2} \\ \hline C_{K1} & C_{K2} & D_K \end{array}\right] \tag{11.141}$$

where A_{Kij}, B_{Ki}, C_{Ki}, D_K $(i, j = 1, 2)$ continuously depend on ϵ and all have finite limits when $\epsilon \downarrow 0$, that is,

$$(A_{Kij}, B_{Ki}, C_{Ki}, D_K) \to (\overline{A}_{Kij}, \overline{B}_{Ki}, \overline{C}_{Ki}, \overline{D}_K) \text{ as } \epsilon \downarrow 0,\ i, j = 1, 2. \tag{11.142}$$

We have the following result.

Theorem 11.6.2 *Assume that the generalized system* $\mathbf{T}$ *in (11.128) satisfies (AD1)–(AD7). Then the controller*

$$\overline{\mathbf{K}} = \left[\begin{array}{c|c} \overline{A}_{K22} & \overline{B}_{K2} \\ \hline \overline{C}_{K2} & \overline{D}_K \end{array}\right], \tag{11.143}$$

of reduced order $n - r$*, where* $\overline{A}_{K22}, \overline{B}_{K2}, \overline{C}_{K2}$*, and* $\overline{D}_K$ *are defined in (11.141), (11.142), is a solution to the optimal DF problem for* $\mathbf{T}$.

Proof. As in the continuous–time case, the proof of the theorem follows immediately once we show that the limit of the closed–loop system $\mathbf{T}_{y_1u_1}$, when $\gamma \downarrow \gamma_{\min}$, is a system, say $\overline{\mathbf{T}}_{y_1u_1}$, which is internally stable and for which $\|\overline{\mathbf{T}}_{y_1u_1}\| = \gamma_{\min}$. The proof of these latter facts implies some tedious manipulations and several auxilliary results on the behavior of the closed–loop system which we defer to the next subsection. ∎

The behavior of the closed–loop system when $\epsilon \downarrow 0$

Let $\gamma > \gamma_{\min}$. We write a realization of the resultant closed–loop system of $\mathbf{T}$ and $\mathbf{K}$ by using (11.131) in (2.44), and get after a similarity transformation and removal of the stable uncontrollable part,

$$\mathbf{T}_{y_1 u_1} = \left[\begin{array}{c|c} A_R & B_R \\ \hline C_R & D_R \end{array}\right] = \left[\begin{array}{c|c} M_\gamma^{-1}A & M_\gamma^{-1}B_1 \\ \hline C_1 - D_{12}B_2^T X_\gamma M_\gamma^{-1}A & D_{11} - D_{12}B_2^T X_\gamma M_\gamma^{-1}B_1 \end{array}\right]. \tag{11.144}$$

Clearly A_R is stable and $\|\mathbf{T}_{y_1u_1}\|_\infty < \gamma$ since $\mathbf{K}$ is a solution to the suboptimal γ–DF problem. Consider the γ–contractiveness Popov triplet of $\mathbf{T}_{y_1u_1}$ given explicitly by

$$\begin{aligned} \Sigma_R &= (A_R, B_R; C_R^T C_R, C_R^T D_R, -\gamma^2 I + D_R^T D_R) \\ &= (M_\gamma^{-1}A, M_\gamma^{-1}B_1; C_1^T C_1 + A^T M_\gamma^{-T} X_\gamma B_2 B_2^T X_\gamma M_\gamma^{-1} A, \\ &\quad C_1^T D_{11} + A^T M_\gamma^{-T} X_\gamma B_2 B_2^T X_\gamma M_\gamma^{-1} B_1, \\ &\quad -\gamma^2 I + D_{11}^T D_{11} + B_1^T M_\gamma^{-T} X_\gamma B_2 B_2^T X_\gamma M_\gamma^{-1} B_1), \end{aligned} \tag{11.145}$$

where for the last equality we have used (AD5). Hence by applying Lemma 7.7.1, we get that the KSPYS$(\Sigma_R, -I)$ given explicitly by

$$\begin{aligned} -\gamma^2 I + D_R^T D_R + B_R^T X_R B_R &= -V_R^T V_R, \\ C_R^T D_R + A_R^T X_R B_R &= -W_R^T V_R, \\ C_R^T C_R + A_R^T X_R A_R - X_R &= -W_R^T W_R, \end{aligned} \tag{11.146}$$

has a stabilizing solution (X_R, V_R, W_R), with $X_R \geq 0$. The result below connects X_R to X_γ and is crucial to understand the behavior of the closed–loop system $\mathbf{T}_{y_1u_1}$.

Proposition 11.6.3 *Assume that the discrete-time generalized system* $\mathbf{T}$ *in (11.128) satisfies (AD1)–(AD5), let* $\gamma > \gamma_{\min}$ *and* $\mathbf{K}$ *be the central controller (11.131) that solves the* γ*–DF problem for* $\mathbf{T}$*. Then the following equality holds*

$$X_\gamma = X_R. \tag{11.147}$$

Proof. The idea of the proof is to relate the ESPs associated with the triplets Σ_R and Σ_c. The ESP(Σ_c), denoted $zM_{\Sigma_c} - N_{\Sigma_c}$, is given by

$$M_{\Sigma_c} = \begin{bmatrix} I & O & O & O \\ O & -A^T & O & O \\ O & -B_1^T & O & O \\ O & -B_2^T & O & O \end{bmatrix}, N_{\Sigma_c} = \begin{bmatrix} A & O & B_1 & B_2 \\ C_1^T C_1 & -I & C_1^T D_{11} & O \\ D_{11}^T C_1 & O & -\gamma^2 I + D_{11}^T D_{11} & O \\ O & O & O & I \end{bmatrix}, \tag{11.148}$$

and the ESP(Σ_R), denoted $zM_{\Sigma_R} - N_{\Sigma_R}$, is given by

$$\begin{aligned} M_{\Sigma_R} &= \begin{bmatrix} I & O & O \\ O & -A^T M_\gamma^{-T} & O \\ O & -B_1^T M_\gamma^{-T} & O \end{bmatrix}, \\ N_{\Sigma_R} &= \begin{bmatrix} M_\gamma^{-1}A & O & M_\gamma^{-1}B_1 \\ C_1^T C_1 + A^T \widetilde{X}_\gamma A & -I & C_1^T D_{11} + A^T \widetilde{X}_\gamma B_1 \\ D_{11}^T C_1 + B_1^T \widetilde{X}_\gamma A & O & -\gamma^2 I + D_{11}^T D_{11} + B_1^T \widetilde{X}_\gamma B_1 \end{bmatrix}, \end{aligned} \tag{11.149}$$

where we have denoted $\widetilde{X}_\gamma := M_\gamma^{-T} X_\gamma B_2 B_2^T X_\gamma M_\gamma^{-1}$. Since the KSPYS($\Sigma_c, J_c$) and the KSPYS($\Sigma_R, -I$) have stabilizing solutions, we can switch to the corresponding DTAREs and get by Theorem 5.5.1,

$$N_{\Sigma_c} V_{\Sigma_c} = M_{\Sigma_c} V_{\Sigma_c} S_{\Sigma_c}, \quad V_{\Sigma_c} = \begin{bmatrix} I \\ X_\gamma \\ F_1 \\ F_2 \end{bmatrix}, \quad \Lambda(S_{\Sigma_c}) \subset \mathbb{D}, \tag{11.150}$$

where $S_{\Sigma_c} := A + BF = A + B_1 F_1 + B_2 F_2$, and

$$N_{\Sigma_R} V_{\Sigma_R} = M_{\Sigma_R} V_{\Sigma_R} S_{\Sigma_R}, \quad V_{\Sigma_R} = \begin{bmatrix} I \\ X_R \\ F_R \end{bmatrix}, \quad \Lambda(S_R) \subset \mathbb{D}, \tag{11.151}$$

where $S_R = M_\gamma^{-1} A + M_\gamma^{-1} B_1 F_R$. Here $F = \begin{bmatrix} F_1 \\ F_2 \end{bmatrix}$ and F_R are the stabilizing Riccati feedbacks of the CTARE(Σ_c) and CTARE(Σ_R), respectively. By eliminating F_2, equation (11.150) reduces to

$$\begin{bmatrix} M_\gamma^{-1} A & O & M_\gamma^{-1} B_1 \\ C_1^T C_1 + X_\gamma B_2^T B_2 X_\gamma & -M_\gamma^T & C_1^T D_{11} \\ D_{11}^T C_1 & O & -\gamma^2 I + D_{11}^T D_{11} \end{bmatrix} \begin{bmatrix} I \\ X_\gamma \\ F_1 \end{bmatrix}$$

$$= \begin{bmatrix} I & O & O \\ O & -A^T & O \\ O & -B_1^T & O \end{bmatrix} \begin{bmatrix} I \\ X_\gamma \\ F_1 \end{bmatrix} M_\gamma^{-1} A_{F_1} \tag{11.152}$$

where $F_2 = -B_2^T X_\gamma M_\gamma^{-1} A_{F_1}$, $A_{F_1} := A + B_1 F_1$, and

$$S_{\Sigma_c} = M_\gamma^{-1} A_{F_1}. \tag{11.153}$$

Further, using (11.153), (11.152) becomes

$$\widetilde{N} \begin{bmatrix} I \\ M_\gamma^T X_\gamma \\ F_1 \end{bmatrix} = \widetilde{M} \begin{bmatrix} I \\ M_\gamma^T X_\gamma \\ F_1 \end{bmatrix} M_\gamma^{-1} A_{F_1} \tag{11.154}$$

where

$$\begin{aligned} \widetilde{M} &= \begin{bmatrix} I & O & O \\ O & -A^T M_\gamma^{-T} & O \\ O & -B_1^T M_\gamma^{-T} & O \end{bmatrix}, \\ \widetilde{N} &= \begin{bmatrix} M_\gamma^{-1} A & O & M_\gamma^{-1} B_1 \\ C_1^T C_1 + X_\gamma B_2 B_2^T X_\gamma & -I & C_1^T D_{11} \\ D_{11}^T C_1 & O & -\gamma^2 I + D_{11}^T D_{11} \end{bmatrix}. \end{aligned} \tag{11.155}$$

Next we defer our attention to the explicit form of (11.151). We perform the following elementary row and column operations: (a) premultiply the first block row by $A^T M_\gamma^{-T} X_\gamma B_2^T B_2 X_\gamma$ and subtract it from the second one; (b) subtract the first block row premultiplied by $B_1 M_\gamma^{-T} X_\gamma B_2 B_2^T X_\gamma$ from the third block row. Finally, one obtains

$$\begin{bmatrix} M_\gamma^{-1}A & O & M_\gamma^{-1}B_1 \\ C_1^T C_1 & -I & C_1^T D_{11} \\ D_{11}^T C_1 & O & -\gamma^2 I + D_{11}^T D_{11} \end{bmatrix} \begin{bmatrix} I \\ X_R \\ F_R \end{bmatrix}$$

$$= \begin{bmatrix} I & O & O \\ -A^T M_\gamma^{-T} X_\gamma B_2 B_2^T X_\gamma & -A^T M_\gamma^{-T} & O \\ -B_1^T M_\gamma^{-T} X_\gamma B_2 B_2^T X_\gamma & -B_1^T M_\gamma^{-T} & O \end{bmatrix} \begin{bmatrix} I \\ X_R \\ F_R \end{bmatrix} S_{\Sigma_R}.$$

Further, by subtracting the second block column, postmultiplied by $X_\gamma^T B_2 B_2^T X_\gamma$, from the first one, we obtain

$$\widetilde{N} \begin{bmatrix} I \\ X_R + X_\gamma B_2 B_2^T X_\gamma \\ F_2 \end{bmatrix} = \widetilde{M} \begin{bmatrix} I \\ X_R + X_\gamma B_2 B_2^T X_\gamma \\ F_R \end{bmatrix} S_{\Sigma_R} \tag{11.156}$$

where the notation (11.155) has been used. Since S_{Σ_c} and S_{Σ_R} (see also (11.153)) are both stable, a comparison of (11.154) and (11.156) reveals that

$$X_R + X_\gamma B_2^T B_2 X_\gamma = M_\gamma^T X_\gamma = X_\gamma + X_\gamma B_2^T B_2 X_\gamma.$$

Hence $X_R = X_\gamma$ and the proof ends. ■

Finally, we look at the behavior of the closed–loop as $\epsilon \downarrow 0$. Using again (11.139) and (11.140) but now in (11.144) one gets

$$\mathbf{T}_{y_1 u_1} = \left[\begin{array}{c|c} A_R & B_R \\ \hline C_R & D_R \end{array}\right] = \left[\begin{array}{cc|c} \epsilon A_{R11} & \epsilon A_{R12} & \epsilon B_{R1} \\ A_{R21} & A_{R2} & B_{R2} \\ \hline C_{R1} & C_{R2} & D_R \end{array}\right] \tag{11.157}$$

where $A_{Rij}, B_{Ri}, C_{Ri}, D_R$ $(i, j = 1, 2)$ continuously depend on ϵ and have all finite limits when $\epsilon \downarrow 0$, that is

$$(A_{Rij}, B_{Ri}, C_{Ri}, D_R) \to (\overline{A}_{Rij}, \overline{B}_{Ri}, \overline{C}_{Ri}, \overline{D}_R) \text{ as } \epsilon \downarrow 0,\ i, j = 1, 2. \tag{11.158}$$

For the closed–loop system we can prove the following key result.

Theorem 11.6.4 *The system*

$$\overline{\mathbf{T}}_{y_1 u_1} = \left[\begin{array}{c|c} \overline{A}_{R22} & \overline{B}_{R2} \\ \hline \overline{C}_{R2} & \overline{D}_R \end{array}\right] \tag{11.159}$$

with $\overline{A}_{R22}, \overline{B}_{R2}, \overline{C}_{R2}$ and $\overline{D}_R$ defined in (11.157), (11.158), is internally stable, i.e.,

$$\Lambda(\overline{A}_{R22}) \subset \mathbb{D}, \tag{11.160}$$

and

$$\|\overline{\mathbf{T}}_{y_1 u_1}\|_\infty = \gamma_{\min}. \tag{11.161}$$

Proof. 1. Introduce C_R and A_R from (11.157) into the last equation (11.146) where $W_R = \left[\begin{array}{cc} W_{R1} & W_{R2} \end{array} \right]$ has been partitioned conformably. According to Proposition 11.6.3, set $X_R = X_\gamma$ and take into account the diagonal structure of X_γ given explicitly in (11.134). Finally, take only the equation in which C_{R1} does not appear, i.e.,

$$C_{R2}^T C_{R2} + A_{R22}^T X_2 A_{R22} - X_2 + \epsilon A_{R12}^T X_1 A_{R12} = -W_{R2}^T W_{R2}. \qquad (11.162)$$

Let us prove (11.160). According to (11.158), the limit on the right–hand side of (11.162) exists and it is positive semidefinite, i.e., $W_{R2}^T W_{R2} \to W_0$ as $\epsilon \downarrow 0$. Therefore, (11.162) reveals that

$$\overline{C}_{R2}^T \overline{C}_{R2} + W_0 + \overline{A}_{R22}^T \overline{X}_2 \overline{A}_{R22} - \overline{X}_2 = 0, \qquad (11.163)$$

where $\overline{X}_2 \geq 0$. Provided the following condition holds:

(AD8) $(\overline{C}_{R2}, \overline{A}_{R22})$ is detectable,

we deduce from (11.163) with Theorem 1.5.5. that (11.160) holds. Notice that (AD8) is generically fulfilled.

The equality (11.161) follows directly from Theorem 2.5.1. ■

11.7 Robust stabilization for normalized coprime factors uncertainties

In this section we solve the optimal robust stabilization problem for the case in which the uncertainties act on normalized coprime factors. Consider the discrete–time system

$$\mathbf{G} = \left[\begin{array}{c|c} A & B \\ \hline C & O \end{array} \right] \qquad (11.164)$$

where $A \in \mathbb{R}^{n\times n}$, $B \in \mathbb{R}^{n\times m}$, $C \in \mathbb{R}^{p\times n}$, the pair (A, B) is stabilizable and the pair (C, A) is detectable. Let $(\widetilde{\mathbf{N}}, \widetilde{\mathbf{M}})$ be a normalized left coprime factorization over RH_+^∞ (see Section 7.9) such that $\mathbf{G} = \widetilde{\mathbf{M}}^{-1}\widetilde{\mathbf{N}}$. For a $\delta > 0$, introduce the class of discrete–time systems

$$\begin{aligned} \mathcal{D}_\delta^{cf} \quad := \quad & \{\mathbf{G}_\Delta^{cf} : \mathbf{G}_\Delta^{cf} = (\widetilde{\mathbf{M}} + \mathbf{\Delta}_M)^{-1}(\widetilde{\mathbf{N}} + \mathbf{\Delta}_N), \\ & \mathbf{\Delta} = \left[\begin{array}{c|c} \mathbf{\Delta}_N & -\mathbf{\Delta}_M \end{array} \right] \in RH_{+,p\times(p+m)}^\infty \text{ and } \|\mathbf{\Delta}\|_\infty < \delta\}. \end{aligned} \qquad (11.165)$$

The main goal of this section is to evaluate the maximum stability margin $\delta = \delta_{\max}$ for which there exists a single discrete–time stabilizing controller $\mathbf{K}$ for all systems in the class $\mathcal{D}_{\delta_{\max}}^{cf}$, and to construct such an optimal controller $\mathbf{K}$.

Consider the standard DTARE for control

$$A^T X A - X - A^T X B (I + B^T X B)^{-1} B^T X A + C^T C = 0 \qquad (11.166)$$

and the standard DTARE for estimation

$$A Y A^T - Y - A Y C^T (I + C Y C^T)^{-1} C Y A^T + B B^T = 0, \qquad (11.167)$$

both associated with the system (11.164). Let X, Y be their positive semidefinite stabilizing solutions which exist according to Corollary 4.8.2. Let $F := -(I +$

$B^TXB)^{-1}B^TXA$ and $K := -AYC^T(I+CYC^T)^{-1}$, $V^TV := I + B^TXB$, $\widehat{V}\widehat{V}^T := I+CYC^T$, $A_F := A+BF$, $A_K := A+KC$. Let $(\widetilde{\mathbf{N}}, \widetilde{\mathbf{M}})$ be a normalized left coprime factorization over RH_+^∞ (see Section 7.9), such that $\mathbf{G} = \widetilde{\mathbf{M}}^{-1}\widetilde{\mathbf{N}}$.

As we have proved in Section 11.1, the robust stabilization problem for the class of systems (11.165) is equivalent to a DAP formulated for the generalized system $\mathbf{T}^{cf}$ given in (11.6). In the particular case at hand in which the coprime factors $\widetilde{\mathbf{M}}$ and $\widetilde{\mathbf{N}}$ are normalized, we use the formulas for the discrete–time normalized coprime factors given in Theorem 7.9.1, and obtain an explicit realization in the form

$$\mathbf{T}^{cf} = \left[\begin{array}{c|c:c} A & B_1 & B_2 \\ \hline C_1 & D_{11} & D_{12} \\ \hdashline C_2 & D_{21} & O \end{array}\right] = \left[\begin{array}{c|c:c} A & -K\widehat{V} & B \\ \hline O & O & I \\ C & \widehat{V} & O \\ \hdashline C & \widehat{V} & O \end{array}\right]. \tag{11.168}$$

It is easy to check with (11.168) that (AD1)–(AD5) are fulfilled and the Popov triplet $\Sigma_c = (A, B; Q_c, L_c, R_c)$ in (11.130) is

$$\Sigma_c = (A, \left[\begin{array}{cc} -K\widehat{V} & B \end{array}\right]; C^TC, \left[\begin{array}{cc} C^T\widehat{V} & O \end{array}\right], \left[\begin{array}{cc} \widehat{V}^T\widehat{V} - \gamma^2 I & O \\ O & I \end{array}\right]). \tag{11.169}$$

Our approach to solving the robust stabilization problem at hand is to apply the results in Section 11.6 for the particular case of $\mathbf{T}^{cf}$.

11.7.1 An evaluation of the maximum stability margin

Theorem 11.7.1 *The maximum stability margin for the discrete–time robust stabilization problem with normalized coprime factors uncertainties formulated for the stabilizable and detectable system (11.164) is given by*

$$\delta_{\max} = (1 + \rho(XY))^{-\frac{1}{2}} \tag{11.170}$$

where X and Y are the unique stabilizing solutions to the DTAREs given in (11.166) and (11.167), respectively.

Proof. In order to explicitate (11.129) in terms of the data (11.168), notice first that

$$\left[\begin{array}{cc} \widetilde{\mathbf{T}}_{11} & \widetilde{\mathbf{T}}_{12} \end{array}\right] = \left[\begin{array}{c|cc} \widetilde{A} & B_1 & B_2 \\ \hline \widetilde{C} & D_{11} & D_{12} \end{array}\right] = \left[\begin{array}{c|cc} A_F & -K\widehat{V} & B \\ \hline F & O & I \\ C & \widehat{V} & O \end{array}\right] = \left[\begin{array}{cc} -\widetilde{\mathbf{W}} & \mathbf{M}V \\ \widetilde{\mathbf{U}} & \mathbf{N}V \end{array}\right] \tag{11.171}$$

where for the last relation we have used the notation (7.73) introduced in Theorem 7.9.1. Since $\widehat{\mathbf{T}}_{12} = \widetilde{\mathbf{T}}_{12}V^{-1}$ we may use the same computations as in the continuous case and arrive at

$$\gamma_{\min} = (1 + \rho(\mathbb{H}_{\mathbf{\Omega}}\mathbb{H}_{\mathbf{\Omega}}^*))^{\frac{1}{2}} \tag{11.172}$$

where

$$\mathbf{\Omega} := -\widetilde{\mathbf{W}}^*\mathbf{M} + \widetilde{\mathbf{U}}^*\mathbf{N}. \tag{11.173}$$

To get an expression for $\gamma_{\min}$ in terms of the original data, we first evaluate a realization for $\boldsymbol{\Omega}$. To this end, by using formulas (7.73), one gets, after some routine computations, that $\boldsymbol{\Omega}$ in (11.173) has the descriptor realisation

$$\begin{aligned} \sigma x &= A_F x + BV^{-1}u, \\ \xi &= (F^T F + C^T C)x + A_F^T \sigma\xi + F^T V^{-1}u, \\ y &= \widehat{V}^T C x - \widehat{V}^T K^T \sigma\xi. \end{aligned} \tag{11.174}$$

As (11.166) is equivalent to

$$A_F^T X A_F - X + FF^T + C^T C = 0,$$

the second equation in (11.174) can be rewritten as

$$\xi = Xx - A_F^T X A_F x + A_F^T \sigma\xi + F^T V^{-1}u,$$

or equivalently as

$$\begin{aligned} \xi - Xx &= -A_F^T X(\sigma x - BV^{-1}u) + A_F^T \sigma\xi + F^T V^{-1}u \\ &= A_F^T \sigma(\xi - Xx) + (A_F^T X B V^{-1} + F^T V^{-1})u, \end{aligned} \tag{11.175}$$

where the first equation in (11.174) has also been used. By an appropriate modification of the last equation in (11.174) one obtains

$$\begin{aligned} y &= (\widehat{V}^T C - \widehat{V}^T K^T X A_F)x - \widehat{V}^T K^T(\sigma\xi - XA_F x) \\ &= (\widehat{V}^T C - \widehat{V}^T K^T X A_F)x - \widehat{V}^T K^T \sigma(\xi - Xx) - \widehat{V}^T K^T X B V^{-1}u, \end{aligned} \tag{11.176}$$

where the first equation in (11.174) has been used again.

Let $\eta := \xi - Xx$. Then by joining the first equation in (11.174) with (11.175) and (11.176) one gets eventually the following realization, in descriptor form, for $\boldsymbol{\Omega}$:

$$\begin{aligned} \sigma x &= A_F x + BV^{-1}u, \\ \eta &= A_F^T \sigma\eta + (A_F^T X B V^{-1} + F^T V^{-1})u, \\ y &= (\widehat{V}^T C - \widehat{V}^T K^T X A_F)x - \widehat{V}^T K^T \sigma\eta - \widehat{V}^T K^T X B V^{-1}u. \end{aligned} \tag{11.177}$$

The system (11.177) is a mixed direct and inverse–time system. However, the direct and inverse–time systems are decoupled, i.e., the system can be written as sum of a direct–time stable system with an inverse–time stable system. Since the causal Hankel norm of an inverse–time stable system is zero (see Section 2.4) we retain for further computations only the direct–time part which is

$$\boldsymbol{\Omega}_s := \left[\begin{array}{c|c} A_F & BV^{-1} \\ \hline \widehat{V}^T C - \widehat{V}^T K^T X A_F & O \end{array}\right]. \tag{11.178}$$

From Corollary 2.8.7 we have

$$\rho(\mathbb{H}_{\boldsymbol{\Omega}} \mathbb{H}_{\boldsymbol{\Omega}}^*) = \rho(\mathbb{H}_{\boldsymbol{\Omega}_s} \mathbb{H}_{\boldsymbol{\Omega}_s}^*) = \rho(P_c P_o) \tag{11.179}$$

where P_c and P_o are the causal controllability and observability Gramians of $\boldsymbol{\Omega}_s$. Combining (11.179) with (11.178) and taking into account that $\delta_{\max} = 1/\gamma_{\min}$, the conclusion is immediate providing we show, as in the continuous–time case, that

$$P_c = Y(I + XY)^{-1} \tag{11.180}$$

and

$$P_o = (I + XY)X, \tag{11.181}$$

respectively. Unlike the continuous–time case, the derivation of (11.180) and (11.181) is not trivial.

We shall prove first that P_c given by (11.180) is the causal controllability Gramian of $\mathbf{\Omega}_s$, i.e., P_c satisfies the symmetric Stein equation

$$A_F P_c A_F^T - P_c + B(V^T V)^{-1} B^T = 0. \tag{11.182}$$

Let $zM - N$ be the ESP corresponding to the DTARE (11.167) where

$$M = \begin{bmatrix} I & O & O \\ O & -A & O \\ O & -C & O \end{bmatrix}, \quad N = \begin{bmatrix} A^T & O & C^T \\ BB^T & -I & O \\ O & O & I \end{bmatrix}.$$

Then we can write

$$\begin{bmatrix} A^T & O & C^T \\ BB^T & -I & O \\ O & O & I \end{bmatrix} \begin{bmatrix} I \\ Y \\ K^T \end{bmatrix} = \begin{bmatrix} I & O & O \\ O & -A & O \\ O & -C & O \end{bmatrix} \begin{bmatrix} I \\ Y \\ K^T \end{bmatrix} A_K^T. \tag{11.183}$$

In both sides of (11.183) perform successively the following elementary row and column operations: (a) add to the second block column the first one postmultiplied by $-X$; (b) subtract from the first block row the last one premultiplied by C^T. We obtain

$$\begin{bmatrix} A^T & -A^T X & O \\ BB^T & -I - BB^T X & O \\ O & O & I \end{bmatrix} \begin{bmatrix} I + XY \\ Y \\ K^T \end{bmatrix} = \begin{bmatrix} I & C^T C - X & O \\ O & -A & O \\ O & -C & O \end{bmatrix} \begin{bmatrix} I + XY \\ Y \\ K^T \end{bmatrix} A_K^T. \tag{11.184}$$

From (11.184) we get

$$\begin{bmatrix} A^T & -A^T X \\ BB^T & -I - BB^T X \end{bmatrix} \begin{bmatrix} I \\ Y(I + XY)^{-1} \end{bmatrix} = \begin{bmatrix} I & C^T C - X \\ O & -A \end{bmatrix} \begin{bmatrix} I \\ Y(I + XY)^{-1} \end{bmatrix} \widehat{A}_K^T \tag{11.185}$$

where $\widehat{A}_K^T = (I + XY) A_K^T (I + XY)^{-1}$. But

$$A_F^T = (I + XY) A_K^T (I + XY)^{-1}. \tag{11.186}$$

With (11.186), (11.185) reads as

$$\begin{bmatrix} A^T & -A^T X \\ BB^T & -I - BB^T X \end{bmatrix} \begin{bmatrix} I \\ Y(I + XY)^{-1} \end{bmatrix} = \begin{bmatrix} A & -A^T X A_F \\ O & A \end{bmatrix} \begin{bmatrix} I \\ Y(I + XY)^{-1} \end{bmatrix} A_F^T \tag{11.187}$$

where (11.166) has been written as

$$A^T X A_F - X + C^T C = 0.$$

In both sides of (11.187) we perform successively the following operations: (a) premultiply the second block row by $(I + BB^T X)^{-1}$; (b) subtract from the first block row the updated last one premultiplied by $A^T X$. We obtain

$$\begin{bmatrix} A_F^T & O \\ (I + BB^T X)^{-1} BB^T & -I \end{bmatrix} \begin{bmatrix} I \\ Y(I + XY)^{-1} \end{bmatrix} = \begin{bmatrix} I & O \\ O & -A_F \end{bmatrix} \begin{bmatrix} I \\ Y(I + XY)^{-1} \end{bmatrix} A_F^T. \tag{11.188}$$

Since $A_F = (I + BB^T X)^{-1} A$ and $(I + BB^T X)^{-1} BB^T = B(I + B^T XB)^{-1} B^T = B(V^T V)^{-1} B$, (11.188) yields

$$A_F Y(I + XY)^{-1} A_F^T - Y(I + XY)^{-1} + B(V^T V)^{-1} B^T = 0. \qquad (11.189)$$

By comparing (11.189) with (11.182) and invoking uniqueness arguments it follows that $P_c = Y(I + XY)^{-1}$.

Let us now prove (11.181). For this, we have to show that the symmetric Stein equation

$$A_F^T P_o A_F - P_o + (C^T \widehat{V} - A_F^T XK\widehat{V})(\widehat{V}^T C - \widehat{V}^T K^T XA_F) = 0 \qquad (11.190)$$

is fulfilled for P_o given by (11.181).

By performing the computations in the left–hand side of (11.190) we get

$$\begin{aligned} A_F^T P_o A_F - P_o + C^T(I + CYC^T)C + A_F^T XAYC^T C + C^T CYA^T XA_F \\ + A_F^T XAYC^T(I + CYC^T)^{-1} CYA^T XA_F \\ = A_F^T P_o A_F - P_o + C^T(I + CYC^T)C + A_F^T XAYC^T C \\ + C^T CYA^T XA_F + A_F^T X(AYA^T - Y + BB^T)XA_F. \end{aligned} \qquad (11.191)$$

In the right–hand side of (11.191), substitute $A_F = (I + BB^T X)^{-1} A$ and $P_o = (I + XY)X$ and then replace $A^T X(I + BB^T X)^{-1} A$ by $X - C^T C$ as directly follows from (11.166). By performing the computation one obtains that the last expression in (11.191) vanishes. Thus the validity of (11.181) is proved. ■

11.7.2 An optimal robustly stabilizing controller

In order to find an optimal robustly stabilizing controller we apply Theorem 11.6.2 in the particular case of the generalized system $\mathbf{T}^{cf}$. But we need first to check the hypotheses (AD1)–(AD7) for $\mathbf{T}^{cf}$. Assumptions (AD1)–(AD5) are obvious, while to check (AD6) we need first the following technical result.

Proposition 11.7.2 *Let $\gamma > \gamma_{\min}$ and Σ_c be the Popov triplet given in (11.169). Then the DTARE(Σ_c) has a stabilizing solution $X_\gamma \geq 0$ which is given by*

$$X_\gamma = -\gamma^2 X[(1 - \gamma^2)I + YX]^{-1}, \qquad (11.192)$$

where $X \geq 0$ and $Y \geq 0$ are the stabilizing solutions to the DTAREs (11.166) and (11.167), respectively.

Proof. Since $\gamma > \gamma_{\min}$, the existence of the stabilizing solution X_γ follows directly from the discrete–time version of Theorem 11.2.1. The rest of the proof makes intense use of manipulations on ESPs.

Let $zM_{\Sigma_c} - N_{\Sigma_c}$ be the ESP(Σ_c), where

$$M_{\Sigma_c} = \begin{bmatrix} I & O & O & O \\ O & -A^T & O & O \\ O & \widehat{V}^T K^T & O & O \\ O & -B^T & O & O \end{bmatrix}, N_{\Sigma_c} = \begin{bmatrix} A & O & -K\widehat{V} & B \\ C^T C & -I & C^T \widehat{V} & O \\ \widehat{V}^T C & O & \widehat{V}^T \widehat{V} - \gamma^2 I & O \\ O & O & O & I \end{bmatrix}. \qquad (11.193)$$

Since X_γ is the stabilizing solution to the DTARE(Σ_c), one can write

$$N_{\Sigma_c} V_{\Sigma_c} = M_{\Sigma_c} V_{\Sigma_c} S_{\Sigma_c}, \quad V_{\Sigma_c} = \begin{bmatrix} I \\ X_\gamma \\ F_1 \\ F_2 \end{bmatrix}, \quad \Lambda(S_{\Sigma_c}) \subset \mathbb{D}. \tag{11.194}$$

Introduce the following two nonsingular matrices:

$$U_0 = \begin{bmatrix} (1-\gamma^{-2})^{-1} I & (1-\gamma^{-2})^{-1} AY & (1-\gamma^2)^{-1} K\widehat{V} & (1-\gamma^2)^{-1} B \\ O & I + (1-\gamma^2)^{-1} C^T C Y & -(1-\gamma^2)^{-1} C^T \widehat{V} & O \\ O & O & O & I \\ O & -(1-\gamma^2)^{-1} CY & (1-\gamma^2)^{-1} \widehat{V} & O \end{bmatrix}$$

and

$$W_0 = \begin{bmatrix} (1-\gamma^{-2}) I & -\gamma^{-2} Y & O & O \\ O & I & O & O \\ \gamma^{-2} \widehat{V}^{-1} C & \gamma^{-2} \widehat{V}^{-1} CY & O & \widehat{V}^{-1} \\ O & O & I & O \end{bmatrix}.$$

Then (11.194) can be rewritten as

$$N_0 V_0 = M_0 V_0 S_0 \tag{11.195}$$

where

$$M_0 = U_0 M_{\Sigma_c} W_0 = \begin{bmatrix} I & O & O & O \\ O & -A^T & O & O \\ O & -B^T & O & O \\ O & O & O & O \end{bmatrix}, \quad N_0 = U_0 N_{\Sigma_c} W_0 = \begin{bmatrix} A & O & B & O \\ C^T C & -I & O & O \\ O & O & I & O \\ O & O & O & I \end{bmatrix},$$

$$V_0 = W_0^{-1} V_{\Sigma_c} = \begin{bmatrix} (\gamma^2 - 1)^{-1} (\gamma^2 I + Y X_\gamma) \\ X_\gamma \\ F_2 \\ \star \end{bmatrix}, \tag{11.196}$$

and $S_0 = S_{\Sigma_c}$. From (11.195) and (11.196) we deduce that

$$\begin{bmatrix} A & O & B \\ C^T C & -I & O \\ O & O & I \end{bmatrix} \begin{bmatrix} I \\ \widehat{X}_\gamma \\ \widehat{F}_2 \end{bmatrix} = \begin{bmatrix} I & O & O \\ O & -A^T & O \\ O & -B^T & O \end{bmatrix} \begin{bmatrix} I \\ \widehat{X}_\gamma \\ \widehat{F}_2 \end{bmatrix} \widehat{S}_0 \tag{11.197}$$

where

$$\begin{aligned} \widehat{X}_\gamma &= (\gamma^2 - 1) X_\gamma (\gamma^2 I + Y X_\gamma)^{-1}, \\ \widehat{F}_2 &= (\gamma^2 - 1) F_2 (\gamma^2 I + Y X_\gamma)^{-1}, \\ \widehat{S}_0 &= (\gamma^2 I + Y X_\gamma) S_0 (\gamma^2 I + Y X_\gamma)^{-1}. \end{aligned} \tag{11.198}$$

The leftmost matrices in both sides of (11.197) define the ESP corresponding to the DTARE (11.166). Hence, by uniqueness arguments, one gets

$$\widehat{X}_\gamma = X, \quad \widehat{F}_2 = F, \quad \widehat{S}_0 = A_F. \tag{11.199}$$

Finally, by combining the first equation in (11.198) with the first equation (11.199), one gets $X = (\gamma^2 - 1)X_\gamma(\gamma^2 I + YX_\gamma)^{-1}$ from which (11.192) follows immediately. ■

We return now to the check of (AD6).

Assume that both the DTAREs (11.166), (11.167) are already in *balanced* form such that

$$X = Y = \begin{bmatrix} \overline{\sigma} I_\gamma & O \\ O & \mathrm{diag}\,(\sigma_{r+1}, \ldots, \sigma_n) \end{bmatrix} \tag{11.200}$$

where $\overline{\sigma} > \sigma_{r+1} \geq \ldots \geq \sigma_n$ and r is the multiplicity of $\overline{\sigma}$. Such a balanced form can be obtained similarly as in the continuous–time case where now we use the standard DTAREs (11.166) and (11.167). Hence $\rho(XY) = \overline{\sigma}^2$ and (11.133) becomes

$$\gamma^2 = \epsilon + 1 + \overline{\sigma}^2. \tag{11.201}$$

With (11.200) substituted in (11.192), X_γ acquires the form (11.134) with

$$X_1 := (\epsilon + 1 + \overline{\sigma}^2)\overline{\sigma} I_r, \quad X_2 := \mathrm{diag}\left(\frac{\epsilon + 1 + \overline{\sigma}^2}{\overline{\sigma}^2 - \sigma_i^2 + \epsilon}\sigma_i\right)_{i=r+1}^{n} \tag{11.202}$$

and

$$\overline{X}_1 = (1 + \overline{\sigma}^2)\overline{\sigma} I_r, \quad \overline{X}_2 = \mathrm{diag}\left(\frac{1 + \overline{\sigma}^2}{\overline{\sigma}^2 - \sigma_i^2}\sigma_i\right)_{i=r+1}^{n} \tag{11.203}$$

where $X_i \to \overline{X}_i$ as $\epsilon \downarrow 0$, $i = 1, 2$. Thus (AD6) holds. Let

$$B = \begin{bmatrix} B_{21} \\ B_{22} \end{bmatrix},$$

be partitioned conformably with $X = Y$ in (11.200), and let

$$\overline{M} = \begin{bmatrix} \overline{M}_{11} & \overline{M}_{12} \\ \overline{M}_{21} & \overline{M}_{22} \end{bmatrix} := \begin{bmatrix} B_{21} B_{21}^T \overline{X}_1 & B_{21} B_{22}^T \overline{X}_2 \\ B_{22} B_{21}^T \overline{X}_1 & I + B_{22} B_{22}^T \overline{X}_2 \end{bmatrix}^{-1}, \ \overline{N} := \begin{bmatrix} O & O \\ \overline{M}_{21} & \overline{M}_{22} \end{bmatrix}. \tag{11.204}$$

Define also

$$\left[\begin{array}{c|c} \overline{N} A_K & -\overline{N} K \\ \hline -B^T \mathrm{diag}\,(\overline{X}_1, \overline{X}_2)\overline{M} A_K & -B^T \mathrm{diag}\,(\overline{X}_1, \overline{X}_2)\overline{M} K \end{array}\right] =: \left[\begin{array}{cc|c} O & O & O \\ \overline{A}_{K21} & \overline{A}_{K22} & \overline{B}_{K2} \\ \hline \overline{C}_{K1} & \overline{C}_{K2} & \overline{D}_K \end{array}\right]. \tag{11.205}$$

Theorem 11.7.3 *Assume that (AD7) and (AD8) hold for* $\mathbf{T}^{cf}$*, and* $\mathbf{G}$ *in (11.164) is in balanced form with respect to the standard DTAREs (11.166) and (11.167). Then*

$$\overline{\mathbf{K}} = \left[\begin{array}{c|c} \overline{A}_{K22} & \overline{B}_{K2} \\ \hline \overline{C}_{K2} & \overline{D}_K \end{array}\right] \tag{11.206}$$

is an optimal robustly stabilizing controller of (reduced) order $n - r$ *for the discrete–time system* $\mathbf{G}$*, with respect to normalized coprime factors uncertainties. Moreover, the maximum stability margin is given by*

$$\delta_{\max} = (1 + \overline{\sigma}^2)^{-\frac{1}{2}}. \tag{11.207}$$

Proof. We have already proved that (AD1)–(AD6) hold for $\mathbf{T}^{cf}$, and together with (AD7) and (AD8) assumed in the statement it follows that Theorem 11.6.2 is applicable for $\mathbf{T}^{cf}$. The formulas (11.206) are a matter of straightforward substitutions by using formulas (11.131) for the actual data given by (11.168). Finally, (11.207) follows by using (11.200) in (11.170). ∎

11.8 Robust stabilization for multiplicative uncertainties

In this section we solve the discrete–time suboptimal robust stabilization problem for the case in which the uncertainties act multiplicative. Consider the discrete–time system

$$\mathbf{G} = \left[\begin{array}{c|c} A & B \\ \hline C & D \end{array}\right], \tag{11.208}$$

where $A \in \mathbb{R}^{n\times n}$, $B \in \mathbb{R}^{n\times m}$, $C \in \mathbb{R}^{p\times n}$, the pair (A, B) is stabilizable and the pair (C, A) is detectable. For a $\delta > 0$, introduce the class of discrete–time systems

$$\mathcal{D}_\delta^m := \{\mathbf{G}_\Delta^m : \mathbf{G}_\Delta^m = (I + \mathbf{\Delta})\mathbf{G}, \mathbf{\Delta} \in RH^\infty_{+,p\times p} \text{ and } \|\mathbf{\Delta}\|_\infty < \delta\}. \tag{11.209}$$

The main goal of this section is to evaluate the maximum stability margin $\delta = \delta_{\max}$ for which there exists a single discrete–time stabilizing controller $\mathbf{K}$ for all systems in the class $\mathcal{D}^m_{\delta_{\max}}$, and to construct a robustly stabilizing controller $\mathbf{K}$ in the suboptimal case in which $\delta < \delta_{\max}$.

Throughout the section we make heavy use of the notation already introduced in Section 11.4 . We skip from the beginning the case in which $\mathbf{G} \in RH^\infty_{+,p\times m}$. We give now the discrete versions of the three sufficient conditions for solving the γ–DAP for the associated discrete–time generalized system

$$\begin{aligned} \mathbf{T}^m &= \left[\begin{array}{c|cc} A & O & B \\ \hline C & O & D \\ C & I & O \end{array}\right] =: \left[\begin{array}{c|c} A & \widehat{B} \\ \hline \widehat{C} & \widehat{D} \end{array}\right], \\ \begin{bmatrix} y_1 \\ y_2 \end{bmatrix} &= \mathbf{T}\begin{bmatrix} u_1 \\ u_2 \end{bmatrix}. \end{aligned} \tag{11.210}$$

(CD1) The DTARE(Σ_c) given by

$$A^T XA - X - (L_c + A^T X\widehat{B})(R_c + \widehat{B}^T X\widehat{B})^{-1}(\widehat{B}^T XA + L_c^T) + Q_c = 0 \tag{11.211}$$

has a positive semidefinite stabilizing solution X and the two sign constraints

$$\begin{aligned} R_{c22} + B^T XB &> 0, \\ \operatorname{sgn}(R_c + \widehat{B}^T X\widehat{B}) = J_c &:= \begin{bmatrix} -I_p & O \\ O & I_m \end{bmatrix}, \end{aligned} \tag{11.212}$$

are fulfilled.

(CD2) The DTARE(Σ_o) given by

$$AYA^T - Y - AY\widehat{C}^T(R_o + \widehat{C}Y\widehat{C}^T)^{-1}\widehat{C}YA^T = 0 \tag{11.213}$$

has a positive semidefinite stabilizing solution Y and the following two sign constraints

$$R_{o22} + CYC^T > 0,$$
$$\operatorname{sgn}(R_o + \widehat{C}Y\widehat{C}^T) = J_o := \begin{bmatrix} -I_p & O \\ O & I_p \end{bmatrix}, \tag{11.214}$$

are fulfilled.

(CD3) The solution X and Y are linked by

$$\rho(XY) < \gamma^2. \tag{11.215}$$

Notice that if the Bernoulli equation (11.213) has a stabilizing solution then we have from Proposition 5.6.1 that A is dichotomic, R_o is nonsingular and

$$\operatorname{sgn}(R_o + \widehat{C}Y\widehat{C}^T) = \operatorname{sgn}(R_o).$$

Since we have $\operatorname{sgn}(R_o) = J_o$, it follows that that the sign constraint (11.214) is fulfilled. Moreover, as $Y \geq 0$ and $R_{o22} = I_p$, the first sign constraint in (11.214) holds too.

We also have

$$R_c + \widehat{B}^T X \widehat{B} = \begin{bmatrix} -\gamma^2 I_p & O \\ O & D^T D + B^T X B \end{bmatrix}. \tag{11.216}$$

Hence X must be such that $D^T D + B^T X B > 0$. However, similarly as in the continuous–time case, we relax this condition by employing singular perturbation techniques and arrive to the discrete–time versions of (11.124) and (11.125) which are

$$A^T X_\epsilon A - X_\epsilon - (A^T X_\epsilon B + C^T D)(\epsilon^2 I_m + D^T D + B^T X_\epsilon B)^{-1}(B^T X_\epsilon A + D^T C) + C^T C = 0 \tag{11.217}$$

and

$$AYA^T - Y - AYC^T[\gamma^2(\gamma^2-1)^{-1} I_p + CYC^T]^{-1} CYA^T = 0, \tag{11.218}$$

respectively. Let $\gamma \in (0,1)$. Provided (11.218) has a positive semidefinite stabilizing solution Y, it follows with $\gamma^2(\gamma^2-1)^{-1} < 0$ and Proposition 5.6.2 that A is stable. As already explained, this case is skipped as trivial.

Let $\gamma > 1$ and assume that A is dichotomic. This assumption is necessary as follows from Proposition 5.6.1. Since $\gamma > 1$ it follows that $\gamma^2(\gamma^2-1) > 0$. Further, since the pair (C, A) is detectable we have, according to Theorem 5.6.3, that (11.218) has a positive semidefinite stabilizing solution Y.

We focus further on the DTARE (11.217). Since the pair (A, B) is stabilizable and the pair (C, A) is detectable, it follows from Proposition 7.8.1 that (11.217) has a positive semidefinite stabilizing solution X_ϵ. Assume further that $\mathbf{G}$ has a stable inverse to the right which is proper. Then it follows from the discrete–time counterpart of Theorem 7.2.3 that

$$\lim_{\epsilon \downarrow 0} X_\epsilon = 0. \tag{11.219}$$

Hence by taking ϵ small enough, condition (11.215) written explicitly as

$$\rho(X_\epsilon Y) < \gamma^2$$

is fulfilled. Thus we conclude that all three conditions (CD1), (CD2), and (CD3) are fulfilled. Meanwhile we have proved the following theorem.

Theorem 11.8.1 *Assume for the system* **G** *given in (11.208) that the pair* (A, B) *is stabilizable, the pair* (C, A) *is detectable,* A *is dichotomic,* D *is full row rank, and there exists an inverse to the right of* D*, say* $D^\dagger$*, such that* $A - BD^\dagger C$ *is stable. Then we have:*

1. For each $\delta \in (0, 1)$ *the controller*

$$\mathbf{K} = \left[\begin{array}{c|c} (A + BF_\epsilon)M_\epsilon^{-1} & (A + BF_\epsilon)M_\epsilon^{-1} Z_\epsilon C^T \\ \hline F_\epsilon M_\epsilon^{-1} & F_\epsilon M_\epsilon^{-1} Z_\epsilon C^T \end{array} \right] \tag{11.220}$$

is a suboptimal robustly stabilizing controller for **G***, with respect to multiplicative uncertainties, where*

$$Z_\epsilon := Y(I - \delta^2 X_\epsilon Y)^{-1}, \quad F_\epsilon := -(\epsilon^2 I + B^T X_\epsilon B + D^T D)^{-1}(B^T X_\epsilon A + D^T C),$$

$$M_\epsilon := I + Z_\epsilon C^T C,$$

with X_ϵ *and* Y *the unique stabilizing solutions to the DTAREs (11.217) and (11.218), respectively, and* ϵ *chosen such that* $\rho(X_\epsilon Y) < \frac{1}{\delta^2}$.

2. The maximum stability margin for the problem of robust stabilization with multiplicative uncertainties is $\delta_{\max} = 1$.

Proof. The proof runs as in the continuous–time case, with the only difference that the controller is now obtained by using the general formulas given in Theorem 10.12.1. ■

Notes and references

The robust stability problem with multiplicative uncertainties has been investigated by Doyle and Stein [DS81]. Stoorvogel [Sto96] has given an elegant solution to this problem by reducing it to a singular H^∞–control problem whose solution is expressed in terms of linear matrix inequalities. The case of additive uncertainties has been studied by Glover [Glo86], and a complete solution to the case of normalized coprime factors uncertainties has been given by Glover and McFarlane [GM89, MG90]. Similar topics may be found in book form in [GL95]. The optimal reduced order robust controller for coprime factors uncertainties has been obtained by Ionescu and Oară [IO94] for discrete–time and by Drăgan, Halanay, and Stoica [DHS95] for continuous–time systems. The evaluation of the least achievable tolerance in the DF problem and the solution to the robust stabilization problem with multiplicative uncertainties as presented in this chapter has been obtained by Ionescu [Ion95, Ion97b]. Singular perturbation techniques used in conjunction with the Popov function approach have the potential to provide optimal solutions to the H^∞–control problem and to the Nehari problem as well.

References

[AAK71] Adamjan V., Arov D., and Krein M. (1971) Analytic properties of Schmidt pairs for a Hankel operator and the generalized Schur–Takagi problem. *Math. USSR Sbornik* 15: 31–73.

[ADD90] Alpay D., Dewilde P., and Dym H. (1990) Lossless inverse scattering and reproducing kernels for upper triangular operators. volume 47 of *Oper. Th: Advances Appl.*, pages 61–135. Birkhäuser Verlag, Basel.

[AM71] Anderson B. and Moore J. (1971) *Linear Optimal Control.* Prentice Hall, Englewood Cliffs, NJ.

[BAGK95] Ben-Artzi A., Gohberg I., and Kaashoek M. (1995) Discrete nonstationary bounded real lemma in indefinite metric. volume 80 of *Oper. Th: Advances Appl.*, pages 49–78. Birkhäuser Verlag, Basel.

[BC86] Bittanti S. and Colaneri P. (1986) Lyapunov and Riccati equations: periodic inertia theorems. *IEEE Trans. Auto. Control* 31: 659–661.

[Bel68] Belevitch V. (1968) *Classical Network Theory.* Holden Day, San Francisco.

[BGK92] Ball J., Gohberg I., and Kaashoek M. (1992) Reduction of the abstract four block problem to a Nehari problem. In Gohberg I. and Kaashoek M. (eds) *Time–Variant Systems and Interpolation*, volume 58 of *Oper. Th: Advances Appl.*, pages 121–141. Birkhäuser Verlag, Basel.

[BH83] Ball J. A. and Helton J. W. (1983) A Beurling–Lax theorem for the Lie group $U(m,n)$ which contains most classical interpolation theory. *J. Op. Theory* 9: 107–142.

[BIdK94a] Barb F., Ionescu V., and de Koning W. (1994) Digital H^∞–control with measurement feedback for Pritchard–Salamon systems. Part I: The equivalent discrete–time system. In *Proc. IEEE Conf. Decision Contr.* Lake Buena Vista, Florida.

[BIdK94b] Barb F., Ionescu V., and de Koning W. (1994) Digital H^∞–control with measurement feedback for Pritchard–Salamon systems. Part II: A Popov theory based solution. In *Proc. IEEE Conf. Decision Contr.* Lake Buena Vista, Florida.

[BIdK95] Barb F., Ionescu V., and de Koning W. (1995) A Popov theory based survey on digital control of infinite dimensional systems with unboundedness. *IMA J. Math. Contr. Inform.* 12: 252–298.

[BLW91] Bittanti S., Laub A., and Willems J. (eds) (1991) *The Riccati Equation.* Springer Verlag.

[BV88] Beelen T. and Van Dooren P. (1988) An improved algorithm for the computation of Kronecker's canonical form of a singular pencil. *Lin. Alg. Appl.* 105: 9–65.

[CF89] Chen T. and Francis B. (1989) Spectral and inner-outer factorization of rational matrices. *SIAM J. Matrix Anal. Appl.* 10(1): 1–17.

[CZ95] Curtain R. F. and Zwart H. (1995) *An Introduction to Infinite–Dimensional Linear Systems Theory.* Springer Verlag, New York.

[DD92] Dewilde P. and Dym H. (1992) Interpolation for upper triangular operators. volume 56 of *Oper. Th: Advances Appl.*, pages 153–260. Birkhäuser Verlag, Basel.

[DGKF89] Doyle J., Glover K., Khargonekar P., and Francis B. (1989) State–space solutions to standard H^2 and H^∞ control problems. *IEEE Trans. Auto. Control* 34(8): 831–847.

[DHI94] Drăgan V., Halanay A., and Ionescu V. (1994) Infinite horizon disturbance attenuation for discrete–time systems. A Popov–Yakubovich approach. *Integr. Equat. Oper. Th.* 19: 153–215.

[DHS95] Drăgan V., Halanay A., and Stoica A. (1995) Remarks on order reduction for a robustly suboptimal controller via singular perturbations. *Syst. Contr. Lett.* 24: 317–320.

[Dou72] Douglas R. G. (1972) *Banach Algebra Techniques in Operator Theory.* Academic Press, New York.

[DS81] Doyle J. and Stein G. (1981) Multivariable feedback design: concepts for a classical/modern synthesis. *IEEE Trans. Auto. Control* 26: 4–16.

[FD87] Francis B. and Doyle J. (1987) Linear control theory with an H^∞ optimality criterion. *SIAM J. Contr. Optim.* 25(4): 815–844.

[Fei98] Feintuch A. (1998) *Robust Control Theory in Hilbert Spaces.* Springer Verlag, New York.

[FF85] Feintuch A. and Francis B. (1985) Uniformly optimal control of linear feedback systems. *Automatica* 21(5): 563–574.

[For75] Forney G. D. (1975) Minimal bases of rational spaces with applications to multivariable linear systems. *SIAM J. Contr. Optim.* 13: 493–520.

[Fra87] Francis B. (1987) *A Course in H^∞ Control Theory*, volume 88 of *Lect. Notes Contr. Inform. Sci.* Springer Verlag, Berlin.

[Gan59] Gantmacher F. (1959) *Theory of Matrices*, volume I and II. Chelsea, New York.

[GF74] Gohberg I. and Feldman I. (1974) *Convolution Equations and Projection Methods for Their Solutions*, volume 41 of *Transl. Math. Monographs.* AMS, Providence, RI.

[GKW91] Gohberg I., Kaashoek M., and Woerdeman H. (1991) Time variant extension problems of Nehari type and the band method. In *H^∞-Control Theory*, volume 1496 of *Lect. Notes Math.* Springer Verlag, Berlin.

[GL95] Green M. and Limebeer D. (1995) *Linear Robust Control.* Prentice Hall, Englewood Cliffs, NJ.

[GLD+91] Glover K., Limebeer D., Doyle J., Kasenally E., and Safonov M. (1991) A characterization of all solutions to the four block general distance problem. *SIAM J. Contr. Optim.* 29: 283–324.

[Glo84] Glover K. (1984) All optimal Hankel–norm approximations of linear multivariable systems and their L^∞–error bounds. *Int. J. Control* 39(6): 1115–1193.

[Glo86] Glover K. (1986) Robust stabilization of linear multivariable systems: Relations to approximation. *Int. J. Control* 43: 741–766.

[GLR82] Gohberg I., Lancaster P., and Rodman L. (1982) *Matrix Polynomials.* Academic Press, New York.

[GLR83] Gohberg I., Lancaster P., and Rodman L. (1983) *Matrices and Indefinite Scalar Product*, volume 8 of *Oper. Th: Advances Appl.* Birkhäuser Verlag, Basel.

[GLR86] Gohberg I., Lancaster P., and Rodman L. (1986) *Invariant Subspaces of Matrices with Applications.* John Wiley, New York.

[GM89] Glover K. and McFarlane D. (1989) Robust stabilization of normalized coprime factor plant descriptions with H^∞–bounded uncertainty. *IEEE Trans. Auto. Control* 34: 821–830.

[GV89] Golub G. and Van Loan C. (1989) *Matrix Computations.* John Hopkins University Press, Baltimore and London.

[HI93] Halanay A. and Ionescu V. (1993) Generalized discrete–time Popov–Yakubovich theory. *Syst. Contr. Lett.* 20: 1–6.

[HI94a] Halanay A. and Ionescu V. (1994) Anticausal stabilizing solution to discrete reverse–time Riccati equations. *Computer Math.* 28: 115–125.

[HI94b] Halanay A. and Ionescu V. (1994) Extended Nehari problem for time–variant discrete systems. *Syst. Contr. Lett.* 22: 195–201.

[HI94c] Halanay A. and Ionescu V. (1994) Properties of some global solutions to the discrete–time Riccati equation associated to a contracting input–output operator. *J. Diff. Eqs. Appl.* 1: 1–14.

[HI94d] Halanay A. and Ionescu V. (1994) *Time-Varying Discrete Linear Systems. Input-Output Operators. Riccati Equations. Disturbance Attenuation*, volume 68 of *Oper. Th: Advances Appl.* Birkhäuser Verlag, Basel.

[HJ85] Horn R. A. and Johnson C. R. (1985) *Matrix Analysis*. Cambridge University Press, Cambridge.

[HJ91] Horn R. A. and Johnson C. R. (1991) *Topics in Matrix Analysis*. Cambridge University Press, Cambridge.

[Igl91] Iglesias P. (1991) *Robust and Adaptive Control for Discrete-Time Systems*. PhD thesis, Cambridge University, Cambridge.

[IO94] Ionescu V. and Oară C. (1994) A discrete-time reduced order controller for robust stabilization of plant in the normalized coprime-factor plant description. *IMA J. Math. Contr. Inform.* 11: 231–252.

[IO96a] Ionescu V. and Oară C. (1996) The 4 block Nehari problem: A Popov function approach. *IMA J. Math. Contr. Inform.* 13(2): 173–194.

[IO96b] Ionescu V. and Oară C. (1996) The class of suboptimal and optimal solutions to the discrete Nehari problem: Characterization and computation. *Int. J. Control* 64(3): 483–509.

[IO96c] Ionescu V. and Oară C. (1996) Existence conditions for the stabilizing solution to the time-varying discrete Riccati equation. In *Proc. IEEE Conf. Decision Contr.*, pages 2263–2264. Kobe.

[IO96d] Ionescu V. and Oară C. (1996) Generalized continuous-time Riccati theory. *Lin. Alg. Appl.* 232: 111–131.

[IO96e] Ionescu V. and Oară C. (1996) Generalized discrete-time Riccati theory. *SIAM J. Contr. Optim.* 34(2): 601–619.

[IO96f] Ionescu V. and Oară C. (1996) Spectral and inner-outer factorizations for discrete-time systems. *IEEE Trans. Auto. Control* 41: 840–845.

[IO96g] Ionescu V. and Oară C. (1996) The time-varying discrete 4 block Nehari problem: A Popov-Yakubovich type approach. *Integr. Equat. Oper. Th.* 26: 404–431.

[IO97] Ionescu V. and Oară C. (1997) Indefinite sign Riccati theory for descriptor systems. In *Proc. European Contr. Conf.* Bruxelles, Belgium. Paper 755.

[IO98a] Ionescu V. and Oară C. (1998) The extended generalized distance problem in discrete-time. *Int. J. Robust Nonlinear Contr.* 8: 523–534.

[IO98b] Ionescu V. and Oară C. (1998) The extended Hankel norm approximation problem for discrete-time descriptor systems. In *Proc. IEEE Conf. Decision Contr.* Tampa, Florida.

[Ion95] Ionescu V. (1995) On the evaluation of the least achievable tolerance in the disturbance feedforward problem. *Syst. Contr. Lett.* 25: 113–119.

[Ion96] Ionescu V. (1996) Reverse discrete-time Riccati equation and extended Nehari's problem. *Lin. Alg. Appl.* 236: 59–94.

[Ion97a] Ionescu V. (1997) On the generalized Popov theory. In *Proc. IFAC Conf. System Struct. Contr.*, volume 2, pages 10–16. Bucharest.

[Ion97b] Ionescu V. (1997) Robust controller for multiplicative uncertainty: the time-varying case. *Int. J. Control* 65: 1–14.

[Ion98a] Ionescu V. (1998) Redheffer's theorem revisited. *Integr. Equat. Oper. Th.* 38: 286–303.

[Ion98b] Ionescu V. (1998) Time-varying discrete Riccati equation in terms of Ben-Artzi Gohberg dichotomy. *Lin. Alg. Appl.* 277: 313–336.

[IOW97] Ionescu V., Oară C., and Weiss M. (1997) General matrix pencil techniques for the solution of the algebraic Riccati equations. *IEEE Trans. Auto. Control* 42: 1085–1097.

[IW92a] Ionescu V. and Weiss M. (1992) The ℓ^2-control problem for time-varying discrete systems. *Syst. Contr. Lett.* 18: 371–381.

[IW92b] Ionescu V. and Weiss M. (1992) On computing the stabilizing solution to the discrete-time Riccati equation. *Lin. Alg. Appl.* 174: 229–238.

[IW93a] Ionescu V. and Weiss M. (1993) The constrained discrete–time algebraic Riccati equation. In *Proc. European Contr. Conf.*, pages 1800–1804. Groningen, The Netherlands.
[IW93b] Ionescu V. and Weiss M. (1993) Continuous and discrete–time Riccati theory: A Popov function approach. *Lin. Alg. Appl.* 193: 173–209.
[IW93c] Ionescu V. and Weiss M. (1993) Two Riccati formulae for the discrete–time H^∞–control problem. *Int. J. Control* 57(1): 141–195.
[JJ88] Juang J. and Jonckheere E. (1988) On computing the spectral radius of Hankel plus Toeplitz operator. *IEEE Trans. Auto. Control* 33: 1053–1059.
[JJS89] Jonckheere E., Juang J., and Silverman L. (1989) Spectral theory of the linear–quadratic and H^∞ problems. *Lin. Alg. Appl.* 122-124: 273–300.
[JS78] Jonckheere E. and Silverman L. (1978) Spectral theory of the linear quadratic optimal control problem: Discrete time single input case. *IEEE Trans. Circuits Syst.* 25: 810–825.
[JS81] Jonckheere E. and Silverman L. (1981) Spectral theory of the linear–quadratic optimal control problem: analytic factorization of rational matrix valued functions. *SIAM J. Contr. Optim.* 19: 262–281.
[JV86] Jonckheere E. and Verma M. (1986) A spectral characterization of H^∞–optimal feedback performance – the multivariable case. *Syst. Contr. Lett.* 8: 277–302.
[Kai80] Kailath T. (1980) *Linear Systems.* Prentice Hall, Englewood Cliffs, NJ.
[Kal60] Kalman R. (1960) Contributions to the theory of optimal control. *Bol. Soc. Math. Mex.* 5: 102–119.
[Kal64] Kalman R. (1964) When is a linear control system optimal? *ASME Trans., Series D: J. Basic Engineering* 86: 51–60.
[Kat66] Kato T. (1966) *Perturbation Theory for Linear Operators.* Springer Verlag, New-York.
[Kat94] Katayama T. (1994) A spectral factorization algorithm for discrete-time descriptor systems via generalized eigenproblems. *Lin. Alg. Appl.* 205-206: 869–891.
[Kim96] Kimura H. (1996) *Chain–Scattering Approach to H_∞-Control.* Birkhäuser Verlag, Basel.
[KS72] Kwakernaak H. and Sivan R. (1972) *Linear Optimal Control Systems.* Wiley–Interscience, New York.
[Kuc72] Kucera V. (1972) State–space approach to discrete linear control. *Kybernetica (Prague)* 8: 233–251.
[Kuc73] Kucera V. (1973) A review of the matrix Riccati equation. *Kybernetica (Prague)* 9: 42–61.
[Lau79] Laub A. (1979) A Schur method for solving algebraic Riccati equations. *IEEE Trans. Auto. Control* 24: 913–921.
[LR95] Lancaster P. and Rodman L. (1995) *The Algebraic Riccati Equation.* Oxford University Press, Oxford.
[Meh91] Mehrmann V. L. (1991) *The Autonomous Linear Quadratic Control Problem*, volume 163 of *Lect. Notes Contr. Inform. Sci.* Springer Verlag, Berlin.
[MG90] McFarlane D. and Glover K. (1990) *Robust Controller Design Using Normalized Coprime Factor Plant Descriptions*, volume 138 of *Lect. Notes Contr. Inform. Sci.* Springer Verlag, Berlin.
[Mol73] Molinari B. (1973) The stabilizing solution of the algebraic Riccati equation. *SIAM J. Contr. Optim.* 11(2): 262–271.
[Neh57] Nehari Z. (1957) On bounded bilinear forms. *Anal. Math.* 65: 153–162.
[NI96] Niculescu S. I. and Ionescu V. (1996) On stability criteria for time–delay linear systems: A matrix pencil approach. In *Proc. IFAC World Congress*, pages 171–175. San Francisco.
[NI97] Niculescu S. I. and Ionescu V. (1997) On delay independent criteria: A matrix pencil approach. *IMA J. Math. Contr. Inform.* 14(2): 299–306.
[Oar93] Oară C. (1993) On computing the stabilizing solution to the constrained algebraic Riccati equation. In *Proc. Int. Conf. Syst. Contr. Inform.*, pages 23–29. Bucharest.

[Oar94] Oară C. (1994) Proper deflating subspaces: properties, algorithms and applications. *Numer. Alg.* 7: 355–373.

[Oar95] Oară C. (1995) *Generalized Riccati Theory: A Popov Function Approach.* PhD thesis, Polytechnic University Bucharest, Faculty of Automatic Control and Computers.

[Oar96a] Oară C. (1996) The signature condition: a necessary and sufficient frequency domain existence condition for the stabilizing solution to the algebraic Riccati equation. In *Proc. Benelux Meeting Syst. Contr.*, pages 229–230. Mierlo, The Netherlands.

[Oar96b] Oară C. (1996) Stabilizing solution to the reverse discrete–time Riccati equation. *Lin. Alg. Appl.* 246: 113–130.

[Oar98] Oară C. (1998) Minimal factorization of rational matrices: the general case. In *Proc. Symp. Math. Theory Network Syst.* Padova, Italy.

[OV97] Oară C. and Van Dooren P. (1997) An improved algorithm for the computation of structural invariants of a system pencil and related geometric aspects. *Syst. Contr. Lett.* 30: 39–48.

[OV99a] Oară C. and Varga A. (1999) Computation of the general inner–outer and spectral factorizations. *IEEE Trans. Auto. Control* to appear.

[OV99b] Oară C. and Varga A. (1999) Minimal degree coprime factorization of rational matrices. *SIAM J. Matrix Anal. Appl.* to appear.

[OS62] Ostrowski A. and Schneider H. (1962) Some theorems on the inertia of general matrices. *J. Math. Anal. Appl.* 4: 72–84.

[Par88] Partington I. (1988) *An introduction to Hankel operators.* Cambridge Press, Cambridge.

[PLS80] Pappas T., Laub A., and Sandell N. (1980) On the numerical solution of the discrete algebraic Riccati equation. *IEEE Trans. Auto. Control* 25: 631–641.

[Pop66] Popov V. (1966) *Hiperstabilitatea Sistemelor Automate.* Ed. Academiei, Bucharest. in Romanian; translated as "Hyperstability of Control Systems" (1973), Springer Verlag, Berlin.

[Ros70] Rosenbrock H. H. (1970) *State–Space and Multivariable Theory.* Wiley, New York.

[Rud66] Rudin W. (1966) *Real and Complex Analysis.* McGraw–Hill, New York.

[Sar67] Sarason D. (1967) Generalized interpolation in H^∞. *Trans. AMS* 127: 179–203.

[Sil76] Silverman L. (1976) Discrete Riccati equations: alternative algorithms, asymptotic properties and system theoretic interpretations. In Leondes C. (ed) *Control and Dynamic Systems: Advances in Theory and Applications*, volume 12. Academic Press.

[SNF70] Szokefalvi-Nagy B. and Foias C. (1970) *Harmonic Analysis of Operators on Hilbert Spaces.* North–Holland, Amsterdam.

[Ste72] Stewart G. (1972) On the sensitivity of the eigenvalue problem $Ax = \lambda Bx$. *SIAM J. Numer. Anal.* 4: 669–686.

[Sto92] Stoorvogel A. (1992) *The* H^∞ *Control Problem: A State–Space Approach.* Prentice Hall, Englewood Cliffs, NJ.

[Sto96] Stoorvogel A. (1996) Robust stabilization of systems with multiplicative perturbations. In *Proc. IEEE Conf. Decision Contr.* Kobe, Japan.

[Tad90] Tadmor G. (1990) Worst–case design in the time domain. The maximum principle and the standard H^∞–problem. *Math. Contr. Signal Syst.* 3: 301–324.

[TV92] Tadmor G. and Verma M. (1992) Factorization and the Nehari theorem in time–varying systems. *Math. Contr. Signal Syst.* 5: 419–452.

[Van79] Van Dooren P. (1979) The computation of Kronecker's canonical form of a singular pencil. *Lin. Alg. Appl.* 27: 103–141.

[Van81] Van Dooren P. (1981) A generalized eigenvalue approach for solving Riccati equations. *SIAM J. Sci. St. Comp.* 2(2): 121–135.

[Van83] Van Dooren P. (1983) *Reducing subspaces: definitions, properties and algorithms*, volume 973 of *Lect. Notes Math.* Springer Verlag, Berlin.

[Var95] Varga A. (1995) On stabilization of descriptor systems. *Syst. Contr. Lett.* 24: 133–138.

[Vid85] Vidyasagar M. (1985) *Control System Synthesis: A Coprime Factorization Approach.* MIT Press.

[VLK81] Verghese G., Lévy B., and Kailath T. (1981) A generalized state–space for singular systems. *IEEE Trans. Auto. Control* 26: 811–831.

[VVK79] Verghese G., Van Dooren P., and Kailath T. (1979) Properties of the system matrix of a generalized state–space system. *Int. J. Control* 30: 235–243.

[Wei94a] Weiss M. (1994) *Riccati Equations in Hilbert Spaces: A Popov Function Approach.* PhD thesis, University of Groningen, Groningen, The Netherlands.

[Wei94b] Weiss M. (1994) Spectral and inner–outer factorization in the general case through the constrained Riccati equation. *IEEE Trans. Auto. Control* 39: 677–681.

[Wei97] Weiss M. (1997) Riccati equation theory for Pritchard–Salamon systems: A Popov function approach. *IMA J. Math. Contr. Inform.* 14: 45–83.

[Wil71] Willems. J. (1971) Least squares stationary optimal control and the algebraic Riccati equation. *IEEE Trans. Auto. Control* 16: 621–634.

[Won85] Wonham V. (1985) *Linear Multivariable Control. A Geometric Approach.* Springer Verlag, New York.

[Zam81] Zames G. (1981) Feedback and optimal sensitivity: model reference transformations, multiplicative seminorms, and approximate inverses. *IEEE Trans. Auto. Control* 26: 301–320.

[ZDG95] Zhou K., Doyle J., and Glover K. (1995) *Robust and Optimal Control.* Prentice Hall, Englewood Cliffs, NJ.

[ZK88] Zhou K. and Khargonekar P. (1988) An algebraic Riccati equation approach to H^∞ optimization. *Syst. Contr. Lett.* 11: 85–91.

Index